Topics in Applied Physics
Volume 95

Topics in Applied Physics

Topics in Applied Physics is a well-established series of review books, each of which presents a comprehensive survey of a selected topic within the broad area of applied physics. Edited and written by leading research scientists in the field concerned, each volume contains review contributions covering the various aspects of the topic. Together these provide an overview of the state of the art in the respective field, extending from an introduction to the subject right up to the frontiers of contemporary research.
Topics in Applied Physics is addressed to all scientists at universities and in industry who wish to obtain an overview and to keep abreast of advances in applied physics. The series also provides easy but comprehensive access to the fields for newcomers starting research.
Contributions are specially commissioned. The Managing Editors are open to any suggestions for topics coming from the community of applied physicists no matter what the field and encourage prospective editors to approach them with ideas.

Franz X. Kärtner (Ed.)

Few-Cycle Laser Pulse Generation and Its Applications

With 209 Figures

Springer

Professor Franz X. Kärtner
Massachusettts Institute of Technology
Deptartment of Electrical Engeneering and Computer Science
and Research Laboratory of Electronics
77 Massachusetts Ave.
Cambridge, MA 02139, USA
kaertner@mit.edu

Library of Congress Control Number: 2004104066

Physics and Astronomy Classification Scheme (PACS):
42.65-k, 42.62.Eh, 42.62.Fi, 42.65.Ky, 39.30.+w

ISSN print edition: 0303-4216
ISSN electronic edition: 1437-0859
ISBN 3-540-20115-7 Springer Berlin Heidelberg New York

Springer is a part of Springer Science+Business

springeronline.com

Printed in Germany

Typesetting: DA-TEX · Gerd Blumenstein · www.da-tex.de
Production: LE-TEX Jelonek, Schmidt & Vöckler GbR, Leipzig
Cover design: *design & production* GmbH, Heidelberg

Printed on acid-free paper 57/3141/YL 5 4 3 2 1 0

To my parents and brothers for their continued support

Preface

The long-standing dream of researchers in ultrafast laser physics and its applications of having few-cycle laser pulses at their disposal came true over the last few years. It took only a decade of extremely rapid developments following the invention of the Ti:sapphire laser and its first successful mode locking, most importantly the discovery of Kerr-lens mode locking. This development ushered in a revolution in various fields, first of all in ultrafast laser technology itself. The Ti:sapphire laser replaced the dye laser in almost all applications within only a few years, delivering pulses directly from the laser with durations that before were only possible by external compression. Even more so, this was possible with about two orders of magnitude higher average power and at the full repetition rate of the laser. Today, the use of external compression of amplified Ti:sapphire laser pulses even approaches the single-cycle regime. External spectral broadening leads to octave-spanning spectra at the full repetition rate of the laser enabling control of the carrier-envelope phase and, therefore, the electric field of the pulse. This concept, while already known in the 1970s, had to await the arrival of few-cycle laser sources to become of use and unfold its full power. Only four years after the first realization of a carrier-envelope, phase-stabilized, mode-locked laser oscillator, this technology truly revolutionized frequency metrology by making the counting of oscillations of the optical field possible using a relatively simple counting mechanism. Many years of building intricate and Ph.D. consuming frequency chains came to an end, and these chains are now replaced by a far more superior and much cheaper technology, a mode-locked laser. Before these developments, mode-locked lasers were known as an unreliable optical source with respect to frequency-resolved spectroscopic measurements because the origin of the frequency comb was not fixed. The stabilization of the carrier-envelope phase removed this ambiguity and made the light emitted from a mode-locked laser to a frequency ruler of unprecedented precision.

The same is reflected in the time domain. Stabilization of the carrier-envelope phase made the electric field of the optical pulse controllable on a subcycle timescale. This achievement led to the generation of isolated electromagnetic pulses with durations in the attosecond range, opening up a new era in ultrafast laser physics.

This book gives an overview of the various techniques developed to generate few-cycle laser pulses, their full characterization with respect to amplitude and phase, as well as their use in applications. Part I of the book is devoted to source development, and the first chapter therein specifically gives an overview of the solid-state materials available for the construction of few-cycle lasers covering the visible to the infrared wavelength range. Some of these materials can be pumped directly with semiconductor diodes, which is important for wide dissemination of ultrafast sources. The second to fourth chapters describe various techniques for generating few-cycle lasers based on these materials either directly from the laser by external compression or by employing parametric processes. Each of these techniques has its strengths and weaknesses, depending on the intended application. Part II of this volume is devoted to the two major pulse characterization techniques enabling the complete reconstruction of the complex pulse envelope of even single-cycle pulses; frequency resolved optical gating (FROG) and spectral interferometry for direct electric field reconstruction (SPIDER). Part III covers some of the key applications of few-cycle sources that use either the temporal or the spectral coherence properties of the emitted pulse train from a mode-locked laser. The first two chapters are devoted to advances in frequency metrology and high-resolution laser spectroscopy using mode-locked and carrier-envelope phase-stabilized lasers and exploiting the frequency-domain properties of a pulse train. In the remaining three chapters, the use of the time-domain properties of few-cycle laser pulses is exploited. Carrier-envelope phase-controlled, high-energy pulses open the door to the generation of subfemtosecond XUV pulses. Low-energy pulses are used to explore phase-sensitive nonlinear optical processes or processes simply happening on a few femtosecond timescale in condensed matter systems. These examples are intended to give a flavor of the wide variety of applications and intricate measurement techniques that are based on few-cycle lasers. Many more will follow over the next years.

Cambridge, Juli 2004 *Franz X. Kärtner*

Contents

Part I Few-cycle Pulse Generation: Theory and Experiments

Part I

Few-cycle Pulse Generation: Theory and Experiments

Solid-State Materials for Few-Cycle Pulse Generation and Amplification

Evgeni Sorokin

Institut für Photonik, Technische Universität Wien
Gusshausstr. 27/387, 1040 Vienna, Austria
e.sorokin@tuwien.ac.at

Abstract. Well-established and prospective broadband solid-state active media are discussed from the point of view of their use in few-cycle oscillators and amplifiers. The physical principles leading to broad gain bandwidth, diode-pumped and effective room temperature operation are considered, and the outlook for materials with even better parameters are given. The most important materials currently used for few-cycle generation in all wavelength regions from ultraviolet to mid–infrared are described in detail.

1 Introduction

For generation of few-cycle light, the material issue is of critical importance. Many breakthroughs in ultrashort-pulse generation have only become possible owing to the specific properties of the materials used, like, e.g., the invention of the colliding-pulse [1] or Kerr-lens mode locking (KLM) [2] techniques. This chapter will review the materials that can be used for the generation and amplification of few-cycle pulses.

Whether used in an oscillator, amplifier, or nonlinear frequency converter, the corresponding materials should provide optical gain over a broad frequency range, corresponding to that of the ultrashort pulse. The *gain bandwidth* is therefore a primary characteristic of the material. Traditionally, the bandwidth of laser materials and devices, $\Delta\lambda$, is defined as full-width at half-maximum (FWHM) of the gain cross-sectional spectrum in the wavelength domain. The bandwidth of the ultrashort optical pulse, $\Delta\nu$, is commonly defined as its intensity FWHM in the frequency domain. Finally, the pulse duration, $\Delta\tau$, is usually referred to as the FWHM of its intensity profile in the time domain, whereas for the purpose of few-cycle pulse generation, the natural measure of pulse duration is the number of optical cycles per pulse $N = \Delta\tau/T_0 = \nu_0\Delta\tau$, where T_0 is the period of the pulse carrier frequency ν_0. The uncertainty relation $\Delta\nu\Delta\tau \simeq 1/\pi$ provides a measure of the minimum frequency bandwidth of the ultrashort pulse. The bandwidth $\Delta\lambda$ required from the amplifying medium depends on the central wavelength as $\Delta\lambda \approx \Delta\nu\lambda_0^2/c$. On the contrary, the *relative bandwidth* $\Delta\lambda/\lambda_0$ provides a more convenient and natural bandwidth measure because it does not depend on the central wavelength, it is the same in wavelength and frequency

F. X. Kärtner (Ed.): Few-Cycle Laser Pulse Generation and Its Applications,
Topics Appl. Phys. **95**, 3–71 (2004)

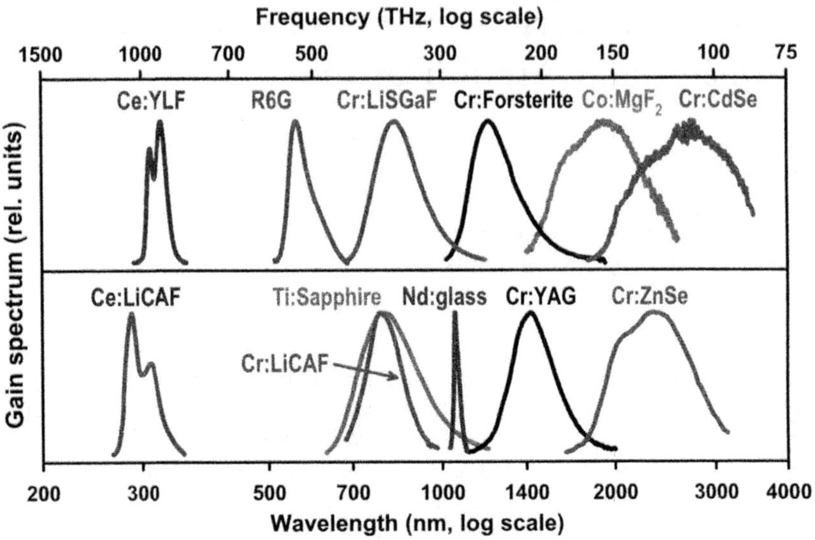

Fig. 1. Overview of broadband laser materials. The wavelength and frequency are plotted on a logarithmic scale to preserve the relative bandwidth $\Delta\lambda/\lambda$. R6G: Rhodamine 6G dye solution in ethanol

domains, and it is directly connected to the number of cycles per pulse: $(\Delta\lambda/\lambda_0)^{-1} \approx (\Delta\nu/\nu_0)^{-1} \propto N$. For this reason, the bandwidth in the rest of this chapter will be used in the sense of the above definition, i.e., as relative bandwidth $\Delta\lambda/\lambda_0$.

Figure 1 compares the bandwidths and wavelength regions covered by different broadband laser and amplifier materials. The wavelength (frequency) is given on a logarithmic scale. This allows a correct comparison in terms of the $\Delta\lambda/\lambda$ ratio of media operating in different wavelength regions.

A number of mechanisms exist for electronic transition line broadening that include vibronic, collision, Raman, Doppler, rotational, natural, and other mechanisms. From these, only a few mechanisms are strong enough to create line broadening comparable to the transition frequency. In gas laser media, even the combined action of Doppler, collisional, and (in molecular gases) rotational broadening turns out to be insufficient to produce large enough $\Delta\lambda/\lambda$ values. Therefore, the gas gain media used in a classical way are not considered prospective for few-cycle pulse generation (but may well be used in nonlinear-optical schemes [3]). The energy levels are subject to the strongest broadening, when the optically active electron directly interacts with the surrounding in condensed matter. This is generally the case for vibrational broadening in transition-metal (TM) doped crystals, dye molecules in solutions or matrices, and color-center crystals (Fig. 2). These materials are considered in detail in Sect. 3. The electrons in inner atomic shells experience much weaker interaction with the surroundings due to shielding by

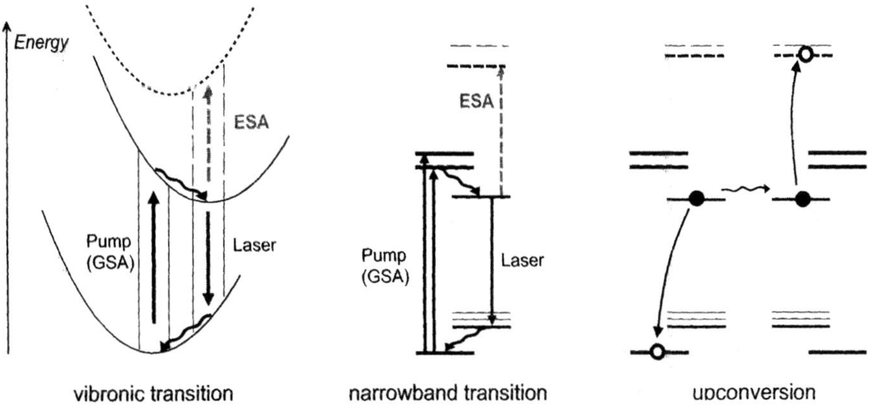

Fig. 2. Schematic diagram of pumping and radiation processes in vibronic-type and rare-earth-based narrowband laser media. Also shown are the processes of excited-state absorption (ESA) and upconversion

the outer shells, and hence broadening of the energy levels is lower. This is the case for rare-earth (RE) lanthanide and actinide ions (Fig. 2). However, the splitting of the levels in solids can be strongly influenced by the surrounding ligand field. In disordered media, ions experience a different local crystalline field, and the laser transition becomes inhomogeneously broadened. The broadband materials operating on this principle are considered in detail in Sect. 4.

It should be mentioned that the terms "homogeneous" and "inhomogeneous" broadening have somewhat different meanings in spectroscopy and laser physics. For lasers and amplifiers, the main feature of a homogeneously broadened transition is the absence of spectral hole burning, i.e., the gain spectrum retains its shape under saturation. Sometimes, inhomogeneously broadened transitions also behave this way. For example, vibronic broadening is, strictly speaking, not homogeneous, yet the high phonon frequencies, fast dephasing, and thermalization rates within vibronic states prevent spectral hole burning.

If the laser (amplifier) is operating in the ultrashort pulse regime, then the situation becomes more complicated. At room temperature, only phonon states with energies below $kT = 210\,\mathrm{cm}^{-1}$ (corresponds to vibrational period of 160 fs) are significantly populated and a few-cycle optical pulse is shorter than the typical period of ion vibrations. During interaction with an ultrashort light pulse, the ions are therefore "frozen" and have fixed (and different) transition wavelengths (Sect. 3, Fig. 3). Thus, the medium behaves as if it were inhomogeneously broadened during single-pulse propagation. At the same time, the medium completely recovers to the moment of the next pulse arrival because the time interval between the consecutive pulses lies in the nanosecond range, whereas thermalization occurs typically on a picosecond timescale. Thus, no spectral hole burning occurs, and, on average, the medium behaves as if it were homogeneously broadened. If the energy flux of an ultrashort pulse is comparable or exceeds the saturation flux of the medium, then the gain saturation will become significant even during a single pass, and the instantaneous inhomogeneity will result in gain spectrum distortion. This situation may happen, e.g., during the final pass through the amplifier, if the pulse duration is of the order of a few picoseconds or less. For oscillators, the pulse flux is comparable to the saturation flux of the gain medium only in systems with the pulse repetition period approaching the gain (absorber) relaxation time (as, e.g., in dye lasers).

This chapter focuses on the broadband active-ion-doped solid-state media. Color-center crystals, also offering broad gain bands in different spectral regions, have now only limited use for ultrashort pulse generation. The reason is the necessity for cumbersome cryogenic apparatus and low saturation intensity that is detrimental for use in amplifiers. Nonlinear-optical conversion techniques (see, e.g., the Chapter by *Kobayashi* et al., and [4,5]) have proved more convenient.

2 Oscillator and Amplifier Operation

In this section, we shall introduce the rate equation model, suitable for description of the broadband oscillators and amplifiers. The model will take into account ground-state absorption (GSA), excited-state absorption (ESA), and the upconversion processes, which will be described in detail in the following paragraphs. To avoid unnecessary details and to simplify the expressions, the spatial effects (transversal as well as the longitudinal inhomogeneity of the inversion and optical field) will be neglected. The balance equation for the population of the upper laser level n that takes into account the upconversion can be written as [6]

$$\frac{\mathrm{d}n}{\mathrm{d}t} = R^{\mathrm{pump}} - F^{\mathrm{las}}\sigma_{\mathrm{em}}n - \frac{n}{\tau} + F^{\mathrm{las}}\sigma^{\mathrm{las}}_{\mathrm{GSA}}(N_{\mathrm{t}} - n) - \alpha n^2\,, \tag{1}$$

where the first three terms stand for pumping rate, stimulated emission, and spontaneous decay, respectively. The last two terms describe the GSA at laser wavelength and the upconversion with the macroparameter α. N_{t} is the total concentration of active ions, n is the excited ion concentration, σ_{em} and $\sigma^{\mathrm{las}}_{\mathrm{GSA}}$ are the emission and GSA cross sections at the laser wavelength, F^{las} is the intracavity emission photon flux, and τ is the lifetime of the upper laser level. Since all known broadband materials employ optical pumping, the pumping rate can be written as

$$R^{\mathrm{pump}} = F^{\mathrm{pump}}\sigma_{\mathrm{abs}}(N_{\mathrm{t}} - n)\,, \tag{2}$$

where σ_{abs} is the pump absorption cross section and F^{pump} is the pump photon flux. The ESA terms do not enter (1) only if the higher lying states decay rapidly back to the upper laser level and therefore ESA does not change the population n. If the upper states have long lifetimes or possess different decay routes, then the laser schemes and rate equations may become quite complicated as, e.g., in the case of Er^{3+} ions [7, 8, 9], but for common broadband materials approximation (1) is valid. However, ESA affects the pumping efficiency, because the incident pump intensity I^{pump} is absorbed by both the ions in the ground state and the ions in the excited state:

$$\frac{\mathrm{d}I^{\text{pump}}}{\mathrm{d}z} = h\nu^{\text{pump}} F^{\text{pump}} \left[\sigma_{\text{abs}}(N_{\text{t}} - n) + \sigma_{\text{ESA}}^{\text{pump}} n\right], \tag{3}$$

where $\sigma_{\text{ESA}}^{\text{pump}}$ is the ESA cross section at the pump wavelength and z is the coordinate along the pump beam. Finally, the rate equation for photon flux at the laser wavelength can be written as

$$\frac{\mathrm{d}F^{\text{las}}}{\mathrm{d}t} = cF^{\text{las}}\sigma_{\text{em}}n - \frac{F^{\text{las}}}{\tau_{\text{c}}} - cF^{\text{las}}\sigma_{\text{ESA}}^{\text{las}}n - cF^{\text{las}}\sigma_{\text{GSA}}^{\text{las}}(N_{\text{t}} - n), \tag{4}$$

where the first two terms on the right-hand side stand for the stimulated emission and loss and the last two terms describe ESA and GSA at the laser wavelength. The photon lifetime in the resonator $\tau_{\text{c}} = l/c(T + L)$, where l is the effective cavity round-trip length and $T+L$ are the logarithmic round-trip losses, consisting of the output coupling T and other losses L.

It should be understood that gain and ESA cross sections, as well as the pulse itself, have broad spectra, and the interaction terms $F^{\text{las}}\sigma_{\text{em}}n$, etc., are averaged overlap integrals of the corresponding spectra. With this understanding, system (1–4) allows us to describe the oscillator and amplifier operation of the medium.

2.1 Oscillator

From the point of view of the laser medium, ultrafast oscillator operation is, in the first approximation, the same as continuous-wave (CW) operation. This is so because the optical pulse repetition period is significantly shorter than the upper laser level lifetime and the change of the population inversion per pass is relatively small. To obtain $\mathrm{d}F^{\text{las}}/\mathrm{d}t \geq 0$ from (4), the population n must exceed the threshold value n_{th} given by

$$n_{\text{th}} = \frac{T + L}{l\left(\sigma_{\text{em}} - \sigma_{\text{ESA}}^{\text{las}}\right)}\left[1 + c\tau_{\text{c}}\sigma_{\text{GSA}}^{\text{las}}(N_{\text{t}} - n_{\text{th}})\right]. \tag{5}$$

Though this is not a closed-form relation (n_{th} is present on both sides), it is convenient to use because the last term in the parenthesis explicitly describes GSA action (quasi-three-level operation) and becomes unity in the four-level scheme. The threshold absorbed pump intensity is then obtained from (1) at

$\mathrm{d}n/\mathrm{d}t \geq 0$ and $F^{\text{las}} = 0$ as

$$I_{\text{th}}^{\text{pump}} = (T+L)\frac{h\nu^{\text{las}}}{\sigma_{\text{em}}\tau}\frac{\lambda^{\text{las}}}{\lambda^{\text{pump}}}\left[1 + c\tau_c\sigma_{\text{GSA}}^{\text{las}}(N_{\text{t}} - n_{\text{th}})\right] \times \left(1 - \frac{\sigma_{\text{ESA}}^{\text{las}}}{\sigma_{\text{em}}}\right)^{-1}\left(1 + \frac{\sigma_{\text{ESA}}^{\text{pump}} n_{\text{th}}}{\sigma_{\text{abs}}(N_{\text{t}} - n_{\text{th}})}\right)(1 + \alpha n_{\text{th}}\tau)\,, \quad (6)$$

where the first row describes the standard result for a quasi-three-level laser and includes the losses $T+L$, the saturation intensity $h\nu/\sigma\tau$, the Stokes shift $\lambda^{\text{las}}/\lambda^{\text{pump}}$, and the GSA term. The last three terms in (6) describe ESA at the laser wavelength, ESA at the pump wavelength, and upconversion, respectively. From the same equations, one can obtain the expression for the slope efficiency with respect to absorbed pump power:

$$\eta_{\text{slope}} = \eta_{\text{pump}}\frac{\lambda^{\text{pump}}}{\lambda^{\text{las}}}\frac{T}{T+L}\left(1 - \frac{\sigma_{\text{ESA}}^{\text{las}}}{\sigma_{\text{em}}}\right)\left(1 + \frac{\sigma_{\text{ESA}}^{\text{pump}} n_{\text{th}}}{\sigma_{\text{ESA}}^{\text{pump}}(N_{\text{t}} - n_{\text{th}})}\right)^{-1}\,, \quad (7)$$

where η_{pump} describes the pump efficiency , the next two terms represent the Stokes shift and output coupling efficiency, and the last two terms stand for the efficiency loss due to ESA at the laser and pump wavelengths, respectively.

2.2 Amplifier

The pumping of an amplifier can be performed in two ways: a relatively slow buildup of the inversion under quasi-CW pumping or by a pumping pulse with duration much shorter than the upper laser level lifetime. In both cases, the inversion density necessary to obtain logarithmic gain $T + L$ is given by (5).

Long-pulse pumping takes place when a pump source with limited instantaneous intensity is used, e.g., laser diodes [10]. In this case, the population buildup is governed by the relaxation process, and the final inversion can be obtained from (1) and (3) in the absence of laser radiation as

$$n = I\frac{\sigma_{\text{abs}}\tau}{1 + \alpha n\tau}\left(1 + \frac{\sigma_{\text{ESA}}^{\text{pump}} n}{\sigma_{\text{abs}}(N_{\text{t}} - n)}\right)^{-1}\,, \quad (8)$$

where I is the absorbed pump intensity. This is again not a closed-form expression because n is found in both sides of (8), but such form allows separating the actions of upconversion and ESA.

In the case of a short-pulsed pump source (e.g., a Q-switched laser), the population buildup during the pump pulse can be obtained by integrating (1) and neglecting the relaxation and upconversion terms:

$$n = N_{\text{t}}\left(1 - \mathrm{e}^{-J/J_{\text{sat}}}\right)\,, \quad (9)$$

where J is the total energy fluence of the pump pulse and $J_{\text{sat}} = h\nu/\sigma_{\text{abs}}$ is the absorption saturation fluence of the medium. The influence of ESA results in a decrease in pumping efficiency and can be estimated by dividing the final energy density stored in the gain medium $h\nu n$ by the energy density absorbed in the medium (3). The exact solution depends on the pump pulse profile and in the case of the rectangular pulse, results in

$$\eta_{\text{pump}} = \left[1 + \frac{\sigma_{\text{ESA}}^{\text{pump}}}{\sigma_{\text{abs}}}\left(\frac{J/J_{\text{sat}}}{1 - \mathrm{e}^{-J/J_{\text{sat}}}} - 1\right)\right]^{-1} . \tag{10}$$

The decrease in pump efficiency is highest when the pumping pulse deeply saturates the medium. In this case, the excited-state population is high, and absorption occurs mostly due to ESA.

Finally, the small-signal gain coefficient during a single pass is obtained from (4) as

$$g = \sigma_{\text{em}} n \left(1 - \frac{\sigma_{\text{ESA}}^{\text{las}}}{\sigma_{\text{em}}}\right) - \sigma_{\text{GSA}}^{\text{las}}(N_{\text{t}} - n) . \tag{11}$$

The effect of ESA and GSA is to reduce the available gain, whereas the upconversion does not affect the gain performance of the amplifier.

2.3 Ideal Broadband Medium

In the ideal case, the gain medium operates in a pure four-level scheme (GSA is absent), there is no excited-state absorption, and upconversion is absent. Still, some fundamental limitations exist, many of which follow from the broadband nature of the medium. Following basic principles [11, 12, 13], we can write the following relation for the total spontaneous decay rate $1/\tau_{\text{rad}}$, valid for any type of optical center:

$$\frac{1}{\tau_{\text{rad}}} = \frac{8\pi}{3hc}\int_0^{\infty} [I_x(\lambda) + I_y(\lambda) + I_z(\lambda)]\,\lambda\,\mathrm{d}\lambda , \tag{12}$$

where $I_j(\lambda)$ is the fluorescence signal from the given level, polarized along the crystalline axis j in watt per unit wavelength interval. It is connected with the emission cross section:

$$I_j(\lambda) = \frac{hc^2 n_j^2}{\lambda^5}\sigma_j(\lambda) , \tag{13}$$

where n_j is the index of refraction in j-polarization. Taking for simplicity the isotropic case, we obtain the following fundamental relation:

$$\frac{1}{\tau_{\text{rad}}} = 8cn^2 \int \frac{\sigma_{\text{em}}(\lambda)\,\mathrm{d}\lambda}{\lambda^4} \approx 8cn^2 \sigma_{\text{em}}(\lambda_0)\frac{1}{\lambda_0^3}\frac{\Delta\lambda}{\lambda_0} . \tag{14}$$

This relation sets the minimum value of the oscillator pump threshold because substituting the $(\sigma_{em}\tau_{rad})$ product from (14) in (6) yields

$$I_{th}^{pump} = 8hc^2(T+L)\left(\frac{\tau}{\tau_{rad}}\right)^{-1}\frac{\lambda}{\lambda^{pump}}\frac{n^2}{\lambda^4}\frac{\Delta\lambda}{\lambda}, \tag{15}$$

where τ/τ_{rad} ratio is the quantum yield of luminescence at working temperature, and it is unity in the ideal case.

It is clearly seen from (15) that the threshold increase is the price paid for the broad bandwidth. As a result of the high threshold intensity, CW oscillation in broadband media could not be obtained under flashlamp pumping, and all successes with CW tunable lasers became possible only with laser pumping: dye lasers under an Ar^+-laser pump [14], color-center laser under a Kr^+laser pump [15], first solid-state Cr:GSGG laser under a Kr^+laser pump [16] followed by Ti:sapphire under an Ar^+-laser pump [17]. For the same reason only a few broadband materials ever allowed diode-pumped operation: Nd:glass [18], Cr:LiSAF [19] and Cr:LiSGaF [6,20], Cr:YAG [21,22, 221], and recently also, Cr:ZnSe [23,24] and Cr:ZnS [25].

A further decrease in the oscillation threshold can be obtained in birefringent crystals, when the emission cross section in one polarization is much higher than that in the other two. In this case, the $(\sigma_{em}\tau_{rad})$ product can be up to three times higher (12–14). It also follows from (15) that the pump threshold is lower in infrared media.

2.4 Real-World Media

The expressions derived in the preceding sections allow analyzing the influence of different spectroscopic and material parameters on the laser and amplifier properties of real-world media. No medium is ever perfect; besides, as we shall see, achieving good values in one set of parameters often requires certain trade-offs in another.

2.4.1 Nonradiative Decay

Nonradiative decay (lifetime quenching) is a main factor limiting the number of useful broadband media at room temperature. The mechanisms of nonradiative decay may vary and will be described in the corresponding sections. In many cases, temperature quenching is the dominant cause and the upper laser level lifetime τ acquires strong dependence on temperature. The nonradiative transition rate W_{nr} results in shortening the lifetime according to the expression,

$$\frac{1}{\tau} = \frac{1}{\tau_{rad}} + W_{nr}\,. \tag{16}$$

The direct consequences of lifetime shortening are an increase in the pump threshold and a decrease in the maximum gain under slow pumping by a factor of $\tau_{\rm rad}/\tau$ (6, 8, 15). Strong temperature quenching limits the usability of many broadband media at room temperature, e.g., color-center crystals and many transition-metal-doped crystals. At the same time, lifetime shortening does not affect slope efficiency or gain spectrum and does not prevent the operation in a short-pulse pumped amplifier, as long as the lifetime still exceeds the pump pulse duration.

2.4.2 Upconversion

Upconversion is an interionic process, taking place between two or more excited optical centers, when the excitation from one of the ions promotes another excited ion to the higher lying energy level. The upconversion term αn^2 enters the basic rate equation (1) as an additional relaxation term, next to the decay term n/τ. Consequently, its action is equivalent to lifetime shortening by a factor $1/(1+\alpha n\tau)$ (6, 8). Upconversion therefore increases the oscillation threshold and decreases the maximum achievable amplifier gain under slow pumping [26, 27], but does not influence the slope efficiency or the gain spectrum. Because upconversion converts one excitation photon into heat (during nonradiative relaxation from the upper lying levels), it produces additional heating of the medium. The extra heating may cause secondary effects, e.g., higher GSA or further reduction of the lifetime due to thermal quenching. These secondary effects sometimes cause dramatic consequences: distortion of the gain profile [28] or catastrophic quenching of the fluorescence and optical bistability [29]. At the same time, since the population inversion is fixed at the threshold value in the oscillators, the heat flow due to upconversion does not grow with the pumping rate in oscillators.

The product $\alpha n\tau$ scales with the upper laser level population and lifetime, and therefore upconversion is less important in four-level and short-lived media than in three-level and long-lived media. As a rule of thumb, upconversion plays a major role in media with millisecond lifetimes but can be neglected in media with lifetimes less than 10 µs.

2.4.3 Ground-State Absorption

GSA (reabsorption) is equivalent to saturable losses, causing an increase in the laser threshold and a decrease in the amplifier gain. It does not, however, reduce the slope efficiency of an oscillator because the populations of the ground and excited states are clumped at the threshold. During the GSA process, the ion is excited to the upper laser level, and thus GSA does not cause additional heating of the medium, unlike all other loss mechanisms. More important is the fact that GSA modifies the gain spectrum. In all practical cases, GSA results in certain gain spectrum narrowing from the blue side.

2.4.4 Stokes Shift

The Stokes shift $\lambda/\lambda^{\text{pump}}$ is the main cause of an active medium heating at high pumping levels. The Stokes loss (energy difference between the pump photon energy and emitted photon energy) is transferred directly to the lattice, and this additional heat load scales with pump power. Obviously, GSA-free four-level laser operation requires a certain minimum value of the Stokes shift, so that the absorption band does not overlap the emission band. Lower Stokes shift losses are realized in quasi-three-level systems, at the expense of higher threshold and spectrum deformation. A low threshold, obtained in four-level schemes is important in small- and medium-power systems. On the contrary, high-power systems benefit more from the reduced heat production in materials with small Stokes shift, at the expense of a higher threshold. A good example is the Yb^{3+} ion (Sect. 4.2), operating in the three-level scheme: low Stokes losses make Yb^{3+}-doped media materials of choice for high average power systems.

2.4.5 Excited-State Absorption

ESA *at pump wavelength* does not modify the gain spectrum but decreases the pump efficiency, therefore affecting the threshold pump intensity and slope efficiency of an oscillator, and reducing the maximum amplifier gain. As a result of the ESA process, pump photons are converted into additional heating of the crystal. This additional heat load grows in proportion to the pump power. The ESA cross section at pump wavelength may have the same order of magnitude as the absorption cross section itself, or even exceed it. However, this does not inhibit oscillator operation, as long as the upper level population in four-level schemes is relatively small, $n \ll N_{\text{t}}$, and the ratio $\sigma_{\text{ESA}}^{\text{pump}} n_{\text{th}}/\sigma_{\text{abs}}(N_{\text{t}} - n_{\text{th}})$ remains small as well. In amplifiers, this type of ESA may be an obstacle to achieving high population densities. The heat load due to ESA at the pump wavelength scales with pump power.

ESA *at laser wavelength* negatively affects all parameters of laser operation and produces an additional heat load proportional to the laser power. ESA at laser wavelength also significantly modifies the gain spectrum. Since ESA and the stimulated emission share the same starting level, the two cross sections enter all expressions as the $(1 - \sigma_{\text{ESA}}^{\text{las}}/\sigma_{\text{em}})$ term, where the two values are compared directly. Therefore, the influence of ESA at laser wavelength does not depend on the pumping level and excited-state population and does not decrease in four-level media as in the case of upconversion or ESA at pump wavelength. ESA at laser wavelength turns out to be the most important factor that renders a wealth of promising broadband materials with excellent parameters unusable (see, e.g.,[30] and references therein).

2.4.6 Host Properties

All processes and parameters mentioned above were related to the active ion, and they directly entered the rate equations. There also exist a number of host-related parameters which significantly influence few-cycle laser or amplifier operation, e.g., the second-order and third-order nonlinear susceptibilities $\chi^{(2)}$ and $\chi^{(3)}$, optical damage threshold, thermal conductivity κ, thermo-optical constant $\mathrm{d}n/\mathrm{d}T$, thermal expansion coefficient α, dispersion, etc. For medium to high average power operation, the thermal and thermo-optical characteristics of the medium play a central role, whereas the value of nonlinear susceptibility $\chi^{(3)}$ is pivotal for Kerr-lens mode-locked and solitary lasers.

2.4.7 Optimal Gain

The threshold condition (6) depends only on a $\sigma_{\mathrm{em}}\tau$ product, and media with short lifetimes and large emission cross sections are in principle as good for obtaining positive gain as long-lived media with small emission cross sections . This is, however, not true if other considerations are taken into account. First of all, the critical population n_{th} is inversely proportional to the emission cross section (5). Therefore, short-lived media with large emission cross sections suffer much less from ESA at pump wavelength and from upconversion. They are also less sensitive to losses and can provide much higher unsaturated gain per pass. At the same time, these media possess low saturation intensity, which limits their use as amplifiers and is detrimental for nonlinear-optical mode-locking schemes [31]. The optimal gain cross section values for few-cycle pulse generation lie typically between $10^{-18}\,\mathrm{cm}^2$ and $10^{-20}\,\mathrm{cm}^2$.

3 Transition-Metal-Doped Materials

The vibronically broadened 3*d*–3*d* transitions of iron group ions proved most successful in providing excellent solid-state materials for few-cycle pulse generation, led by Ti^{3+}:sapphire. The spectroscopy of these ions in a crystalline environment is well understood, so that now it is possible to speak about "engineering" of transition-metal (TM)-doped laser materials [32]. The theory of 3*d*-electron energy levels in a crystalline environment was developed by *Tanabe* and *Sugano* in 1954 [33,34] and turned out to be very useful because it allowed a complete qualitative description of the spectroscopic properties of all possible 3*d*-ions in four-fold (tetrahedral), six-fold (octahedral) and eight-fold (cubic) coordination on the basis of just nine diagrams (Tanabe–Sugano diagrams [32,33,34,35,36]). Using only two fitting parameters Dq/B to characterize ligand field strength and $\gamma = C/B$ to characterize the ion, one can use Tanabe–Sugano diagrams for quantitative description as well [35,36]. The Tanabe–Sugano diagrams are usually drawn for an octahedral environment. The diagrams for the $3d^n$ ion in a tetrahedral or cubic field can then be obtained from the octahedral $3d^{10-n}$ diagram and by correcting the field strength according to the rule [35],

$$Dq(\text{tetrahedral}) = -4/9Dq(\text{octahedral}) = 1/2Dq(\text{cubic})\,. \tag{17}$$

Table 1. Overview of $3d$ transition metal ions. Ions, where tunable generation has been demonstrated, are underlined. (t): tetrahedral coordination, (o): octahedral coordination

$3d^1$	$3d^2$	$3d^3$	$3d^4$	$3d^5$	$3d^6$	$3d^7$	$3d^8$	$3d^9$
$\underline{Ti^{3+}}$(o)	Ti^{2+}							
V^{4+}	V^{3+}	$\underline{V^{2+}}$(o)						
Cr^{5+}	$\underline{Cr^{4+}}$(t)	$\underline{Cr^{3+}}$(o)	$\underline{Cr^{2+}}$(t)					
Mn^{6+}	$\underline{Mn^{5+}}$(t)	Mn^{4+}	Mn^{3+}	Mn^{2+}				
	Fe^{6+}		Fe^{4+}	Fe^{3+}	$\underline{Fe^{2+}}$(t)			
					Co^{3+}	$\underline{Co^{2+}}$(o)		
						Ni^{3+}	$\underline{Ni^{2+}}$(o)	Ni^{+}
							Cu^{3+}	Cu^{2+}

In all known TM-doped laser crystals, the ions occupy sites with either octahedral or tetrahedral symmetry. The important difference is that the octahedral sites are centrosymmetric, whereas tetrahedral sites are not. Strictly speaking, the transitions within the the $3d^n$ configurations are transitions of the same parity and obey the *Laporte* rule [37], i.e., are electric-dipole forbidden. This selection rule can be weakened as a result of the influence of odd-parity crystal-field terms, which mix the d and p orbitals. At octahedral sites, this admixture can occur due to the symmetry distortions or as a result of the odd-parity ion vibrations, when the ion is shifted away from the central position. At tetrahedral noncentrosymmetric sites, the d–p admixture is stronger and always present. The result is the generally higher transition cross section and correspondingly lower radiative lifetimes observed for TM ions at tetrahedral sites. Compare, e.g., the octahedrally coordinated Cr^{3+}:LiCAF [74] (emission cross-section $\sim 10^{-20}\,\text{cm}^2$, $\tau = 170\,\mu\text{s}$) with the tetrahedrally coordinated Cr^{2+}:ZnSe or Cr^{2+}:ZnS (emission cross section $\sim 10^{-18}\,\text{cm}^2$, $\tau \sim 5\,\mu\text{s}$).

Another result of the stronger d–p admixture at a tetrahedral site is the lower crystal-field stabilization of ions [38]. During optical transitions, the charge from the impurity-centered orbitals (i.e., d-like e states) is redistributed to ligand-centered orbitals (i.e., the p–d hybridized t_2 states) [39,40]. The transitions are therefore not purely internal, and ions at tetrahedral sites are prone to change valence and build charge-transfer bands. Such charge-

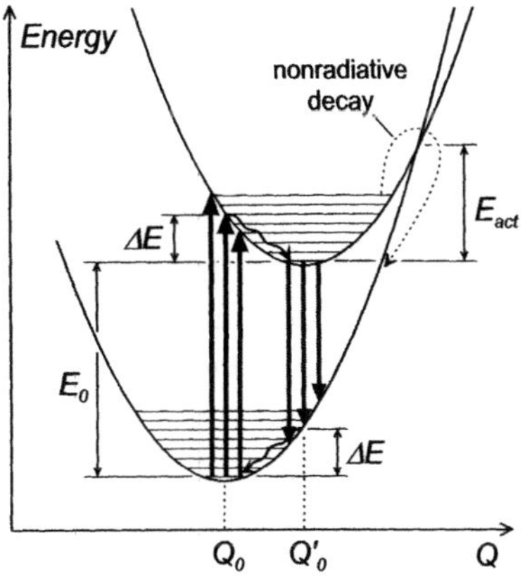

Fig. 3. Energy level diagram of the vibronic transition in a single-coordinate approximation

transfer processes in some cases may affect the laser performance of tunable CW lasers with active ions at tetrahedral sites [21, 24, 25, 41].

In addition to the main crystal-field splitting as described by the Tanabe–Sugano diagram, there exist also additional fine structures caused by local field distortions and low-energy interactions. This fine structure can be seen in the low-temperature spectra or in the narrowband transitions within the same electronic configuration (e.g., R-lines in ruby). However, the linewidth broadening caused by these interactions is much lower than the electron–phonon interaction, which is most interesting from the point of view of few-cycle generation.

For the analysis of line bandwidth, an ion in a crystalline environment can be approximated by a harmonic oscillator. In Fig. 3, the potential energy of the ion in different electronic states is shown as parabolae along a generalized configurational coordinate, corresponding to one of the vibrational modes of the lattice. In principle, corresponding diagrams should be drawn for every vibrational mode, but in practice only one mode, corresponding to the strongest vibronic coupling, is of importance. The electronic wavefunctions in the upper and lower states are different, and hence the equilibrium positions of the ion along the vibrational coordinate are not the same. However, since both states belong to the same d state, the curvatures of the parabolae can be considered equal with high accuracy [35]. Electronic transitions occur on a timescale much faster than vibrational movement of heavy ions (the *Frank–Condon* principle, vertical arrows in Fig. 3), so that absorption and emission of photons leave the ion in an excited vibrational state. The excess energy ΔE is transferred nonradiatively to the phonons of the lattice, producing a Stokes shift $\approx 2\Delta E$ between absorption and emission photon energies. For laser purposes, it is advantageous to have $\Delta E \gg kT$ so that the thermal population of the lower laser level is low (four-level laser scheme). The observed lineshapes of both absorption and emission depend on the temperature and at $T = 0$ are described by a set of lines corresponding to the transitions between individual phonon states with cross sections [35]

$$\sigma(E_0 + m\hbar\omega) \propto \frac{\exp(-S)S^m}{m!}, \tag{18}$$

where $\hbar\omega$ is the phonon energy, and $S = \Delta E/\hbar\omega$ is the Huang–Rhys parameter [42]. The bandwidth of this lineshape is equal to $\Gamma(0) = 2.35\hbar\omega\sqrt{S}$. At higher temperatures, the expression for the bandshape becomes more complex and approaches the Gaussian form with an energy bandwidth given by [35]

$$\Gamma(T)^2 = \Gamma(0)^2 \coth\left(\frac{\hbar\omega}{2kT}\right), \tag{19}$$

which translates to the relative bandwidth

$$\frac{\Delta\nu}{\nu_0} \approx \frac{\Gamma(T)}{E_0} = 2.35\sqrt{\frac{\Delta E}{E_0}\frac{\hbar\omega}{E_0}\coth\left(\frac{\hbar\omega}{2kT}\right)} \leq 4.7\sqrt{\frac{\Delta E}{E_0}\frac{kT}{E_0}}, \tag{20}$$

using $\hbar\omega/2kT \leq 2$ for typical energies of coupling phonons at room temperature. It follows from (20) that a large Stokes shift $\Delta E/E_0$ is favorable for the broader transition bandwidth. However, large $\Delta E/E_0$ also means that the crossing point of the two parabolic potential curves is located lower on the energy scale. The crossing is responsible for thermally activated nonradiative decay with activation energy E_{act} (dashed arrow in Fig. 3), which inversely scales with the Stokes shift $E_{\mathrm{act}} \propto E_0(E_0/\Delta E)$. It can be easily derived from Fig. 3 that when $\Delta E/E_0 > 1/3$, the activation energy E_{act} is less than ΔE and transition becomes nonradiative, i.e., luminescence is quenched at any temperature. In practice, nonradiative decay sets up earlier through tunneling and W_{nr} depends strongly on temperature, having an exponential asymptotic of $\sim \exp(-E_{\mathrm{act}}/kT)$ at higher temperatures [35, 43, 44, 45, 46]. Using $\Delta E/E_0 \leq 1/4$ as a reasonable condition in (20), we obtain

$$\frac{\Delta\nu}{\nu_0} \leq 2.3\sqrt{\frac{kT}{h\nu_0}}. \tag{21}$$

For example, in Ti:sapphire, $\nu_0 = 16\,200\,\mathrm{cm}^{-1}$, and $\Delta\nu_0 = 3200\,\mathrm{cm}^{-1}$ [13], resulting in $\Delta\nu/\nu_0 = 0.2$; estimation by formula (21) gives $\Delta\nu/\nu_0 \leq 0.25$. Ti:sapphire has practically the maximum possible bandwidth in its wavelength range.

To complete the discussion of the spectroscopic features of TM-doped crystals, let us mention the Jahn–Teller effect [47,48,49]. It results in an additional spontaneous local distortion of the lattice, which removes electronic degeneracy. All triply (T) and doubly (E) degenerate terms therefore acquire additional splitting. This may significantly change the lineshape of the optical spectra. For example, Jahn–Teller splitting of the ^2E upper level in

Ti:Sapphire causes the double-peaked absorption band (Fig. 6) [13], whereas Jahn–Teller splitting of the $^5T_2$ ground state of Cr:ZnSe and Cr:ZnS results in the blue-side shoulder on the emission spectrum (Fig. 17) and an additional absorption in the mid-infrared [50, 51, 52, 53, 54, 55]. The Jahn–Teller splitting of the fluorescence band can significantly increase the bandwidth beyond vibronic broadening (e.g., in Cr:ZnSe and Cr:ZnS, see Fig. 17). Looking for media with large Jahn–Teller splitting is one of the possible ways to create new superbroadband materials [56].

3.1 Ti:sapphire

In the 1980s, for about a decade, ultrafast (sub-100 fs) technology was based on the dye solution and colliding-pulse mode-locking [1] technique that was invented about 10 years after the dye laser itself [14]. In the 1990s and up to date, analogous role is being played by the Ti:sapphire laser and the KLM technique [2] that was also introduced about 10 years after the Ti^{3+}:sapphire laser was invented by Moulton in 1982.

Ti^{3+}:sapphire possesses a favorable combination of properties, which were for many years the best among all known broadband laser materials and only recently has become matched by Cr^{2+}:ZnSe (see Table 2 and the next section). Sapphire Al_2O_3 has high thermal conductivity of $28\,W/(m \cdot K)$, exceptional chemical inertness and mechanical resistance. In sapphire, a Ti^{3+} ion substitutes for an Al^{3+} ion at an octahedral site. The introduction of Ti^{3+} in Al_2O_3 presents an interesting contrast to the introduction of Cr^{3+} in ruby or garnet, as the ionic radius of Ti^{3+} (0.67 A) is 26% larger than that of an Al^{3+} ion. Introducing the Ti^{3+} ion thus creates a strong perturbation in the lattice and leads to Ti ions in the wrong valence state, defect, or interstitial sites. As a result, inhomogeneous broadening [57, 58] and parasitic absorption at 800 nm [59, 60] are observed, especially at high Ti concentrations. After-growth annealing returns some of these parasitic ions to the correct valence and sites [60]. The progress in growth techniques and postgrowth processing resulted in significant improvement in the quality of commercial crystals. In the best crystals nowadays, the figure of merit achieves ~ 300.

Today, Ti^{3+}:sapphire is the most widely used commercially available tunable laser. The CW tuning curve of such a laser spans the wavelength range of over 400 nm between ~ 670 nm and 1080 nm [61] and is now duly obtained in commercial products (Fig. 4). The large Ti^{3+} ion in a dense environment experiences strong overlapping of its electron shell with ligands, causing strong electron–phonon coupling to the lattice (large Huang–Rhys parameter S), which is the key to the ultrabroadband operation of this laser and of its overwhelming success in few-cycle generation, as evidenced in many chapters in this volume. For an extensive review of the spectroscopic properties of Ti:sapphire, the reader is referred to the books [32, 36]. Here, only the most important features of this material will be outlined.

Table 2. Spectroscopic and material parameters of broadband transition-metal-doped laser materials. All parameters refer to room temperature. $\Delta\lambda/\lambda_0$: relative emission bandwidth; λ_0 (nm): maximum gain wavelength; $\Delta\lambda$ (nm): gain bandwidth (FWHM); σ_{em} (10^{-20} cm^2): emission cross section at λ_0; τ (μs): upper laser level lifetime; n_2 (10^{-16} cm^2/W): nonlinear refractive index at 1.06 μm, unless specified otherwise; κ $\left(W/(m\cdot K)\right)$: thermal conductivity (for colquiriites, the main value corresponds to the c-axis, value in parenthesis to a-axis); dn/dT (10^{-6} K^{-1}): thermo-optic coefficient (for colquiriites, the main value corresponds to the c-axis, value in parenthesis to a-axis)

Material	$\frac{\Delta\lambda}{\lambda_0}$	λ_0	$\Delta\lambda$	Pump source	σ_{em}	τ	n_2	κ	$\frac{dn}{dT}$	Ref.
Ce^{3+}:LiCAF	0.14	290	40	Nd-, Yb-lasers (4ω) Excimer	600	0.03	0.54[a]	5.1 (4.6)	−4.6 (−4.2)	[278, 279, 358, 359]
$Ti^{3+}:Al_2O_3$	0.3	790	230	Ar^+-laser, Nd-, Yb-lasers (2ω)	39	3.2	3.2	28	12.6	[13, 90]
Cr^{3+}:LiCAF	0.16	780	125	Kr^+-laser, diodes	1.3	170	0.4	5.1 (4.6)	−4.6 (−4.2)	[74, 75, 101, 102, 278]
Cr^{3+}:LiSGaF	0.21	842	180	Kr^+-laser, diodes	3.3	88	1.2	3.6 (3.4)	−1.8 (−2.7)	[101, 103, 112]
Cr^{3+}:LiSAF	0.22	855	190	Kr^+-laser, diodes	4.8	67	0.8	3.3 (3.0)	−4.0 (−2.5)	[101, 102, 112]
$Cr^{4+}:Mg_2SiO_4$ (Cr:forsterite)	0.2	1250	250	Nd-, Yb-lasers	14.5	2.7	2	5	3.8	[30, 183]
Cr^{4+}:YAG	0.2	1450	300	Nd-, Yb-lasers, diodes	33	4	6.2	10	8.9	[30, 90]
$Co^{2+}:MgF_2$	0.3	2050	650	Nd-laser (1.3 μm)	0.11	∼ 40	0.8	3.1	1	[90, 267]
Cr^{2+}:ZnS	0.38	2350	890	Tm-lasers, Er-fiber, diodes	140	4.3	90[b]	27	46	[254, 280]
Cr^{2+}:ZnSe	0.37	2450	900	Tm-lasers, Er-fiber, diodes	130	5.5	120[b] 110[c]	18	70	[225, 280, 281]

[a] at 532 nm [b] at 1.3 μm [c] at 1.55 μm

The lasing ion in Ti:sapphire is an octahedrally coordinated Ti^{3+} ion in $3d^1$ configuration. This is the simplest electronic system containing a single electron in the outer shell. In an octahedral field, the $3d$ level splits into two: a lower energy threefold degenerate $^2T_2$ level and the higher energy twofold degenerate ^2E level. These spin-doublet levels are the only two states in the octahedral symmetry (Fig. 5). This is a remarkable feature because in this case there are no higher excited states, to which ESA would be possible. The upper ^2E level is further split into two levels due to combined action of a trigonal crystal-field distortion, spin-orbit coupling, and the Jahn–Teller effect. The ground-state absorption into these two levels occurs in the blue–green

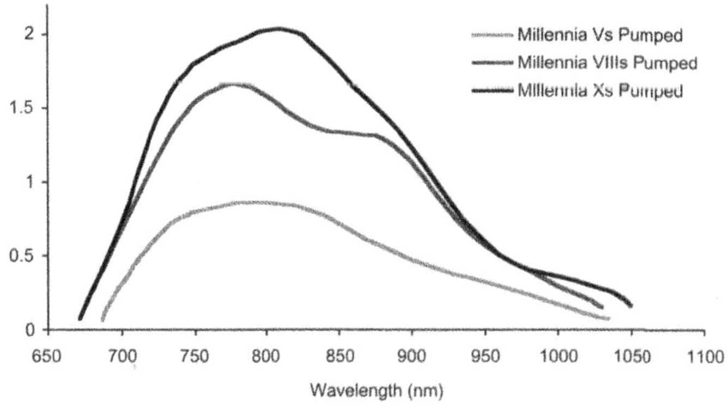

Fig. 4. The tuning curve of a commercial Ti:sapphire laser (adapted from a datasheet of a 3900S laser, Spectra-Physics, Inc.)

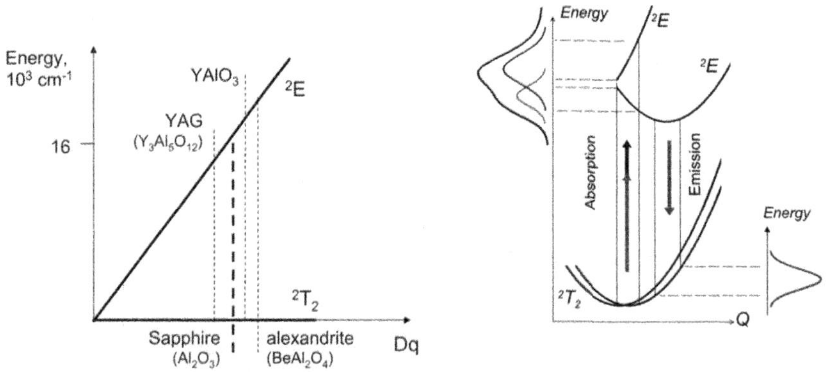

Fig. 5. *Left graph:* The Tanabe–Sugano diagram of an octahedral $3d^1$ ion (Ti^{3+}). *Vertical lines* denote crystal field values corresponding to YAG, sapphire, YALO, and alexandrite. *Right graph:* Coordinational diagram description of Ti:sapphire energy levels. Strong Jahn–Teller splitting of the upper ^2E state results in a double-peaked absorption; splitting of the lower $^2T_2$ state is hidden under the vibronic band

wavelength range peaking at 555 nm and 488 nm, while the luminescence on the $^2E \rightarrow ^2T_2$ transition occurs in a broad band around 800 nm (Fig. 6) with a peak emission cross section of $4 \times 10^{-19}\,\mathrm{cm}^2$ and a slightly quenched lifetime of approx. 3 µs at room temperature [13]. The Ti^{3+}:sapphire laser can be pumped by several pump sources, such as flashlamps, an Ar-ion laser, and frequency-doubled neodymium or ytterbium lasers. Direct diode-pumping of Ti:sapphire cannot be realized due to the lack in the foreseeable future of pump diodes, which would be able to create the necessary high pumping intensity to reach the threshold. The intrinsically high threshold of this material is connected with the large gain bandwidth and short central wavelength (15).

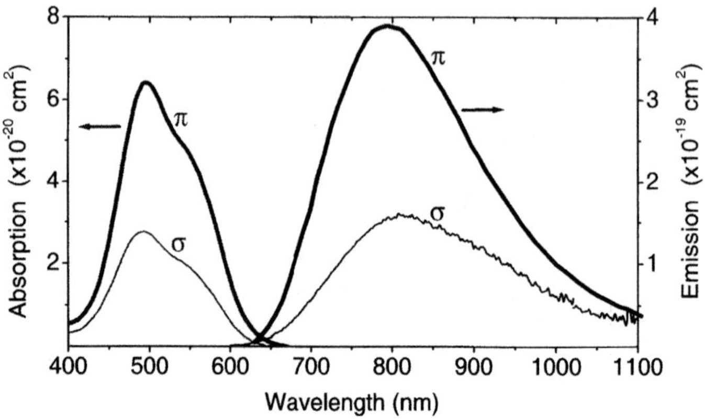

Fig. 6. Absorption and emission cross sections of Ti:sapphire. Note the fivefold vertical scale difference between absorption and emission

However, using very tight pumping, a 14 fs oscillator using relatively inexpensive low-power pump lasers could be realized [62].

Lasers using a number of other Ti^{3+}-doped crystals have been reported including alexandrite [63, 64, 65, 66], YAG [67], and $YAlO_3$ [68, 69], but none of them provided any advantages over Ti:sapphire, primarily due to the ESA into the conduction band (charge-transfer ESA) [30, 70].

3.2 Cr^{3+}-Doped Crystals

Numerous Cr^{3+}-doped crystals, disordered crystals as well as glasses, have been investigated to find an optimum laser medium. For an extensive review of all these materials, the reader is referred to [30, 32, 36]. Here, we briefly discuss the most important materials, which proved suitable for broadly tunable CW and ultrashort-pulsed operation.

The $3d^3$ Cr^{3+}-ions prefer to occupy octahedral sites in all known crystals. The absorption spectra consist of the two broad bands, corresponding to ${}^4A_2 \rightarrow {}^4T_1$ and ${}^4A_2 \rightarrow {}^4T_2$ transitions and filling practically the whole visible range (Fig. 7). The excited state of Cr^{3+}-ion consists of a superposition of two states, which are in thermal equilibrium with each other (Fig. 7): the metastable ${}^2E$ state and the relatively short-living ${}^4T_2$ state. The narrowband transition from ${}^2E$ to the ground state is parity and spin forbidden, whereas the broadband transition from the ${}^4T_2$ state is spin allowed [36]. The energy gap ΔE_{TE} between ${}^4T_2$ and ${}^2E$ levels plays an important role in making the laser transition broadband because it determines the population degree of the ${}^4T_2$ level. If the gap is negative (${}^4T_2$ lies below ${}^2E$) or is of the order of kT at room temperature, then the emission occurs from the ${}^4T_2$ level with a lifetime of the order of tens of microseconds. This is the case, for example, in such low crystalline field active media as Cr^{3+}-doped scandium garnets (e.g., Cr:GSGG, Cr:YSGG, Cr:GSAG) or colquiri-

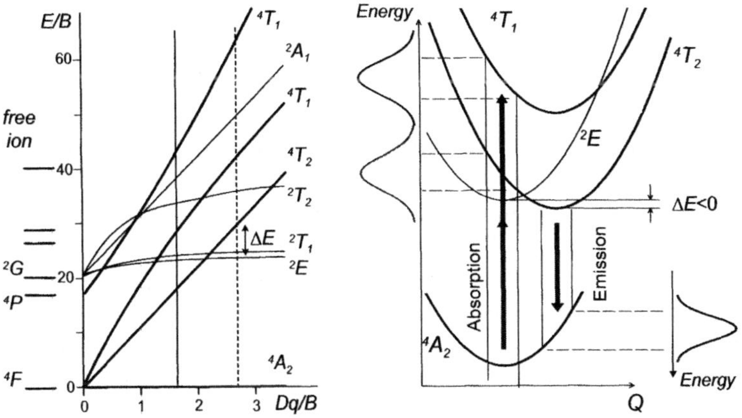

Fig. 7. *Left graph:* Simplified Tanabe–Sugano diagram of the Cr^{3+} ion at an octahedral site. *Thick lines* show quartet terms. The *dashed vertical line* corresponds to the high-field hosts (e.g., YAG, ruby, alexandrite) with $\Delta E > 0$. The *solid vertical line* corresponds to the low-field broadband media (GSGG, colquiriites) with $\Delta E < 0$. *Right graph:* The configurational coordinate diagram of energy levels of Cr^{3+} ion in the octahedral field (low-field case, e.g., LiCAF)

ites (Cr:LiSAF, Cr:LiSGaF, Cr:LiCAF). The opposite takes place in ruby ($\Delta E_{\mathrm{TE}} \sim 2300\,\mathrm{cm}^{-1}, \tau = 3\,\mathrm{ms}$ [36]) or Cr^{3+}:YAG ($\Delta E_{\mathrm{TE}} \sim 1500\,\mathrm{cm}^{-1}, \tau = 1.52\,\mathrm{ms}$ [71]), where only narrowband luminescence due to R-lines corresponding to the ${}^2E \rightarrow {}^4A_2$ transitions can be observed [72, 73].

From the large pool of Cr^{3+}-doped crystals (see, e.g., Table 4 in [30]), only few materials so far proved suitable for broadband tunable CW and ultrashort-pulsed operation. According to the Tanabe–Sugano diagram (Fig. 7), transition wavelengths of low crystalline field media are generally longer than those of high crystalline field media. Besides the central wavelength shift, Cr^{3+}-doped materials demonstrate other tendencies: lower field materials possess broader transitions, a larger Stokes shift, a higher nonradiative decay rate, and a higher ESA. Thus, shorter wavelength materials demonstrate better laser properties at the expense of the bandwidth (compare, e.g., otherwise analogous Cr:LiCAF and Cr:LiSAF [74, 75, 76, 77, 78], Fig. 8), and all practically important Cr^{3+}-doped crystals operate at wavelengths below 1 µm.

Successfully working Cr^{3+}-doped laser crystals include Cr:$BeAl_2O_4$ (alexandrite or chromium-doped chrysoberil), various Cr-doped scandium garnets, and Cr-doped colquiriites.

Cr:$BeAl_2O_4$ is the third material after ruby and YAG, in which Cr^{3+} was reported to lase and the first to tune [79]. Before the appearance of Ti:sapphire, alexandrite was the most popular and commercially available tunable laser. Good optical and thermo-optical properties (thermal conductivity $23\,\mathrm{W/(m \cdot K)}$), high quantum yield of 95% [32], and the absence of the concentration quenching of the lifetime, which is 260 µs at room temperature, allow flashlamp-pumped [80] and even sun-pumped operation [81]. CW operation was demonstrated with Ar^+ [82], Kr^+ [83], and diode pumping [84, 85]. One should also mention the n_2 value of $3.52 \times 10^{-16}\,cm^2/W$ at 1.06 µm [89, 90], low material dispersion near the gain maximum, and the gain cross section of $\sim 7 \times 10^{-21}\,cm^2$ [95], allowing high-power amplifier operation. In spite of these attractive features and room-temperature tunability [86], alexandrite has not found use as an ultrafast laser. This is mostly due to the narrower bandwidth ($\leq 1500\,cm^{-1}$) in comparison with Ti:sapphire in the same wavelength range. The attempts to mode-lock alexandrite lead to generation of actively mode-locked 35 ps pulses with a limited tunability of 30 nm around 750 nm [93] as well as passively mode-locked 10 ps pulses using various saturable absorbers [94]. The shortest passively mode-locked pulses reported so far were 700 fs to 800 fs [97].

An interesting property of alexandrite that might be useful for some special applications is that the gain in this material increases with temperature and therefore cooling is not required. The nature of this phenomenon lies in the fact that alexandrite has a positive $\Delta E_{TE} \sim 800\,cm^{-1}$ and the thermal population of the $^4T_2$ level increases with temperature. This leads to a temperature-dependent gain spectrum [91]. The authors of [92] reported a factor of 4 increase in output energy with a temperature increase from 38 °C to 225 °C.

The second developed class of tunable Cr^{3+}-doped crystals was scandium-containing garnets: Cr:GSGG (gadolinium scandium gallium garnet), Cr:YSGG with yttrium instead of gadolinium, and Cr:GSAG with aluminum instead of gallium. All of these crystals are low crystal field media and possess relatively broad gain ($\sim 1700\,cm^{-1}$) centered between 700 nm and 800 nm. Using flash-lamp pumping, continuous tuning was observed between 765 nm and 801 nm [96]. Using longitudinal Kr-laser pumping, CW operation tunable between 740 nm and 840 nm at 200 mW output power was achieved in both Cr:GSGG [98] and Cr:GSAG [99]. The main problem of chromium-doped garnets remains significant ESA at the pump wavelength. In spite of the fairly high third-order nonlinearity ($(9$ to $12) \times 10^{-16}\,cm^2/W$ at 1.06 µm [90]) and the theoretically achievable approx. three optical cycle duration, mode locking has not yet been demonstrated in these crystals.

Most successful Cr^{3+}-doped materials for few-optical-cycle generation are colquiriites. The colquiriite itself, $Cr:LiCAlF_6$ (Cr:LiCAF), and its derivatives, $Cr:LiSrAlF_6$ (Cr:LiSAF) and $Cr:LiSrGaF_6$ (Cr:LiSGaF), as well as their solid solutions, were first proposed at the end of the 1980s by the group of scientists at the LLNL in collaboration with *Chai* from CREOL [74, 76]. They were initially supposed to be employed in the Inertial Confinement Fusion Program.

Unlike Ti:sapphire that operates in the same wavelength region, these crystals have longer upper laser level lifetime (e.g., 67 µs in Cr:LiSAF and 88 µs in Cr:LiSGAF [100, 101]) and therefore can be flashlamp pumped. But

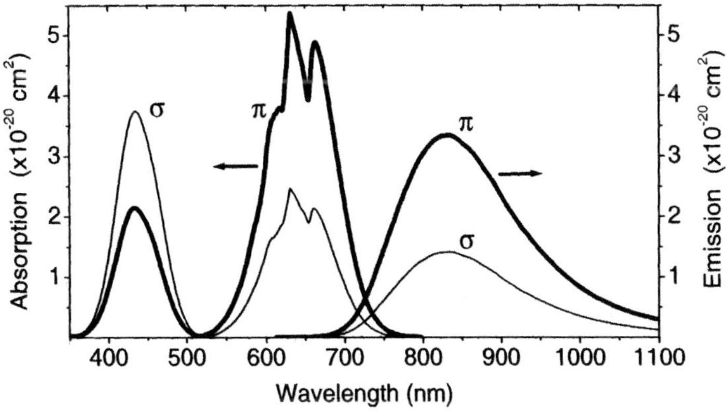

Fig. 8. Absorption and emission spectra of Cr^{3+}:LiSGaF.

even more important is the fact that they can be directly diode pumped by red laser diodes [102] in the 630 nm to 680 nm range (Fig. 8). This caused increased attention to this class of crystals as a possible compact and cost-effective alternative to Ti:sapphire. The development of compact directly diode-pumped, few-optical-cycle sources in the 800 nm region relies currently on Cr^{3+}-doped colquiriites.

Colquiriites are birefringent materials, which is advantageous for laser operation. The typical absorption and emission spectra are shown for Cr:LiSGaF in Fig. 8. The oscillator strength of Cr:LiCAF [77, 101] is approximately a factor of 2 lower than that of Cr:LiSAF and Cr:LiSGaF. The larger transition strengths of the latter arise from the stronger odd-parity T_{2u} symmetry distortion [104]. This determines a generally lower threshold in these crystals in comparison to Cr:LiCAF. On the other hand, due to the higher crystalline field (Fig. 7) Cr:LiCAF has an advantage of a smaller ESA in laser operation. Combined with weak thermal quenching at 300 K, this accounts for the highest efficiency among the colquiriites in Cr:LiCAF. Passive mode-locking was obtained in KLM Cr:LiCAF using both the prism- as well as mirror-dispersion compensation technique [105, 106]. The latter configuration produces pulses down to 52 fs at 75 mW output power from a diode-pumped cavity and even sub-30 fs pulses at low output coupling [106]. The solid solution of LiCAF and LiSAF, i.e., Cr:$LiSr_{0.8}Ca_{0.2}F_6$, has also been successfully mode-locked [107].

Albeit small differences in the spectroscopic properties of all colquiriite crystals, they have considerable similarity: they have broad emission and absorption spectra and high intrinsic slope efficiencies in excess of 50%. This makes these materials suitable for few-optical-cycle generation. As a result, many compact ultrashort pulse sources based on Cr^{3+}:LiSAF [19, 108, 109, 110, 111] have been developed during the last decade. Next to Cr:LiSAF, the attention of researchers was drawn to Cr:LiSGaF, which has a factor of 1.5 higher third-order nonlinearity ($n_2 \approx 1.2 \times 10^{-16}\,\mathrm{cm^2/W}$ [112]) than

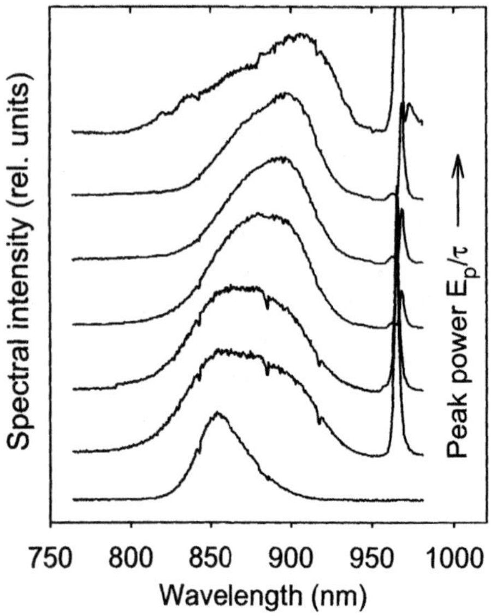

Fig. 9. Self-frequency shift in a mode-locked Cr:LiSGaF laser with changing intracavity pulse peak power [130]

that of Cr:LiSAF ($n_2 \approx 0.8 \times 10^{-16}\,\mathrm{cm}^2/\mathrm{W}$ [113]). The first femtosecond laser operation in LiSGaF was demonstrated in 1994 [114]. In comparison to Cr:LiSAF, this crystal has better power handling capability [101, 115, 116]. This can be explained by the reduced thermal quenching of luminescence and the lower thermal expansion coefficients, as well as the decreased anisotropy of thermal expansion of Cr:LiSGaF in comparison to Cr:LiSAF. In [115], up to 900 mW was demonstrated in the CW regime without special cooling techniques. This inspired the whole series of further works on diode pumping and passive mode locking of this crystal [20, 117, 118, 119, 120, 121, 122], including using saturable semiconductor absorber mirrors (SESAM [123]).

Along with diode pumping and implementation of SESAM, the prismless mirror-dispersion-control (MDC) approach plays a considerable role in making the resonator design of Cr:colquiriites really compact and stable [122]. The first implementation was based on Gires–Tournois mirrors in Cr:LiSAF and Cr:LiSGaF lasers, delivering sub-40 fs pulses at $\sim$ 200 mW output power [122]. The MDC approach led finally to the generation of only five optical-cycle pulses ($\sim$ 14 fs) from both Cr:LiSAF and Cr:LiSGAF lasers at $\sim$ 100 mW output power [124] using a Kr-laser pump. The latter was later substituted by the red diode-lasers to yield even shorter (12 fs) pulses in Cr:LiSAF at 6.2 mW output power [125].

It is interesting to note that at short pulse durations and megawatt intracavity pulse peak powers, a considerable pulse self-frequency shift (Fig. 9) was observed in both Cr:LiSAF and Cr:LiSGaF lasers [124]. Induced Raman scattering was identified as a mechanism causing this shift. Based on this observation, a general theory of a Raman pulse shift in ultrashort-pulsed crystalline lasers has been developed [126, 127]. Later, induced Raman scat-

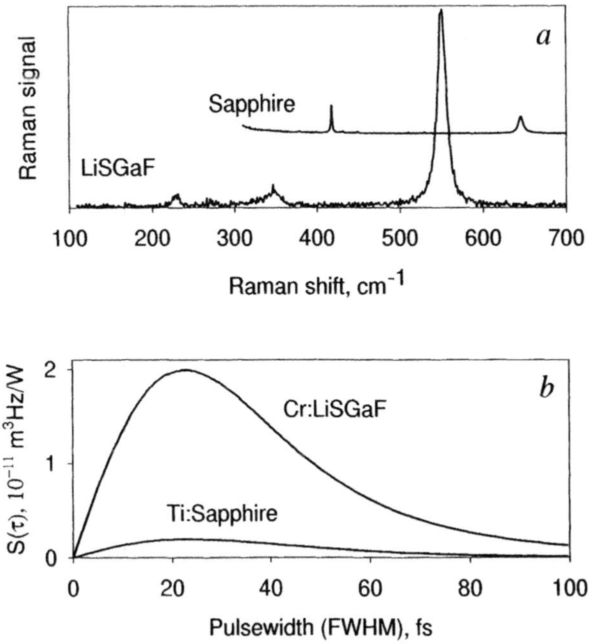

Fig. 10. Analysis of Raman self-frequency shift in ultrashort-pulse oscillators for Ti:sapphire and Cr:LiSGaF as active media [127]. (**a**) Raman spectra of Ti:sapphire and Cr:LiSGaF. (**b**) Parameter $S(\tau)$, characterizing the strength of Raman self-action for the given medium

tering in these materials has been also studied experimentally [128, 129] and theoretically [130, 131]. The fact that few-cycle Cr:colquiriites exhibit the remarkably large frequency shift is explained by the strong and broad Raman lines in colquiriites [127, 128, 132] and high intracavity power in the low-gain Cr:colquiriites. The main characteristic feature of the Raman effect in crystalline mode-locked lasers, which distinguishes it from the well-known Raman self-frequency shift in glass fibers, is the existence of the maximum Raman response function at a certain pulse duration (usually between 20 fs and 30 fs depending upon the material) (Fig. 10). This means that at significantly short pulses, the self-frequency shift again becomes smaller. Similar phenomena have been observed in another low-gain femtosecond material, Cr:YAG, in the 1.5 μm wavelength region [133].

The current state of the art in few-cycle-pulse generation from Cr^{3+}-doped colquiriites employs diode pumping and doubly chirped mirrors [134], possibly in combination with prisms. Four optical cycle pulses (10 fs) at 40 mW output power can be generated in Cr:LiCAF using KLM [135] (see also the Chapter by *Kärtner*). Cr:LiSAF crystal serves as a base for entirely portable, self-contained, and battery-powered sub-100 fs Cr:LiSAF lasers with up to 20 mW of output power and with an electrical-to-optical efficiency of almost 4 [136, 137, 138, 139]. These results make Cr:colquiriites a useful alternative to Ti:sapphire in many real-world applications, especially in the compact low-power sector.

3.3 Cr^{4+}-Doped Crystals

A separate broad class of crystalline media, which is suitable for broadly tunable CW and femtosecond pulse operation in the near infrared region between approximately 1.1 μm and 1.7 μm, consists of Cr^{4+}-doped crystals [140]. These include various garnets (e.g., $Y_3Al_5O_{12}$ or YAG) [141], $Lu_3Al_5O_{12}$ or LuAG [142], YSAG [143], etc.), silicates (e.g., Mg_2SiO_4 or forsterite [144, 145], Y_2SiO_5 [146, 147], tephroite Mn_2SiO_4, monticellite $MgCaSiO_4$ and diopside $MgCaSi_2O_6$ [148]), germanites (e.g., cunyite Ca_2GeO_4 [149], akermanite $B_2MgGe_2O_7$ [150], melilite $Ca_2Ga_2SiO_7$ [151]), and apatites (e.g., $Ca_{10}(PO_4)_6F_2$ [152]). Cr^{4+}-doped glasses [153,154,155] have been also investigated [153,155,156]. Despite the large variety to date of existing Cr^{4+}-doped materials, the emphasis in this section will be on two Cr^{4+}-doped crystals, with which few-optical-cycle operation has been obtained: Cr^{4+}:YAG and Cr^{4+}:forsterite [157, 158, 159].

The discovery of the Cr^{4+} ion at tetrahedral sites goes back to 1986, when a broad band of parasitic absorption centered around 1.06 μm [160] was observed in Cr,Nd:GSGG and Cr,Nd:YSGG containing trace amounts of Fe, Mg, or Ca impurities. This saturable absorption was used to self-Q-switch laser oscillation at the 1.06 μm transition of neodymium [160].

Since chromium preferably enters octahedral sites in the trivalent state, the optical center was initially ascribed to the Cr^{3+}–oxygen vacancy complex and only in 1988, *Petricevic* et al. suggested the model of the Cr^{4+} ion at distorted tetrahedral sites [161]. This hypothesis was confirmed in numerous investigations, including the electron-paramagnetic resonance and site-selective spectroscopy [162, 163, 164, 165, 166, 167]. The confusion about the Cr^{4+} ion in YAG is partially due to the fact that chromium is aliovalent and can exist in several valence states and coordinations, whereas the garnet structure offers three sites with different coordination. Although only Cr^{4+} ions at tetrahedral sites are responsible for laser action [168], large concentrations of parasitic Cr^{4+} and Cr^{3+} ions at octahedral sites [169, 170, 171], as well as the presence of Cr^{6+} at tetrahedral sites [172], have been reported. Additionally, due to the relatively large concentration, up to $10:1$, of Mg^{2+} or Ca^{2+} charge-compensating ions with respect to Cr^{4+}, there is a high probability of ME^{2+}–Cr^{4+} pair center formation. All this contributes to an increase in the threshold and reduction of the efficiency in this material due to possible interactions between Cr^{4+} and Cr in other valency states [169, 172], as well as to the slow bleaching due to charge transfer and color-center formation [41, 173], especially in highly concentrated crystals [21]. The relative amount of each type of existing chromium centers depends on crystal-growth conditions and postgrowth treatments. This explains the variation in spectral properties from crystal to crystal. Indeed, there is a significant spread in the literature of reported values for absorption cross section as well as for excited-state absorption cross section, varying by nearly an order of magnitude [175]. In fact, the relative amount of the Cr^{4+} lasing ion to the overall chromium ion content is usually as low as 1.5% to 2% [176] and reaches 5% only in highly concentrated samples [21].

Research on this type of tunable room-temperature laser started simultaneously with two active media: Cr:YAG [141, 179] operating between 1.35 μm and 1.65 μm, and Cr:$MgSiO_4$ (forsterite) [144, 177, 178] lasing between 1.1 μm and 1.37 μm. From the discussion above, it is clear, that an ideal host for the Cr^{4+} ion would be a crystal where only tetrahedral sites for a fourfold charged ion of roughly the same ionic size as Cr^{4+} state exist. One such host is the cunyite Ca_2GeO_4, and others can probably be found. Cr^{3+} ions were absent in cunyite; Cr^{4+} ions demonstrated high slope efficiency of 21 [180] and allowed direct diode pumping [181]. Passive mode locking by a SESAM yielded 60 ps pulses so far [180] that can be certainly improved by proper dispersion compensation. Yet the best results have been obtained with Cr:YAG and Cr:forsterite because of their availability and developed technology.

Generally, Cr:YAG has a greater potential for applications given better thermo-optical and thermomechanical properties than forsterite. More than 2 W in the CW regime is accessible from Cr:YAG [159]. The small-signal figure of merit value of Cr:YAG crystals was measured to be between 60 and 120 [21], whereas the same parameter does not usually exceed 20 to 30 in Cr:forsterite [182].

Direct diode pumping has been attempted in various Cr^{4+}-doped crystals. Diode pumping of Cr:forsterite with a 1.5 W AlGaInP diode at 670 nm was demonstrated [183], yielding 5 mW of output power. However, this excitation route is not prospective due to high ESA and low efficiency. The excitation route via 1 μm is more difficult due to lower absorption in this wavelength range and the challenges connected with growing highly concentrated crystals (see above). As a result, Cr:forsterite and Cr:Ca_2GeO_4 can only be pumped using the TEM_{00} beam of a rather complex and very expensive master-oscillator-power-amplifier (MOPA) diode laser source [181]. With the progress in high-brightness laser diodes, the first directly diode-pumped room-temperature Cr:YAG laser was realized [21], yielding more than 200 mW of output power tunable over 120 nm with the electrical-to-optical efficiency of approx. 2%. Although this value might seem rather small, it is still higher by a factor of 3 to 4 than the corresponding values estimated for the Cr^{4+}:forsterite laser, if pumped by a double-clad fiber laser [159], which is an alternative approach to the development of compact Cr^{4+}-doped systems. Recently with the availability of laser diodes with even higher brightness, broad tunability of $\sim$ 200 nm between 1350 nm and 1550 nm was achieved [184].

The Tanabe–Sugano diagram (Fig. 11) describes the energy level scheme of the Cr^{4+} ion. An example of the absorption spectrum is shown in Fig. 12.

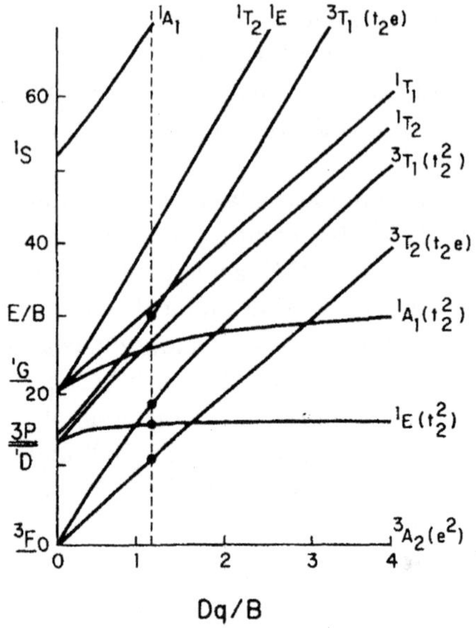

Fig. 11. Tanabe–Sugano diagram of the Cr^{4+} ion. The *vertical line* and energy level positions correspond to Cr:forsterite. From [174]

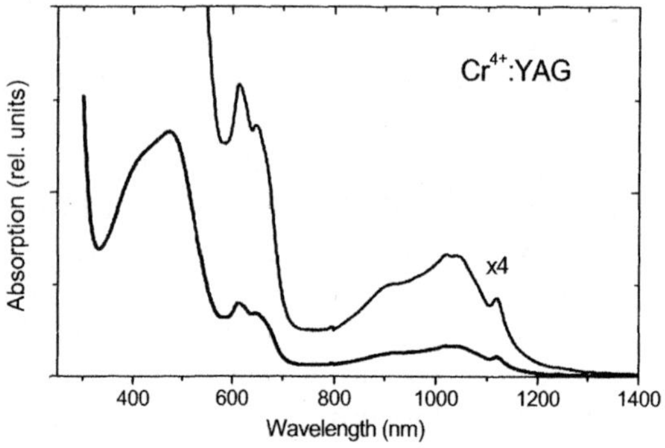

Fig. 12. Absorption spectrum of Cr^{4+}:YAG

The strong broad absorption bands around 620 nm and 1 μm belong to the ^3E and ^3A crystal field components of the ^3F level, respectively [176]. Such large splitting is explained by the strong distortion of the tetrahedron. The broad absorption feature around 480 nm is due to the absorption of octahedrally coordinated Cr^{4+} ions and tetrahedrally coordinated Cr^{6+} ions. The strong and broad absorption over the entire visible wavelength range determines the dark brown coloration of Cr^{4+}-doped crystals. Calculation of the crystal field parameters is complicated and not unambiguous. The latest results yield the energy level scheme shown in Fig. 13 [185]. According to this scheme, the absorption band of interest around 1 μm is assigned to the $^3A_2 \rightarrow {}^3T_2$ transition (the authors of [176] assigned it to the $^3A_2 \rightarrow {}^3T_1$ transition),

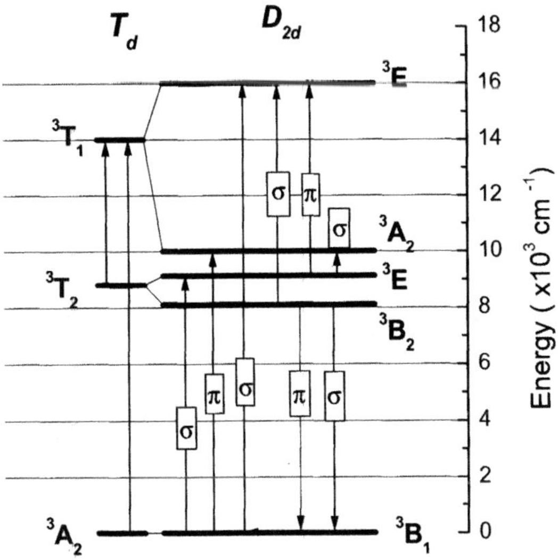

Fig. 13. Energy level scheme of the tetrahedrally coordinated Cr^{4+} ion in a YAG crystal. *Left side* shows the levels in cubic site symmetry (group T_d), and the *right side* gives the energy levels in the distorted tetrahedra (group D_{2d}). Electronic transitions are shown by vertical arrows, π and σ denote polarization of radiation relative to the S_4 axis of the distorted tetrahedron. From [185], courtesy of A. Okhrimchuk

and the band around 620 nm is assigned to the $^3A_2 \rightarrow {}^3T_1$ transition. The corresponding emission spectra consist of a single band due to a $^3T_2 \rightarrow {}^3A_2$ transition, which extends from $\sim 1.1\,\mu$m to $1.7\,\mu$m in different hosts [30].

The luminescence is broadened by electron–phonon coupling, and its lifetime is usually quenched at room temperature. In Cr-doped garnets, the room temperature lifetime of $2\,\mu$m to $5\,\mu$m is shorter by a factor of ~ 10 than the corresponding radiative lifetime. According to the Tanabe–Sugano diagram, the fluorescence bands shift to the red with a decreasing crystalline field [186]. Simultaneously, as for Cr^{3+}-doped crystals, ESA increases, reducing the laser efficiency. Cr:YAG already shows reduced efficiency, and other known hosts with lower crystal field demonstrate even lower values. The longest central wavelength obtained in Cr^{4+}-doped media was 1628 nm in the pulsed regime from the YAG:YSAG solid solution ($Y_3Sc_{0.5}Al_{4.5}O_{12}$)[143]. For more details on the spectroscopy of Cr^{4+} ions in alternative crystals, the reader is referred to [30]. Here, we outline the four main crystallographic and spectroscopic features, which distinguish Cr^{4+} systems and determine their laser properties: 1. multisite and multivalent nature of chromium in the majority of the crystals studied; in garnets, there are three tetrahedral sites for Cr^{4+} ions that are distorted along the three orthogonal crystallographic axes; as a result, only one-third of the available tetrahedral Cr^{4+} ions can effectively participate in lasing when the pump and the laser waves are polarized along one

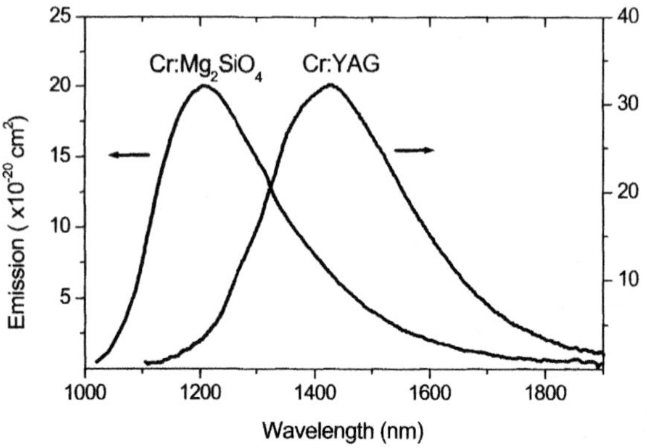

Fig. 14. Fluorescence spectra of Cr^{4+}:forsterite and Cr^{4+}:YAG. Note the vertical scale difference

of these axes; 2. high ($\sim 10^{-18}\,\mathrm{cm}^2$) transition cross sections that are typical of noncentrosymmetric tetrahedral sites, enabling operation at very low concentration levels of lasing ions; 3. excited-state absorption at the pump wavelength [175, 187]; and 4. nonradiative decay resulting in the decreased quantum efficiency ($\sim$ 14% in Cr:YAG at room temperature [186]).

The broad bandwidth of the Cr^{4+} laser transition ($\Delta\lambda/\lambda \sim 0.20$), combined with the high nonlinearity of YAG and forsterite (n_2 is $6.2\times10^{-16}\,\mathrm{cm}^2/\mathrm{W}$ in YAG [77] and $2\times10^{-16}\,\mathrm{cm}^2/\mathrm{W}$ in forsterite [188]) made these materials attractive for passive mode locking and ultrashort pulse generation. Passive mode locking has been achieved using various techniques such as Kerr-lens mode locking (KLM) in Cr^{4+}:YAG [189, 190, 191, 192, 193, 194, 195, 196, 197] and in Cr:forsterite [198, 199, 200, 201, 202, 203, 204], regeneratively initiated mode locking in Cr:YAG [190] and in Cr:forsterite [205, 206], synchronous pumping of Cr:YAG [192], as well as using a semiconductor saturable absorber (SESAM) in Cr^{4+}:YAG [207, 208, 209, 210, 211, 212] and in Cr:forsterite [213, 214, 215]. High repetition rate (up to 4 GHz) femtosecond Cr^{4+}:YAG lasers have also been demonstrated [139, 212, 216, 217, 218]. Pulses as short as 43 fs at 200 mW of average power were demonstrated in Cr^{4+}:YAG [194, 195] and 25 fs at 300 mW in Cr:forsterite [188], results that have remained unsurpassed for several years. The pulse duration was limited in Cr^{4+}:YAG by high third-order dispersion (TOD), which amounted to $\sim 9000\,\mathrm{fs}^3$ [194].

Recently, the parameters of both systems have been significantly improved. Using the negative nonlinear phase shift in a lithium triborate crystal, 13 fs have been demonstrated with a Cr:forsterite crystal [219]. Later, 14 fs was demonstrated in Cr:forsterite [157] and 20 fs in Cr:YAG [220] at 400 mW of output power, using only chirped mirrors for dispersion compensation in a KLM scheme. This corresponds to about four optical cycles. A self-starting KLM Cr^{4+}:YAG laser was realized [159], generating near transform-lim-

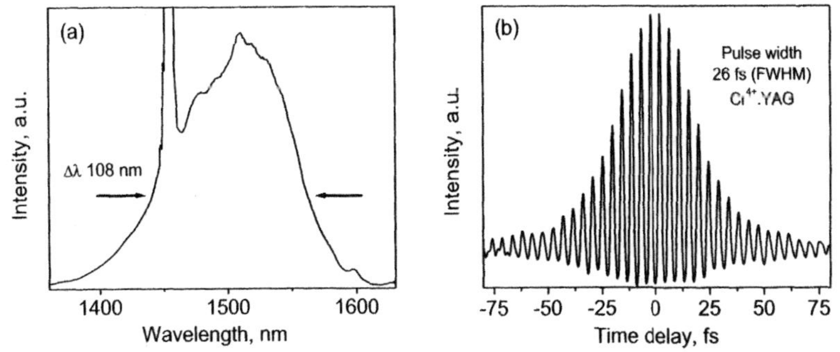

Fig. 15. Spectrum (**a**) and autocorrelation (**b**) of the self-starting five-optical-cycle pulses from a Cr^{4+}:YAG laser [159]

ited five-optical-cycle (26 fs) pulses at 250 mW to 450 mW output power. The maximum average output power in Cr:YAG is 600 mW at 55 fs [159] and 800 mW with sub-100 fs pulses in Cr:forsterite [215]). Finally, a directly diode-pumped KLM Cr:YAG laser has been realized that delivers 65 fs pulses at 30 mW average power [221].

Various SESAM schemes have been realized. With nanocrystalline InAs-doped silica films, 25 fs pulses have been demonstrated in Cr:forsterite [222]. In Cr^{4+}:YAG using the broadband gold-reflector-based SESAM, 44 fs pulses generated at 65 mW at 1520 nm [210, 211]. Recently, ultra-broadband saturable Bragg reflectors consisting of a GaAs/Al_xO_y Bragg mirror and an InGaAs/InP quantum well were used to start 36 fs pulses in a chirped-mirror-compensated Cr:YAG laser [223]. An alternative approach using an ultrabroadband hybrid dielectric-InGaAs/InP saturable mirror [159] allowed generation of the shortest so far of SESAM-initiated KLM Cr:YAG laser pulses (27 fs) of only five optical cycles in duration (Fig. 15). The same mirror implemented in the directly diode-pumped cavity enabled realization of the first directly diode-pumped Cr:YAG laser [22], yielding 62 fs pulse duration at 15 mW.

For the latest achievements, the reader is referred also to the Chapter by *Kärtner* in this volume. Given the ongoing high interest, all results will undoubtedly be improved in the near future.

3.4 Cr^{2+}-Doped Crystals

One of the most important developments of the last decade in broadband diode-pumped lasers was the discovery of the new class of Cr^{2+}-doped zinc chalcogenides [224,225,226], and in particular of Cr:ZnSe as a broadly tunable continuous-wave laser operating around 2.5 μm [227, 228, 229]. This is so because the previously existing tunable mid–infrared (mid-IR) lasers had either only limited tunability, e.g., various Tm-doped lasers operating around 2 μm or Er-doped lasers around 3 μm, or had to be cryogenically cooled, like e.g., the $Co:MgF_2$ laser operating between 1.6 μm and 2.4 μm. The majority of the known vibronically broadened laser transitions in the infrared are quenched at room temperature due to the high probability of multiphonon relaxation processes. The ultrabroad gain bandwidth, which is comparable and even exceeds that of Ti:sapphire (Fig. 2) of the new class of laser crystals allows generation of ultimately short pulses of only few optical cycles. Such pulses in the mid–infrared spectral region can be used as unique diagnostic tools for investigation of numerous transient processes on the femtosecond scale. The broadly tunable ultrashort pulsed lasers in this wavelength range are also attractive for such applications as remote sensing, environmental monitoring, mid-IR free-space communications, optical frequency standards as well as optical coherence tomography [230], and ophthalmology and dermatology in medicine [231, 232]. They can be also be used for pumping mid-IR optical parametric oscillators to produce even longer wavelengths [233].

The remarkable properties of the ultrabroadband Cr^{2+}:ZnSe crystal, such as high emission cross section $\sigma_{em} = 1.3 \times 10^{-18}\,cm^2$, negligibly low ESA [23], fairly good chemical and mechanical stability, and thermal conductivity close to that of sapphire gives this material enormous potential as a laser medium for diode-pumped few-cycle mid-IR lasers and amplifiers. Cr^{2+}:ZnSe as well as other Cr^{2+}-doped II–VI compounds possess two important features: 1. the existence of chemically stable divalent Cr^{2+} dopant ions, with no need for charge compensation and 2. a tendency to crystallize in tetrahedral coordination. As explained above, tetrahedral coordination results in high transition cross section, short radiative lifetime, and smaller crystal field splitting, placing the central wavelength further into the IR.

A number of other features makes Cr^{2+}-doped II–VI compounds special. First of all, the low nonradiative decay rate that allows room-temperature operation. This is due to two factors: 1. the low-phonon nature of such host materials as chalcogenides, providing low energy of the optical phonon cutoff ($\sim 250\,cm^{-1}$) and 2. the nonradiative decay is small in comparison with the short radiative lifetime (16). Second is the absence of ESA. In a tetrahedral crystal field, the two lowest levels, $^5T_2$ and ^5E, originate from the crystal-field splitting of the ^5D ground state of the $3d^4$ ion, which is the only quintet state (Fig. 16). Since all higher lying states are singlets or triplets, ESA transitions from the upper state are spin forbidden. One can thus consider only quintet states (thick lines in Fig. 16), and the resulting energy level diagram becomes very simple, as in Ti:sapphire (Fig. 5). A 200 nm to 300 nm broad absorption band is centered around 1.8 μm [234, 235, 236], and the emission is located between 2 μm and 3 μm [236, 237, 238, 239, 240] (Fig. 17). The energy level scheme resembles that of Ti:sapphire (Fig. 6). Jahn–Teller splitting is stronger in the ground state; fluorescence is now double-peaked, and absorption is single-peaked (Fig. 17). This adds to the fluorescence bandwidth of Cr^{2+}-doped materials, resulting in the highest $\Delta\lambda/\lambda$ values of more than 0.35 (Table 2).

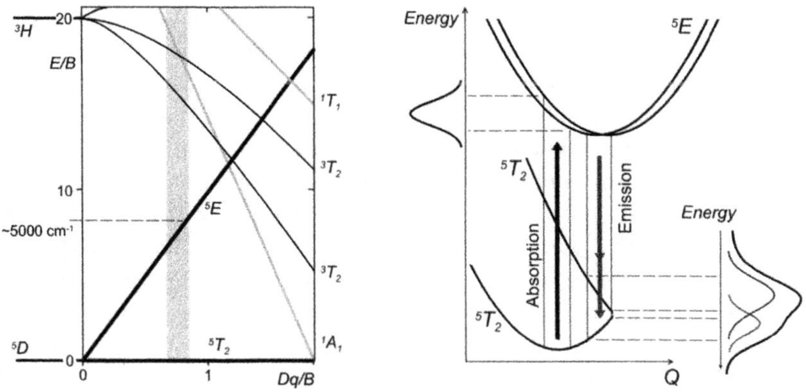

Fig. 16. *Left graph:* Tanabe–Sugano diagram of the Cr^{2+} ion at a tetrahedral site. The *vertical gray bar* shows the range of field parameters that are realized in Zn and Cd chalcogenides. *Right graph:* The coordinational diagram of the laser transition in Cr^{2+} ions, including Jahn–Teller splitting

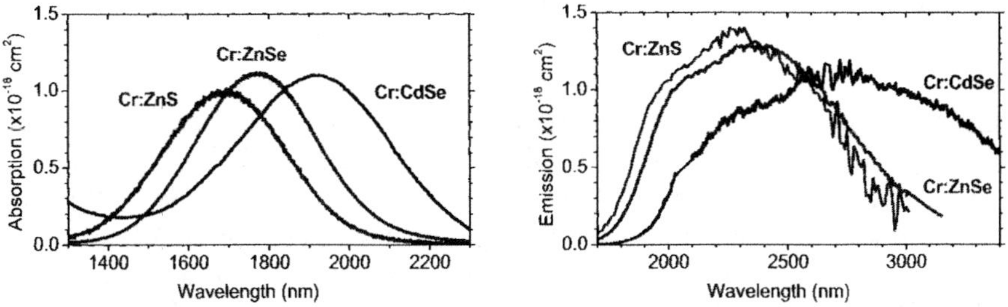

Fig. 17. Absorption and emission cross sections of some Cr^{2+}-doped materials

As discussed in Sect. 3.3, chromium is aliovalent, also in II–VI compounds [24, 25, 242]. Opposite to Cr^{4+}-lasers, this does not lead to deterioration of the output characteristics of Cr^{2+}-lasers because there are simply no other than divalent tetrahedral sites in II–VI compounds. For more details on the spectroscopic and laser properties of Cr^{2+}-doped crystals, the reader is referred to the specialized reviews [30, 280].

Since the first experimental demonstration [225, 226], the laser performance of Cr:ZnSe has greatly improved, including demonstration of direct diode pumping [23, 244, 245, 246], tunable diode-pumped operation [24], active [247, 248] and passive [248] mode locking, continuous-wave operation spanning several hundreds of nanometers in the mid-IR [226, 229] at close to the quantum limit slope efficiency over 60% [23, 227, 228, 229, 244], and power levels in the CW regime in excess of 1.8 W [249]. Recently, up to 18 W of output power was achieved in the pulsed gain-switched mode [250]. Based on the analysis of the mechanical, thermal, spectroscopic, and laser properties of Cr:ZnSe, output power over 10 W in the CW regime and several watt in the mode-locked regime are envisaged. To this point, the following record parameters were also demonstrated from this laser: 1. the broadest tuning bandwidth of 1100 nm between 2000 nm and 3100 nm in the CW regime using three different mirror sets [243] and as much as 850 nm in a single setup (Fig. 18) 2. a 350 nm tuning range at 65 mW output power in the diode-pumped regime [24] (450 nm in Cr:ZnS [25]), and 3. active [247, 248] and passive mode locking [248] with pulses as short as 4 ps at 80 mW and 400 mW of output power, respectively. Though the material parameters are somewhat different from the common femtosecond laser parameter set, theoretical analysis renders generation down to a few optical pulses possible [251]. The n_2 value of $1.7 \times 10^{-14}\,\mathrm{cm^2/W}$ at 1.8 µm in Cr:ZnSe is higher by a factor of 50 than that in Ti:sapphire. This suggests very exciting physics and new opportunities, e.g., for intracavity ultrabroad spectra generation. Opposite to most other crystalline media, II–VI compounds are noncentrosymmetric and possess strong second-order nonlinearity. The interaction of cascaded second-order nonlinearity with third-order nonlinearity is another interesting field of investigation [280].

The most impressive results have been obtained so far using Cr^{2+}:ZnSe crystals. However, there exist other promising Cr^{2+}-doped crystals. One of these is Cr^{2+}:ZnS [225, 226]. This crystal has remained less studied as a laser material due to the lack of good optical quality single crystals. Having spectroscopic properties similar to Cr:ZnSe, Cr:ZnS crystal is known to have a larger bandgap (compare 3.84 eV in ZnS and 2.83 eV in ZnSe [252]); better hardness, a higher thermal shock parameter (compare $7.1\,\mathrm{W/m^{1/2}}$ and $5.3\,\mathrm{W/m^{1/2}}$ in Cr:ZnS and Cr:ZnSe, respectively [225]); and a lower $\mathrm{d}n/\mathrm{d}T$ ($46 \times 10^{-6}\,\mathrm{K^{-1}}$ vs. $70 \times 10^{-6}\,\mathrm{K^{-1}}$ [225]) than those in Cr:ZnSe. Therefore, the power handling capability of this material should be higher than that of Cr:ZnSe, making Cr:ZnS attractive for high-power applications. Pulsed laser operation of a Cr:ZnS laser was first reported in [225, 226, 253]. Spectroscopic study and the first continuous-wave operation was recently reported in [254]. Using Er-fiber pumping up to 700 mW, room-temperature CW operation tunable over 700 nm (between 2.1 µm and 2.8 µm) was demonstrated [25]. Tuning over 400 nm between 2250 nm and 2650 nm in a directly diode-pumped configuration [25] as well as an Er-fiber pumped CW microchip laser at 2320 nm were recently demonstrated [255]. A certain advantage of Cr:ZnS is the shift of absorption by about 100 nm to the blue (Fig. 17), allowing convenient pumping of this material with available 1.6 µm telecommunication diodes [25].

Another important issue remains the extension of the operating range of Cr^{2+}-doped lasers beyond the 3 µm wavelength. This could be obtained by using other II–VI compounds with larger lattice constants and hence lower crystalline fields (Fig. 16). For example, hosts like CdSe [256, 257], CdTe [258], and CdMnTe [259, 260, 261] also allow room-temperature operation with the

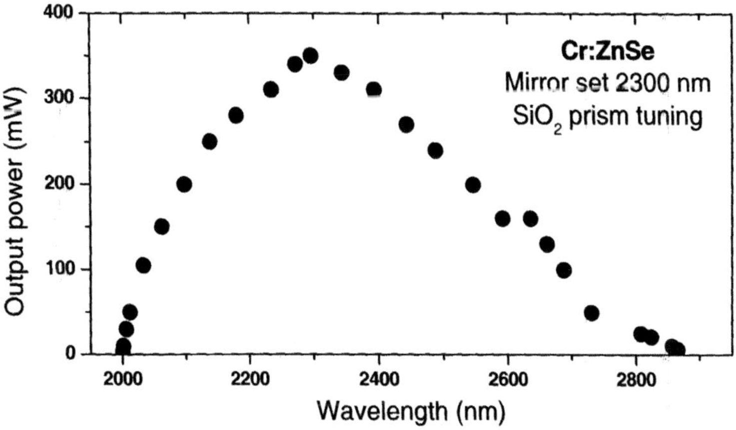

Fig. 18. Tuning range of Cr^{2+}:ZnSe laser. Pump source is Er:fiber laser, mirror set is centered at 2300 nm

Cr^{2+} ion. Tuning up to record 3.4 μm in the pulsed regime was demonstrated in Cr:CdSe [262].

Finally, maybe one of the most interesting advantages of Cr:ZnSe is the availability of technologically developed and low-cost polycrystalline material. The existing technologies of producing ceramic ZnSe, such as chemical vapor deposition (CVD) or hot pressing of powders result in high optical quality low-cost substrates of arbitrary size. Recently, the first ceramic directly diode-pumped, CW tunable, and actively mode-locked lasers were developed [263, 264]. Further optimization should lead to the generation of few optical cycle pulses from the most practical directly diode-pumped ceramic lasers, operating between 2 μm and 3 μm.

3.5 Other TM-Ion Doped Crystals

As follows from the Table 1, there exist a number of other TM ions; some of them are isoelectronic with those described above. For example, V^{4+} has the same electron configuration as Ti^{3+}, Mn^{3+} is analogous to the Cr^{2+} ion, etc. As a general rule, one should expect that one and the same transition will have a shorter wavelength in isoelectronic ions with higher ionization states (e.g., V^{4+} vs. Ti^{3+}, Cr^{4+} vs. V^{3+}). On a Tanabe–Sugano diagram, this corresponds to higher values of the Dq/B parameter.

From this point of view, the V^{4+}-doped system represents a particular interest, as it would provide the gain band at even shorter than the Ti^{3+} wavelength, yet retaining the advantages of the Ti^{3+} – a simple and well-known energy scheme, the absence of ESA channels, and the possibility to use an efficient pump source like the second-harmonic of Yb^{3+}-based lasers. For this purpose, V^{4+}-doped Al_2O_3, YAG, and $YAlO_3$ have been thoroughly investigated [265], but laser oscillation could not be obtained. The reasons were ESA to the conduction band (charge-transfer ESA) and unstable valency of the V^{4+} ion, causing color-center formation. Both problems might be overcome by an extensive search for a host with a larger bandgap (e.g., fluorides) and providing higher crystal-field stabilization of the V^{4+} ion. Thus, the possibility of finding a working V^{4+}-based configuration still exists, but requires a significant effort that could be justified only if a compelling need exists for an ultrabroadband system with a center wavelength between 600 nm and 650 nm.

Considerable effort has been invested in the development and characterization of various TM^{3+}-doped crystals, including Cr^{5+}, Mn^{6+}, Ni^{2+}, Mn^{3+} Co^{2+}, Fe^{2+}, V^{2+}, Mn^{4+}, V^{3+}, Mn^{5+}, and Fe^{6+} [30]. Only few of these systems have produced stimulated emission, the main problem being ESA at the laser wavelength. For the purposes of few-cycle pulse generation, probably only Co^{2+}:MgF_2 may present interest for amplifier systems in the otherwise hardly accessible wavelength region centered at 2 µm, as will be discussed below.

First suggested as a laser material in 1964 [266], Co^{2+}:MgF_2 has two main points of interest: a broad fluorescence band centered at 2 µm and a pump band around 1.35 µm (Fig. 19). Besides the usual vibronic broadening, the fluorescence band is additionally broadened by about 1300 cm^{-1} due to the splitting of the lower $^4T_1$ level [267], resulting in a 650 nm bandwidth with $\Delta\lambda/\lambda = 0.33$ (Fig. 19), on a par with such materials as Ti:sapphire and Cr:ZnSe. The position of the absorption band is another important advantage of this material, as it allows pumping with Nd lasers at 1.32 µm to 1.34 µm that are readily available in pulsed and CW modes with high power levels and even diode-pumped [268]. Laser operation of Co^{2+}:MgF_2 has been demonstrated in CW [267,269,270], pulsed [267,271,272,273], Q-switched [267,274], and actively mode-locked [275] regimes. Tuning was observed from 1.6 µm to beyond 2.4 µm.

The spectroscopy of the Co^{2+} ion is well understood [30, 36, 243, 266, 267, 276]; however, there still remains some discrepancy in the polarization data because spectroscopic measurements suggest roughly equal gain in π and σ polarizations (Fig. 19 [267]), whereas laser experiments demonstrate a clear preference for σ polarization for laser purposes [273, 277]. In MgF_2, the cobalt ion occupies a highly symmetrical octahedral position resulting in a relatively low transition cross section of the order of 10^{-21} cm^2 and a long radiative lifetime of 1.85 ms at 20 K. Compared to this value, even a moderate nonradiative decay rate causes severe thermal quenching of the lifetime (16). At room temperature, the lifetime is $\sim$ 37 µs, and only pulsed operation is possible. This value, however, is more than sufficient for realizing a multipass or regenerative amplifier.

The relatively low cross section results in a very high saturation fluence of $\sim 40\ J/cm^2$ but also in low gain values. With a typical 1 wt% doped Co:MgF_2 crystal (Co^{2+} ion density 2×10^{20} cm^{-3}) and using a saturating pump fluence,

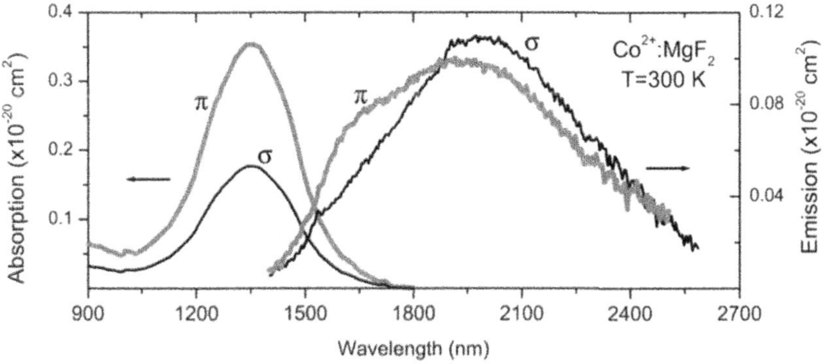

Fig. 19. Absorption and emission spectra of Co^{2+}:MgF_2 at room temperature. Data courtesy of P. F. Moulton, Qpeak Inc.

one could reach a small-signal gain coefficient of $\sigma_{em}n = 0.3\,\text{cm}^{-1}$. However, ESA at the pump wavelength is up to $\sim$ 60% of the pump absorption [276], thus decreasing pumping efficiency (10). The really usable gain coefficient thus reduces to about $0.1\,\text{cm}^{-1}$ to $0.2\,\text{cm}^{-1}$, but this is compensated for by the possibility of growing long crystals with very high optical quality (losses $\sim$ 0.3 %/cm [272]). Another approach would be to use crystals with higher Co^{2+} concentration. For power handling, a moderate thermal conductivity of 3.1 W/(m · K) is compensated for by a very low $\mathrm{d}n/\mathrm{d}T \approx 1 \times 10^{-6}\,\text{K}^{-1}$, allowing CW TEM_{00} operation at 4 W of average output power [271] and 6.5 W of average output power in pulsed mode at room temperature, both values not yet at the limit.

Another possible host for Co^{2+} ions includes various fluorides like ZnF_2 [266], $KMgF_3$, $KZnF_3$ [266, 276], and oxides like MgO [276], but so far alternative hosts did not show any advantages over MgF_2.

4 Rare-Earth Doped Materials

Lanthanide rare-earth (RE) atoms possess filled 4*d*, 5*s*, 5*p*, and 6*s* shells. In crystals, the rare-earth ions tend to enter the trivalent state and can be easily incorporated in any type of position, octahedral, dodecahedral, etc. Optically active electrons originate from the partially (except La^{3+} and Lu^{3+}) filled 4*f* shell which is shielded from the surrounding crystalline environment by the filled spherically symmetrical 5*s* and 5*p* shells. The spin-orbit coupling is the most significant interaction that determines the free-ion multiplets that are then further split by the crystalline field (Stark splitting).

As can be seen in Fig. 20, the magnitudes of the Coulomb and the spin-orbit couplings are approximately equal and are of the order of a few thousand wave numbers. Stark splitting due to the crystalline field is much smaller and is of the order of hundreds of wavenumbers. The energy levels of rare-earth

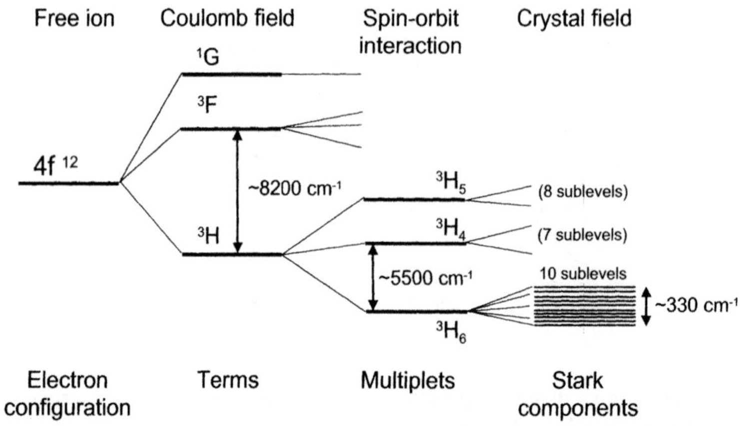

Fig. 20. The effect of the crystalline field on energy-level splitting in lanthanide ions in the $4f$ electron shell. The terms and numbers on the graph correspond to a Tm^{3+} ion in $GdVO_4$ crystal

ions change only slightly from host to host. Therefore, using the well-known and very useful Dieke diagram [282] to estimate the relative positions of multiplets and the widths of the Stark splitting in crystals is justified. More details can be found in [282, 283, 284, 285, 286].

Due to the predictability of their spectroscopic properties, rare-earth ions have produced by far the largest number of operating lasers both in crystalline and in glass environments [286, 292]. The search has been concentrated on finding the best hosts from the point of view of various secondary factors: thermal and mechanical properties, particular spectral properties, production convenience, etc.

From the point of view of broadband generation, the structure of the rare-earth ions is unfavorable; due to shielding by the filled electron shells, the ions have a very small shift of their electronic energy parabolae along the configurational coordinate, rendering vibronic broadening ineffective. Broadband spectra of rare-earth ions therefore have been primarily connected with inhomogeneous broadening that occurs when ions experience deviations of the local crystalline field. The corresponding changes in energy levels can only be a fraction of the crystalline splitting itself, i.e., of the order of $100\,\mathrm{cm}^{-1}$ to $300\,\mathrm{cm}^{-1}$. This is about as much as can be obtained in the rare-earth-doped glasses, e.g., Nd:glass or Er:glass, and even this is obtained at a price of significantly worse thermo-optical parameters in glasses, compared to crystals.

Inhomogeneous broadening in crystals can be for two different reasons: due to irregularities of the local surrounding, i.e. disordering, and because the active ion can be found at different crystallographic sites in a regular lattice. In the latter case, the result is the appearance of separate sets of spectral lines, whereas disordering causes appearance of closely-spaced replicas of the original lines approaching to a glass-type broadening of every transition line.

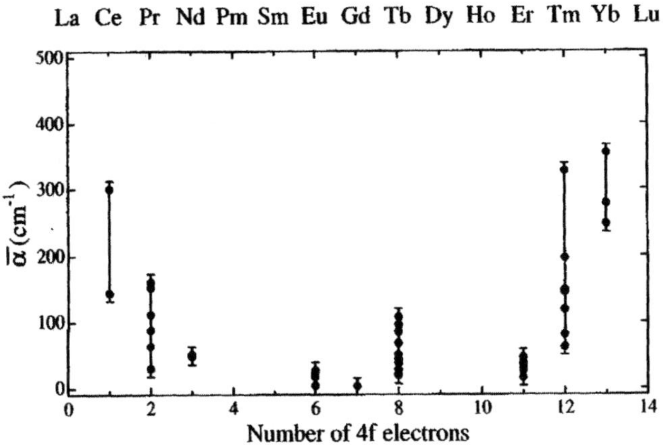

Fig. 21. The electron–phonon coupling parameter $\bar{\alpha}$ of trivalent lanthanide ions in YLF ($LiYF_4$) crystal. From [294]

From the point of view of ultrafast pulse amplification and mode locking, it is preferable to have smooth gain spectra, as created by disordering. At the same time, strong disordering in crystals impedes their thermo-optical parameters, reducing the main advantages of crystalline media.

Yet ways exist to obtain larger gain bandwidths from rare-earth-doped materials, even in perfect crystals. The absence of vibronic (Sect. 3) bands does not mean that electron–phonon coupling does not exist at all. In particular, electron–phonon coupling defines the homogeneous linewidths of the individual transitions between Stark sublevels. Among the various mechanisms of line broadening in rare-earth ions in crystals, Raman-type processes have been found dominant [35]. The linewidth in this case has the functional form

$$\Delta\nu = \bar{\alpha}\left(\frac{T}{T_{\mathrm{D}}}\right)^7 \int_0^{T_{\mathrm{D}}/T} \frac{x^6 \mathrm{e}^x}{(\mathrm{e}^x - 1)^2}\,\mathrm{d}x = \bar{\alpha}\cdot F(T/T_{\mathrm{D}}), \tag{22}$$

where T_{D} is the Debye temperature of the crystal [35,293,294]. For common laser matrices, $T_{\mathrm{D}} = 200\,\mathrm{K}$ to $500\,\mathrm{K}$, giving typical values of $F(T/T_{\mathrm{D}}) = 0.4$ to 0.05 at room temperature. Besides the difference due to the matrix, it has been found that rare-earth ions demonstrate significant variations in the electron–phonon coupling constant $\bar{\alpha}$, as illustrated by Fig. 21. From this figure, we can expect Yb^{3+}, Tm^{3+}, and Ce^{3+} ions to have much larger homogeneous linewidths than. e.g., the Nd^{3+} ion in the same crystal. If the ion has a number of transitions between Stark sublevels with comparable cross sections separated by some tens of reciprocal centimeter, then the homogeneous broadening may be sufficient to produce a smooth spectrum from a set of individual lines. In this case, the overall width of the spectrum may become even larger than that obtained by purely inhomogeneous broadening of one or two strongest lines, as in Nd:glass. Such conditions can be realized in Tm^{3+} and Yb^{3+} crystals, as will be shown in the next sections.

The *actinide* rare-earth ions have partially filled $5f$ shells that are more exposed to the crystalline environment than $4f$ shells of the lanthanides. As a result, these ions demonstrate much stronger electron–phonon broadening than lanthanides. Because of the radioactivity of most actinides, only trivalent uranium has been tested as a laser material in fluorides: CaF_2 [287, 288], BaF_2 [289], SrF_2 [290], and YLF [291]. The relatively smooth emission spectrum extends between 2.2 and 2.8 µm, and laser action has been observed at various wavelengths in this range, depending upon the host. In 1994, room-temperature, CW laser operation at 2.8 µm was demonstrated in U^{3+}:YLF [291]. However, interest in this laser did not result in further development, in particular due to the discovery in 1995 of Cr^{2+}-doped chalcogenides, operating in the same wavelength range (Sect. 3.4).

4.1 Nd^{3+}-Doped Materials

Lasers based on Nd^{3+} ion have been for about 40 years the most widely used laser types, second only to semiconductor lasers. The reasons for this are that Nd^{3+} operates in a purely four-level scheme even at elevated temperatures, can directly absorb flashlamp pump light, and the lifetime of its upper laser level is in the 150 µs to 500 µs range, well suited for pulsed flashlamp pumping. Radiative transitions occur from the metastable $^4F_{3/2}$ doublet to the $^4I_{15/2}$ (1.8 µm), $^4I_{13/2}$ (1.3 – –1.44 µm), $^4I_{11/2}$ (1 µm), and $^4I_{9/2}$ (0.9 µm, ground state) multiplets. Transitions into the $^4I_{11/2}$ state concentrate typically about 50% of the integral cross section, and within this group there are one or two narrow lines with much higher peak cross-sections, enabling low threshold operation. ESA and upconversion are negligible in most hosts. Thus, Nd^{3+}-doped media at 1 µm transition are practically ideal materials both as oscillators and as amplifiers. However, the existence of a single transition line with much higher cross-section than others means that only disordering can produce smooth broad spectra, as required for ultrashort-pulse generation.

Nd^{3+}-doped glasses have for a long time been the only rare-earth-doped materials used for ultrashort-pulse oscillators and amplifiers. The flashlamp-pumped Nd:glass laser was mode locked as early as in 1966 [295] and since that time, Nd-doped glass materials have been implemented in many configurations: pulsed flashlamp-pumped oscillators and amplifiers, CW-mode-locked diode-pumped oscillators, and CW-mode-locked fiber lasers. The typical spectra of Nd^{3+} ions in silicate and phosphate glasses with a 25 nm to 30 nm bandwidth ($\Delta\lambda/\lambda = 0.025$) are shown in Fig. 22. With the availability of the laser diodes, directly diode-pumped operation of Nd:glass materials has become possible [18]. Based on improvements in laser diode parameters and the semiconductor saturable absorber (SESAM) technique, 60 fs pulses

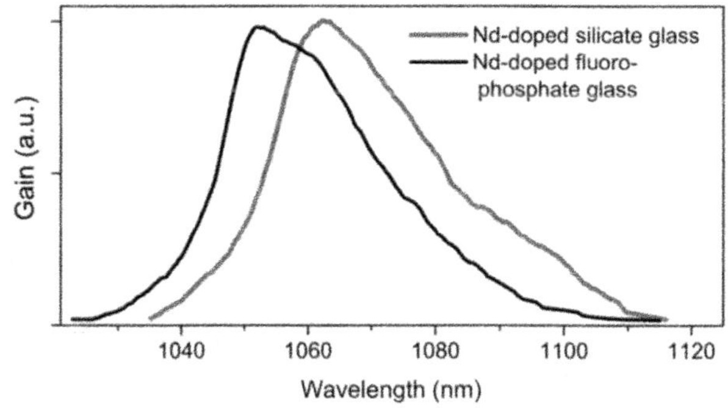

Fig. 22. Fluorescence spectra of Nd^{3+} ions in silicate and fluorophosphate glasses

have been demonstrated using a fluorophosphate glass [296], as well as mode-locked average power of over 1 W [297]. The spectral bandwidth of the 60 fs pulse was 21.6 nm [296], which is close to the limit set by the gain spectrum width. Further improvement in pulse duration could be reached using two different glass plates un the same resonator, exploiting the fact that Nd spectra in silicate and phosphate glasses are wavelength-shifted. In this way, pulses as short as 38 fs with 41 nm bandwidth ($\Delta\lambda/\lambda = 0.04$) have been demonstrated [298], representing the shortest achievement to date from Nd-doped bulk material. This configuration has the potential to be diode-pumped. The prospects for Nd^{3+}:glass oscillators in the 1 µm range are not clear because of the competition from Yb^{3+}-doped glasses, where similar parameters have been obtained (see Sect. 4.2).

Another important application area of Nd^{3+}:glass is the construction of high-energy amplifiers. With a saturation fluence of 3 J/cm^2 to 4 J/cm^2 and the possibility of preparing large elements of arbitrary shape, Nd:glass still dominates systems with pulse energies above 10 J. While classic systems employed direct flashlamp pumping, now progress in high-power diodes allows constructing diode-pumped amplifiers using glasses with a longer upper laser level lifetime. Here, however, Nd:glass systems are meeting increased competition from Yb:glass systems that could not be used before with the flashlamp technology. Typical pulse durations are of the order of 0.5 ps to 1 ps due to the initial bandwidth limitation and gain narrowing. To increase the bandwidth, one can mix silicate and phosphate glasses, as in the Nd:glass oscillators described above [299]. The pulse duration improvement compared to a single glass system is expected to be about 1.5 [300], as demonstrated in the oscillators [296, 298]. The pulse duration is not likely to be further improved because of fundamental bandwidth limitations.

Nd^{3+}:glass has also been successfully operated as a fiber laser. Introduced into the core of a single-mode silica fiber, Nd^{3+} ions exhibit a gain spectrum typical of that of silicate glass and are capable of supporting pulses as short as 38 fs [301] around 1060 nm using the cross-phase modulation mechanism.

The excellent gain properties of the single-mode fiber also allow demonstration of 53 fs pulses at 920 nm (a three-level transition to the ground state in Nd^{3+}) [302]. However, all these systems required large and bulky free-space dispersion compensation of the large positive dispersion of silica at 1 μm, making the use of the fiber almost senseless. The attempt to create an all-fiber source using chirped gratings resulted in picosecond pulses due to the excessive dispersion of chirped gratings [303, 304]. Recently, the techniques of microstructured and photonic crystal fibers made possible to control the dispersion of silica fibers, setting a zero point anywhere in the near-infrared region, including the 920 nm and 1060 nm wavelengths. This might revive an interest to all-fiber Nd-based femtosecond oscillators.

In Nd^{3+}-doped crystalline materials, the first subpicosecond pulse operation has been demonstrated using a solid solution of GSAG and YSGG garnets [305] and by using the LMA ($LaMgAl_{11}O_{19}$) crystal [306]. The spectrum broadening in the Nd:GSAG:YSGG material is due to disordering [307]; in the Nd:LMA crystal, it is due to the existence of three different crystallographic sites [308], amounting to $\Delta\lambda/\lambda = 0.005$ to 0.01. A number of materials with a somewhat broader fluorescence spectrum have been suggested [309, 310], but none of them has the potential to generate pulses significantly shorter than 260 fs, demonstrated in the Nd:GSAG:YSGG mixed garnet [307]. At the same time, progress in diode-pumped Nd:glass lasers has been quite impressive, reaching pulse durations down to 60 fs [296]. It is doubtful that Nd-doped crystals with much broader fluorescence spectra will be found, so that the 260 fs result [307] is likely to remain the shortest femtosecond pulse obtained from Nd-doped crystalline media.

4.2 Yb^{3+}-Doped Materials

The Yb^{3+} ion has long been in the shadows of the laser community's interests because of its spectroscopic features: there are only two multiplets that have relatively small Stark splitting of the order of 500 cm^{-1} to 800 cm^{-1}, so that the ion operates in a quasi-three level scheme at room temperature. Besides, the absorption lies in the 900 nm to 980 nm range, and thus the Yb^{3+} ion can absorb only a very small fraction of flashlamp radiation. Unlike other rare-earth ions, Yb^{3+} does not possess higher lying groups of lines in the visible that could better absorb flashlamp light. Thus, low absorption in combination with the high threshold due to the three-level nature made the Yb^{3+} ion unsuitable for flashlamp pumping. The situation changed radically in 1991, when *Lacovara* et al. demonstrated the first diode-pumped Yb:YAG laser [311]. This was made possible by the progress in InGaAs laser diodes in the 900 nm region. In the following decade, many features of Yb^{3+} that were considered drawbacks, turned out to be advantages. The absence of higher lying levels excluded ESA and upconversion losses, the small Stark splitting meant low Stokes heating in high-power operation, and the high threshold due to the quasi-three-level scheme did not turn out to be a big problem with high-brightness laser diodes. As an extra bonus, InGaAs diodes in the 900 nm range proved to possess the highest brightness and best reliability of all high-power laser diodes, as a result of their development for the telecom market. By the end of the decade, diode-pumped Yb^{3+}-doped media became widely accepted as the best high-power laser materials.

From the point of view of ultrashort-pulse generation, the Yb ion has the remarkable feature of relatively high electron–phonon coupling and hence large homogeneous broadening of the transitions between Stark sublevels, about 30 cm^{-1} to 70 cm^{-1} (see Fig. 21). In all site symmetries lower than octahedral, the upper $^7F_{5/2}$ level is split into three components, and the lower $^7F_{7/2}$ into four. Thus, the absorption and emission spectra are typically an approx. 1000 cm^{-1} broad group of 12 spectral lines with a typical distance of 70 cm^{-1} to 100 cm^{-1}. The exact spectral positions and intensity distribution between the lines depend on the crystal. In many materials, the lines have comparable intensity, are more or less evenly spaced, and can build up a relatively smooth spectrum. This is illustrated in Fig. 23. The spectral bandwidths of Yb-doped crystalline materials compare or even exceed those of Yb-doped glass. To increase the bandwidth further, one should look for crystals with larger Stark splitting (stronger crystalline field). In this case, the homogeneous broadening may be insufficient to fill the gaps between the lines. One looks therefore for crystals with different crystallographic positions suitable for the Yb ion to achieve additional inhomogeneous broadening. As already mentioned above, such inhomogegeous broadening is preferable over disorder-type broadening from the point of view of thermo-optical parameters. A persistent search in this direction resulted in some impressive products, e.g., Yb:GdCOB [312], Yb:BOYS [313] and Yb:SYS [314] that generated pulses down to 69 fs, almost as short as from Yb:glass (60 fs [315]). The comparative merits of some Yb-doped crystals can be seen in Table 3.

Table 3. Parameters and short-pulse performance of different Yb-doped materials

	$\Delta\lambda$ nm	σ_{em}, peak 10^{-20} cm^2	Lifetime ms	κ (undoped) W/(m·K)	τ_{pulse} fs	Ref.
Yb:SYS	73	0.4	1.1	2.8	94	[314, 321]
Yb:BOYS	60	0.2	1.1	1.4	69	[313]
Yb:GdCOB	44	0.35	2.6	2.1	89	[312]
Yb:LSB	40	0.13	1.7	3.52		[316]
Yb:glass	35	0.05	1.3	0.85	60	[315]
Yb:KGW	20	2.8	0.6	3.3	112	[322]
Yb:NaGW	18	2.0	0.34			[318]
Yb:KYW	16	3.0	0.6	3.3	71	[317]
Yb:YAG	9	2.2	0.95	13	340	[315]

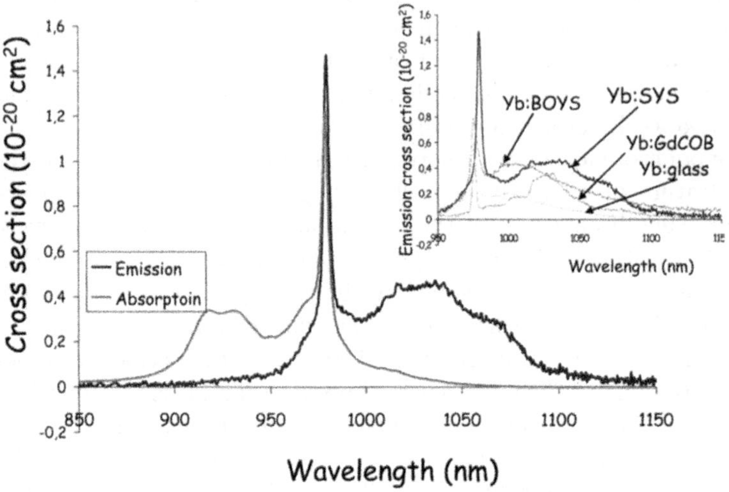

Fig. 23. Absorption (*gray line*) and emission (*black line*) cross-sectional spectra of the Yb:SYS crystal. The *inset* compares the gain spectra of different Yb^{3+}-doped materials. From [314] (courtesy of P. Georges)

The enormous power capability of Yb-doped materials allows using a different approach to shorter pulses by using the chirp-compressor sequence. Using an 800 fs Yb:YAG thin-disk laser with average power of 60 W and pulse energy of 1.75 μJ directly from the oscillator, *Brunner* et al. demonstrated almost 20-fold compression down to 43 fs, retaining as much as 24 W of the average power (0.7 μJ pulse energy). This is, by at least an order of magnitude, more than anything obtained from an oscillator in the sub-100 fs regime, and it is about 15 optical cycles. It is also possible to achieve even shorter pulses, because one can use other media with broader bandwidths in the high-power thin-disk configuration, like, e.g., Yb:KYW [319]. In this case, one can obtain 240 fs pulses at an average power of 22 W directly from the oscillator [319], and probably even shorter pulses with other Yb-doped crystalline media, that would then allow further compressing [320].

The long lifetime ($\sim$ 1 ms) allows efficient energy storage in Yb ions, making Yb-doped media excellent candidates for amplifiers, with the bandwidth exceeding that of Nd:glass in the same spectral region. Some of the novel media have already been successfully implemented as high repetition rate and high average power amplifiers [321].

Summarizing, the bright prospects of Yb-doped media in the field of few-cycle optical generation lie in the possibility of generating pulses with high average power directly from diode-pumped oscillators, making the Yb-based system probably the most efficient source of medium-energy pulses. The pulsewidth of the Yb-based system may not qualify as few-cycle, but can be further compressed or used for pumping optical parametric oscilla-

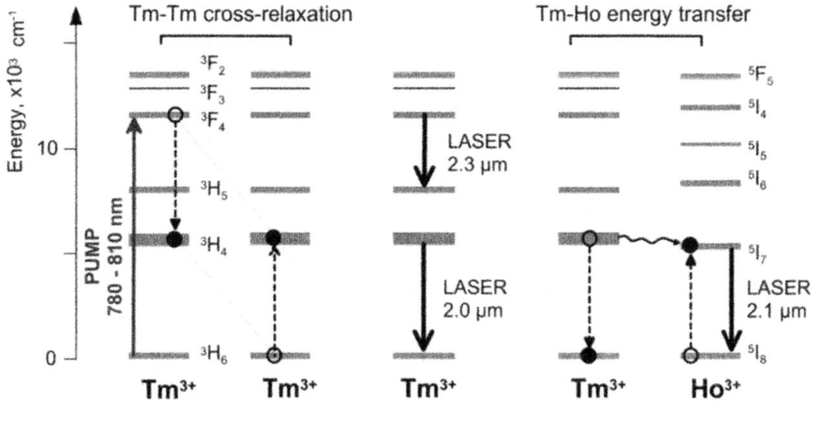

Fig. 24. Pump schemes and laser transitions of Tm^{3+}- and Ho^{3+}-doped crystals

tors using the chirped-pulse amplification technique (see also the Chapter by *Kobayashi* et al.).

4.3 Tm^{3+}- and Ho^{3+}-Doped Crystals

Tm- and Ho-doped crystals have been known to lase since 1962 [323,324,325], but it was not until 1986 that the first room-temperature CW operation of a Cr,Ho,Tm-doped YSAG and YSGG [326] under a Kr^+-ion laser was demonstrated using a codoping scheme suggested by *Antipenko* et al. [327]; this was followed immediately by diode-pumped operation of Tm,Ho:YAG [328, 329, 330], Tm:YAG [331], single-frequency monolithic Tm:YAG [332] and the tunable Tm,Ho:YLF [333, 334].

At the moment, 2 μm lasers based on Tm-, Ho-, and Tm,Ho-doped crystals are well-developed and available commercially. Both ions operate in overlapping spectral regions, but their spectroscopic features differ dramatically. The Ho^{3+} ion demonstrates sharp well-separated lines on the 2 μm $^5I_7$–$^5I_8$ transition, enabling low-threshold operation at low dopant concentrations. The Tm^{3+} ion operates at a quasi-three-level scheme, with relatively broad lines (according to Fig. 21) but allows high-power operation of up to 115 W in diode-pumped CW operation [335]. This is possibly due to the remarkable feature of the Tm^{3+} ion, where one pump photon at $\sim$ 800 nm can produce two excited Tm^{3+} ions (quantum efficiency $\sim$ 2 in a cross-relaxation process see Fig. 24), thus reducing Stokes losses to the minimum. In Tm,Ho-codoped systems, the excitation from the $^3H_4$ level of the Tm^{3+} ion is additionally transferred to the $^5I_7$ level of the Ho^{3+} ion (Fig. 24). The lifetimes of Tm^{3+} and Ho^{3+} ions lie typically in the millisecond range.

If Ho^{3+} is the lasing ion, then obtaining a broad continuous spectrum should rely on inhomogeneous broadening, which should be sufficient to fill the gaps between spectral lines. An example is the Cr,Tm,Ho:YSGG:GSAG

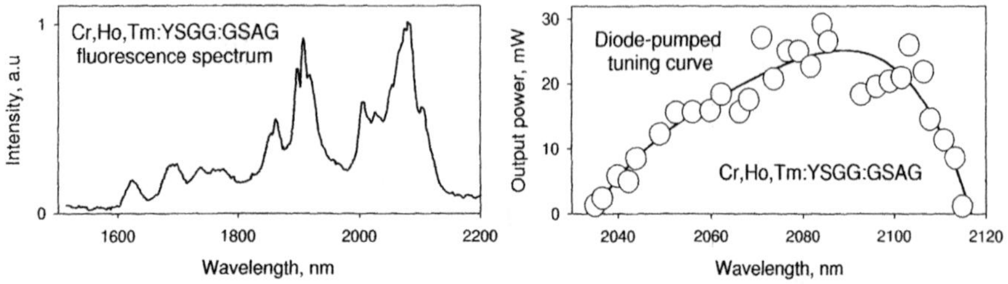

Fig. 25. Fluorescence spectrum and tuning curve of the Cr,Tm,Ho:YSGG:GSAG laser. From [336]

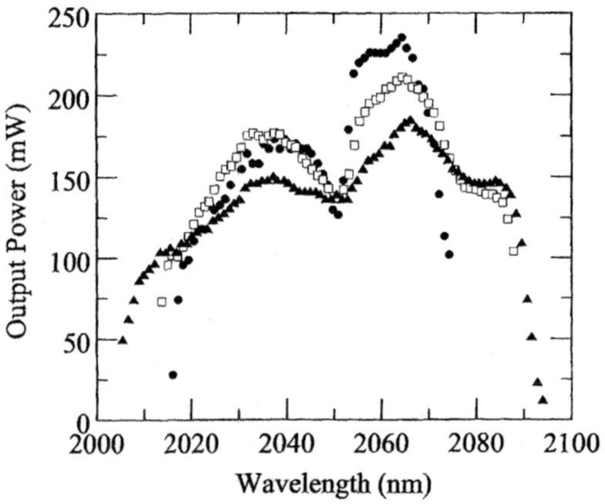

Fig. 26. Tuning curve of the Tm,Ho:BaY_2F_8 laser. Output coupler transmission is 1% (*triangles*), 2% (*squares*), and 4% (*circles*). From [338]

crystal, where the inhomogeneous broadening is of the disordering type, similar to the Cr,Nd:YSGG:GSAG crystal mentioned in Sect. 4.1. The resulting inhomogeneous broadening allows as much as 80 nm of continuous tuning (Fig. 25)[336, 337]. Recently, a continuous tuning range of 89 nm has also been achieved with a multiple-site Tm,Ho:BaY_2F_8 crystal [338], corresponding to $\Delta\lambda/\lambda = 0.04$ (Fig. 26).

Tm-doped crystalline lasers can provide tunable operation in two spectral regions: 1.9 µm to 2 µm using the ${}^3H_4$–${}^3H_6$ transition and 2.2 µm to 2.4 µm on the ${}^3F_4$–${}^3H_5$ transition (Fig. 24). Continuous tuning has been demonstrated on both transitions mentioned. In the 1.9 µm to 2 µm range, tuning has been demonstrated in Tm:YAG and TmYSGG (1.87 µm to 2.16 µm and 1.84 µm to 2.14 µm, respectively) [339], Tm:$YAlO_3$ (1.93 µm to 2.00 µm) [340] in Tm:Y_2O_3 and Tm:Sc_2O_3 (1.93 µm to 2.09 µm and 1.93 µm to 2.16 µm, respectively) [341], Tm:YLF (1.91 µm to 2.07 µm) [342], Tm:BaY_2F_8 (1.85 µm to 2.06 µm) [343], and in Tm:$GdVO_4$ (1.86 µm to 1.99 µm) [344]. The latter crystal [345, 346, 347, 348] is especially attractive because of its high

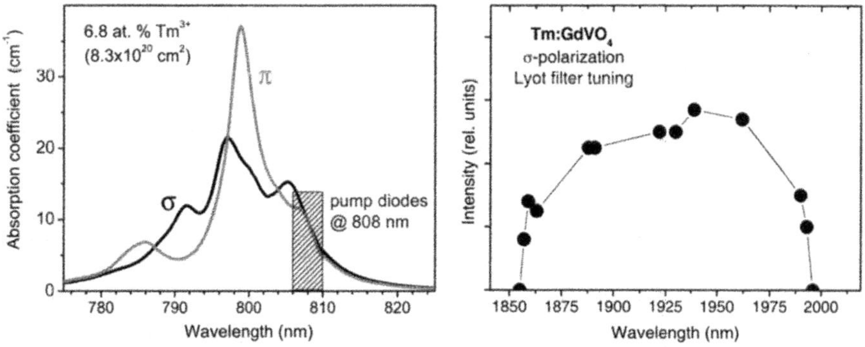

Fig. 27. *Left graph:* Absorption spectrum of Tm:$GdVO_4$. *Right graph:* Free-running oscillation and tuning range of the Tm:$GdVO_4$ laser

absorption at 808 nm (Fig. 27). All previously mentioned Tm- and Tm,Ho-doped crystals possess usable pumping bands in the 780 nm to 795 nm region. However, the pumping diodes in this range are more expensive and possess lower brightness than diodes in the 805 nm to 810 nm range. The extremely high absorption coefficient of the Tm:$GdVO_4$ crystal (and of its analogue Tm:YVO_4 [349, 350]) at typical working concentrations also allows pumping in the 805 nm to 810 nm range. The demonstrated smooth tuning of the Tm:$GdVO_4$ crystal covers the 1.86 nm to 1.99 μm range (Fig. 27).

In the 2.3 μm range ($^3F_4$–$^3H_5$ transition), tuning has been demonstrated in Tm:YLF (2.2 μm to 2.46 μm) [351] and in Tm:GSGG (2.2 μm to 2.37 μm) [352]. It is not possible to realize 2 μm and 2.3 μm operation in the same crystal because two transitions require different Tm^{3+} ion concentrations. Pumping the $^3H_4$–$^3H_6$ transition relies on the cross-relaxation process (Fig. 24) that requires a *high* concentration of Tm^{3+} ions, typically $> 5\%$. This process, however, would be detrimental to operation on the $^3F_4$–$^3H_5$ transition because cross-relaxation will depopulate the upper laser level $^3F_4$ (Fig. 24). Therefore, operation in the 2.3 μm range requires *low* concentrations of Tm^{3+} ions, typically $< 2\%$.

From the point of view of ultrashort-pulse operation, Tm^{3+}- and Ho^{3+}-doped materials are interesting mostly because of the otherwise hardly accessible spectral region of operation. Active mode-locking experiments have been performed with both Tm- and Ho-based lasers in the picosecond range [336, 337, 353, 354, 355]. Tm^{3+}-doped lasers are more prospective because of the larger bandwidth of up to $\Delta\lambda/\lambda > 0.1$ and high-power operation capability. Long lifetime and low cross section allow using these materials as amplifiers. However, because both Tm^{3+} and Ho^{3+} ions exhibit upconversion losses due to many upper lying levels (Fig. 24), they should be driven at relatively low population densities.

4.4 Ce^{3+}- and Other Rare-Earth-Doped Crystals

Ce^{3+}-doped materials represent the shortest wavelength-tunable, solid-state laser materials demonstrated so far. The most successful fluoride Ce^{3+}-doped materials operate between 266 nm and 335 nm, with a fluorescence band of about 30 nm to 35 nm, corresponding to $\Delta\lambda/\lambda \sim 0.1$. Note that the Ce^{3+} spectrum has a different nature than other rare-earth ions considered here. The Ce^{3+} ion has a $4f^1$ electronic configuration, with a $^2F_{5/2}$ ground state and a $^2F_{7/2}$ excited state, which is only 2000 cm^{-1} above the ground state [282]. The transitions observed in the visible and UV are due to the next higher excited state, corresponding to the $5d^1$ configuration (f–d transitions). As in transition metals, the $5d$ electron is then strongly exposed to the crystalline environment, and its states are described by a configurational diagram typical of the transition-metal ion (Fig. 2a). The main difference from the d–d transitions is that the ground and upper state parabolae now have different curvatures and formulas (20, 21) do not apply. Formulas (18, 19) are still valid, but the bandshapes (and Huang–Rhys parameter S) are no longer the same for the absorption and emission transitions.

Another important difference with the inner shell transitions is that the f and d states are of different parity and that the Laporte rule does not apply. The transitions are therefore electric-dipole allowed. In Ce^{3+}, the $4f^1$–$5d^1$ transitions are also spin-allowed, resulting in high cross-sectional values ($\sim 10^{-17}$ cm^{-1}) and a short lifetime (tens of nanosecond).

The $5d^1$ electron state resembles that of the $3d^1$ electron of the Ti^{3+} ion and in the octahedral field also has a lower $^2T_2$ and higher lying ^2E states [30]. In sites with lower symmetry, the degeneracy is lifted, and the absorption spectrum of Ce^{3+} can contain up to five bands, corresponding to the transitions from the $^2F_{5/2}$ to each of the split $5d$ states. The fluorescence of the Ce^{3+} ion consists of two partially overlapping bands, corresponding to the transitions from the lower $5d$ state to the $^2F_{7/2}$ and $^2F_{5/2}$ states. The bands are separated by $\sim$ 2000 cm^{-1}. The Figure 28 shows spectra typical of Ce^{3+}-doped fluorides.

As with other broadband materials, ESA is the main problem plaguing the operability of many Ce^{3+}-doped materials. To avoid ESA into the conduction band, materials with the broadest possible energy gap should be used. This is one of the reasons why Ce^{3+}-doped fluorides like Ce:YLF [356], Ce:LaF_3 [357], Ce:LiCaF [358, 359], Ce:LiBAF [361], Ce:LiSAF [360], Ce:SAF [362] and Ce:$LuLiF_4$ [363] have been successful, whereas, e.g., Ce^{3+}:YAG failed [364, 365]. ESA to the conduction band is present in all materials, but in YAG, it overlaps with the emission band; in fluorides, it lies higher on the energy scale, overlapping only the pump bands. Data on spectroscopic features of more than 300 different Ce^{3+}-doped compounds can be found in [366]. UV operation is undoubtedly the most interesting feature of the Ce^{3+} ion, but the fluorescence may go as far as 650 nm, depending on the host [366]. Among Ce^{3+}-doped materials that actually worked as laser media, Ce^{3+}:LiCAF has

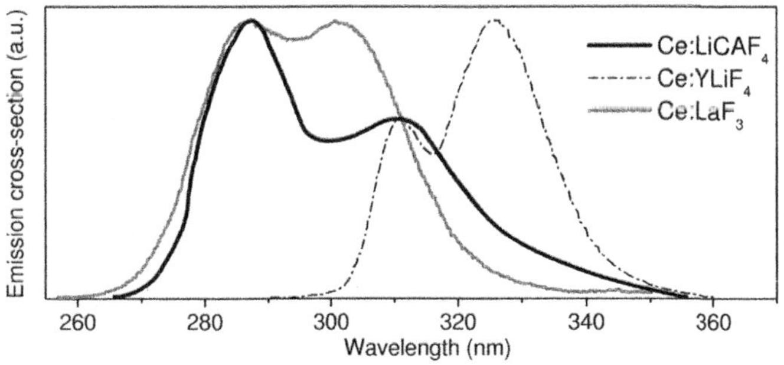

Fig. 28. Emission cross-sectional spectra of Ce^{3+}:LiCAF, Ce^{3+}:YLF, and $Ce^{3+}:LaF_3$ [356, 357, 358]

the broadest bandwidth of over 45 nm ($\Delta\lambda/\lambda \sim 0.15$) due to the combination of vibronic and inhomogeneous broadening [359]. Ce:LiCAF, as well as other Ce^{3+}-doped colquiriites, allows convenient pumping around 260 nm, the fourth harmonic of Nd- and Yb-based lasers. High quality crystals are now available with diameters up to 7 cm [367]. Of all colquiriites, LiCAF exhibits the highest thermal conductivity of $\sim 5\,\mathrm{W/(cm\cdot K)}$.

For pumping, excimer lasers and fourth harmonics of Nd and Yb lasers can be used, depending on the absorption bands of the particular materials. Obtaining CW operation is extremely difficult because of fundamental reasons: as follows from (15), the threshold pump intensity scales as λ^{-4}. Taking for comparison Cr^{3+}:LiSAF (Sect. 3.2, $\lambda = 860$ nm, $\Delta\lambda/\lambda = 0.2$, minimum threshold power ~ 30 mW in a tightly focused TEM_{00} beam), we obtain for $\lambda = 290$ nm and $\Delta\lambda/\lambda = 0.14$ an estimate of 1.5 W of the minimum threshold pump power using the scaling rule (15). This power level lies on the verge of feasibility with fourth-harmonic Nd lasers [368], but current implementation is impractical for applications due to fast degradation of the nonlinear crystal. Thus, obtaining CW operation of Ce^{3+}-doped crystals requires a considerable improvement in pump source technology.

Under pulsed pumping, Ce^{3+}-doped fluorides demonstrate 25% to 50% slope efficiency [30, 369], and up to 100 mJ output energies [370]. Extensive reviews on pulsed Ce^{3+}-doped lasers and amplifiers can be found in [371, 372, 373].

From the point of view of ultrashort-pulse generation, Ce^{3+}-doped fluorides are interesting mostly as broadband amplifiers in the ultraviolet. Recently, Ce:LiCaF has been used in a chirped-pulse amplifier scheme at 290 nm, delivering 6 mJ pulses of 115 fs duration [374]. Shorter pulses and energies reaching TW levels are feasible. The same wavelength region can be accessed by frequency-tripling of the high-energy Ti:sapphire amplifier running at 870 nm, but direct generation of UV pulses from a Ce^{3+}-doped amplifier has significantly higher potential. The main problems associated with the 3ω Ti:sapphire route are the difficulty of making large-aperture BBO

(β-BaB_2O_4) crystals and lower efficiency of the broadband femtosecond 3ω conversion ($\eta < 10\%$), compared with the narrowband nanosecond 2ω process using, e.g., a newly developed $Li_2B_4O_7$ crystal ($\eta > 30\%$) [375]. For higher pump energies, combining multiple laser sources to pump a Ce-doped amplifier crystal is possible.

Other trivalent rare-earth ions also exhibit f–d transitions. Since all of them lie even higher on the energy scale (deeper in the UV) than Ce^{3+} [282], none of them could be used as laser transitions in common host materials due to charge-transfer ESA. Some of the *divalent* rare-earth ions also exhibit broadband interconfigurational spectra, e.g., Sm^{2+}, Eu^{2+}, and Yb^{2+}. However, these materials suffer from ESA to higher lying $4f$ states. Only Sm^{2+}:CaF_2 was shown to operate at temperatures below 210 K [376].

5 Conclusion

This chapter has surveyed current and prospective solid-state laser materials for few-cycle pulse generation and amplification. The emphasis in the discussion has been to connect the spectroscopic and physical properties of different active media with their performance in the ultrashort-pulse regime. Understanding this connection helps to assess critically the emerging laser materials and to perform directed search for required parameters. More than 40 years after the first solid-state laser was operated, research on novel laser media continues to bring often unanticipated fruits, and some of them will certainly advance few-cycle generation further toward performance and better user-friendliness.

This review deliberately avoided brandmarking the materials as "good" or "bad", "prospective" or "useless." The advances in pulse generation techniques and pump sources may turn a well forgotten material into a fresh starter, as happened with Yb:YAG. As ultrashort-pulse generation becomes more and more application-oriented (one may recall, e.g., femtochemistry, optical coherence tomography, and metrology), the requirements of corresponding sources will be reevaluated, calling for more variety in wavelength ranges and output parameters. The demand for new materials will probably only increase in the foreseeable future.

Acknowledgements

It is my pleasure to thank Mark Dubinski (Army Research Lab), Patrick Georges (Institut d'Optique), and Peter Moulton (QPeak Inc.) for providing their results and figures used in this review. I would like to thank Irina Sorokina, Manfred Ober, Ernst Wintner, Angelika Unterhuber, Sergei Naumov, Vladimir Kalashnikov, with whom I had the privilege to work at the Technical University of Vienna. Some of the results were obtained in collaboration with Alberto Di Lieto and Mauro Tonelli at Pisa University as well as with Andrei Alpat'ev, Hermann Haus, Andrei Okhrimchuk, and Ralf Page during their visits in Vienna. All of them deserve my sincere gratitude.

References

[1] R. L. Fork, B. I. Greene, C. V. Shank: Generation of optical pulses shorter than 0.1 psec by colliding pulse mode locking, Appl. Phys. Lett. **38**, 671–672 (1981)

[2] D. E. Spence, P. N. Kean, W. Sibbett: 60-fs pulse generation from a self-mode-locked Ti:sapphire laser, Opt. Lett. **16**, 42–44 (1991)

[3] A. V. Sokolov: Single-cycle optical pulses produced by coherent molecular oscillations, in Ultrafast Optics IV **95** (Springer, New York 2004)

[4] M. Ebrahimzadeh: Mid–infrared ultrafast and continuous-wave optical parametric oscillators, in *Solid-State Mid–Infrared Laser Sources*, Top. Appl. Phys. **79** (Springer, Berlin, Heidelberg 2003) pp. 179–218

[5] Ultrafast Optics IV, **95** (Springer, New York, 2004)

[6] I. T. Sorokina, E. Sorokin, E. Wintner, A. Cassanho, H. P. Jenssen, M. Noginov: Efficient continuous-wave TEM_{00} and femtosecond Kerr-lens mode-locked Cr:LiSrGaF laser, Opt. Lett. **21**, 204–206 (1996)

[7] J. Koetke, G. Huber: Infrared excited-state absorption and stimulated-emission cross sections of Er^{3+}-doped crystals, Appl. Phys. B **61**, 151 (1995)

[8] M. A. Noginov, V. A. Smirnov, I. A. Shcherbakov: Non-linear population processes of Er^{3+} laser levels in chromium-doped garnet crystals, Opt. Quantum Electron. **22**, S61–S74 (1990)

[9] F. Sanchez, A. Kellou: Laser dynamics with escited-state absorption, J. Opt. Soc. Am. B **14**, 209–213 (1997)

[10] R. Mellish, S. C. W. Hyde, N. P. Barry, R. Jones, P. M. W. French, J. R. Taylor, C. J. van der Poel, A. Valster: All-solid-state diode-pumped Cr:LiSAF femtosecond oscillator and regenerative amplifier, Appl. Phys. B **65**, 221–226 (1997)

[11] D. E. McCumber: Theory of phonon-terminated optical masers, Phys. Rev. **134**, A299–A306 (1964)

[12] D. E. McCumber: Einstein relations connecting broadband emission and absorption spectra, Phys. Rev. **136**, A954–A957 (1964)

[13] P. F. Moulton: Spectroscopic and laser characteristics of Ti:Al_2O_3, J. Opt. Soc. Am. B **3**, 125–132 (1986)

[14] O. G. Peterson, S. A. Tuccio, B. B. Snavely: CW operation of an organic dye solution laser, Appl. Phys. Lett. **17**, 245–247 (1970)

[15] L. F. Mollenauer, D. H. Olson, A broadly tunable cw laser using color centers, Appl. Phys. Lett. **24**, 386 (1974)

[16] B. Struve, G. Huber, V. V. Laptev, I. A. Scherbakov, E. V. Zharikov: Tunable room-temperature cw laser action in Cr^{3+}:GdScGa-garnet, Appl. Phys. B **30**, 117–120 (1983)

[17] P. Albers, E. Stark, G. Huber: Continuous-wave laser operation and quantum efficiency of titanium-doped sapphire, J. Opt. Soc. Am. B **3**, 134–139 (1986)

[18] S. Basu, R. L. Byer: Continuous-wave mode-locked Nd:glass laser pumped by a laser diode, Opt. Lett. **13**, 458–460 (1988)

[19] R. Mellish, N. P. Barry, S. C. W. Hyde, R. Jones, P. M. W. French, C. J. Van der Poel, A. Valster: Diode-pumped Cr:LiSAF all-solid-state femtosecond oscillator and regenerative amplifier, Opt. Lett. **20**, 2312–2314 (1995)

[20] V. P. Yanovsky, F. W. Wise, A. Cassanho, H. P. Jenssen: Femtosecond diode-pumped Cr:LiSGAF lasers, IEEE J. Quantum Elect. **2**, 465 (1996)

[21] I. T. Sorokina, S. Naumov, E. Sorokin, E. Wintner, A. V. Shestakov: Directly diode-pumped tunable continuous-wave room-temperature Cr^{4+}:YAG laser, Opt. Lett. **24**, 1578–1580 (1999)

[22] S. Naumov, E. Sorokin, I. T. Sorokina: Directly diode-pumped femtosecond Cr^{4+}:YAG laser, TOPS **83** (OSA, Washington 2003) pp. 163–166

[23] A. V. Podlipensky, V. G. Shcherbitsky, N. V. Kuleshov, V. I. Levchenko, V. N. Yakimovich, M. Mond, E. Heumann, G. Huber, H. Kretschmann, S. Kück: Efficient laser operation and continuous-wave diode pumping of Cr^{2+}: ZnSe single crystals, Appl. Phys. B. **72**, 253–255 (2001)

[24] E. Sorokin, I. T. Sorokina: Tunable diode-pumped continuous-wave Cr^{2+}:ZnSe laser, Appl. Phys. Lett. **80**, 3289–3291 (2002)

[25] I. T. Sorokina, E. Sorokin, S. Mirov, V. Fedorov, V. Badikov, V. Panyutin, K. Schaffers: Broadly tunable compact continuous-wave Cr^{2+}:ZnS laser, Opt. Lett. **27**, 1040–1042 (2002)

[26] F. Balembois, F. Druon, F. Falcoz, P. Georges, A. Brun: Comparison between Cr:LiSAF and Cr:LiSGaF for cw diode-pumped Q-switched operation, TOPS **10** (OSA, Washington 1997) pp. 170–173

[27] S. Guy, C. L. Bonner, D. P. Shepherd, D. C. Hanna, A. C. Tropper, B. Ferrand: High-inversion densities in Nd:YAG-upconversion and bleaching, IEEE J. Quantum Elect. **34**, 900–909 (1998)

[28] F. Balembois, F. Falcoz, F. Kerboull, F. Druon, P. Georges, A. Brun: Theoretical and experimental investigations of small-signal gain for a diode-pumped Q-switched Cr:LiSAF laser, IEEE J. Quantum Elect. **33**, 269–278 (1997)

[29] M. A. Noginov, M. Vondrova, B. D. Lucas: Thermally induced optical bistability in Cr-doped colquiriite crystals, Phys. Rev. B **65**, 035112 (2001)

[30] S. Kück: Laser-related spectroscopy of ion-doped crystals for tunable solid-state lasers, Appl. Phys. B. **72**, 515–562 (2001)

[31] E. P. Ippen, L. Y. Lui, H. A. Haus: Self-starting condition for additive-pulse mode-locked lasers, Opt. Lett. **15**, 183 (1990)

[32] B. Henderson, R. H. Bartram: *Crystal-Field Engineering of Solid-State Laser Materials*, (Cambridge Univ. Press, Cambridge, New York 2000), p. 398

[33] Y. Tanabe, S. Sugano: On the absorption spectra of complex ions, Part I, J. Phys. Soc. Japan **9**, 753–766 (1954); Part II, J. Phys. Soc. Japan **9**, 766–779 (1954)

[34] S. Sugano, Y. Tanabe, H. Kamimura: *Multiplets of Transition-Metal Ions in Crystals* (Acad. Press, New York 1970)

[35] B. Henderson, G. F. Imbusch: *Optical Spectroscopy of Inorganic Solids* (Clarendon Pr., Oxford 1989)

[36] R. C. Powell: *Physics of Solid-State Laser Materials* (Springer, New York 1998)

[37] O. Laporte: Z. Physik, **23**, 135 (1924)

[38] L. E. Orgel: *An Introduction to Transition-Metal Chemistry Ligand-Field Theory* (Methuen, London 1966)

[39] A. Zunger, U. Lindefelt: Electronic structure of transition-atom impurities in semiconductors: substitutional 3d impurities in silicon, Phys. Rev. B **27**, 1191 (1983)
[40] V. Singh, A. Zunger: Electronic structure of transition-atom impurities in GaP, Phys. Rev. B **31**, 3729 (1985)
[41] I. T. Sorokina, S. Naumov, E. Sorokin, A. G. Okhrimchuk: The mechanisms of slow bleaching in YAG:Cr^{4+} under cw pumping, in *Laser Optics 2000: Solid-State Lasers*, Proc. SPIE **4350** (2001) pp. 99–105
[42] K. Huang, A. Rhys: P. Roy. Soc. Lond. A Mat. **204**, 406 (1950)
[43] W. H. Fonger, C. W. Struck: Unified model of energy transfer for arbitrary Franck-Condon offset and temperature, J. Lumin. **17**, 241–261 (1978)
[44] M. Yamaga, Y. Gao, F. Rasheed, K. P. O'Donnel, B. Henderson, B. Cockayne: Radiative and non-radiative decays from the excited state of Ti^{3+} ions in oxide crystals, Appl. Phys. B **51**, 329 (1990)
[45] M. D. Sturge: Temperature dependence of multiphonon nonradiative decay at an isolated impurity center, Phys. Rev. B **8**, 6–14 (1973)
[46] R. Englman, J. Jortner: Mol. Phys. **18**, 145 (1970)
[47] H. A. Jahn, E. Teller: P. Roy. Soc. Lond. A Mat. **161**, 220 (1937)
[48] R. Engleman: *The Jahn–Teller Effect in Molecules and Crystals* (Willey-Interscience, New York 1972)
[49] I. B. Bersuker: *The Jahn–Teller Effect and Vibronic Interactions in Quantum Chemistry* (Plenum, New York 1984)
[50] M. Kaminska, J. M. Baranovski, S. M. Uba, J. T. Vallin: Absorption and luminescence of $Cr^{2+}(d^4)$ in II-VI compounds, J. Phys. C Solid State **12**, 2197–2214 (1979)
[51] J. T. Vallin, G. A. Slack, S. Roberts, A. E. Hughes: Infrared absorption in some II-VI compounds doped with Cr, Phys. Rev. B **2**, 4313–4333 (1970)
[52] S. W. Biernacki: Jahn-Teller coupling of Cr^{2+} ion in ZnS, ZnSe and ZnTe, Phys. Stat. Solidi (b) **87**, 607–612 (1978)
[53] A. L. Natadze, A. I. Ryskin: Jahn-Teller coupling of Cr^{2+} ion with degenerate modes in ZnS, ZnSe, and ZnTe crystals: microscopic treatment, Phys. Stat. Sol. (b) **97**, 175–185 (1980)
[54] G. Goetz, H. Zimmermann, H. J. Schulz: Jahn-Teller interaction at $Cr^{2+}(d^4)$ centres in tetrahedrally coordinated II-VI lattices studied by optical spectroscopy, Z. Physik-B **91**, 429–436 (1993)
[55] I. T. Sorokina, E. Sorokin, A. Di Lieto, M. Tonelli; R. H. Page, K. I. Schaffers: Efficient broadly tunable continuous-wave Cr^{2+}:ZnSe laser, J. Opt. Soc. Am. B **18**, 926–930 (2001)
[56] E. V. Pestryakov, V. V. Petrov, V. I. Trunov, A. V. Kirpichnikov, A. G. Volkov, A. I. Alimpiev, A. Ja. Rodionov: $Be_3Al_2Si_6O_{18}$ crystals doped with Jahn-Teller ions: a promising active media for superbroadband lasers, Laser Phys. **11**, 1138–1141 (2001)
[57] P. Lacovara, L. Esterowitz, M. Kokta: Growth, spectroscopy and lasing of titanium-doped sapphire, IEEE J. Quantum Elect. **21**, 1614-1618 (1985)
[58] J. M. Eggleston, L. G. Deshazer, K. W. Kangas: Characteristics and kinetics of laser-pumped Ti:sapphire oscillators, IEEE J. Quantum Elect. **24**, 1009–1015 (1988)
[59] P. Albers, E. Stark, G. Huber: Continuous-wave laser operation and quantum efficiency of titanium doped sapphire, J. Opt. Soc. Am. B **3**, 134–139 (1986)

[60] R. L. Aggarwal, A. Sanchez, M. M. Stuppi, R. E. Fahey, A. J. Strauss, W. R. Rapoport, Ch. Khattak: Residual infrared absorption in As-grown and annealed crystals of Ti:Al_2O_3, IEEE J. Quantum Elect. **24**, 1003–1008 (1988)
[61] T. Izawa, M. Maeda, N. Yamamura, R. Uchimura, R. Ikeda, S. Kimura, T. Yakuoh, N. Sarukura, Z. Liu, Y. Segawa, H. Tashiro: Full-range tunable operation of a cw Ti-sapphire laser with a single set of extrmely broad-band, low-loss mirrors, in OSA Proc. ASSL **24**, 257–261 (1995)
[62] A. M. Kowalevicz, Jr., T. R. Schibli, F. X. Kärtner, J. G. Fujimoto: Ultralow-threshold Kerr-lens mode-locked Ti:Al_2O_3 laser, Opt. Lett. **27**, 2037–2039 (2002)
[63] A. I. Alimpiev, G. V. Bukin, V. N. Matrosov, E. V. Pestryakov, V. P. Solntsev, V. I. Trunov, E. G. Tsvetkov, V. P. Chebotaev: Tunable $BeAl_2O_4$:Ti^{3+} laser, Kvantovaya Elektronika **13**, 885–886 (1986); English translation in Sov. J. Quantum Elect. **16**, 579–580 (1986)
[64] E. V. Pestryakov, V. I. Trunov, A. I. Alimpiev: Generation of tunable radiation in a $BeAl_2O_4$:Ti^{3+} laser subjected to pulsed coherent pumping at a high repetition frequency, Kvantovaya Elektronika **14**, 919–922 (1987); English translation in Sov. J. Quantum Elect. **17**, 585–586 (1987)
[65] Y. Segawa, A. Sugimoto, P.H. Kim, S. Namba, K. Yamagishi, Y. Anzai, Y. Yamaguchi: Optical properties and lasing of Ti^{3+} doped $BeAl_2O_4$, Jpn. J. Appl. Phys. **24**, L291–L292 (1987)
[66] A. Sugimoto, Y. Segawa, Y. Anzai, K. Yamagishi, P. H. Kim, S. Namba: Flash-lamp-pumped tunable Ti:$BeAl_2O_4$ laser, Jpn. J. Appl. Phys. **29**, L1136–L1137 (1990)
[67] T. A. Driscoll, M. Peressini, R. E. Stone, G. Hansen, H. J. Hoffman: CLEO (OSA, Washington DC 1986) p. 106, paper TUK29
[68] J. Kvapil, M. Koselja, Ji. Kvapil, B. Perner, V. Skoda, J. Kubelka, K. Hamel, V. Kubecak: Growth and stimulated emission of YAP:Ti, Czech. J. Phys. B **38**, 237–240 (1988)
[69] J. Kvapil, M. Koselja, Ji. Kvapil, K. Hamel: CLEO **11** (OSA, Washington DC 1989) p. 6, paper MC2
[70] T. Danger, K. Petermann, G. Huber: Polarized and time-resolved measurements of excited-state absorption and stimulated emission in Ti:$YAlO_3$ and Ti:Al_2O_3, Appl. Phys. A **57**, 309–313 (1993)
[71] Z. J. Kiss, R. C. Duncan: Cross-pumped Cr^{3+}-Nd^{3+}:YAG laser system, Appl. Phys. Lett. **5**, 200–202 (1964)
[72] T. H. Maiman, Stimulated optical radiation in ruby, Nature **187**, 493–494 (1960)
[73] B. K. Sevast'yanov, Kh. S. Bagdasarov, L. B. Pasternak, S. Yu. Volkov, V. P. Orekhova: Optical generation on Cr^{3+} ions in yttrium-aluminium garnet crystals, JETP Lett. **17**, 47 (1973)
[74] S. A. Payne, L. L. Chase, H. W. Newkirk, L. K. Smith, W. F. Krupke: $LiCaAlF_6$:Cr^{3+}: a promising new solid-state laser material, IEEE J. Quantum Elect. **24**, 2243–2252 (1988)
[75] P. Beaud, M. C. Richardson, Y.-F. Chen, B. H. T. Chai: Optical amplification characteristics of Cr:LiSAF and Cr:LiCAF under flashlamp-pumping, IEEE J. Quantum Elect. **30**, 1259–1266 (1994)

[76] L. L. Chase, S. A. Payne, J. E. Marion, L. K. Smith, B. H. T. Chai, H. W. Newkirk, W. Kway, L. Davis, M. D. Shinn, B. Woods: Flashlamp-pumped performance of the solid state laser material $LiCaAlF_6$:Cr^{3+}, CLEO (1989) p. 111, paper TUJ59
[77] S. A. Payne, L. L. Chase, G. D. Wilke: Optical spectroscopy of the new laser materials, $LiSrAlF_6$:Cr^{3+} and $LiCaAlF_6$:Cr^{3+}, J. Lumin. **44**, 167 (1089)
[78] S. A. Payne, L. L. Chase, L. K. Smith, W. Kway, H. W. Newkirk: Laser performance of Cr:LiSAF, J. Appl. Phys. **66**, 1051–1054 (1989)
[79] J. C. Walling, H. P. Jenssen, R. C. Morris, E. W. O'Dell, O. G. Peterson: Tunable laser performance in $BeAl_2O_4$:Cr^{3+}, Opt. Lett. **4**, 182–184 (1979)
[80] J. C. Walling, D. F. Heller, H. Samelson, D. J. Harter, J. A. Pete, R. C. Morris: Tunable alexandrite lasers: development and performance, IEEE J.Quantum Elect. **21**, 1568 (1985)
[81] M. Lando, Y. Shimony, R. M. J. Benmair, D. Abramovich, V. Krupkin, A. Yogev: Visible solar-pumped lasers, Opt. Mater. **13**, 111 (1999)
[82] S. T. Lai, M. L. Shand: High efficiency cw laser-pumped tunable alexandrite laser, J. Appl. Phys. **54**, 56642 (1983)
[83] M. L. Shand, D. J. Harter, S. T. Lai, H. Samelson: in Proc. Int. Conf. Lasers, 161 (1984)
[84] R. Scheps, B. M. Gately, J. F. Myers, J. S. Krasinski, D. F. Heller: Alexandrite laser pumped by semiconductor lasers, Appl. Phys. Lett. **56**, 2288 (1990)
[85] R. Scheps, J. F. Myers, T. R. Glesne, H. B. Serreze: Monochromatic end-pumped operation of an alexandrite laser, Opt. Commun. **97**, 363 (1993)
[86] J. C. Walling, O. G. Peterson, H. P. Jenssen, R. C. Morris, E. W. O'Dell: Tunable alexandrite lasers, IEEE J. Quantum Elect. **16**, 1302 (1980)
[87] D. Spence, P. Kean, W. Sibbett: 60 fs pulse generation from a self-mode-locked Ti:sapphire laser, Opt. Lett. **16**, 41 (1991)
[88] M. Piche: Beam reshaping and self-mode-locking in nonlinear laser resonators, Opt. Commun. **86**, 156 (1991)
[89] M. J. Weber, D. Milam, W. L. Smith: Nonlinear refractive index of glasses and crystals, Opt. Eng. **17**, 463–469, (1978)
[90] R. Adair, L. L. Chase, S. A. Payne: Nonlinear refractive index of optical crystals, Phys. Rev. B **39**, 3337 (1989)
[91] C. L. Sam, J. C. Walling, H. P. Jenssen, R. C. Morris, E. W. O'Dell: Characteristics of alexandrite lasers in Q-switched and tuned operations, Proc. Soc. Phot. Opt. Inst. Eng., **247**, 130 (1980)
[92] C. Guch Jr., C. E. Jones: Alexandrite laser performance, Opt. Lasers (Dec. 1982)
[93] R. Frey, F. De-Rougemont, C. H. Lee: An actively mode-locked continuous wave alexandrite laser, Opt. Commun. **73**, 282 (1989)
[94] F. Volker, Q. Lu, H. Weber: Passive mode-locking of an alexandrite laser for picosecond pulse generation, J. Appl. Phys. **69**, 3432 (1991)
[95] M. L. Shand, H. P. Jenssen: Temperature dependence of the excited state absorption of alexandrite, IEEE J. Quantum Elect. **19**, 480 (1983)
[96] J. V. Meier, N. P. Barnes, D. K. Remelius, M. R. Kokta: Flashlamp-pumped Cr^{3+}:GSAG laser, IEEE J. Quantum Elect. **22**, 2058–2064 (1986)
[97] V. N. Lisitsyn, V. N. Matrosov, V. P. Orekhova, E. V. Pestryakov, B. K. Sevast'yanov, V. I. Trunov, V. N. Zenin, Yu. L. Renigallo: Generation of

0.7-0.8 picosecond pulses in an alexandrite laser with passive mode-locking, Sov. J. Quantum Elect., 368–370, (1982)

[98] B. Struve, G. Huber, V. V. Laptev, I. A. Shcherbakov, E. V. Zharikov: Tunable room-temperature cw laser action in Cr^{3+}:GdScGa-garnet, Appl. Phys. B **30**, 117–120 (1983)

[99] J. Drube, B. Struve, G. Huber: Tunable room-temperature cw laser action in Cr^{3+}:GdScAl-garnet, Opt. Commun. **50**, 45 (1984)

[100] L. K. Smith, S. A. Payne, L. L. Chase, W. Kway, B. H. T. Chai: Cr^{3+}:$LiSrGaF_6$ – a new material of the colquiriite structure, CLEO (OSA, Washington DC 1991) pp. 388–390, paper CThH1

[101] L. K. Smith, S. A. Payne, W. Kway, L. L. Chase, B. H. T. Chai: Investigation of the laser properties of Cr^{3+}:$LiSrGaF_6$, IEEE J. Quantum Elect. **28**, 2612 (1992)

[102] R. Scheps: Cr-doped solid state lasers pumped by visible laser diodes, Opt. Mater. **1**, 1 (1992)

[103] R. D. Peterson, A. T. Pham, H. P. Jenssen, A. Cassanho, V. Castillo: Thermo-optical Comparison of LiSAF, LiCAF, and LiSGaF, TOPS **26** (OSA, Washington DC 2003) pp. 647–651

[104] M. Yamaga, B. Henderson, K. P. O'Donnell: Polarization spectroscopy of Cr^{3+} ions in laser host crystals: II. The broadband transitions, J. Lumin. **46**, 397 (1990)

[105] P. Li Kam Wa, B. H. T. Chai, A. Miller, Self-mode-locked Cr^{3+}:$LiCalF_6$ laser, Opt. Lett. **17**, 1438–1440 (1992)

[106] K. M. Gäbel, P. Rußbüldt, R. Lebert, A. Valster: Diode-pumped Cr^{3+}:$LiCalF_6$ femtosecond laser, Opt. Commun. **157**, 327 (1998)

[107] H. S. Wang, P. Li Kam Wa, J. L. Lefaucheur, B. H. T. Chai: CW and self-modelocking performance of a red pumped Cr:$LiSr_{0.8}Ca_{0.2}F_6$ laser, Opt. Commun. **110**, 679–688 (1994)

[108] N. H. Rizvi, P. M. W. French, J. R. Taylor: 50-fs pulse generation from a self-starting cw passively mode-locked Cr:$LiSrAlF_6$ laser, Opt. Lett. **17**, 877–879 (1992)

[109] N. H. Rizvi, P. M. W. French, J. R. Taylor: Generation of 33-fs pulses from a passively mode-locked Cr3+:$LiSrAlF_6$ laser, Opt. Lett. **17**, 1605–1607 (1992)

[110] M. J. P. Dymott, A. I. Ferguson: Pulse duration limitations in a diode-pumped femtosecond Kerr-lens mode-locked Cr:LiSAF laser, Appl. Phys. B. **65**, 227 (1997)

[111] D. Kopf, A. Prasad, G. Zhang, M. Moser, U. Keller: Broadly tunable femtosecond Cr:LiSAF laser, Opt. Lett. **22**, 621–623 (1997)

[112] E. Van Stryland, unpublished result, private communication

[113] A. Miller, P. Li Kam Wa, B. H. T. Chai, E. W. Van Stryland: Generation of 150-fs tunable pulses in Cr:$LiSrAlF_6$, Opt. Lett. **17**, 195–197 (1992)

[114] I. T. Sorokina, E. Sorokin, E. Wintner, A. Cassanho, H. P. Jenssen: A new cw self-mode-locked Cr:LiSGaF femtosecond laser, in CLEO (1994) p.369, paper CThN3

[115] I. T. Sorokina, E. Sorokin, E. Wintner, A. Cassanho, H. P. Jenssen, M. A. Noginov: Efficient continuous-wave TEM_{00} and femtosecond Kerr-lens mode-locked Cr:LiSrGaF laser, Opt. Lett. **21**, 204–206 (1996)

[116] F. Balembois, F. Druon, F. Falcoz, P. Georges, A. Brun: Performances of Cr$LiSrAlF_6$ and Cr$LiSrGaF_6$ for continuous-wavediode-pumped Q-switched operation, Opt. Lett. **22**, 387–389 (1997)

[117] I. T. Sorokina, E. Sorokin, E. Wintner, A. Cassanho, H. P. Jenssen, M. Noginov: All-solid-state diode-pumped cw and mode-locked Cr:LiSrGaF laser, OSA Proc. ASSL **24** (1995) pp. 146–149
[118] V. P. Yanovsky, F. W. Wise, A. Cassanho, H. P. Jenssen: Kerr-lens mode-locked diode-pumped Cr:LiSGAF laser, Opt. Lett. **20**, 1304–1306 (1995)
[119] I. T. Sorokina, E. Sorokin, E. Wintner, M. A. Noginov, R. Szipöcs, A. Cassanho, H. P. Jenssen: Femtosecond pulses from the novel solid-state laser source Cr:LiSGaF, Laser Phys. **7**, 187–195 (1997)
[120] K. M. Gäbel, P. Rußbüldt, R. Lebert, P. Loosen, R. Poprawe, H. Heyer, A. Valster: Diode pumped, chirped mirror dispersion compensated, fs-laser, Opt. Commun. **153**, 275 (1998)
[121] F. H. Loesel, C. Horvath, F. Grasbon, M. Jost, M. H. Niemz: Selfstarting femtosecond operation and transient dynamics of a diode-endpumped Cr:LiSGaF laser with a semiconductor saturable absorber mirror, Appl. Phys. B **65**, 783–787 (1997)
[122] I. T. Sorokina, E. Sorokin, E. Wintner, A. Cassanho, H. P. Jenssen, R. Szipöcs: Sub-20 fs pulse generation from the mirror dispersion controlled Cr:LiSGaF and Cr:LiSAF lasers, Appl. Phys. B **65**, 245 (1997)
[123] U. Keller, K. J. Weingarten, F. X. Kärtner, D. Kopf, B. Braun, I. D. Jung, R. Fluck, C. Hönninger, N. Matuschek, J. Aus der Au: IEEE J. Sel. Top. Quant. **2**, 435 (1996)
[124] I. T. Sorokina, E. Sorokin, E. Wintner, A. Cassanho, H. P. Jenssen, R. Szipöcs: 14-fs pulse generation in Kerr-lens mode-locked prismless Cr:LiSGaF and Cr:LiSAF lasers: observation of pulse self-frequency shift, Opt. Lett. **22**, 1716 (1997)
[125] S. Uemura, K. Torizuka, Generation of 12-fspulses from a diode-pumped Kerr-lens mode-locked CrLiSAF laser, Opt. Lett. **24**, 780–782 (1999)
[126] H. A. Haus, I. T. Sorokina, E. Sorokin: Raman induced red-shift of ultrashort mode-locked laser pulses, J. Opt. Soc. Am. B **15**, 223–231 (1998)
[127] I. T. Sorokina, E. Sorokin, E. Wintner, A. Cassanho, H. P. Jenssen: Raman induced pulse self-frequency shift in sub-20 fs Kerr-lens mode-locked Cr:LiSGaF and Cr:LiSAF lasers, in W. R. Bosenberg and M. M. Fejer (Eds.): TOPS **19** (OSA, Washington DC 1998), pp. 359–362
[128] S. Smolorz, F. Wise: Time-resolved nonlinear refraction in femtosecond laser gain media, Opt. Lett. **23**, 1381–1383 (1998)
[129] K. M. Gäbel, R. Labert, A. Valster: Signature of the Raman self-frequency shift on the autocorrelation of sub-20-fs pulses from colquiriite lasers, TOPS **39** (2000), p. 483, paper CThM48
[130] V. I. Kalashnikov, E. Sorokin, I. T. Sorokina: Mechanisms of spectral shift in ultrashort-pulse laser oscillators, J. Opt. Soc. Am. B **18**, pp. 1732–1741 (2000)
[131] Z. Li, L. Li, H. Tian, G. Zhou, K. H. Spatschek: Chirped femtosecond solitonlike laser pulse form with self-frequency shift, Phys. Rev. Lett. **89**, 263901 (2002)
[132] S. Smolorz, F. Wise: Femtosecond two-beam coupling energy transfer from Raman and electronic nonlinearities, J. Opt. Soc. Am. B **17**, 1636–1644 (2000)
[133] V. L. Kalashnikov, E. Sorokin, S. Naumov, I. T. Sorokina: Spectral properties of the Kerr-lens mode-locked Cr^{4+}:YAG laser, J. Opt. Soc. Am B **20**, 2084–2092 (2003)

[134] F. X. Kärtner, N. Matuschek, T. Schibli, U. Keller, H. A. Haus, C. Heine, R. Morf, V. Scheuer, M. Tilsch, T. Tschudi: Design and fabrication of double-chirped mirrors, Opt. Lett. **22**, 831–833 (1997)

[135] P. Wagenblast, R. Ell, U. Morgner, F. Grawert, F. Kärtner: 10-fs, diode-pumped Cr:LiCAF laser, in Ultrafast Optics IV **95** (Springer, New York 2004) pp. 55–60

[136] D. Burns, M. P. Critten, W. Sibbett: Low-threshold diode-pumped femtosecond Cr^{3+}:$LiSrAlF_6$ laser, Opt. Lett. **21**, 477–479 (1996)

[137] G. J. Valentine, J. M. Hopkins, P. Loza-Alvarez, G. T. Kennedy, W. Sibbett, D. Burns, A. Valster: Ultralow-pump-threshold, femtosecond $Cr^{3+}LiSrAlF_6$ laser pumped by a single narrow-stripe AlGaInP laser diode, Opt. Lett. **22**, 1639–1641 (1997)

[138] J. Hopkins, G. J. Valentine, B. Agate, A. J. Kemp, U. Keller, W. Sibbett: Highly compact and efficient femtosecond Cr:LiSAF lasers, IEEE J. Quantum Elect. **38**, 360 (2002)

[139] C. G. Leburn, A. A. Lagatsky, C. T. Brown, W. Sibbett: Femtosecond Cr^{4+}:YAG laser with a 4 GHz pulse repetition rate, Adv. Solid-State Phot. Top. Meeting (2004), paper WE4

[140] C. R. Pollock, D. B. Barber, J. L. Mass, S. Markgraf: Cr^{4+}-lasers: present performance and prospects for new host lattices, IEEE J. Sel. Top. Quant. **1**, 62–66 (1995)

[141] N. B. Angert, N. I. Borodin, V. M. Garmash, V. A. Zhitnyuk, A. G. Okhrimchuk, O. G. Siyuchenko, A. V. Shestakov: Lasing due to impurity color centers in yttrium alluminum garnet crystals in the range 1.35-145 μm, Sov. J. Quantum Elect. **18**, 73-74 (1988)

[142] H. Eilers, U. Hömmerich, S. M. Jacobsen, W. M. Yen, M. Kokta: Spectroscopic properties of Cr^{4+}:$Lu_3Al_5O_{12}$, Opt. Lett. **18**, 1928-1930 (1993)

[143] S. Kück, K. Petermann, U. Pohlmann, U. Schönhoff, G. Huber: Tunable room-temperature laser action of Cr^{4+} -doped $Y_3Sc_xAl_{5-x}O_{12}$, Appl. Phys. B **58**, 153 (1994)

[144] V. Petricevic, S. K. Gayen, R. R. Alfano, K. Yamagashi, H. Anzai, K. Yamagashi, H. Anzai, Y. Yamaguchi: Laser action in chromium-doped forsterite, Appl. Phys. Lett. **52**, 1040 (1988)

[145] V. Petricevic, S. K. Gayen, R. R. Alfano: Near infrared tunable operation of chromium doped forsterite laser, Appl. Opt. **28**, 1609 (1988)

[146] J. Koetke, S. Kück, K. Petermann, G. Huber, G. Cerullo, M. Danailov, V. Magni, L. F. Qian, O. Svelto: Quasi-continuous wave laser operation of Cr^{4+}-doped Y_2SiO_5 at room-temperature, Opt. Commun. **101**, 195 (1993)

[147] A. G. Avanesov, V. A. Lebedev, B. Denker, B. I. Galagan, V. V. Osiko, S. E. Sverchkov, A. V. Shestakov: Room-temperature laser action of Y_2SiO_5:Cr^{4+} crystal, in OSA Proc. ASSL **20** (1994) pp. 185–187

[148] H. Eilers, U. Hömmerich, S. M. Jacobsen, W. M. Yen: The near infrared emission of Mn_2SiO_4 and $MgCaSiO_4$, Chem. Phys. Lett. **212**, 109–112 (1993)

[149] V. Petricevic, A. B. Bykov, J. M. Evans, R. R. Alfano: Room-temperature near-infrared tunable laser operation of Cr^{4+}: Ca_2GeO_4, Opt. Lett. **21**, 1750–1752 (1996)

[150] M. H. Garrett, V. H. Chan, H. P. Jenssen, M. H. Whitmore, A. Sacra, D. Singel, D. J. Simkin: Comparison of Cr-doped forsterite and akermanite laser host crystals, in OSA Proc. ASSL **10** (1991) pp. 76–81

[151] L. D. Merkle, T. H. Alik, B. H. T. Chai: Crystal growth and spectroscopic properties of Cr^{4+} in $Ca_2Al_2SiO_7$ and $Ca_2Ga_2SiO_7$, Opt. Mater. **1**, 91–100 (1992)

[152] M. A. Scott, T. P. J. Han, H. G. Gallagher, B. Henderson: Near-infrared laser crystals based on $3d^2$ ions, spectroscopic studies of $3d^2$ ions in oxide, melilite and apatite crystals, J. Lumin. **72-74**, 260–262 (1997)

[153] K. Cerqua-Richardson, B. Peng, T. Izumitani: Spectroscopic investigation of Cr^{4+}-doped glasses, in OSA Proc. ASSL **13** (1992) pp. 52–55

[154] U. Hömmerich, H. Eilers, W. M. Yen, J. S. Hayden, M. K. Aston: Near infrared emission at 1.35 μm in Cr doped glass, J. Lumin. **60-61**, 119–122 (1994)

[155] T. Izumitani, B. Peng, K. Cerqua-Richardson, Spectroscopic investigation of Cr-doped silicate and aluminate glasses, Rev. Laser Eng. **24**, 689–694 (1996)

[156] S. Tanabe, X. Feng: Temperature variation of near-infrared emission from Cr^{4+} in alluminate glass for broadband telecommunication, Appl. Phys. Lett. **77**, 818 (2000)

[157] C. Chudoba, J. G. Fujimoto, E. P. Ippen, H. A. Haus, U. Morgner, F. X. Kärtner, V. Scheuer, G. Angelow, T. Tschudi: All-solid-state Cr:forsterite laser generating 14-fs pulses at 1.3 μm, Opt. Lett. **26**, 292–294 (2001)

[158] D. J. Ripin, C. Chudoba, J. T. Gopinath, J. G. Fujimoto, E. P. Ippen, U. Morgner, F. X. Kärtner, V. Scheuer, G. Angelow, T. Tschudi: Generation of 20-fs pulses by a prismless Cr^{4+}:YAG laser, Opt. Lett. **27**, 61–63 (2002)

[159] S. Naumov, E. Sorokin, V. L. Kalashnikov, G. Tempea, I. T. Sorokina: Self-starting five optical cycle pulse generation in Cr^{4+}:YAG laser, Appl. Phys. B **76**, 1–11 (2003)

[160] E. V. Zharikov, A. M. Zabaznov, A. M. Prokhorov, A. P. Shkadarevich, I. A. Shcherbakov: Use of GSGG:Cr:Nd crystals with photochromic centers as active elements in solid lasers, Sov. J. Quantum Elect. **16**, 1552–1554 (1986)

[161] V. Petricevic, S. K. Gayen, R. R. Alfano: Laser action in chromium-activated forsterite for near-infrared excitation: is Cr^{4+} the lasing ion? Appl. Phys. Lett. **53**, 2590–2592 (1988)

[162] N. I. Borodin, V. A. Zhitnyuk, A. G. Okhrimchuk, A. V. Shestakov: Oscillation of a $Y_3Al_5O_{12}$:Cr^{4+} laser in wavelength region of 1.35–1.6 μm, Bull. Acad. Sci. USSR, Phys. Ser. **54**, 1500–1506 (1990)

[163] K. R. Hoffman, J. Casas-Gonzalez, S. M. Jacobsen, W. M. Yen: Electron-paramagnetic-resonance and fluorescence-line-narrowing measurements of the lasing center in Cr-doped forsterite, IEEE J. Quantum Elect. **44**, 12589 (1991)

[164] R. Moncorge, G. Cormier, D. J. Simkin, J. A. Capobianco: Fluorescence analysis of chromium-doped forsterite (Mg_2SiO_4), IEEE J. Quantum Elect. **27**, 114–120 (1991)

[165] N. I. Borodin, A. G. Okhrimchuk, A. V. Shestakov: Polarizing spectroscopy of $Y_3Al_5O_{12}$, $SrAl_2O_4$, $CaAl_2O_4$ crystals containing Cr, in OSA Proc. ASSL **13** (1992) 42–46

[166] H. Eilers, U. Hömmerich, S. M. Jacobsen, W. M. Yen: On the optical center in Cr, M (M=Ca or Mg) doped YAG, in OSA Proc. ASSL **15**, 437–440 (1993)

[167] H. Eilers, W. M. Dennis, W. M. Yen, S. Kück, K. Petermann, G. Huber, W. Jia: Performance of a Cr:YAG laser, IEEE J. Quantum Elect. **29**, 2508 (1993)

[168] S. Kück, K. Petermann, G. Huber: Spectroscopic investigation of the Cr^{4+}-center in YAG, OSA Proc. ASSL **10** (1991) p. 92

[169] A. G. Okhrimchuk, A. V. Shestakov: Performance of YAG:Cr^{4+} laser crystal, Opt. Mat. **3**, 1–13 (1994)

[170] L. I. Kurutova, A. V. Lukin, V. A. Sandulenko, E. A. Sidorova, V. M. Solntsev: Phototropic centers in chromium-doped garnets, Opt. Spectr. **63**, 693–695 (1987)

[171] K. R. Hoffman, U. Hömmerich, S. M. Jacobsen, W. M. Yen: On the emission and excitation spectrum of the NIR laser center in Cr:YAG, J. Lumin. **52**, 277 (1992)

[172] A. Sugimoto, Y. Nobe, K. Yamagishi: Crystal growth and optical characterization of Cr,Ca:$Y_3Al_5O_{12}$, J. Cryst. Growth **140**, 349–354 (1994)

[173] I. T. Sorokina, S. Naumov, E. Sorokin, W. Ching, Y. Xu, B. Brickeen: Anomalous absorption saturation in Cr,Mg:YAG and Cr,Ca:YAG, CLEO (2000) p. 108, paper CtuK43

[174] W. Jia, H. Liu, S. Jaffe, W. M. Yen, B. Denker: Spectroscopy of Cr^{3+} and Cr^{4+} ions in forsterite, Phys. Rev. B **43**, 5234–5242 (1991)

[175] Z. Burshtein, P. Blau, Y. Kalisky, Y. Simony, M. R. Kokta: Excited-state absorption studies of Cr^{4+} ions in several garnet host crystals, IEEE J. Quantum Elect. **34**, 292 (1998)

[176] H. Eilers, U. Hömmerich, S. M. Jacobsen, W. M. Yen, K. R. Hoffman, W. Jia: Spectroscopy and dynamics of Cr^{4+}:$Y_3Al_5O_{12}$, Phys. Rev. B **49**, 15505–15513 (1994)

[177] V. Petricevic, S. K. Gayen, R. R. Alfano: Near infrared tunable operation of chromium doped forsterite laser, Appl. Opt. **28**, 1609 (1989)

[178] V. Petricevic, S. K. Gayen, R. R. Alfano: Continuous-wave laser operation of chromium-doped forsterite, Opt. Lett. **14**, 612–614 (1989)

[179] G. M. Zverev, A. V. Shestakov: Tunable near-infrared oxide crystal lasers, in OSA Proc. ASSL **5** (1989) pp.66–70

[180] B. Xu, J. M. Evans, V. Petricevic, S. P. Guo, O. Maksimov, M. C. Tamargo, R. R. Alfano: Continuous-wave and passively mode-locked operation of a cunyite (Cr^{4+}:Ca_2GeO_4) laser, Appl. Opt. LP-**39**, 4975–4978 (2000)

[181] J. M. Evans, V. Petricevic, A. B. Bykov, R. R. Alfano: Direct diode-pumped continuous-wave near-infrared tunable laser operation of Cr^4+:forsterite and Cr^{4+}:Ca_2GeO_4, Opt. Lett. **22**, 1171–1173 (1997)

[182] A. Agnesi, E. Piccinini and G. C. Reali, Influence of thermal effects in Kerr-lens mode-locked femtosecond Cr^{4+}:forsterite lasers, Opt. Commun. **135**, 77 (1997)

[183] L. F. Qian, X. Liu, F. Wise: Cr:forsterite laser pumped by broad-area laser diodes, Opt. Lett. **22**, 1707–1709 (1997)

[184] A. J. Alcock, P. Scorah, K. Hnatovsky: Broadly tunable continuous-wave diode-pumped Cr^{4+}:YAG laser, Opt. Commun. **215**, 153 (2003)

[185] A. G. Okhrimchuk, A. V. Shestakov: Absorption saturation mechanism for YAG:Cr^{4+} crystals, Phys. Rev. B **61**, 988–995 (2000)

[186] S. Kück, K. Petermann, U. Pohlmann, G. Huber: Near-infrared emission of Cr^{4+}-doped garnets: lifetimes, quantum efficiencies, and emission cross sections, Phys. Rev. B **51**, 17323–17331 (1995)

[187] S. Kück, K. L. Schepler, S. Hartung, K. Petermann, G. Huber: Excited state absorption and its influence on the laser behaviour of Cr^{4+}-doped garnets, J. Lumin. **72-74**, 222–223 (1997)

[188] V. Yanovsky, Y. Pang, F. Wise, B. I. Minkov: Generation of 25-fs pulses from a self-mode-locked Cr:forsterite laser with optimized group-delay dispersion, Opt. Lett. **18**, 1541 1543 (1993)

[189] P. M. W. French, N. H. Rizvi, J. R. Taylor, A. V. Shestakov: Continuous-wave mode-locked Cr^{4+}:YAG laser, Opt. Lett. **18**, 39–41 (1993)

[190] A. Sennaroglu, C. R. Pollock, H. Nathel: Continuous-wave self-mode-locked operation of a femtosecond Cr^{4+}:YAG laser, Opt. Lett. **19**, 390 (1994)

[191] P. J. Conlon, J. P. Tong, P. M. French, J. R. Taylor, A. V. Shestakov: Passive mode locking and dispersion measurement of sub-100 fs Cr^{4+}:YAG laser, Opt. Lett. **19**, 1468–1469 (1994)

[192] P. J. Conlon, J. P. Tong, P. M. French, J. R. Taylor, A. V. Shestakov: Femtosecond pulse generation from synchronously pumped self-mode-locked Cr^{4+}:YAG, J. Mod. Opt. **42**, 723–726 (1995)

[193] J. Ishida, K. Naganuma: Compact diode-pumped all-solid-state femtosecond Cr^{4+}:YAG laser, Opt. Lett. **21**, 51–53 (1996)

[194] Y. P. Tong, P. M. W. French, J. R. Taylor, J. O. Fujimoto: All-solid-state sources in the near infrared, Opt. Comm. **136**, 235–238 (1997)

[195] Y. P. Tong, L. M. Sutherland, P. M. W. French, J. R. Taylor, A. V. Shestakov, B. H. T. Chai: Self-starting Kerr-lens mode-locked femtosecond Cr^{4+}YAG and picosecond Pr^{3+} YLF solid-state lasers, Opt. Lett. **21**, 644–646 (1996)

[196] Y. Ishida, K. Naganuma: Characteristics of femtosecond pulses near 1.5 mm in self-mode-locked Cr $^{4+}$:YAG laser, Opt. Lett.**19**, 2003–2005, 1994

[197] J. Theimer, M. Hayduk, M. F. Krol, J. W. Haus: Mode-locked Cr $^{4+:}$YAG laser: model and experiment, Opt. Comm. **142**, 55–60 (1997)

[198] Y. Pang, V. Yanovsky, F. Wise, B. I. Minkov: Self-mode-locked Cr:forsterite laser, Opt. Lett. **18**, 1168–1170 (1993)

[199] A. Seas, V. Petricevic, R. R. Alfano: Generation of sub-100-fs pulses from a cw mode-locked chromium-doped forsterite laser, Opt. Lett. **17**, 937–939, 1992

[200] A. Seas, V. Petricevic, R. R. Alfano: Self-mode-locked chromium-doped forsterite laser generates 50-fs pulses, Opt. Lett. **18**, 891–893 (1993)

[201] E. Slobodchikov, J. Ma, V. Kamalov, K. Tominaga, K. Yoshihara, Cavity-dumped femtosecond Kerr-lens mode locking in a chromiumdoped forsterite laser, Opt. Lett. **21**, 354–356 (1996)

[202] X. Liu, L. Qian, F. Wise, Zh. Zhang, T. Itatani, T. Sugaya, T. Nakagawa, K. Torizuka, Femtosecond Cr:forsterite laser diode pumped by a double-clad fiber, Opt. Lett. **23**, 129 (1998)

[203] A. Robertson, H. Fuchs, U. Ernst, R. Wallenstein, V. Scheuer, T. Tschudi, J. Opt. Soc. Am. B **17**, 668 (2000)

[204] T. Miura, K. Kobayashi, Zh. Zhang, K. Torizuka, F. Kannari: Stable mode-locking operation in a Cr:forsterite laser with a five-mirror cavity, Opt. Lett. **24**, 554–556 (1999)

[205] A. Sennaroglu, C. R. Pollock, H. Nathel: Generation of 48-fs pulses and measurement of crystal dispersion by using a regeneratively initiated self-mode-locked chromium-doped forsterite laser, Opt. Lett. **18**, 826–828 (1993)

[206] A. Sennaroglu, C. R. Pollock, H. Nathel: Generation of tunable femtosecond pulses in the 1.21–1.27 μm and 605–635 nm wavelength region by using a regeneratively initiated self-mode-locked Cr:forsterite laser, IEEE J. Quantum Elect. **30**, 1851–1861 (1994)

[207] B. C. Collings, J. B. Stark, S. Tsuda, W. H. Knox, J. E. Cunningham, W. Y. Jan, R. Pathak, K. Bergmann: Saturable Bragg reflector self-starting passive mode-locking of a Cr^{4+}:YAG laser pumped with a diode-pumped Nd:YVO_4 laser, Opt. Lett. **21**, 1171–1173 (1996)
[208] S. Spaelter, M. Boehm, M. Burk, B. Mikulla, R. Fluck, I. D. Jung, G. Zhang, U. Keller, A. Sizmann, G. Leuchs: Self-starting soliton-modelocked femtosecond Cr^{4+}:YAG laser using an antiresonant Fabry-Perot saturable absorber, Appl. Phys. B **65**, 335–338 (1997)
[209] M. J. Hayduk, S. T. Johns, M. F. Krol, C. R. Pollock, R. P. Leavitt: Self-starting mode-locked tunable femtosecond Cr^{4+}:YAG laser using a saturable-absorber mirror, Opt. Comm. **137**, 55–58 (1997)
[210] Z. Zhang, T. Nakagawa, K. Torizuka, T. Sugaya, K. Kobayashi: Self-starting mode-locked Cr^{4+}:YAG laser with a low-loss broad-band semiconductor saturable-absorber mirror, Opt. Lett. **24**, 1768–1770 (1999)
[211] Z. Zhang, T. Nakagawa, K. Torizuka, T. Sugaya, K. Kobayashi: Gold-reflector based semiconductor saturable absorber mirror for femtosecond mode-locked Cr^{4+}:YAG lasers, Appl. Phys. B **70**, S59–S62 (2000)
[212] B. C. Collings, K. Bergman, W. H. Knox: True fundamental solitons in a passively mode-locked short-cavity Cr^{4+}:YAG laser, Opt. Lett. **22**, 1098–1100 (1997)
[213] Z. Zhang, K. Torizuka, T. Itatani, K. Kobayashi, T. Sugaya, T. Nakagawa: Self-starting mode-locked femtosecond forsterite laser with a semiconductor saturable-absorber mirror, Opt. Lett. **22**, 1006–1008 (1997)
[214] Z. Zhang, K. Torizuka, T. Itatani, K. Kobayashi, T. Sugaya, T. Nakagawa: Femtosecond Cr:forsterite laser with mode locking initiated by a quantum-well saturable absorber, IEEE J. Quantum Elect. **33**, 1975 (1997)
[215] V. Petrov, V. Shcheslavskiy, T. Mirtchev, F. Noack, T. Itatani, T. Sugaya, T. Nakagawa: High-power self-starting femtosecond Cr:forsterite laser, Electron. Lett. **34**, 559 (1998)
[216] R. Mellish, S. V. Chernikov, P. M. French, J. R. Taylor: All-solid-state compact high repetition rate modelocked Cr^{4+}:YAG laser, Electron. Lett. **34**, 552–553 (1998)
[217] T. Tomaru, H. Petek: Femtosecond Cr^{4+}:YAG laser with an L-fold cavity operating at a 1.2-GHz repetition rate, Opt. Lett. **25**, 584–586 (2000)
[218] T. Tomaru: Two-element-cavity femtosecond Cr^{4+}:YAG laser operating at a 2.6-GHz repetition rate, Opt. Lett. **26**, 1439–1441 (2001)
[219] L. J. Qian, X. Liu, F. W. Wise: Femtosecond Kerr-lens mode locking with negative nonlinear phase shifts, Opt. Lett. **24**, 166–169 (1999)
[220] D. J. Ripin, C. Chudoba, J. T. Gopinath, J. G. Fujimoto, E. P. Ippen, U. Morgner, F. X. Kärtner, V. Scheuer, G. Angelow, T. Tschudi: Generation of 20-fs pulses by a prismless Cr^{4+}:YAG laser, Opt. Lett. **27**, 61–63 (2002)
[221] S. Naumov. E. Sorokin, I. T. Sorokina: Directly diode-pumped Kerr-lens mode-locked Cr^{4+}:YAG laser, Opt. Lett. **29**, 1276 (2004)
[222] R. P. Prasankumar, C. Chudoba, J. G. Fujimoto, P. Mak, M. F. Ruane: Self-starting mode locking in a Cr:forsterite laser by use of non-epitaxially-grown semiconductor-doped silica films, Opt. Lett. **27**, 1564–1566 (2002)
[223] D. J. Ripin, J. T. Gopinath, H. M. Shen, A. A. Erchak, G. S. Petrich, L. A. Kolodziejski, F. X. Kärtner, E. P. Ippen: Oxidized GaAs/AlAs mirror with a quantum-well saturable absorber for ultrashort-pulse Cr^{4+}:YAG laser, Opt. Commun. **214**, 285–289 (2002)

[224] R. H. Page, L. D. DeLoach, G. D. Wilke, S. A. Payne, R. J. Beach, W. F. Krupke: Cr^{2+}-doped II-VI crystals: new widely tunable, room-temperature mid-IR lasers, Proc. IEEE 8th Lasers Elect.-Opt. Soc., vol. 1 (1995) pp. 449–450

[225] L. D. DeLoach, R. H. Page, G. D. Wilke, S. A. Payne, W. P. Krupke: Transition metal-doped zinc chalcogenides: spectroscopy and laser demonstration of a new class of gain media, IEEE J. Quantum Elect. **32**, 885–895 (1996)

[226] R. H. Page, K. I. Schaffers, L. D. DeLoach, G. D. Wilke, F. D. Patel, J. B. Tassano, S. A. Payne, W. F. Krupke, K. T. Chen, A. Burger: Cr^{2+}-doped zinc chalcogenides as efficient widely tunable midinfrared lasers, IEEE J. Quantum Elect. **33**, 609–619 (1997)

[227] G. J. Wagner, T. J. Carrig, R. H. Page, K. I. Schaffers, J. Ndap, X. Ma, A. Burger: Continuous-wave broadly tunable Cr^{2+}:ZnSe laser, Opt. Lett. **24**, 19–21 (1999)

[228] A. Sennaroglu, A. Ozgun Konza, C. R. Pollock: Continuous-wave power performance of a 2.47-μm Cr^{2+}:ZnSe laser: experiment and modelling, IEEE J. Quantum Elect. **36**, 1199–1205 (2000)

[229] I. T. Sorokina, E. Sorokin, A. DiLieto, M. Tonelli, R. H. Page, K. I. Schaffers: Efficient broadly tunable continuous-wave Cr^{2+}:ZnSe laser, J. Opt. Soc. Am. B **18**, 926–930 (2001)

[230] U. Morgner, W. Drexler, F. X. Kärtner, X. D. Li, C. Pitris, E. P. Ippen, J. G. Fujimoto: Spectroscopic optical coherence tomography, Opt. Lett. **25**, 111–113 (2000)

[231] F. K. Tittel, D. Richter, A. Fried: Mid–infrared laser applications in spectroscopy, in *Solid-State Mid–Infrared Laser Sources*, Top. Appl. Phys. **79** (Springer, Berlin, Heidelberg 2003) pp. 445–510

[232] B. Jean, T. Bende: Mid-IR Laser Applications in Medicine, in *Solid-State Mid–Infrared Laser Sources*, Top. Appl. Phys. **79** (Springer, Berlin, Heidelberg 2003) pp. 511–544

[233] W. Pelouch, G. J. Wagner, T. J. Carrig, W. J. Scharpf: Mid-wave ZGP OPO pumped by a Cr:ZnSe laser, TOPS **50** (OSA, Washington DC 2001) pp. 670–674

[234] J. T. Vallin, G. A. Slack, S. Roberts, A. E. Hughes: Near and far infrared absorption in Cr doped ZnSe, Sol. State Comm. **7**, 1211 (1969)

[235] J. T. Vallin, G. A. Slack, S. Roberts, A. E. Hughes: Infrared absorption in some II-VI compounds doped with Cr, Phys. Rev. B **2**, 4313 (1970)

[236] M. Kaminska, J. M. Baranowski, S. M. Uba, J. T. Vallin: Absorption and luminescence of $Cr^{2+}(d^4)$ in II-VI compounds, J. Phys. C Solid State **12**, 2197–2214 (1979)

[237] H. Nelkowski, G. Grebe: IR-luminescence of ZnS:Cr, J. Lumin. **1/2**, 88–93 (1970)

[238] G. Grebe, H. J. Schulz: Interpretation of excitation spectra of ZnS:Cr^{2+} by fitting the eigenvalues of the Tanabe-Sugano matrices, Phys. Stat. Solidi B **54**, 69-72 (1972)

[239] G. Grebe, H. J. Schulz: Luminescence of Cr^{2+} centers and related optical interactions involving crystal field levels of chromium ions in ZnS, Z. Naturf. **29a**, 1803 (1974)

[240] G. Grebe, G. Roussos, H. J. Schulz: Infrared luminescence of ZnSe:Cr crystals, J. Lumin. **12/13**, 701 (1976)

[241] A. Fazzio, M. J. Caldas, A. Zunger: Many-electron multiplet effects in the spectra of 3-d impurities in heteropolar semiconductors, Phys. Rev. B **30**, 3430 (1984)

[242] I. T. Sorokina, E. Sorokin: Sensitization of the induced radiation in Cr^{2+}:ZnSe and Cr^{2+}:ZnS lasers, in OSA TOPS **79** (2002) pp. 167–169

[243] I. T. Sorokina, Crystalline mid–infrared lasers, in *Solid-State Mid–Infrared Laser Sources*, Top. Appl. Phys. **79** (Springer, Berlin, Heidelberg 2003) pp. 255–350

[244] R. H. Page, J. A. Skidmore, K. I. Schaffers, R. J. Beach, S. A. Payne, W. F. Krupke: Demonstration of diode-pumped and grating tuned ZnSe: Cr^{2+}laser, TOPS (OSA, Washington DC 1997) pp. 208–210

[245] E. Sorokin, I. T. Sorokina, R. H. Page, Room-temperature cw diode-pumped Cr^{2+}:ZnSe laser, TOPS **46** (OSA, Washington 2001) pp. 101–105

[246] M. Mond, D. Albrecht, E. Heumann, G. Huber, S. Kück, V. I. Levchenko, V. N. Yakimovich, V. G. Shcherbitsky, V. E. Kisel, N. V. Kuleshov, M. Rattunde, J. Schmitz, R. Kiefer, J. Wagner: 1.9-μm and 2.0-μm laser diode pumping of Cr^{2+}:ZnSe and Cr^{2+}:CdMnTe, Opt. Lett. **27**, 1034–1036 (2002)

[247] T. J. Carrig, G. J. Wagner, A. Sennaroglu, J. Y. Jeong, C. R. Pollock: Mode-locked Cr^{2+}:ZnSe laser, Opt. Lett. **25**, 168–170 (2000)

[248] I. T. Sorokina, E. Sorokin, A. Di Lieto, M. Tonelli, R. H. Page, K. I. Schaffers: Active and passive mode-locking of Cr^{2+}:ZnSe laser, OSA TOPS **46** (OSA, Washington DC 2001) pp. 157–161

[249] G. J. Wagner, T. J. Carrig: Power scaling of Cr^{2+}: ZnSe lasers, TOPS **50** (OSA, Washington DC 2001) pp. 506–510

[250] T. J. Carrig, G. J. Wagner, W. J. Alford, A. Zakel: Chromium-doped chalcogenide lasers, Photonics Europe 2004, Proc. SPIE 5460-11

[251] V. L. Kalashnikov, E. Sorokin, I. T. Sorokina: Multipulse operation and limits of the Kerr-lens mode locking stability, IEEE J. Quantum Elect. **39**, 323–336 (2003)

[252] B. Segall, D. T. F. Marple: in M. Aven, J. S. Prener (eds.): *Physics and chemistry of II-IV compounds*, (North Holland, Amsterdam 1967) pp. 319–381

[253] K. Graham, S. Mirov, V. Fedorov, M. E. Zvanut, A. Avanesov, V. Badikov, B. Ignat'ev, V. Panutin, G. Shevirdyaeva: Spectroscopic characterization and laser performance of diffusion doped Cr^{2+}:ZnS, TOPS **46** (OSA, Washington DC 2001) pp. 561–567

[254] I. T. Sorokina, E. Sorokin, S. Mirov, V. Fedorov, V. Badikov, V. Panyutin, A. Di Lieto, M. Tonelli: Continuous-wave tunable Cr^{2+}:ZnS laser, Appl. Phys B **74**, 607-611 (2002)

[255] S. B. Mirov, V. V. Fedorov, K. Graham, I. Moskalev, V. Badikov, V. Panyutin: Er-fiber laser pumped continuous-wave microchip Cr^{2+}:ZnS and Cr^{2+}:ZnSe lasers, Opt. Lett. **27**, 909–911 (2002)

[256] K. L. Schepler, S. Kueck, L. Shiozawa: Cr^{2+} emission spectroscopy in CdSe, J. Lumin. **72–74**, 116–117 (1997)

[257] J. McKay, K. L. Schepler, G. C. Catella: Efficient grating-tuned midinfrared Cr^{2+}:CdSe laser, Opt. Lett. **24**, 1575–1577 (1999)

[258] A. G. Bluiett, U. Hömmerich, J. T. Seo, R. Shah, S. B. Trivedi, S. W. Kutcher, C. C. Wang, P. R. Boyd: Observation of lasing from Cr^{2+}:CdTe and compositional effects in Cr^{2+} doped II-VI semiconductors, J. Electron. Mater. **31**, 806-810 (2002)

[259] J. T. Seo, U. Hömmerich, S. B. Trivedi, R. J. Chen, S. Kutcher: Slope efficiency and tunability of a Cr^{2+}- doped $Cd_{0.85}Mn_{0.15}Te$ midinfrared laser, Opt. Comm. **153**, 267–270 (1998)
[260] J. T. Seo, U. Hömmerich, H. Zong, S. B. Trivedi, S. W. Kutcher, C. C. Wang, R. J. Chen: Midinfrared lasing from a novel optical material: Chromium-doped $Cd_{0.55}Mn_{0.45}Te$, Phys. Stat. Solidi A **175**, R 3 (1999)
[261] U. Hömmerich, J. T. Seo, M. Turner, A. Bluiett, S. B. Trivedi, H. Zong, S. Kutcher, C. C. Wang, R. J. Chen: Midinfrared laser development based on transition metal doped cadmium manganese telluride, J. Lumin. **87–89**, 1143–1145 (2000)
[262] J. McKay, W. B. Roh, K. L. Schepler: Extended mid-IR tuning of a Cr^{2+}:CdSe laser, OSA TOPS **68** (2002) pp. 371–373
[263] I. T. Sorokina, E. Sorokin, A. Di Lieto, M. Tonelli, R. H. Page: Tunable diode-pumped continuous-wave single crystal and ceramic Cr^{2+}:ZnSe lasers, CLEO (2001) paper WJPD-1
[264] E. Sorokin, I. T. Sorokina, A. Di Lieto, M. Tonelli, P. Minguzzi: Mode-locked ceramic Cr^{2+}:ZnSe Laser, TOPS **83** (OSA, Washington DC 2003) pp. 227–230
[265] J.-P. Meyn, T. Danger, K. Petermann, G. Huber: Spectroscopic characterization of V^{4+}-doped Al_2O_3 and $YAlO_3$, J. Lumin. **55**, 55–62 (1993)
[266] L. F. Johnson, R. E. Dietz, H. J. Guggenheim: Spontaneous and stimulated emission from Co^{2+} ions in MgF_2 and ZnF_2, Appl. Phys. Lett. **5**, 21-22 (1964)
[267] P. F. Moulton: An investigation of the Co:MgF_2 laser system, IEEE J. Quantum Elect. **21**, 1582–1595 (1985)
[268] A. Di Lieto, P. Minguzzi, A. Pirastu, S. Sanguinetti, A. Toncelli: High-power diffraction-limited diode-pumped Nd:YVO_4 cw laser at 1.34 μm, TOPS **68** (OSA, Washington DC 2003) pp. 570–574
[269] P. F. Moulton, A. Mooradian: Broadly tunable cw operation of Ni:MgF_2 and Co:MgF_2 lasers, Appl. Phys. Lett. **35**, 838 (1979)
[270] A. Di Lieto: Development of a cw Co:MgF_2 laser, Opt. Laser Eng. **39/3**, 305–308 (2003)
[271] D. Welford, P. F. Moulton: Room-temperature operation of a Co:MgF_2 laser, Opt. Lett. **13**, 975–977 (1988)
[272] D. M. Rines, P. F. Moulton, D. Welford, G. A. Rines: High-energy operation of a Co:MgF_2 laser, Opt. Lett. **19**, 628–630 (1994)
[273] Z. M. Zhang, Y. B. Cui, F. L. Li, G. B. Zhang, Q. R. Pu, G. J. Xu: Polarization characteristic of a room-temperature Co:MgF_2 laser, Appl. Opt. **41**, 1071–1074 (2002)
[274] S. Løvold, P. F. Moulton, D. K. Killinger, N. Menyuk: Frequency tuning characteristics of a Q-switched Co:MgF_2 laser, IEEE J. Quantum Elect. **21**, 202–208 (1985)
[275] B. C. Johnson, P. F. Moulton, A. Mooradian, Mode-locked operation of Co:MgF_2 and Ni:MgF_2 lasers, Opt. Lett. **10**, 116–118 (1984)
[276] H. Manaa, Y. Guyot, R. Moncorge: Spectroscopic and laser properties of Co^{2+}-doped single crystals, Phys. Rev. B **48**, 3633 (1993)
[277] J. Harrison, D. Welford, P. F. Moulton: Threshold analysis of pulsed lasers with application to a room-temperature Co:MgF_2 laser, IEEE J. Quantum Elect. **25**, 1708–1711 (1989)

[278] B. W. Woods, S. A Payne, J. E. Marion, R. S. Hughes, L. E. Davis: Thermo-mechanical and thermo-optical properties of the $LiCaAlF_6$:Cr^{3+} laser material, J. Opt. Soc. Am. **8**, 970–977 (1991)

[279] G. Toci, M. Vannini, R. Salimbeni, M. A. Dubinskii, E. Giorgetti: Z-scan measurements on crystal hosts for ultraviolet laser emission, Proc. Lasers (STS Press, Mclean 2000) pp. 18–21

[280] I. T. Sorokina: Cr^{2+}-doped II-VI materials for lasers and nonlinear optics, Opt. Mater., doi:10.1016/j.optmat.2003.12.025

[281] A. A. Major, F. Yoshino, J. S. Aitchison, P. W. E. Smith, I. T. Sorokina, E. Sorokin: Ultrafast nonresonant third-order optical nonlinearities in ZnSe for photonic switching at telecom wavelengths, NLGW (2004) paper TuC13

[282] G. H. Dieke: *Spectra and Energy Levels of Rare-Earth Ions in Crystals* (Wiley-Intersci., New York 1968)

[283] B. R. Judd: Optical absorption intensities of rare-earth ions, Phys. Rev. **127**, 750–761 (1962)

[284] B. G. Wybourne: *Spectroscopic Properties of Rare-Earths* (Wiley-Intersci., New York 1965)

[285] S. Hufner: *Optical Spectra of Transparent Rare-Earth Compounds* (Acad. Press, New York 1978)

[286] A. A. Kaminskii: *Laser Crystals* (Springer, Berlin, Heidelberg 1990) p. 465

[287] P. P. Sorokin, M. J. Stevenson: Stimulated infrared emission from trivalent uranium, Phys. Rev. Lett. **5**, 557–559 (1960)

[288] G. D. Boyd, R. J. Collins, S. P. S. Porto, A. Yariv, W. A. Hargreaves: Excitation, relaxation and continuous maser action in the 2.613-micron transition of CaF_2:U^{3+}, Phys. Rev. Lett. **8**, 269–272 (1962)

[289] S. P. S. Porto, A. Yariv: Optical maser characteristics of BaF_2:U^{3+}, Proc. IRE **50**, 1542–1543 (1962)

[290] S. P. S. Porto, A. Yariv: Excitation, relaxation and optical maser action at 2.407 micron in SrF_2:U^{3+}, Proc. IRE **50**, 1543–1544 (1962)

[291] D. Meichenin, F. Auzel, S. Hubert, E. Simoni, M. Louis, J. Y. Gesland: New room-temperature cw laser at 2.82 μm: $U^{3+}/LiYF_4$, Electron. Lett. **30**, 1309 (1994)

[292] W. Koechner: *Solid-State Laser Engineering* (Springer, Berlin, Heidelberg 1999) p. 746

[293] A. Ellens, H. Andres, A. Meijerink, G. Blasse: Spectral-line-broadening study of the trivalent lanthanide-ion series. I. Line broadening as a probe of the electron-phonon coupling strength, Phys. Rev. B **55**, 173–179 (1997)

[294] A. Ellens, H. Andres, M. L. H. ter Heerdt, R. T. Wegh, A. Meijerink, G. Blasse: Spectral-line-broadening study of the trivalent lanthanide-ion series. II. The variation of the electron-phonon coupling strength through the series, Phys. Rev. B **55**, 180–186 (1997)

[295] A. J. DeMaria, D. A. Stetser, A. Heynau: Appl. Phys. Lett. **8**, 174 (1966)

[296] J. Aus der Au, D. Kopf, F. Morier-Genoud, M. Moser, U. Keller: 60-fs pulses from a diode-pumped Nd:glass laser, Opt. Lett. **22**, 307–309 (1997)

[297] J. Aus der Au, F. H. Loesel, F. Morier-Genoud, M. Moser, U. Keller: Femtosecond diode-pumped Nd:glass laser with more than 1 W of average output power, Opt. Lett. **23**, 271–273 (1998)

[298] S. Han, W. Lu, B. Y. Sheh, L. Yan, M. Wraback, H. Shen, J. Pamulapati, P. G. Newman: Generation of sub-40fs pulses from a mode-locked dual-gain-media Nd:glass laser, Appl. Phys. B **74**, s177–s179 (2002)

[299] N. Blanchot, C. Rouyer, Sauteret, A. Migus: Amplification of sub-100-TW femtosecond pulses by shifted amplifying Nd:glass amplifiers: theory and experiments, Opt. Lett. **20**, 395–397 (1995)
[300] I. N. Ross, M. Trentelman, C. N. Danson: Optimization of a chirped-pulse amplification Nd:glass laser, Appl. Opt. LP-**36**, 9348–9358 (1997)
[301] M. Hofer, M. H. Ober, F. Haberl, M. E. Fermann: Characterization of ultra-short pulse formation in passively mode-locked fiber lasers, IEEE J. Quantum Elect. **28**, 720–728 (1992)
[302] R. Hofer, M. Hofer, G. A. Reider, M. Cernusca, M. H. Ober: Modelocking of a Nd-fiber laser at 920 nm, Opt. Comm. **140**, 242–244 (1997)
[303] M. Hofer, M. H. Ober, R. Hofer, M. E. Fermann, G. Sucha, D. Harter, K. Sugden, I. Bennion, C. A. C. Mendonca, T. H. Chiu: High-power neodymium soliton fiber laser that uses a chirped fiber grating, Opt. Lett **20**, 1710–1703 (1995)
[304] M. Hofer, M. H. Ober, R. Hofer, G. A. Reider, K. Sugden, I. Bennion, M. E. Fermann, G. Sucha, D. Harter, C. A. C. Mendonca, T. H. Chiu: Monolithic polarization-insensitive passively mode-locked fiber laser, Opt. Fiber Comm., 4–5 (1996)
[305] M. H. Ober, E. Sorokin, I. Sorokina, F. Krausz, E. Wintner, I. A. Shcherbakov: Subpicosecond mode-locking of a Nd^{3+}-doped garnet laser, Opt. Lett. **17**, 1364–1366 (1992)
[306] M. W. Philips, J. R. M. Barr, D. W. Hughes, D. C. Hanna, Z. Chang, C. N. Danson, C. B. Edwards: Self-starting additive-pulse mode locking of a Nd:LMA laser, Opt. Lett. **17**, 453–455 (1992)
[307] E. Sorokin, M. H. Ober, I. Sorokina, E. Wintner, A. J. Schmidt, A. I. Zagumennyi, G. B. Loutts, E. W. Zharikov, I. A. Shcherbakov: Femtosecond solid-state lasers using Nd^{3+}-doped mixed scandium garnet, J. Opt. Soc. Am. B **10**, 1436–1442 (1993)
[308] L. D. Schearer, M. Leduc, D. Vivien, A.-M. Lejus, J. Thery: LNA: a new cw Nd laser tunable around 1.05 and 1.08 μm, IEEE J. Quantum Elect. **22**, 713–717 (1986)
[309] Yu. K. Voronko, A. A. Sobol, A. Ya. Karasik, N. A. Eskov, P. A. Rabochkina, S. N. Ushakov: Calcium niobium gallium and calcium lithium niobium gallium garnets doped with rare earth ions effective laser media, Opt. Mater. **20**, 197–209 (2002)
[310] V. V. Petrov, E. V. Pestryakov, I. N. Nyushkov, V. I. Trunov, A. V. Kirpichnikov, A. I. Alimpiev: $Nd:BeLaAl_{11}O_{19}$: a promising new laser medium for helium optical pumping at 1080 nm, Laser Phys. **12**, 586–590 (2002)
[311] P. Lacovara, H. K. Choi, C. A. Wang, R. L. Aggarwal, T. Y. Fan: Room-temperature diode-pumped Yb:YAG laser, Opt. Lett. **16**, 1089–1091 (1991)
[312] F. Druon, F. Balembois, P. Georges, A. Brun, A. Courjaud, C. Hönninger, F. Salin, A. Aron, F. Mougel, G. Aka, D. Vivien: Generation of 90-fs pulses from a mode-locked diode-pumped $Yb^{3+}:Ca_4GdO(BO_3)_3$ laser, Opt. Lett. **25**, 423–425 (2000)
[313] F. Druon, S. Chénais, P. Raybaut, F. Balembois, P. Georges, R. Gaumé, G. Aka, B. Viana, S. Mohr, D. Kopf: Diode-pumped $YbSr_3Y(BO_3)_3$ femtosecond laser, Opt. Lett. **27**, 179–181 (2002)
[314] F. Druon, S. Chénais, P. Raybaut, F. Balembois, P. Georges, R. Gaumé, G. Aka, B. Viana, D. Vivien, S. Dhellemmes, V. Ortiz, C. Larat: Apatite-

structure crystal, Yb^{3+}:$SrY_4(SiO_4)_3O$, for the development of diode-pumped femtosecond lasers, Opt. Lett. **27**, 1914–1916 (2002)

[315] C. Hönninger, R. Paschotta, M. Graf, F. Morier-Genoud, G. Zhang, M. Moser, S. Biswal, J. Nees, A. Braun, G. A. Mourou, I. Johannsen, A. Giesen, W. Seeber, U. Keller: Ultrafast ytterbium-doped bulk lasers and laser amplifiers, Appl. Phys. B **69**, 3–17 (1999)

[316] J. J. Romero, K. Petermann, G. Huber, E. Heumann: Spectroscopy and continuous-wave laser action of Yb^{3+}-doped lanthanum scandium borate, ASSP (2004) paper WB3

[317] H. Liu, J. Nees, G. Mourou: Diode-pumped Kerr-lens mode-locked Yb:$KY(WO_4)_2$ laser, Opt. Lett. **26**, 1723–1725 (2001)

[318] Yu. K. Voronko, E. V. Zharikov, D. A. Lis, A. A. Sobol, K. A. Subbotin, S. N. Ushakov, V. E. Shukshin, S. Dröge: Growth and luminescent properties of $NaGd(WO_4)_2$:Yb^{3+} crystals, Neorganicheskie Materialy **39**, 1509-1516 (2003), English translation in: Inorg. Mater. **39**, 1308-1314 (2003)

[319] F. Brunner, T. Südmeyer, E. Innerhofer, F. Morier-Genoud, R. Paschotta, V. E. Kisel, V. G. Shcherbitsky, N. V. Kuleshov, J. Gao, K. Contag, A. Giesen, U. Keller: 240-fs pulses with 22-W average power from a mode-locked thin-disk $YbKY(WO_4)_2$ laser, Opt. Lett. **27**, 1162–1164 (2002)

[320] F. Druon, P. Georges: Pulse-compression down to 20 fs using a photonic crystal fiber seeded by a diode-pumped Yb:SYS laser at 1070 nm, Opt. Express **12**, 3383 (2004)

[321] P. Raybaut, F. Druon, S. Chénais, F. Balembois, P. Georges, R. Gaumé, B. Viana, D. Vivien, S. Dhellemmes, V. Ortiz, C. Larat: Diode-pumped 100-fs lasers based on a new apatite-structure crystal: Yb^{3+}:$SrY_4(SiO_4)_3O$, ASSP (2003) pp. 158–161, paper TuA8

[322] F. Brunner, G. J. Spühler, J. Aus der Au, L. Krainer, F. Morier-Genoud, R. Paschotta, N. Lichtenstein, S. Weiss, C. Harder, A. A. Lagatsky, A. Abdolvand, N. V. Kuleshov, U. Keller: Diode-pumped femtosecond $YbKGd(WO_4)_2$ laser with 1.1-W average power, Opt. Lett. **25**, 1119–1121 (2000)

[323] L. F. Johnson, G. D. Boyd, K. Nassau: Optical maser characteristics of Ho^{3+} in $CaWO_4$, Proc. IRE **50**, 87 (1962)

[324] L. F. Johnson, G. D. Boyd, K. Nassau: Optical maser characteristics of Tm^{3+} in $CaWO_4$, Proc. IRE **50**, 86 (1962)

[325] L. F. Johnson: Optical maser characteristics of rare-earth ions in crystals, J. Appl. Phys. **34**, 897–909 (1963)

[326] E. W. Duczynski, G. Huber, V. G. Ostroumov, I. A. Shcherbakov: CW double cross pumping of the $^5I_7$-$^5I_8$ laser transition in Ho^{3+}-doped garnets, Appl. Phys. Lett. **48**, 1562–1563 (1986)

[327] B. M. Antipenko: Cross-relaxation schemes for pumping laser transitions, Zh. Tekh. Phys. **54**, 385–388 (1984), English translation: Sov. Phys.–Tech. Phys. **29**, 228–230 (1984)

[328] L. Esterowitz, R. Allen, L. Goldberg, J. F. Weller, M. Storm, I. Abella, Diode-pumped 2 µm holmium laser, in *Tunable Solid-State Lasers II*, Springer Ser. Opt. Sci. **52** (Springer, Berlin, Heidelberg 1986) p. 291–292

[329] G. J. Kintz, L. Esterowitz, R. Allen: CW diode-pumped Tm^{3+}, Ho^{3+}:YAG 2.1 µm room-temperature laser, Electron. Lett. **23**, 616 (1987)

[330] T. Y. Fan, G. Huber, R. L. Byer, P. Mitzscherlich: Continuous-wave operation at 2.1 μm of a diode-laser-pumped, Tm-sensitized Ho:$Y_3Al_3O_{12}$ laser at 300K, Opt. Lett. **12**, 678–680 (1987)
[331] G. J. Kintz, R. Allen, L. Esterowitz: CW laser emission at 2.02 μm from diode-pumped Tm^{3+}:YAG at room temperature, CLEO (OSA, 1988) paper FB2
[332] T. J. Kane, T. S. Kubo: Diode-pumped single-frequency lasers and Q-switched laser using Tm:YAG and Tm,Ho:YAG, ASSL **6** (OSA, 1990) pp. 136–139
[333] B. T. McGuckin, R. T. Menzies: Efficient cw diode-pumped Tm, Ho:YLF laser with tunability near 2.067 μm, IEEE J. Quantum Elect. **28**, 1025–1028 (1992)
[334] A. Di Lieto, P. Minguzzi, A. Toncelli, M. Tonelli, H. P. Jenssen: A diode-laser-pumped tunable Ho:YLF laser in the 2 μm region, Appl. Phys. B **57**, 3172–3175 (1993)
[335] E. C. Honea, R. J. Beach, S. B. Sutton, J. A. Speth, S. C. Mitchell, J. A. Skidmore, M. A. Emanuel, S. A. Payne: 115-W Tm:YAG diode-pumped solid-state laser; IEEE J. Quantum Elect. **33**, 1592–1600 (1997)
[336] E. Sorokin, I. T. Sorokina, A. Unterhuber, E. Wintner, A. I. Zagumenny, I. A. Shcherbakov, V. Carozza, A. Toncelli, M. Tonelli: A novel cw tunable and mode-locked 2 μm Cr,Tm,Ho:YSGG:GSAG laser, in W. R. Rosenberg, M. M. Fejer (Eds.): TOPS **19** (OSA, Washington DC 1998) pp. 197–200
[337] E. Sorokin, I. T. Sorokina, A. Unterhuber, E. Wintner, A. I. Zagumenny, I. A. Shcherbakov, V. Carozzo, M. Tonelli: A novel cw mode-locked 2 μm Cr,Tm,Ho:YSGG:GSAG laser, CLEO (1998) p. 296, paper CWM1
[338] F. Cornacchia, E. Sani, A. Toncelli, M. Tonelli, M. Marano, S. Taccheo, G. Galzerano, P. Laporta: Optical spectroscopy and diode-pumped laser characteristics of codoped Tm-Ho:YLF and Tm-Ho:BaYF: a comparative analysis, Appl. Phys. B **75**, 817–822 (2002)
[339] R. C. Stonemann, L. Esterowitz: Efficient, broadly tunable, laser-pumped Tm:YAG and Tm:YSGG cw lasers, Opt. Lett. **15**, 486–488 (1990)
[340] R. C. Stonemann, L. Esterowitz: Efficient 1.94-μm Tm:YALO laser, IEEE J. Sel. Top. Quant. **1**, 78–80 (1995)
[341] L. Fornasiero, N. Berner, B.-M. Dicks, E. Mix, V. Peters, K. Petermann, G. Huber: Broadly tunable laser emission from Tm:Y_2O_3 and Tm:Sc_2O_3 at 2 μm, TOPS **26** (1999) pp. 450–453
[342] A. Dergachev, K. Wall, P. F. Moulton: A cw side-pumped Tm:YLF laser, OSA TOPS **68** (2002) pp. 343–350
[343] G. Galzerano, D. Parisi, A. Toncelli, M. Tonelli, P. Laporta, to be published
[344] E. Sorokin, A. N. Alpatiev, I. T. Sorokina, A. I. Zagumennyi, I. A. Shcherbakov: Tunable efficient continuous-wave room-temperature Tm^{3+}:$GdVO_4$ laser, OSA TOPS **68** (2002) pp. 347–350
[345] P. A. Studenikin, A. I. Zagumennyi, Yu. D. Zavartsev, P. A. Popov, I. A. Shcherbakov: $GdVO_4$ as a new medium for solid-state lasers: some optical and thermal properties of crystals doped with Nd^{3+}, Tm^{3+} and Er^{3+} ions, Quantum Elect. **25/12**, 1162–1165 (1995)
[346] V. A. Mikhailov, Yu. D. Zavartzev, A. I. Zagumennyi, V. G. Ostroumov, P. A. Studenikin, E. Heumann, G. Huber, I. A. Shcherbakov: Tm^{3+}: $GdVO_4$ – a new efficient medium for diode-pumped 2 μm lasers, Quantum Elect. **27/1**, 13–14 (1997)

[347] Ch. P. Wyss, W. Lüthy, H. P. Weber, V. I. Vlasov, Yu. D. Zavartsev, P. A. Studenikin, A. I. Zagumennyi, I. A. Shcherbakov: Emission properties of a Tm^{3+}: $GdVO_4$ microchip laser at 1.9 µm, Appl. Phys. B **67**, 1–4 (1998)
[348] A. I. Zagumennyi, V. A. Mikhailov, Yu. D. Zavartsev, S. A. Koutovoi, F. Zerrouk, I. A. Shcherbakov: A diode-pumped 0.6 W cw Tm:$GdVO_4$ microchip laser, CLEO (2001) p. 99, paper C-PSL 129
[349] H. Saito, S. Chaddha, R. S. F. Chang, N. Djeu: Efficient 1.94-μm Tm^{3+} laser in YVO_4 host, Opt. Lett. **17**, 189–191 (1992)
[350] K. Ohta, H. Saito, M. Obara, N. Djeu: Characterization of a longitudinally pumped cw, room-temperature operation of Tm^{3+}:YVO_4 laser, Jpn. J. Appl. Phys. **32**, 1651–1657 (1993)
[351] J. F. Pinto, L. Esterowitz, G. H. Rosenblatt: Tm^{3+}:YLF laser continuously tunable between 2.20 and 2.46 µm, Opt. Lett **19**, 883–885 (1994)
[352] G. H. Rosenblatt, J. F. Pinto, R. C. Stonemann, L. Esterowitz: Continuously tunable 2.3 µm Tm:GSGG laser, IEEE LEOS (1993) pp. 689–690
[353] J. F. Pinto, L. Esterowitz, G. H. Rosenblatt: Continuous-wave mode-locked 2-µm Tm:YAG laser, Opt. Lett. **17**, 731–732 (1992)
[354] F. Heine, E. Heumann, G. Huber, K. L. Schepler: Mode locking of room-temperature cw thulium and holmium lasers, Appl. Phys. Lett. **60**, 1161–1162 (1992)
[355] K. L. Schepler, B. D. Smith, F. Heine, P. A. Budni: Mode-locking of a diode-pumped Tm,Ho:YLF, OSA ASSL **20** (1994) pp. 257–259
[356] D. J. Ehrlich, P. F. Moulton, R. M. Osgood, Jr.: Ultraviolet solid-state Ce:YLF laser at 325 nm, Opt. Lett. **4**, 184-186 (1979)
[357] D. J. Ehrlich, P. F. Moulton, R. M. Osgood, Jr.: Optically pumped Ce:LaF_3 laser at 286 nm, Opt. Lett. **5**, 339–341 (1980)
[358] M. A. Dubinskii, V. V. Semashko, A. K. Naumov, R. Y. Abdulsabirov, S. L. Korableva: Spectroscopy of a new active medium of a solid-state UV laser with broadband single-pass gain, Laser Phys. **3**, 216–217 (1993)
[359] V. V. Semashko, M. A. Dubinskii, R. Yu. Abdulsabirov, A. K. Naumov, S. L. Korableva: Investigation of multisite activation in $LiCaAlF_6$:Ce^{3+} Crystals using stimulated quenching of luminescence, Laser Phys. **5**, 69–72 (1995)
[360] C. D. Marshall, J. A. Speth, S. A. Payne, W. F. Krupke, G. J. Quarles, V. Castillo, B. H. T. Chai: Ultraviolet-laser emission properties of Ce^{3+}-doped $LiSrAlF_6$ and $LiCaAlF_6$, J. Opt. Soc. Am. B **11**, 2054–2065 (1994)
[361] M. A. Dubinskii, K. L. Schepler, V. V. Semashko, A. K. Naumov, R. Yu. Abdulsabirov, S. L. Korableva: Ce^{3+}:$LiBaF_3$ as new prospective active material for tunable UV laser with direct UV pumping, in C. R. Pollock, W. R. Bosenberg (Eds.): TOPS ASSL **10** (OSA, Washington DC 1997) pp. 30–34
[362] M. A. Dubinskii, K. L. Schepler, V. V. Semashko, R. Yu. Abdulsabirov, S. L. Korableva, A. K. Naumov: Ce^{3+}:$SrAlF_5$, a new material for all-solid-state tunable UV lasing, CLEO (OSA, Washington DC 1997) p. 405
[363] M. A. Dubinskii, V. V. Semashko, A. K. Naumov, R. Yu. Abdulsabirov, S. L. Korableva: A new active medium for a tunable solid-state UV laser with an excimer pumping, Laser Phys. **4**, 480–484 (1994)
[364] J. F. Owen, P. F. Dorain, T. Kobayashi: Excited-state absorption in Eu^{2+}:CaF_2 and Ce^{3+}:YAG single crystals at 298 and 77 K, J. Appl. Phys **52**, 1216–1223 (1980)

[365] D. S. Hamilton, S. K. Gayen, G. J. Pogatshnik, R. D. Ghen: Optical-absorption and photoionization measurements from the excited states of Ce^{3+}:$Y_3Al_5O_{12}$, Phys. Rev. B **39**, 8807-8815 (1989)

[366] P. Dorenbos: The 5d level positions of the trivalent lanthanides in inorganic compounds, J. Lumin. **91**, 155–176 (2000)

[367] K. Shimamura, S. L. Baldochi, N. Mujilatu, K. Nakano, Zh. Liu, N. Sarukura, T. Fukuda: Growth of Ce-doped $LiCaAlF_6$ and $LiSrAlF_6$ single crystals by the Czochralski technique under CF_4 atmosphere, J. Cryst. Growth **211**, 302–307 (2000)

[368] E. Zanger, R. Müller, B. Liu, M. Kötteritzsch, W. Gries: Diode-pumped cw all solid-state laser at 266 nm, OSA TOPS **26** (1999) pp. 104–111

[369] A. J. Bayramian, C. D. Marshall, H. Wu J., J. A. Speth, S. A. Payne, G. J. Quarles, V. K. Castillo: Ce:$LiSrAlF_6$ laser performance with antisolarant pump beam, J. Lumin. **69**, 85–94 (1996)

[370] S. Ono, Y. Suzuki, T. Kozeki, H. Murakami, H. Ohtake, N. Sarukura, H. Sato, S. Machida, K. Shimamura, T. Fukuda: High-energy, all-solid-state, ultraviolet laser power-amplifier module design and its output-energy scaling principle, Appl. Opt. LP **41**, 7556–7560 (2002)

[371] N. Sarukura, M. A. Dubinskii, Zh. Liu, V. V. Semashko, A. K. Naumov, S. L. Korableva, R. Y. Abdulsabirov, K. Edamatsu, Y. Suzuki, T. Ito, Y. Segava: Ce^{3+}-activated fluoride crystals as prospective active media for widely tunable ultraviolet ultrafast lasers with direct 10-nanosecond pumping, IEEE J. Sel. Top. Quant. **1**, 792–804 (1995)

[372] Z. Liu, N. Sarukura, M. A. Dubinskii, R. Yu. Abdulsabirov, S. L. Korableva: All-solid-state subnanosecond tunable ultraviolet laser sources based on Ce^{3+}-activated fluoride crystals, J. Nonlinear Opt. Phys. **8**, 41–54 (1999)

[373] Zh. Liu, N. Sarukura, M. A. Dubinskii: All-solid-state, short-pulse, tunable ultraviolet laser sources based on Ce^{3+}-activated fluoride crystals, in P. Misra, M. A. Dubinskii (Eds.): *Ultraviolet Spectroscopy and UV Lasers* (Marcel Dekker, New York 2002) pp. 397–450

[374] Z. Liu, T. Kozeki, Y. Suzuki, N. Sarukura, K. Shimamura, T. Fukuda, M. Hirano, H. Hosono: Chirped-pulse amplification of ultraviolet femtosecond pulses by use of Ce^{3+}:$LiCaAlF_6$ as a broadband, solid-state gain medium, Opt. Lett. **26**, 301–303 (2001)

[375] Y. Suzuki, S. Ono, H. Murakami, T. Kozeki, H. Ohtake, N. Sarukura, G. Masada, H. Shiraishi, I. Sekine: 0.43 J, 10 Hz fourth harmonic generation of Nd:YAG laser using large $Li_2B_4O_7$ crystals, Jpn. J. Appl. Phys **41**, L823–L824 (2002)

[376] Yu. S. Vagin, V. M. Marchenko, A. M. Prokhorov: On the spectrum of a laser based on electron-vibrational transitions in a CaF_2:Sm^{2+} crystal, Zh. Eksp. Teor. Fiziki **55**, 1717–1726 (1968); English translation in Sov. Phys. JETP-USSR **28**, 902 (1969)

Few-Cycle Pulses Directly from a Laser

Franz X. Kärtner[1], Uwe Morgner[2], Thomas Schibli[1], Richard Ell[2],
Hermann A. Haus[1], James G. Fujimoto[1], and Erich P. Ippen[1]

[1] Massachusetts Institute of Technology, Cambridge, MA 02139, USA
kaertner@mit.edu
[2] Universität Karlsruhe, 76131 Karlsruhe, Germany

Abstract. Advances in ultrashort-pulse laser technology have led to the generation of few-cycle optical pulses directly from oscillators at high repetition rates, in contrast to external compression schemes. In this chapter, we review the technique of mode locking to obtain few-cycle pulses and describe the major developments leading to sub-two-cycle pulse generation from a Ti:sapphire laser. The master equation approach is used in a consistent way to analyze the nonlinear dynamics of such a laser as well as the noise and tuning properties of its frequency mode comb. Control over the mode combs of a few-cycle Ti:sapphire laser and a Cr:forsterite laser enables their coherent superposition in a long-term stable arrangement. An approach toward synthesizing single-cycle pulses using this technique is presented.

1 Introduction

Over the past three decades, remarkable progress has been achieved in short pulse generation directly from a laser. The shortest pulses are generated by excitation and phase coherent locking of many of the longitudinal modes of a laser. In standard lasers with a typical repetition rate of 100 MHz, one is now able to lock approximately 2 million of these modes, which results in pulses as short as 5 fs – shorter than two optical cycles – at a center wavelength of 800 nm. Several key developments have led to this result. As discussed in the Chapter by *Sorokin*, broadband laser materials with a relative bandwidth of $|\Delta\lambda/\lambda_0| = |\Delta f/f_0| = 0.25$ to 0.3 have been developed in the near- to mid-infrared region to support the generation of few-cycle pulses directly from a laser. The most important materials today are Ti:sapphire, the colquiriite family Cr:LISAF/LICAF/LISGAF, Cr:forsterite, and Cr:YAG (see Fig. 1 of the Chapter by *Sorokin* in this volume). To generate the ultrashort pulses permitted by such broadband gain media, a nonlinear mechanism must be placed inside the laser to favor pulsed operation over continuous wave operation of the laser. This technique of forming short pulses has been termed mode locking, since it effectively results in the phase coherent excitation of many longitudinal modes of the laser.

The concept of mode locking and the first experimental realizations of this principle already appeared in the work of several researchers in the early 1960s [1]. But the result that introduced the prospect of generating truly short pulses was that of *DeMaria* et al. in 1965 [2]. With a saturable absorber dye

F. X. Kärtner (Ed.): Few-Cycle Laser Pulse Generation and Its Applications,
Topics Appl. Phys. **95**, 73–135 (2004)

in the cavity of a flashlamp-pumped Nd:glass laser, they observed bursts of pulses too short to be measured by electronic detection. Thus began a decade of activity directed toward measuring, manipulating, and applying ultrashort pulses [3]. Many of the techniques that are still used today, including nonlinear autocorrelation, pulse compression, and pump–probe experiments, had their origin during this period. The pulses produced by the flashlamp-pumped Nd:glass laser had durations in the 5 ps to 10 ps range, and experiments were limited to slow repetition rates (1 Hz to 10 Hz).

Dramatic reductions in pulse duration and equally dramatic increases in repetition rate were ushered in by the achievement of passive mode locking with continuous wave (CW) dye lasers [4]. This led to the generation of the first subpicosecond pulses in 1974 [5] and, over the next decade, ultimately to pulses as short as 6 fs [6]. During this period, the theory of passive mode locking was also greatly extended to useful analytical descriptions of the "slow saturable absorber" (dye lasers and semiconductor lasers) [7, 8], the "fast saturable absorber" (solid-state lasers) [9], and the effects of self-phase modulation and dispersion [10].

A second revolution came in the late 1980s with the discovery of two ultrafast passive mode locking techniques that worked with CW solid-state lasers. The first came from experiments with a coupled cavity "soliton" laser [11]. It was shown [12] that the interference between two pulses experiencing different degrees of self-phase modulation produced pulse shortening. This is now called additive pulse mode locking (APM) [13] or coupled-cavity mode locking (CCM) [14]. The APM principle, in particular, has been useful in the development of ultrashort-pulse fiber lasers [15], in which it manifests itself as the nonlinear interference of two polarization states and as nonlinear polarization rotation.

An even more dramatic component of this ultrafast solid-state revolution was the femtosecond-pulse-producing self-mode-locking discovery of the *Sibbett* group in 1991 [16]. Explained as a consequence of self-focusing inside the laser [17,18,19], the technique has become known as Kerr-lens mode locking (KLM). It is now the basis for femtosecond-pulse generation in a wide variety of solid-state lasers (Fig. 1).

The effect of self-focusing in the KLM laser, in the presence of aperturing due to either the size of the gain spot or some other physically introduced aperture, can cause pulse shortening in precisely the manner described by "fast saturable absorber" mode locking theory [9]. In addition, simultaneously occurring self-phase modulation (SPM), in the presence of group velocity dispersion (GVD) in the resonator, provides a strong soliton-like shaping to the pulses. This can, in fact, be the dominant pulseshaping effect, but KLM is necessary to suppress the growth of noise between the pulses and to stabilize the mode locking. An important variant of soliton shaping occurs when the (GVD) alternates between positive and negative values as the pulse propagates. The result can be the formation of "dispersion-managed

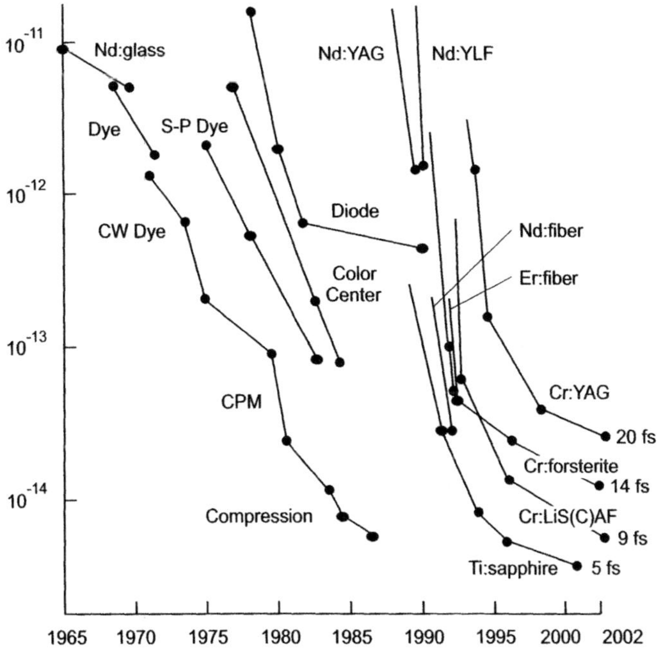

Fig. 1. The history of mode locking shows years in which shorter and shorter pulses were achieved. Distinct branches pertaining to different laser types are shown. The earliest successful generation of subpicosecond pulses was achieved with dye lasers. Then came the mode locking of semiconductor lasers. Finally, the shortest pulses were generated with solid-state systems, in particular with Ti:sapphire

solitons" when the average GVD in the resonator is zero or net normal [20]. Such dispersion management has important application in "stretched-pulse" fiber lasers [21] and plays an important role in very short-pulse Ti:sapphire lasers [22].

Ultrashort-pulse generation has been pursued by university groups in Washington [23, 24], Vienna [25, 26, 27, 28], Zurich [29, 30, 31], Cambridge, USA [22, 32], and Karlsruhe [33, 34], as well as many others. The Washington group noticed first the limitations to pulse shortening by higher order dispersion. The Vienna group introduced chirped mirrors for dispersion compensation, which were further refined at ETH-Zurich with the concept of a double-chirped mirror design. Dispersion compensation over one octave, ultimately achieved by the use of double-chirped mirror pairs at MIT and the University of Karlsruhe, led to the generation of 5 fs pulses that are comprised of less than two optical cycles. Dispersion compensating mirrors are now widely used to exploit the full bandwidth of several other broadband laser materials. Figure 2, right scale, shows how the fractional bandwidth of ultrashort pulses has increased over the last decade. The increase in spectral bandwidth has been more dramatic than the decrease in pulsewidth over this period. This was only possible due to the development of a novel chirped

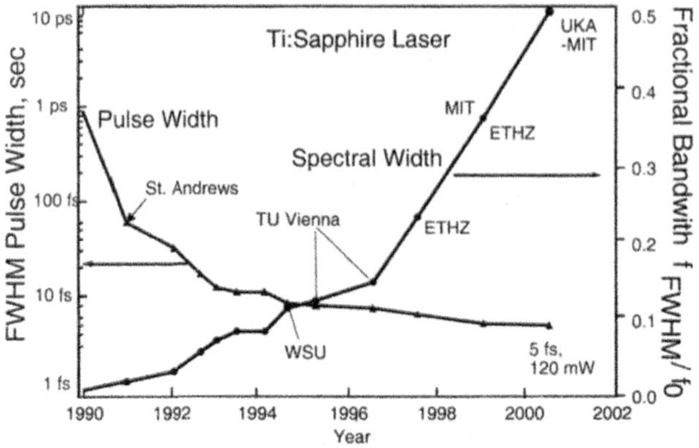

Fig. 2. Evolution of pulsewidth and spectral width of ultrashort pulses from the Ti:sapphire laser achieved over the years. WSU: Washington State University; ETHZ: Eidgenössisch Technische Hochschule Zürich; UKA: University of Karlsruhe; MIT: Massachusetts Institute of Technology

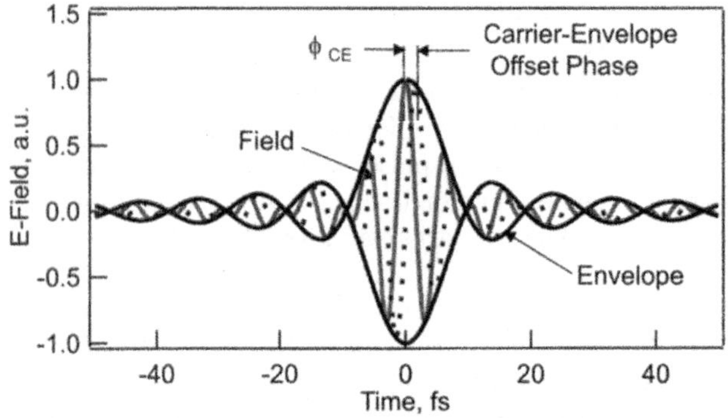

Fig. 3. Electric field of a pulse with an octave-spanning rectangularly shaped spectrum centered at 800 nm, with zero carrier–envelope phase shift (*solid line*) and with $\pi/2$-carrier–envelope phase shift (*dashed line*). The envelope is a sinc function

mirror technology. The achievement of an octave broadband spectrum, or even two-thirds of an octave, has important implications for the control of the electric field waveform of the pulse. It has been recognized by *Xu* et al. [35] that for few-cycle pulses, the maximum electric field in a pulse depends quantitatively on the phase relationship between the carrier wave and the maximum of the envelope (see Fig. 3, and for more details, Chapters by *Udem* et al. and *Jones* et al.).

It is obvious that nonlinear optical processes, which depend on electric field, must depend on the carrier–envelope phase for very short pulses. This is visualized in Fig. 4 by sketching the output spectrum of instantaneous $\chi^{(2)}$- and $\chi^{(3)}$-processes for a chirp-free pulse with a rectangularly shaped spec-

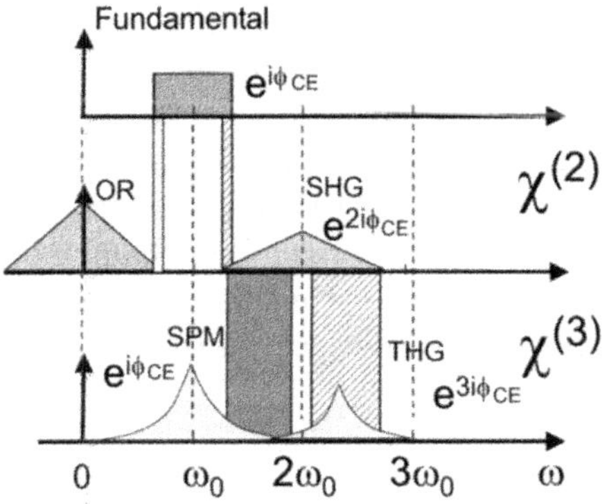

Fig. 4. Fundamental spectrum of an octave-spanning pulse centered at ω_o; below are spectra generated by an instantaneous second- and third-order nonlinearity due to optical rectification (OR), second-harmonic generation (SHG), self-phase modulation (SPM), and third-harmonic generation (THG)

trum covering one octave. For a one-octave pulse, the spectra of the second-harmonic and optical rectification terms overlap the fundamental spectrum.

Interference in the overlap regions will be constructive or destructive, depending on the carrier–envelope phase. If higher order processes are employed, such as a $\chi^{(3)}$-nonlinearity, there are similar interference terms between the $\chi^{(2)}$ and $\chi^{(3)}$ terms. Then, phase-dependent interference is already possible even if the pulse spans only two-thirds of an octave. All of these interferences have been used to detect and control the carrier–envelope phase. A special case, cited in [36] and discussed in the Chapter by *Baltuška* et al., is that optical rectification or difference-frequency mixing of an optical pulse can be used to produce a phase-independent infrared pulse. Currently, control of the carrier–envelope phase for a variety of applications is one of the strongest motivations for the development of few-cycle- or even single-cycle-pulse lasers.

This chapter reviews the understanding and technology needed to achieve few-cycle pulses directly from a laser, starting with a short overview of the principles of mode locking. The chapter is then organized as follows. In Sect. 2, we review passive mode locking based solely on loss mechanisms within the framework of the master equation of mode locking. In Sect. 3, we add solution-like pulseshaping to these schemes, which allows us to shorten the pulses further below the limits achievable with loss modulation alone. Extending this treatment to the formation of dispersion-managed solitons, we discuss the ultimate case when the pulses are so short or intense that the nonlinear effects and spectral breathing of the pulses become large within the laser elements themselves. This leads to effective rescaling of the soliton dynamics, but, otherwise, the basic phenomena and their interplay stay the same. Soliton perturbation theory will give us a natural basis for the discussion of fluctuations in pulse amplitude, phase, timing, and center frequency.

That, in turn, enables us to analyze, in Sect. 4, the dynamics of lasers influenced by noise and the control of the carrier–envelope phase. In Sect. 5, we discuss the main technical challenges that must be solved to generate few-cycle pulses; that is, the optimization of KLM and self-phase modulation and the design of broadband dispersion compensating laser mirrors. Section 6 is devoted to the experimental results obtained with various broadband gain media. In Sect. 7, we explore the possibility of single-cycle-pulse synthesis by coherently superimposing the mode combs of two few-cycle lasers. Section 8 concludes and summarizes this Chapter.

2 The Master Equation of Mode Locking

Over the past three decades, a unified picture of ultrashort-pulse generation in terms of the master equation of mode locking has been developed. This section gives a brief review of these developments. Limits on the pulsewidth achievable, given the most important laser parameters, are presented. Although the master equation is the basis for active mode locking and slow absorber mode locking as well, we limit ourselves here to fast saturable absorber mode locking that is applicable to ultrashort-pulse, solid-state lasers.

A schematic of a simple passively mode-locked laser is shown in Fig. 5. Between the two mirrors forming the cavity are elements of gain, loss, and saturable absorption. We assume that the pulse beginning one round-trip of the cavity may be described by an amplitude $a(t)$, where t is the local time expressed around the center of the pulse. After one round-trip, the pulse is $a(t + T_\mathrm{r})$, where T_r is the round-trip time of the cavity. The round-trip loss of the laser cavity (which we assume is frequency independent and includes the transmission of the mirrors) is determined by multiplying by the factor ℓ_c. We assume that the laser is in the steady state and that the gain has a homogeneously saturated value with parabolic frequency dependence near its peak of the form

$$\Delta g(\Omega) = \frac{g}{1 + \left(\frac{\Omega}{\Omega_\mathrm{g}}\right)^2} = g\left[1 - \left(\frac{\Omega}{\Omega_\mathrm{g}}\right)^2\right]. \tag{1}$$

There are no significant temporal dynamics of the gain on the timescale of the pulse. The gain saturates on average power. We may then determine the change in $a(t)$ with each round-trip by Fourier transforming to $\tilde{a}(\Omega)$, then multiplying by $g(\Omega)$ and taking the inverse transform. The result is $a(t)$ is

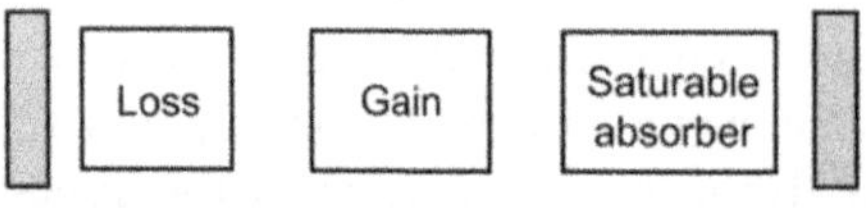

Fig. 5. Schematic of a simple, passively mode-locked laser

multiplied each round-trip by

$$g\left(1+\frac{1}{\Omega_{\mathrm{g}}^{2}}\frac{\partial^{2}}{\partial t^{2}}\right). \tag{2}$$

The round-trip change (decrease) in $a(t)$ due to the saturable absorber is given by $q(t)a(t)$, where the absorption $q(t)$ is itself rapidly modulated by the intensity $I(t)$ of the pulse. For the fast (instantaneously responding) saturable absorber, we write

$$q(t)=\frac{q_{0}}{1+I(t)/I_{\mathrm{sat}}}, \tag{3}$$

where q_0 is the unsaturated loss, $I(t)$ is the time-dependent intensity, and I_{sat} is the saturation intensity of the absorber. If the saturation is relatively weak, expression (3) can be expanded to give

$$q(t)=q_{0}-q_{0}I(t)/I_{\mathrm{sat}}. \tag{4}$$

The intensity $I(t)$ multiplied by the effective area of the mode A_{eff} gives the power in the mode. We normalize the mode amplitude so that $|a(t)|^2 = \text{power}$. Then, the transmission can be written

$$q(t)=q_{0}-\frac{q_{0}|a(t)|^{2}}{I_{\mathrm{sat}}A_{\mathrm{eff}}}\equiv q_{0}-\gamma|a(t)|^{2}, \tag{5}$$

where γ is the self-amplitude modulation (SAM) coefficient. Finally, we have the master equation of passive mode locking with a fast saturable absorber:

$$a(t+T_{\mathrm{r}})-a(t)=(g-\ell)a+D_{\mathrm{f}}\frac{\partial^{2}}{\partial t^{2}}a+\gamma|a|^{2}a, \tag{6}$$

where we have introduced the total frequency-independent loss $\ell = \ell_{\mathrm{c}} + q_0$ and the loss (or gain) curvature $D_{\mathrm{f}} = g/{\Omega_{\mathrm{g}}}^2$. In the steady state, $a(t+T_{\mathrm{r}}) - a(t) = 0$, and there is an analytical solution for $a(t) = 0$. It is a simple hyperbolic secant

$$a_{0}(t)=A_{0}\operatorname{sech}(t/\tau), \tag{7}$$

with

$$\frac{1}{\tau^{2}}=\frac{\gamma A_{0}^{2}}{2D_{\mathrm{g}}}=\frac{\gamma A_{0}^{2}{\Omega_{\mathrm{g}}}^{2}}{2g} \tag{8}$$

and

$$\ell-g=\frac{D_{\mathrm{g}}}{\tau^{2}}. \tag{9}$$

The net gain is negative preceding and following the pulse. At the pulse peak, the net gain is positive due to the bleaching of the saturable absorber. Another interesting feature emerges: a hyperbolic secant has exponential tails. Such tails are mandatory in passive mode locking; the system behaves linearly in the tails since the intensity is small. The second-order differential equation dictates exponential solutions for any bounded pulse. The ultimate pulsewidth achieved for a fully saturated absorber, i.e., $q_0 = \gamma A_0^2$, is then given by

$$\tau = \frac{1}{\Omega_{\mathrm{g}}} \sqrt{\frac{g}{q_0}} \,. \tag{10}$$

Note that the saturated gain level is not independent of the unsaturated value of the saturable absorber. The final pulsewidth achieved depends on the saturated gain level that arranges itself, which dependd also on the residual losses of the absorber saturated by the pulse. The solution (7) is not stable unless steady-state, average intensity gain saturation is explicitly included. The solution is also stabilized if the full saturation behavior (3) is heeded, but no closed-form solutions have been found.

3 Soliton Mode Locking

So far, mode locking using saturable absorbers has been discussed. As shorter and shorter pulses were generated by dye lasers, it was realized that the group delay dispersion (GDD) and self-phase-modulation (SPM) caused by the Kerr effect of the laser elements could not be ignored. Group delay was controlled by the insertion of prism pairs into the laser resonator [37]. These new effects become important for subpicosecond pulses and especially for few-cycle lasers, due to their enormous bandwidth and short pulsewidth, i.e., high peak power.

3.1 Soliton Formation in Mode-Locked Lasers

Here, we discuss only the extension of fast saturable absorber mode locking, including GDD and SPM. The master equation (6) is modified to

$$\frac{1}{T_{\mathrm{r}}} \frac{\partial}{\partial T} a = (g - \ell) a + (D_{\mathrm{f}} - \mathrm{i} D) \frac{\partial^2}{\partial t^2} a + (\gamma + \mathrm{i}\delta) |a|^2 a \,. \tag{11}$$

Here, D is the group velocity dispersion parameter, D_{f} the gain curvature, and the associated filtering action is represented by $D_{\mathrm{f}} \frac{\partial^2}{\partial t^2} a$. In a medium of length L, with a propagation constant whose second derivative is β''; the parameter D is $D = \beta'' L/2$. The Kerr coefficient is $\delta = (2\pi/\lambda) n_2 L / A_{\mathrm{eff}}$, where λ is the carrier wavelength, n_2 is the nonlinear index in centimeter per watt, and A_{eff} is the effective mode cross-sectional area in square centimeter. The gain is taken as time independent as applicable for a gain medium with a long relaxation time. The bandwidth is assumed to be limited by a filter of bandwidth Ω_f. This equation has a simple steady-state solution [10]:

$$a(t) = A_0 \,\mathrm{sech}^{(1-\mathrm{i}\beta)}\left(\frac{t}{\tau}\right). \tag{12}$$

The presence of group velocity dispersion raises the possibility of unequal phase and group velocities. When this happens, the carrier phase may slip with respect to the envelope in one round-trip time. This means that $\frac{1}{T_r}\frac{\partial a}{\partial T} = j\psi a$. Introducing this ansatz into (12) and balancing terms, one obtains two complex equations:

$$\ell - \frac{(1-\mathrm{i}\beta)^2}{\tau^2}(D_\mathrm{f} - \mathrm{i}D) = g + \mathrm{i}\psi\,, \tag{13}$$

$$\frac{1}{\tau^2}(D_\mathrm{f} - \mathrm{i}D)(2 - 3\mathrm{i}\beta - \beta^2) = (\gamma + \mathrm{i}\delta)A_0^2 = 0\,. \tag{14}$$

We investigate how the pulse parameters vary as one adjusts the group velocity dispersion by a pair of prisms or the SAM coefficient by changes in the APM or KLM setup. In such a case, the energy W is fixed by the pump level. By taking the real and imaginary parts of (13) and (14), one obtains four real equations for the four parameters: 1. net gain, 2. the phase shift, 3. the pulsewidth, and 4. the chirp parameter.

The real part and imaginary part of (14) give

$$\frac{1}{\tau^2}\left[D_\mathrm{f}(2-\beta^2) - 3\beta D\right] = \gamma A_0^2\,, \tag{15}$$

$$\frac{1}{\tau^2}\left[D(2-\beta^2) + 3\beta D_\mathrm{f}\right] = -\delta A_0^2\,. \tag{16}$$

We introduce the normalized dispersion, $D_\mathrm{n} = D/D_\mathrm{f}$, and the pulsewidth of the system without GDD and SPM, i.e., the width of the purely saturable absorber mode-locked system, $\tau_0 = 4D_\mathrm{f}/(\gamma W)$. Dividing (15) by (16) and introducing the normalized nonlinearity $\delta_\mathrm{n} = \delta/\gamma$, we obtain a quadratic equation for the chirp. With the parameter,

$$\chi = \frac{2-\beta^2}{3\beta} = \frac{\delta_\mathrm{n} D_\mathrm{n} - 1}{\delta_\mathrm{n} + D_\mathrm{n}}\,, \tag{17}$$

closely related to the chirp, the solution to this quadratic equation can be written as

$$\beta = -\frac{3}{2}\chi \pm \sqrt{\left(\frac{3}{2}\chi\right)^2 + 2}\,. \tag{18}$$

Knowing the chirp, we obtain from (15) the pulsewidth

$$\tau = \frac{\tau_0}{2}\left(2 - \beta^2 - 3\beta D_\mathrm{n}\right), \tag{19}$$

which can also be written as

$$\tau = \frac{3\tau_0}{2}\beta\left(\chi - D_{\rm n}\right). \tag{20}$$

To be physically meaningful, the pulsewidth must be a positive number, i.e., the product $(\chi - D_{\rm n})$ must always be greater than 0, which determines the root in (18) according to

$$\beta = \begin{cases} -\frac{3}{2}\chi + \sqrt{\left(\frac{3}{2}\chi\right)^2 + 2} & \text{for } \chi > D_{\rm n}\,, \\ -\frac{3}{2}\chi - \sqrt{\left(\frac{3}{2}\chi\right)^2 + 2} & \text{for } \chi < D_{\rm n}\,. \end{cases} \tag{21}$$

A necessary, but not, sufficient criterion for the stability of pulses is that there must be a net loss leading and following the pulse. From the real part of (13), we obtain for the net gain leading and following the pulse,

$$g - \ell = -\frac{1}{\tau^2}\left[(1-\beta^2)D_{\rm f} + 2\beta D\right] < 0\,. \tag{22}$$

The pulsewidth as a function of dispersion is plotted in Fig. 6a. For nonzero SPM, the shortest pulses are obtained with negative dispersion. The pulses are always longer with positive dispersion. The chirp parameter is plotted in Fig. 6b. A combination of negative dispersion with finite SPM can find a zero chirp solution. For a small SAM coefficient, weak filtering, and negative values of D, one finds that the pulse is chirp-free. In this case, the pulse is soliton-like, a solution of the nonlinear Schrödinger equation, an approximation to (11):

$$\frac{1}{T_{\rm r}}\frac{\partial}{\partial T}a = -{\rm i}D\frac{\partial^2}{\partial t^2}a + {\rm i}\delta|a|^2 a\,, \tag{23}$$

with the chirp-free hyperbolic secant "soliton" solution:

$$a(T,t) = A_0\,{\rm sech}\left(\frac{t}{\tau}\right){\rm e}^{{\rm i}\delta|A_0|^2 T/2T_{\rm r}}\,. \tag{24}$$

The pulse is continuously phase shifted by the Kerr effect. The amplitude and pulsewidth obey the "area theorem"

$$|A_0|\tau = \sqrt{2|D|/\delta}\,. \tag{25}$$

The pulse forms through the balance of GVD and SPM. This behavior is ideally reached in fiber ring lasers with uniform dispersion $(D < 0)$, but is also characteristic of solid-state lasers. The pulse in the laser may be described as a soliton weakly perturbed by SAM and filtering. The pulseshaping may be attributed to the soliton process. SAM action is required solely for stabilization of the pulse against noise buildup in the intervals between pulses. Filtering selects the pulsewidth by causing a monotonic increase in loss with

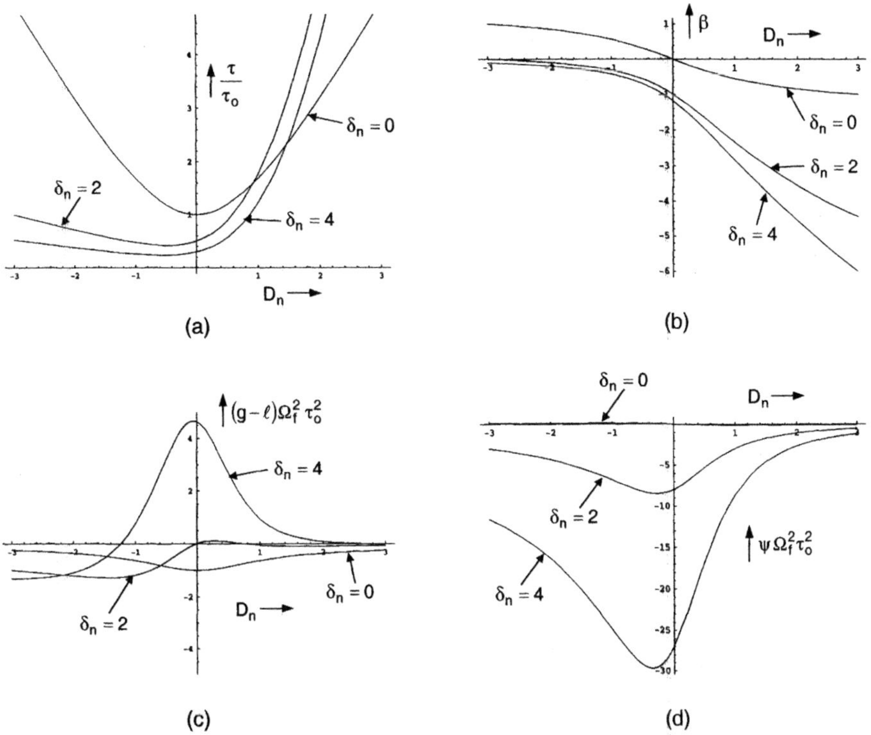

Fig. 6. (**a**) Pulsewidth. (**b**) Chirp parameter. (**c**) Net gain. (**d**) Phase shift per pass

decreasing pulsewidth. Figure 6c plots the parameter g–ℓ, which must be negative if the pulse train is to be stable against buildup of noise between pulses. We see that excessive SPM can lead to instability near zero dispersion and for positive dispersion.

The pulse can also be made chirp-free by a proper balance between the SAM coefficient and the filtering action. The pulsewidth increases rapidly with increasing positive values of D. However, the pulse with the broadest bandwidth is found for slightly positive D. When compressed by grating pairs external to the laser, the shortest pulses can be generated from a given system.

Figure 6d plots the phase shift ψ per pass. If a phase is added upon each pass, the phase increases linearly in time, which corresponds to a frequency shift. Fluctuations of ψ that may be induced by noise, cause the spectral lines of the modes to acquire finite linewidth. Further, the phase shift per pass causes a change in phase velocity. The phase velocity dropped out of the analysis when we resorted to the slowly varying envelope approximation suppressing the carrier. The carrier $\exp(\mathrm{i}\omega_\mathrm{c}t - \mathrm{i}\beta_0 z)$ does not appear explicitly. This carrier is multiplied by the envelope function. The carrier slips underneath the envelope within one round-trip, due to the difference in phase and group velocity

$$\frac{1}{v_\mathrm{p}} = \frac{\beta_\mathrm{c}}{\omega_\mathrm{c}}\,, \tag{26}$$

$$\frac{1}{v_\mathrm{g}} = \left.\frac{\partial\beta(\omega)}{\partial\omega}\right|_{\omega=\omega_\mathrm{c}} \tag{27}$$

by a phase

$$\Delta\Phi = \omega_\mathrm{c}\left(\frac{1}{v_\mathrm{g}} - \frac{1}{v_\mathrm{p}}\right)L\,, \tag{28}$$

where $\beta(\omega)$ is the dispersion relation for the linear waves in the resonator. Normally, one does not care when such a slip occurs, unless the pulse is only a few cycles long. Then, such a slip leads to a change in the electric field pulse-shape from pulse to pulse and to other effects that will be studied in later sections and chapters. It should be noted that the total inverse group velocity and phase velocity of the pulse inside the laser is also affected by the Kerr effect, thus leading to an energy-dependent contribution, which will be discussed in the framework of soliton perturbation theory in Sect. 4.

There is another mechanism that is very important in limiting the pulse shortening. These are the parasitic sidebands first described and explained by *Kelly* [38]. The soliton is periodically perturbed by the gain, loss, filtering, and SAM action. In the process, it radiates (generates continuum). If the continuum generated by the soliton is phase matched from pulse to pulse, its energy can build up and drain the soliton. Figure 7 shows a phase-matching diagram in the frequency domain. A spectral component of the continuum at a frequency deviation Ω_s from the carrier frequency is phase delayed in one pass through the resonator by $D\Omega_\mathrm{s}^2$, and phase advanced if the dispersion is anomalous ($D < 0$). The soliton pulse and spectrum experience the phase delay $\Phi_0 = \delta|A_0|^2$. The excitation of the continuum is matched if the sum of delay and advance is equal to a multiple of 2π. When this happens, sidebands appear in the pulse spectrum at position

$$\Omega_\mathrm{s} = \pm\frac{1}{\tau}\sqrt{\frac{2m\pi}{\Phi_0} - 1}\,, \qquad m = 1, 2, 3, \ldots\,. \tag{29}$$

One such experimental trace is shown in Fig. 8 for a fiber laser [21].

3.2 Dispersion-Managed Soliton Formation

The nonlinear Schrödinger equation describes pulse propagation in a medium with continuously distributed dispersion and self-phase-modulation. It was first pointed out by *Spielmann* et al. that large changes in the pulse occur within one round-trip and that the ordering of the pulseshaping elements within the cavity has a major effect on pulse formation [25]. For pulses as short as 10 fs and as intense as those in typical femtosecond lasers, the discrete action of linear dispersion in the arms of the laser resonator, and the discrete,

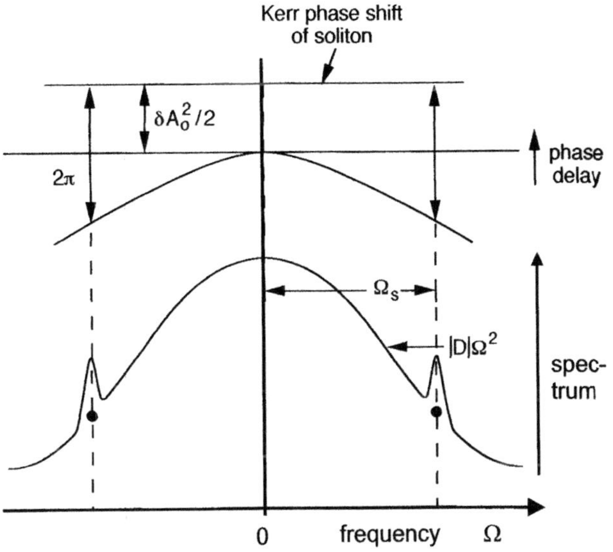

Fig. 7. Phase matching of parasitic sidebands

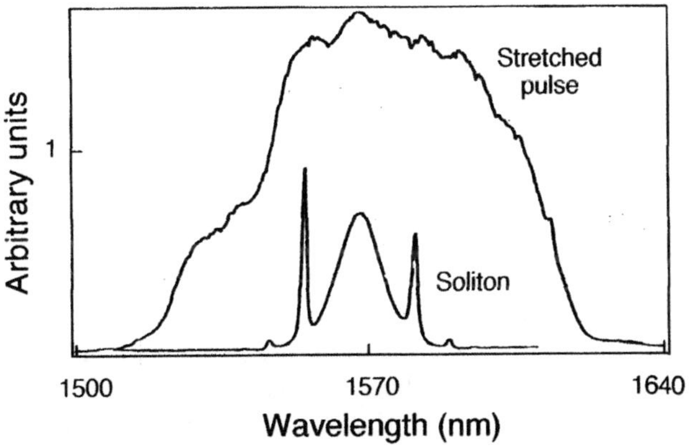

Fig. 8. Spectra of a soliton laser and of a stretched-pulse laser

but simultaneous, action of positive SPM and positive GDD in the laser crystal cannot be neglected any longer. The importance of the strong variations in dispersion first discovered in a fiber laser was called stretched-pulse mode locking [15]. The positive dispersion in the Er-doped fiber section of a fiber ring laser was balanced by a negative dispersive passive fiber. The pulse circulating in the ring was stretched and compressed by a factor of as much as 20 in one round-trip. One consequence of this behavior was a dramatic decrease in nonlinearity and, thus, increased stability against SPM-induced instabilities. No Kelly sidebands were observed (see Fig. 8). The energy of the output pulses could be increased 100-fold. The minimum pulsewidth was

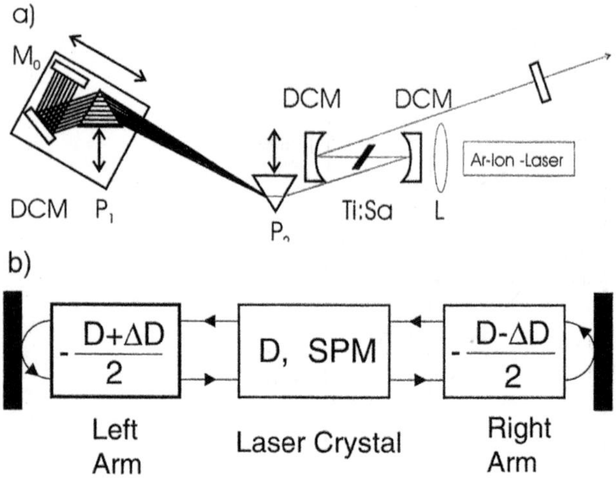

Fig. 9. (**a**) Schematic of a Kerr-lens mode-locked Ti:sapphire laser: Ps, prisms; L, lens; DCMs, double-chirped mirror; TiSa, Ti:sapphire. (**b**) Correspondence with dispersion-managed fiber transmission

63 fs, about twice as broad as the gain bandwidth of the Er-doped fiber [39]. Figure 8 clearly shows the spectral enhancement of the fiber laser in the dispersion-managed regime.

The generation of ultrashort pulses from Ti:sapphire lasers has progressed over the past decade and has led to the generation of pulses as short as 5 fs directly from the laser. The spectra generated by these lasers are not of simple shape [24, 34, 35], and various models for their explanation have been presented. *Christov* et al. have shown by computation that, when the second-order group velocity is balanced, fourth-order dispersion with gain bandwidth filtering and KLM action can lead to steady-state pulse generation [40]. Extensive three-dimensional simulations have been carried out [41, 42, 43].

Here, we want to consider the impact on the spectral shape of such laser pulses, due to the discrete action of dispersion and nonlinearity in femtosecond lasers, which become important in the sub-10 fs range in Ti:sapphire lasers along the lines of the stretched-pulse fiber laser. A mode-locked Ti:sapphire laser producing ultrashort pulses consists of at least a gain medium (Ti:sapphire crystal) and dispersion-balancing components, e.g., mirrors, prism pairs (see Fig. 9a). The system can be decomposed into the resonator arms and the crystal (see Fig. 9b). To achieve ultrashort pulses, the dispersion-balancing components should produce close to zero net dispersion, whereas the dispersion element(s) individually produce significant group delay over the broad bandwidth of the laser pulse. This fact suggests an analogy with dispersion-managed pulse propagation along a dispersion-managed fiber transmission link [20]. A system with sufficient variation of dispersion can support waves called nonlinear Bloch waves [44]. One can show that the

Kerr nonlinearity produces a self-consistent nonlinear scattering potential that permits the formation of a periodic solution with a simple phase factor in a system with zero net dispersion. It has been shown that nonlinear propagation along a dispersion-managed fiber near zero net GDD possesses a narrower spectrum in the segment of positive dispersion than that in the segment of negative dispersion [20, 44]. Thus, the effect of negative dispersion is greater than that of positive dispersion, and it imparts an effective net negative dispersion to the pulse. This effective negative dispersion can balance the Kerr-induced phase, thus leading to steady-state pulses at zero net dispersion. This is true even when there is no nonlinearity in the negative dispersion segment. The pulses are analogous to solitons in that they are self-consistent solutions of the Hamiltonian (lossless) problem, as in the conventional solitons discussed above. But they are not secant hyperbolic in shape.

Figure 10 shows a numerical simulation of a self-consistent solution of the Hamiltonian pulse-propagation problem in a linear medium of negative dispersion and subsequent propagation in a nonlinear medium of positive dispersion with positive self-phase modulation, following the equation,

$$\frac{\partial}{\partial z}a = -\mathrm{i}D(z)\frac{\partial^2}{\partial t^2}a + \mathrm{i}\delta(z)|a|^2 a\,. \tag{30}$$

The dispersion map $D(z)$ is shown as an inset in Fig. 11. Note that the dispersion coefficient $D(z)$ and the nonlinear coefficient $\delta(z)$ are now defined per unit length, in contrast to (23). In Fig. 10, the steady-state intensity profiles are shown at the center of the negative dispersion segment over 1000 round-trips. In addition, we can include in the model the saturable gain and Lorentzian gain filtering. The KLM can be modeled by a fast saturable absorber. Figure 11 shows the behavior in one period (one round-trip through the resonator), including these effects. The response of the absorber is $q(a) = q_0/(1+|a|^2/P_\mathrm{A})$, with $q_0 = 0.01\,\mathrm{mm}$ and $P_\mathrm{A} = 1\,\mathrm{MW}$. The bandwidth-limited gain is modeled by the Lorentzian profile with gain bandwidth $\Omega_\mathrm{g} = 2\pi \times 43\,\mathrm{THz}$. The filtering and saturable absorber reduce the spectral and temporal sidelobes of the Hamiltonian problem, as inferred from Fig. 11. The steady-state pulse formation can be understood in the following way. By symmetry, the pulses are chirp-free in the middle of the dispersion cells. A chirp-free pulse starting in the center of the gain crystal, i.e., the nonlinear segment, is spectrally broadened by the SPM and disperses in time due to the GVD, which generates a mostly linear chirp over the pulse. After the pulse leaves the crystal, it experiences GVD in the arms of the laser resonator, which compresses the positively chirped pulse to its transform limit at the end of each arm, where an output coupler can be placed. Backpropagation toward the crystal imposes a negative chirp, generating the time-reversed solution of the nonlinear Schrödinger equation (30). Therefore, subsequent propagation in the nonlinear crystal is compressing the pulse spectrally and temporally to its initial shape in the center of the crystal. The spectrum is narrower in

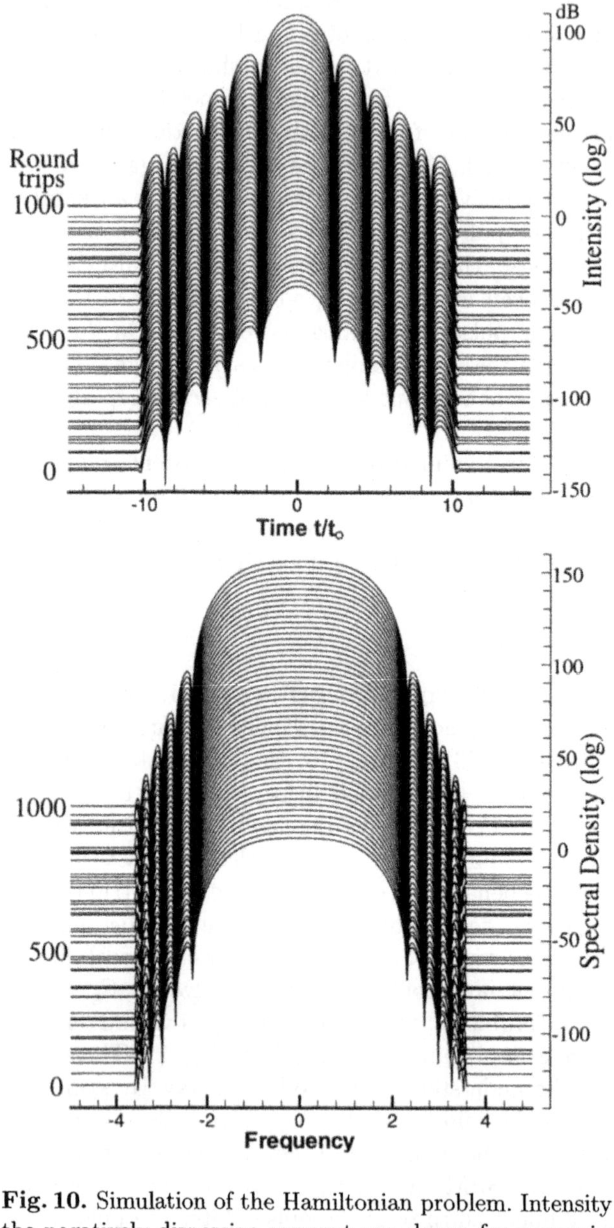

Fig. 10. Simulation of the Hamiltonian problem. Intensity profiles at the center of the negatively dispersive segment are shown for successive round-trips. The total extent in 1000 round-trips. $D = D^{(\pm)} = \pm 60\,\mathrm{fs}^2/\mathrm{mm}$, segment of crystal length $L = 2\,\mathrm{mm}, \tau_{\mathrm{FWHM}} = 5.5\,\mathrm{fs}, \delta = 0$ for $D < 0$, $\delta = 1\,\mathrm{MW}^{-1} \cdot \mathrm{mm}^{-1}$ for $D > 0$

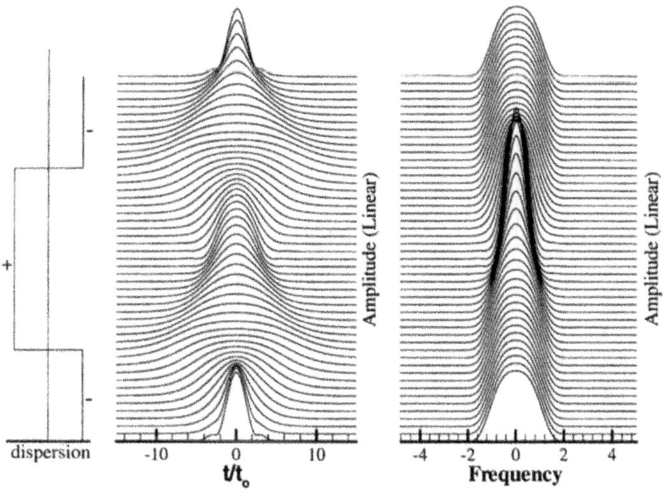

Fig. 11. Pulseshaping in one round-trip. The negative segment has no nonlinearity

the crystal than in the negative dispersion sections because it is negatively prechirped before it enters the SPM section, and spectral spreading occurs again only after the pulse has been compressed. This result further explains that, in a laser with a linear cavity for which the negative dispersion is located in only one arm of the laser resonator (i.e., in the prism pair and no use of chirped mirrors), the spectrum is widest in the arm that contains the negative dispersion [25].

In a laser with a linear cavity for which the negative dispersion is equally distributed in both arms of the cavity, the pulse runs through the dispersion map twice per round-trip. The pulse is short at each end of the cavity, and, most importantly, the pulses are identical in all paths through the crystal, which exploits the KLM action twice per round-trip [32], in contrast to an asymmetrical dispersion distribution in the resonator arms. Thus, a symmetrical dispersion distribution may lead to an effective saturable absorption that is twice as strong as an asymmetrical dispersion distribution, which results in substantially shorter pulses. Furthermore, the dispersion swing between the negative and positive dispersion sections is only half, which allows for shorter dispersion-managed solitons operating at the same average power level. To illustrate further the efficiency of dispersion-managed soliton formation, we present a series of simulations that start with a linear segment of negative dispersion and a nonlinear segment of positive dispersion of the same magnitude, KLM action, and filtering.

Figure 12 shows the pulseshape for this arrangement (*dashed curves*). We obtained the other traces by increasing the SPM while keeping the energy fixed through adjustment of the gain. As one can see, increasing the SPM permits shorter pulses. The shortest pulse generated can be up to three times

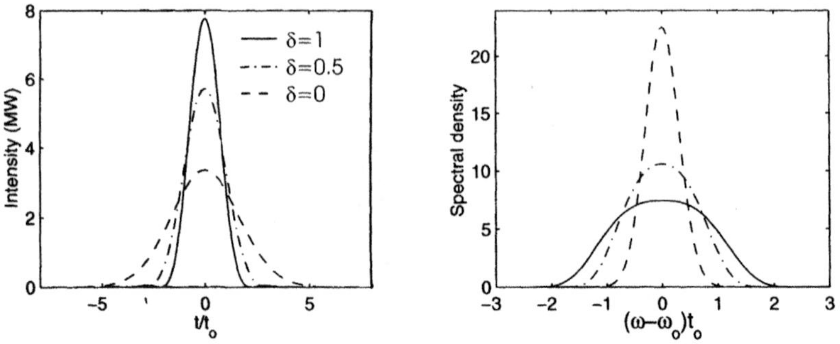

Fig. 12. Sequence of pulse profiles in the center of the negatively dispersive segment for three magnitudes of SPM. $t_0 = 3\,\mathrm{fs}$, with *solid curves* (5.5 fs) for $\delta = 1\,\mathrm{MW}^{-1}\cdot\mathrm{mm}^{-1}$, *dashed-dotted curve* (7 fs) for $\delta = 0.5\,\mathrm{MW}^{-1}\cdot\mathrm{mm}^{-1}$, and *dashed curves* for no SPM of $\delta = 0$. The dispersion map is of Fig. 11. The output coupler loss is 3%

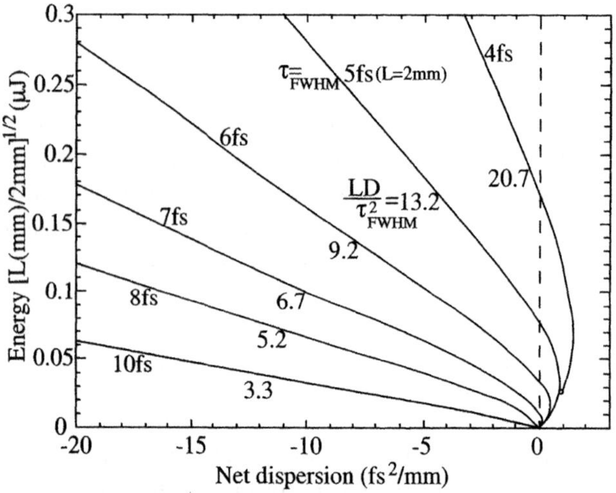

Fig. 13. Energy of the pulse in a lossless dispersion-managed system with stretching $S = LD/\tau_{\mathrm{FWHM}}^2$ or for a fixed crystal length L and pulsewidth as parameters; $D = 60\,\mathrm{fs}^2/\mathrm{mm}$ for Ti:sapphire at 800 mm

shorter than a pulse without SPM. The parameters chosen for the simulations are listed in the figure captions. In this respect, the behavior is similar to that of other soliton lasers [45]. A major difference in conventional soliton formation is illustrated in Fig. 13. The figure shows the parameter ranges for a dispersion-managed Hamiltonian system (no gain, no loss, no filtering) that is unbalanced, which results in the net dispersion that serves as the abscissa of the figure. Each curve gives the locus of energy versus net cavity dispersion for a stretching ratio $S = LD/\tau_{\mathrm{FWHM}}^2$ (or pulsewidth with fixed crystal length L). One can see that, for a pulsewidth longer than 8 fs with crystal length $L = 2\,\text{mm}$, no solution of finite energy exists in the dispersion-managed system for zero or positive net dispersion. Pulses longer than 8 fs require net negative dispersion. Hence, one can reach ultrashort dispersion-managed operation at zero net dispersion only by first providing the system with negative dispersion. At the same energy, one can form a shorter pulse by reducing the net dispersion, provided that the 8 fs threshold has been passed. For a fixed dispersion swing $\pm D$, the stretching increases quadratically with the spectral width of the inverse pulsewidth. Long pulses with no stretching have a sech shape. For stretching ratios of 3 to 10, the pulses are Gaussian shaped. For even larger stretching ratios, the pulse spectra become increasingly more flattopped, as shown in Fig. 10.

For gaining insight into laser dynamics and later in noise and tuning behavior, a master equation approach that was already developed for the operation of the stretched-pulse laser [46] will be introduced. Note that the Kerr phase-shift is produced by a pulse of varying amplitude and width as it circulates around the ring. The Kerr phase shift for a pulse of constant width, $\delta|a|^2$, had to be replaced by a phase profile that mimicked the average shape of the pulse, weighted by its intensity. Therefore, the SPM action of (11) is replaced by

$$\delta|a|^2 = \delta_0|A_0|^2\left(1-\mu\frac{t^2}{\tau^2}\right), \tag{31}$$

where A_0 is the pulse amplitude at the position of minimum width. The Kerr phase profile is expanded to second order in t. The coefficient δ_0 and μ are evaluated variationally. The SAM action is similarly expanded. Finally, the net intracavity dispersion acting on average on the pulse is replaced by the effective dispersion D_{net} in the resonator within one round-trip. The master equation becomes

$$\begin{aligned}\frac{1}{T_\text{r}}\frac{\partial}{\partial T}a = (g-\ell)a + (D_\text{f} - \mathrm{i}D_\text{net})\frac{\partial^2}{\partial t^2}a \\ + (\gamma_0 + \mathrm{i}\delta_0)|A_0|^2\left(1-\mu\frac{t^2}{\tau^2}\right)a\,.\end{aligned} \tag{32}$$

This equation has Gaussian-pulse solutions. Since the approximations made in arriving at (32) are not applicable to the wings of the pulse, the wings are not Gaussian. However, especially in few-cycle lasers, the wings are not very well described by the simple parabolic gain profile either. Most often, the mirrors and output couplers form a hard spectral filter. If the saturable absorber action is strong enough to stabilize the generation of spectrum beyond the high reflectivity bandwidth of the output coupler, the output spectrum may show strong wings, due to the enhanced output coupling. The master equation (32) is a patchwork; it is not an ordinary differential equation. The coefficients in the equation depend on the pulse solution and eventually must be found iteratively. Nevertheless, the equation accounts for the pulse-shaping in the system in an analytic fashion. It will allow us to extend the conventional soliton perturbation theory to dispersion-managed solitons. The master equation (32) for the stretched-pulse laser also permits steady-state solutions without filtering and SAM, as long as $D_{\rm net} < 0$. Again, one may look at this operation as a perturbed dispersion-managed soliton; the SAM and filtering are small perturbations that stabilize against noise buildup and set the pulsewidth, respectively. The stretching and compression lead to a reduction of the nonlinear phase shift during propagation of the pulse in the cavity. Operations at close to zero average dispersion per round-trip also lead to a reduced conversion of frequency to timing fluctuations.

4 Noise, Stabilization, and Tuning of Mode-Locked Lasers

So far, we have considered only deterministic steady-state pulse formation in ultrashort-pulse laser systems due to the most important pulseshaping mechanisms prevailing in today's femtosecond lasers. Due to the recent interest in the phase of mode-locked lasers and the use of mode-locked lasers in high-resolution laser spectroscopy, as well as phase-sensitive nonlinear optics, a unified approach for predicting the noise and tuning properties of mode-locked lasers shall be presented here. It is built upon soliton-perturbation theory, which has been successfully used to predict the noise behavior of many solid-state and fiber laser systems [47], as well as changes in both group and phase velocity in mode-locked lasers due to intracavity nonlinear effects [48]. We start by reconsidering the derivation of the master equation for describing the pulseshaping effects in a mode-locked laser. We assume that, in steady state, the laser generates at some position z (for example, at the point of the output coupler) a periodic sequence of pulses inside the laser:

$$
\begin{aligned}
A(T,t) = & \sum_{m=-\infty}^{+\infty} a\left[T, t - mT_{\rm r}(T)\right] \\
& \cdot \exp\left[-{\rm i}\left(\omega_{\rm c}(T)\left\{\left[t - mT_{\rm r}(T) + \left(\frac{1}{v_{\rm g}} - \frac{1}{v_{\rm p}}\right) 2mL\right] + \theta_0\right\}\right)\right]
\end{aligned} \tag{33}
$$

at a repetition rate $f_{\rm r} = 1/T_{\rm r}$ and center frequency $\omega_{\rm c}$. Both are, in general, subject to slow drifts, due to mirror vibrations, thermal drifts, changes, in intracavity pulse energy that may be converted to phase changes, and changes in center frequency from nonlinearities or dispersion. Note that the center frequency and repetition rate are defined only for times long compared to the round-trip time in the laser. Usually, they change only on a timescale three orders of magnitude longer than the expectation value of the repetition rate. θ_0 is an arbitrary but constant contribution to the carrier–envelope phase of the pulse. To obtain self-consistent equations for the changes in repetition rate, center frequency, and other pulse parameters, we employ soliton-perturbation theory [49]. This is justified when the steady-state pulse is close to a soliton. For fast saturable absorber mode locking, this is the chirp-free solution, occurring when the ratio of gain filtering to dispersion is equal to the ratio of SAM action to self-phase modulation; see (24). Then the pulse solution in the mth round-trip is a solution of the nonlinear Schrödinger equation stabilized by the irreversible dynamics and is subject to additional perturbations described by the term L_{pert}:

$$\frac{1}{T_\text{r}}\frac{\partial}{\partial T}a = -\text{i}D\frac{\partial^2}{\partial t^2}a + \text{i}\delta|a|^2a + (g-\ell)a + D + \frac{\partial^2}{\partial t^2}a + \gamma|a|^2a + L_{\text{pert}} \,. \quad (34)$$

Due to the irreversible processes and the perturbations, the solution at (34) is a soliton-like pulse with perturbations

$$a(t,T) = \left[A_0 \operatorname{sech}\left(\frac{t-t_0}{\tau}\right) + \Delta a(T, t-t_0)\right] \text{e}^{\phi_\text{s}T/T_\text{r} + \text{i}\theta_0} \,, \quad (35)$$

with soliton phase shift

$$\phi_\text{s} = \frac{1}{2}\delta A_0^2 = \frac{|D|}{\tau^2} \,, \quad (36)$$

and pulse energy $W_0 = 2A_0^2\tau$.

The perturbations cause fluctuations in amplitude, phase, center frequency, and timing of the soliton and generate possible background radiation, i.e., continuum

$$\Delta a(T, t-t_0) = \Delta W(T) f_W(t-t_0) + \Delta\theta(T) f_\theta(t-t_0) + \Delta\omega(T) f_\omega(t-t_0) + \Delta t(T) f_\text{t}(t-t_0) + \Delta a_\text{c}(T, t-t_0) \,, \quad (37)$$

where the basis functions for the perturbations in energy phase, frequency and timing can be derived by building proper derivatives of the steady-state solutions (for more details see [47]):

$$f_W(t) = \frac{1}{W_0}\left[1 - \frac{t}{\tau}\tanh\left(\frac{t}{\tau}\right)\right] a_\text{s}(t) \,, \quad (38)$$

$$f_\theta(t) = \text{i}a_\text{s}(t) \,, \quad (39)$$

$$f_\omega(t) = \text{i}t a_\text{s}(t) \,, \quad (40)$$

$$f_\text{t}(t) = \frac{1}{\tau}\tanh\left(\frac{t}{\tau}\right) a_\text{s}(t) \,. \quad (41)$$

The dynamics of the pulse parameters, due to the perturbed nonlinear Schrödinger equation (34), can be projected from the perturbation using the adjoint basis and the orthogonality relation,

$$\bar{f}_W(t) = 2a_\mathrm{s}(t)\,, \tag{42}$$

$$\bar{f}_\theta(t) = 2\mathrm{i}\frac{1}{W_0}\left[1 - \frac{t}{\tau}\tanh\left(\frac{t}{\tau}\right)\right] a_\mathrm{s}(t)\,, \tag{43}$$

$$\bar{f}_\omega(t) = \mathrm{i}\frac{2}{W_0\tau}\tanh\left(\frac{t}{\tau}\right) a_\mathrm{s}(t)\,, \tag{44}$$

$$\bar{f}_\mathrm{t}(t) = \frac{2}{W_0} t a_\mathrm{s}(t)\,, \tag{45}$$

$$\mathrm{Re}\left[\int_{-\infty}^{+\infty} \bar{f}_i^*(t) f_j(t)\,\mathrm{d}t\right] = \delta_{i,j}\,, \tag{46}$$

respectively. We obtain

$$\frac{\partial}{\partial T}\Delta W(T) = -\frac{1}{\tau_W}\Delta W + \frac{1}{T_\mathrm{r}}\mathrm{Re}\left[\int_{-\infty}^{+\infty} \bar{f}_W^*(t) L_\mathrm{pert}(T,t)\,\mathrm{d}t\right], \tag{47}$$

$$\frac{\partial}{\partial T}\Delta\theta(T) = \frac{2\phi_\mathrm{s}}{T_\mathrm{r}}\frac{\Delta\omega}{W_0} + \frac{1}{T_\mathrm{r}}\mathrm{Re}\left[\int_{-\infty}^{+\infty} \bar{f}_\theta^*(t) L_\mathrm{pert}(T,t)\,\mathrm{d}t\right], \tag{48}$$

$$\frac{\partial}{\partial T}\Delta\omega(T) = -\frac{1}{\tau_\omega} + \frac{1}{T_\mathrm{r}}\mathrm{Re}\left[\int_{-\infty}^{+\infty} \bar{f}_\omega^*(t) L_\mathrm{pert}(T,t)\,\mathrm{d}t\right], \tag{49}$$

$$\frac{\partial}{\partial T}\Delta t(T) = \frac{-2|D|}{T_\mathrm{r}}\Delta\omega + \frac{1}{T_\mathrm{r}}\mathrm{Re}\left[\int_{-\infty}^{+\infty} \bar{f}_\mathrm{t}^*(t) L_\mathrm{pert}(T,t)\,\mathrm{d}t\right], \tag{50}$$

with the normalized decay constants for energy and center frequency fluctuations

$$\frac{1}{\tau_W} = (2g_\mathrm{d} - 2\gamma A_0^2)\,, \tag{51}$$

$$\frac{1}{\tau_\omega} = \frac{4}{3}\frac{g_\mathrm{s}}{\Omega_\mathrm{g}^2\tau^3}\,. \tag{52}$$

Here, g_s is the saturated gain, and g_d is related to the differential gain,

$$g_\mathrm{s} = \frac{g_0}{1 + W_0/W_\mathrm{s}}\,, \tag{53}$$

$$g_\mathrm{d} = \frac{\mathrm{d}g_\mathrm{s}}{\mathrm{d}W_\mathrm{s}} W_0\,. \tag{54}$$

Note, in this model, we assumed that the gain instantaneously follows the intracavity average power or pulse energy, which is not true in general. W_s is the saturation energy of the gain medium. When necessary, it is straightforward to include the relaxation of the gain by adding a dynamic gain model to the perturbation equations. For simplicity, we restrict ourselves to this simple gain model.

4.1 Noise in Mode-Locked Lasers

Within this framework, the response of the laser to noise can easily be included. The spontaneous emission noise due to the amplifying medium, with saturated gain g_s and excess noise factor Θ, leads to additive white noise in the perturbed master equation (34) with $L_{\text{pert}} = \xi(t,T)$, where ξ is a white noise source with autocorrelation function,

$$\langle \xi(t',T')\xi(t,T)\rangle = T_r^2 P_n \delta(t-t')\delta(T-T'), \tag{55}$$

where the spontaneous emission noise energy $P_n \cdot T_r$ with

$$P_n = \Theta \frac{2g_s}{T_r}\hbar\omega_c = \Theta \frac{\hbar\omega_c}{\tau_c} \tag{56}$$

is added to the pulse within each round-trip in the laser. Here, τ_c is the cavity decay time, and Θ is the excess noise factor of the gain medium. Projecting the equations of motion for the pulse parameters in the presence of this noise according to (37–50), we obtain the following power spectra for the amplitude and center frequency fluctuations, as well as phase and timing jitter due to the amplifier noise:

$$\left|\frac{\Delta W(\Omega)}{W_0}\right|^2 = \frac{4}{1/\tau_W^2 + \Omega^2}\frac{P_n}{W_0}, \tag{57}$$

$$|\Delta\omega(\Omega)\tau|^2 = \frac{1}{1/\tau_\omega^2 + \Omega^2}\frac{4}{3}\frac{P_n}{W_0}, \tag{58}$$

$$|\Delta\theta(\Omega)|^2 = \frac{1}{\Omega^2}\frac{4}{3}\left(1+\frac{\pi^2}{12}\right)\frac{P_n}{W_0} + \frac{16}{(1/\tau_W^2+\Omega^2)}\frac{\phi_s^2}{T_r^2}\frac{P_n}{W_0}, \tag{59}$$

$$\left|\frac{\Delta t(\Omega)}{\tau}\right|^2 = \frac{1}{\Omega^2}\frac{\pi^2}{3}\frac{P_n}{W_0} + \frac{1}{\Omega^2(1/\tau_\omega^2+\Omega^2)}\frac{4}{3}\frac{4|D|^2}{T_r^2\tau^4}\frac{P_n}{W_0}. \tag{60}$$

These equations indicate, that the energy and center frequency fluctuations become stationary with mean-square fluctuations

$$\left\langle \left(\frac{\Delta W}{W_0}\right)^2 \right\rangle = 2\frac{P_n\tau_W}{W_0}, \tag{61}$$

$$\langle(\Delta\omega\tau)^2\rangle = \frac{2}{3}\frac{P_n\tau_\omega}{W_0}, \tag{62}$$

whereas the phase and timing jitter undergo a random walk by following the rules,

$$\sigma_\theta(T) = \left\langle [\Delta\theta(T) - \Delta\theta(0)]^2 \right\rangle$$
$$= \frac{4}{3}\left(1 + \frac{\pi^2}{12}\right)\frac{P_\mathrm{n}}{W_0}|T| + 16\frac{\phi_\mathrm{s}^2}{T_\mathrm{r}^2}\frac{P_\mathrm{n}}{W_0}\tau_W^3\left[\exp\left(-\frac{|T|}{\tau_W}\right) - 1 + \frac{|T|}{\tau_W}\right], \quad (63)$$

$$\sigma_\mathrm{t}(T) = \left\langle \left(\frac{\Delta t(T) - \Delta t(0)}{\tau}\right)^2 \right\rangle$$
$$= \frac{\pi^2}{3}\frac{P_\mathrm{n}}{W_0}|T| + \frac{4}{3}\frac{4|D|^2}{T_\mathrm{r}^2\tau^4}\frac{P_\mathrm{n}}{W_0}\tau_\omega^3\left[\exp\left(-\frac{|T|}{\tau_\omega}\right) - 1 + \frac{|T|}{\tau_\omega}\right]. \quad (64)$$

Thus, the spontaneous emission leads to diffusion processes of phase and pulse timing. The phase noise produces the fundamental finite width of every line of the comb in the optical domain. The timing jitter leads to a finite linewidth of the detected microwave signal, which is equivalent to the laser's fundamental fluctuations in repetition rate.

4.1.1 The Optical Spectrum

Taking only the phase and timing jitter into account, the optical field emitted by a noisy laser is proportional to

$$A(t) = \sum_{m=-\infty}^{+\infty} a\left[t - mT_\mathrm{r} - \Delta t(mT_\mathrm{r})\right]$$
$$\cdot \exp\left[-\mathrm{i}\omega_\mathrm{c}(t - mT_\mathrm{r})\right]\exp\left[-\mathrm{i}\Delta\theta(mT_\mathrm{r}) - \mathrm{i}\Phi_\mathrm{CE}(mT_\mathrm{r})\right], \quad (65)$$

with the carrier–envelope phase of the mth pulse defined by

$$\Phi_\mathrm{CE}(mT_\mathrm{r}) = \omega_\mathrm{c}\left(1 - \frac{v_\mathrm{g}}{v_\mathrm{p}}\right)mT_\mathrm{r} - \phi_\mathrm{s}mT_\mathrm{r} + \theta_0 = \omega_\mathrm{CE}mT_\mathrm{r} + \theta_0. \quad (66)$$

The carrier–envelope phase shift from pulse to pulse in the pulse train causes a frequency shift, the carrier–envelope offset frequency

$$\omega_\mathrm{CE} = \omega_\mathrm{c}\left(1 - \frac{v_\mathrm{g}}{v_\mathrm{p}}\right) - \phi_\mathrm{s}. \quad (67)$$

Note that ϕ_s is the nonlinear phase shift of the steady-state pulse per round-trip, and v_g and v_p are the average group and phase velocity of the pulse per round-trip, respectively. The nonlinear phase shift is actually a modification of the phase velocity of the pulse due to the nonlinearity. As we will encounter later, there is also a modification of the group velocity of the pulse due to nonlinearity.

A deterministic shift of the pulse envelope Δt per round-trip leads to a change in the group velocity of the pulse and in the repetition rate of the laser.

We assume a stationary process, so that the optical power spectrum can be computed from

$$S(\omega) = \lim_{\substack{T=N \\ T_r \to \infty}} \frac{1}{2T} \langle \hat{A}_T^*(\omega) \hat{A}_T(\omega) \rangle \tag{68}$$

with the spectra defined over a finite time interval

$$\begin{aligned} \hat{A}_T(\omega) &= \int_{-T}^{T} A(t) e^{i\omega t} \, dt \\ &= \hat{a}(\omega - \omega_c) \sum_{m=-N}^{N} e^{-i\Phi_{CE}(mT_r) - \omega m T_r} e^{-i[(\omega-\omega_c)\Delta t(mT_r) - \Delta\theta(mT_r)]} , \end{aligned} \tag{69}$$

where $\hat{a}(\omega)$ is the Fourier transform of the steady-state pulse

$$\hat{a}(\omega) = \int_{-\infty}^{\infty} a(t) e^{i\omega t} \, dt . \tag{70}$$

From (68), the optical spectrum of the laser using the statistical independence of the phase and timing jitter follows,

$$\begin{aligned} S(\omega) = \lim_{N \to \infty} & \frac{|\hat{a}(\omega - \omega_c)|^2}{2NT_r} \sum_{m'=-N}^{N} \sum_{m=-N}^{N} e^{i(\omega + \omega_{CE}) T_r (m - m')} \\ & \cdot \left\langle e^{-i(\omega-\omega_c)\left[\Delta t(mT_r) - \Delta t(m'T_r)\right]} \right\rangle \left\langle e^{i\left[\Delta\theta(mT_r) - \Delta\theta(m'T_r)\right]} \right\rangle . \end{aligned} \tag{71}$$

The stationarity of the noise processes ensures that the phase and timing differences depend only on $k = m - m'$. Then, the sum over the index m in (71) can be carried out in the limit of large N, and we obtain

$$\begin{aligned} S(\omega) = & \frac{|\hat{a}(\omega - \omega_c)|^2}{T_r} \sum_{k=-\infty}^{\infty} e^{i(\omega + \omega_{CE}) T_r k} \\ & \cdot \left\langle e^{-i(\omega-\omega_c)[\Delta t(kT_r) - \Delta t(0)]} \right\rangle \cdot \left\langle e^{i[\Delta\theta(kT_r) - \Delta\theta(0)]} \right\rangle . \end{aligned} \tag{72}$$

The expectation values in (72) are averages of a complex phasor $e^{i\phi}$ with a Gaussian-distributed phase. It is well known that this average decays exponentially with the square of the variance of the phase, which leads to

$$\langle e^{-i\phi} \rangle = e^{-\frac{1}{2}\sigma_\phi} \tag{73}$$

with

$$\sigma_\phi = \langle \phi^2 \rangle . \tag{74}$$

With (72) and (73), by summing equal terms in the limit of large N,

$$S(\omega) = \frac{|\hat{a}(\omega - \omega_c)|^2}{2\pi T_r} \cdot \sum_{k=-\infty}^{\infty} e^{i(\omega+\omega_{CE})T_r k} e^{-\frac{1}{2}\sigma_\theta(kT_r)} e^{-\frac{1}{2}\{[(\omega-\omega_c)\tau]^2\sigma_t(kT_r)\}} . \quad (75)$$

Frequently we are interested in the noise very close to the lines, i.e., at frequency offsets much smaller than the inverse energy and frequency relaxation times τ_W and τ_ω. This part of the spectrum is determined by the long-term behavior of the variances of timing and phase, which grow linearly with time T:

$$\sigma_\theta(T) = \frac{4}{3}\left(1 + \frac{\pi^2}{12} + 16\phi_s^2 \frac{\tau_W^2}{T_r^2}\right) \frac{P_n}{W_0} |T| = 2\Delta\omega_\Phi |T| , \quad (76)$$

$$\sigma_t(T) = \frac{1}{3}\left(\pi^2 + 16\frac{|D|^2}{\tau^4} \frac{\tau_\omega^2}{T_r^2}\right) \frac{P_n}{W_0} |T| = 2\Delta\omega_t |T| , \quad (77)$$

where $\Delta\omega_\Phi$ and $\Delta\omega_t$ are related to the optical and microwave linewidth of the laser process, as we will see below. Applying Poisson's sum formula

$$\sum_{k=-\infty}^{+\infty} h(k) e^{ikx} = \sum_{n=-\infty}^{+\infty} G(x + 2n\pi) \quad (78)$$

with

$$G(x) = \int_{-\infty}^{+\infty} h(k) e^{ikx} \, dk \quad (79)$$

to (75), we finally arrive at the optical line spectrum of the mode-locked laser,

$$S(\omega) = \frac{|\hat{a}(\omega - \omega_c)|^2}{T_r^2} \sum_{n=-\infty}^{+\infty} \frac{2\Delta\omega_n}{(\omega - \omega_n)^2 + \Delta\omega_n^2} , \quad (80)$$

which are Lorentzian lines at the comb positions

$$\omega_n = \omega_{CE} + n\omega_r \quad (81)$$

with ω_{CE} according to (67) and a half-width at half-maximum of

$$\Delta\omega_n = \Delta\omega_\phi + [\tau(\omega_n - \omega_c)]^2 \Delta\omega_t . \quad (82)$$

Estimating the number of modes M included in the comb by

$$M = \frac{T_r}{\tau} , \quad (83)$$

we notice that the contribution of the timing fluctuations to the linewidth of the comb lines is negligible in the center of the comb. The timing jitter may lead to additional line broadening in the wings of the comb. The linewidth in the comb center is given by (76):

$$\Delta\omega_{\Phi} = \left[\frac{2}{3}\left(1+\frac{\pi^2}{12}\right) + 16\phi_s^2\frac{\tau_W^2}{T_r^2}\right]\frac{\Theta}{N_0\tau_c}, \tag{84}$$

where $N_0 = w_0/\hbar\omega_c$ is the number of photons in the cavity and τ_c is the cavity decay time. Note that this result for a mode-locked laser is close to the Schawlow–Townes linewidth of a single-mode laser, which is $\Delta\omega = \Theta/(N_0\tau_c)$. For a solid-state laser approximately 1 μm with the following parameters: an intracavity pulse energy of 50 nJ corresponding to $N_0 = 2.5\times 10^{11}$ photons, 100 MHz repetition rate, an output coupler of 10% and an excess noise figure of $\Theta = 2$, a linewidth of $\Delta f_\phi = 8$ μHz without any amplitude to phase conversion, i.e., $\phi_s = 0$, is obtained from (84). These intrinsic fluctuations occur on a timescale much shorter than the cavity round-trip time. For sub-10 fs lasers, the spectra fill the full gain bandwidth, and KLM is rather strong. As a result, the amplitude and center frequency relaxation times are on the order of 10 to 100 cavity round-trips, and the nonlinear phase shift, ϕ_0, approaches one radian per round-trip. Thus, most fluctuations are due to amplitude fluctuations that are converted into phase jitter. This contribution may increase the linewidth by a factor of 100 to 10 000. If the amplitude fluctuations are not damped fast enough, e.g., by feedback, the amplitude-to-phase conversion may bring the linewidth to the hertz level.

4.1.2 The Microwave Spectrum

Photodetection of an ultrafast pulse train converts optical power into microwave power. In this way, highly precise microwave signals can be generated from the optical mode comb. The linewidth of the microwave comb is of interest. Again, neglecting amplitude noise, we find from (65) for the photodetector current,

$$\begin{aligned} I(t) &= \eta\frac{e}{\hbar\omega_c}\left|a(T,t)\right|^2 \\ &= \eta\frac{eW_0}{\hbar\omega_c\tau}\sum_{m=-\infty}^{+\infty}\frac{1}{2}\operatorname{sech}^2\left(\frac{t-mT_r-\Delta t(mT_r)}{\tau}\right), \end{aligned} \tag{85}$$

where η is the quantum efficiency. For simplicity, we neglect the amplitude noise and consider only the consequences due to the timing jitter. Then, for the Fourier transform of the photocurrent, we obtain

$$\hat{I}_T(\omega) = \eta\frac{eW_0}{\hbar\omega_c}\left|a_0\right|^2(\omega)\sum_{m=-N}^{+N} e^{i\omega[mT_r+\Delta t(mT_r)]} \tag{86}$$

with

$$|a_0|^2(\omega) = \int_{-\infty}^{\infty} \frac{1}{2} \operatorname{sech}^2(x) e^{i\omega\tau x} \, dx \tag{87}$$

and its power spectrum according to (68)

$$\begin{aligned} S_I(\omega) &= \frac{(\eta e N_0)^2}{T_r} \left| |a_0|^2(\omega) \right|^2 \sum_{k=-\infty}^{+\infty} e^{i\omega k T_r} \left\langle e^{i\omega[\Delta t(kT_r) - \Delta t(0)]} \right\rangle \\ &= \frac{(\eta e N_0)^2}{T_r} \left| |a_0|^2(\omega) \right|^2 \sum_{k=-\infty}^{+\infty} e^{i\omega k T_r} e^{-\frac{1}{2}\left[(\omega\tau)^2 \sigma_t(kT_r)\right]}, \end{aligned} \tag{88}$$

where N_0 is again the photon number. Using the Poisson formula results in

$$S_I(\omega) = \frac{(\eta e N_0)^2}{\pi T_r^2} \left| |a_0|^2(\omega) \right|^2 \sum_{n=-\infty}^{+\infty} \frac{\Delta\omega_{I,n}}{(\omega - n\omega_r)^2 + \Delta\omega_{I,n}^2} \tag{89}$$

with the linewidth $\Delta\omega_{I,n}$ of the nth harmonic

$$\Delta\omega_{I,n} = \left(2\pi n \frac{\tau}{T_r}\right)^2 \Delta\omega_t = \left(\frac{2\pi n}{M}\right)^2 \Delta\omega_t \tag{90}$$

with

$$\Delta\omega_t = \frac{1}{6}\left(\pi^2 + \frac{|D|^2}{\tau^4} \frac{\tau_\omega^2}{T_r^2}\right) \frac{\Theta}{N_0 \tau_c} . \tag{91}$$

The fundamental line ($n = 1$) of the microwave spectrum has a width on the order M^2 times smaller than the optical linewidth. For a 10 fs laser with a 100 MHz repetition rate, the number of modes M is approximately a million. This result indicates that a perfectly stabilized femtosecond laser may have a microwave linewidth of the order of $\Delta f_t = 10^{-17}$ Hz.

These rather small optical and microwave linewidths are difficult to observe because they are swamped by technical noise such as fluctuations in pump power, which may cause gain fluctuations, mirror vibrations, air-density fluctuations and thermal drifts, which directly cause changes in the laser's repetition rate. However, these noise sources can be greatly reduced in principle by careful design and feedback stabilization.

One way to gain control over the optical phase and the timing of the pulses is by direct manipulation of the group and phase velocities of the intracavity pulse. Therefore, it is important to understand the dependence of the group and phase velocities on the intracavity power or pulse energy, at least within the current basic model. Additional linear and nonlinear effects due to higher order linear dispersion or nonlinearities may cause additional changes in group and phase velocities, which might also create unusual dependencies of group and phase velocities on intracavity pulse energy. Here we discuss, as an example, the impact of the instantaneous Kerr effect on the group and phase velocities of a pulse.

4.2 Group- Versus Phase- Velocity of Solitons and Control of the Mode Comb

The Kerr effect leads to a change in the phase velocity of a pulse, thus resulting in a self-phase shift of the soliton, ϕ_s, per round-trip. A change in group velocity does not appear explicitly in the solution of the NLSE. Self-steepening, which becomes important for ultrashort pulses, leads to an additional term in the NLSE and, therefore, to an additional term in the master equation (92)

$$L_{\text{pert}} = -\frac{\delta}{\omega_c}\frac{\partial}{\partial t}\left[|a(T,t)|^2 a(T,t)\right] . \tag{92}$$

The impact of this term is expected to be small, on the order of $1/(\omega_c \tau)$, and, therefore, only important for few-cycle pulses. However, this term alters the group velocity of a soliton-like pulse twice as strongly as the nonlinear phase shift alters the phase velocity. We take this term into account in form of a perturbation. This perturbation generates an odd and real contribution in the master equation and therefore leads only to a timing shift, when substituted in (50):

$$T_r \frac{\partial \Delta t(T)}{\partial T}\bigg|_{\text{self-steepening}} = \frac{\delta}{\omega_c} A_0^2 = \frac{2\phi_s}{\omega_c} . \tag{93}$$

This timing shift or group delay per round-trip, together with the nonlinear phase shift and linear dispersion, leads to a phase change between carrier and envelope per round-trip given by

$$\begin{aligned}
\Delta\phi_{\text{CE}} &= -\phi_s + \omega_c T_r\left(1 - \frac{v_g}{v_p}\right) + \omega_c\, T_r \frac{\partial}{\partial T}\Delta t(T)\bigg|_{\text{self-steepening}} \\
&= -\frac{1}{2}\delta A_0^2 + \delta A_0^2 + \omega_c T_r\left(1 - \frac{v_g}{v_p}\right) \\
&= \frac{1}{2}\delta A_0^2 + \omega_c T_r\left(1 - \frac{v_g}{v_p}\right) .
\end{aligned} \tag{94}$$

The compound effect of this phase delay per round-trip in the carrier versus envelope leads to a new overall carrier–envelope frequency,

$$\omega_{\text{CE}} = \frac{\Delta\phi_{\text{CE}}}{T_r} = \phi_s f_r + \omega_c\left(1 - \frac{v_g}{v_p}\right) . \tag{95}$$

Note that the group delay due to self-steepening reverses the sign of the carrier–envelope frequency shift as a function of the nonlinear phase shift. The group delay also changes the optical cavity length of the laser and, therefore, alters the repetition rate according to

$$\Delta f_r = -f_r^2 \Delta t(T)\big|_{\text{self-steepening}} = -2\phi_s \frac{f_r}{\omega_c} f_r = -\frac{2\phi_s}{2\pi m_0} f_r , \tag{96}$$

where $m_0 = \omega_c/\omega_r$ is the mode number of the carrier wave. Equations (95) and (96) together determine the shift of the mth line of the optical comb, $f_m = f_{CE} + mf_R$, due to an intracavity pulse energy modulation assuming an energy-independent carrier frequency. In addition, allowing for a change in cavity length leads to a total change in the mth mode frequency by

$$\Delta f_m = \Delta f_{CE} + m\Delta f_r = \frac{\phi_s}{2\pi} f_r \left(1 - \frac{2m}{m_0}\right) \frac{\Delta W}{W_0} - m f_r \frac{\Delta L}{L_0} \tag{97}$$

due to an incremental energy and cavity length change. If the carrier frequency is energy-dependent, too, then the second term in (96) also must be taken into account; see [50]. This formula can be used to control an optical comb by regulating the cavity length and the intracavity pulse energy through the pump power. This will be explored in more detail in the Chapters by *Udem* et al. and *Jones* et al.

Specifically, (97) predicts that the mode with number $m = m_0/2$, i.e., the mode at half the center frequency, does not change its frequency as a function of intracavity pulse energy due to the interplay of the intensity-dependent group and the phase velocities of soliton dynamics in the presence of an instantaneous Kerr effect.

One must remember that this model is based on self-phase modulation and that self-steepening comes from an instantaneous Kerr effect. There may be other mechanisms that cause a power-dependent carrier envelope offset frequency. One such effect is the group delay caused by the laser gain medium; another is the carrier–envelope phase change due to a change in carrier frequency, which gives strong additional dependence on pump power [50].

5 Technical Challenges

To realize the strongly idealized and, so far, purely temporal pulse dynamics discussed in the previous sections, the proper saturable absorbers and laser optics must be provided. The technical challenge to design and fabricate these important components determines how short a pulse can be supported. As mentioned in the introduction, the fast saturable absorber mode-locking model was not employed until the late 1980s and the early 1990s because of the lack of a suitable fast saturable absorber and the success of dye lasers. The discovery of artificial fast saturable absorber mechanisms realized in additive pulse mode locking and Kerr-lens mode locking revolutionized short-pulse generation. The latter technique is used for few-cycle pulse generation.

5.1 Kerr-Lens Mode Locking

In lasers with free-space propagation within a Fabry–Pérot resonator, the Kerr effect can be used to produce intensity-dependent focusing. This way of producing an artificial fast saturable absorber is called Kerr-lens mode

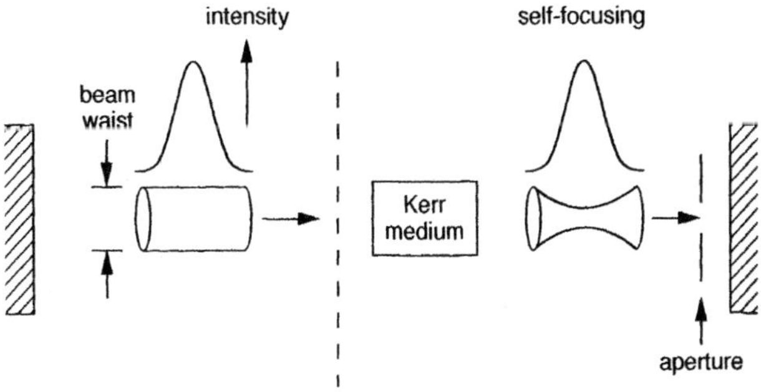

Fig. 14. Artificial saturable absorber realized with Kerr-lens mode locking

locking (KLM). As mentioned earlier, KLM was first accomplished by *Sibbett* and his group [16].

The Kerr effect imposes a spatially varying phase profile upon the beam during propagation through the gain crystal (Fig. 14). This phase profile leads to intensity-dependent beam diameter variations in a Fabry–Pérot resonator with spherical mirror(s) that has been designed to operate close to its instability regime. If the beam diameter is decreased at high intensities within a hard aperture or a spatially varying gain profile, such as that produced by a Gaussian pump beam, the high intensities experience less loss or equivalently higher gain. This reduction of loss or enhancement of gain with increasing intensity is equivalent to saturable absorber action (Fig. 14).

The proper resonator design for optimum KLM has received a great deal of attention and is ongoing. The earliest approaches toward an understanding and optimization of the self-focusing dynamics in a KLM laser were based on an *ABCD*- or *q*-parameter analysis, by modeling the Kerr lens by an intensity-dependent lens, and by using an iterative solution scheme [51,52,53]. Later, other effects like gain guiding were included [54, 55]. More detailed investigations followed. *Kalosha* et al. [56] simulated laser dynamics using a Gaussian approximation for the transverse beam profile, and recently, complete numerical simulations of the full spatiotemporal dynamics of a KLM laser have also been presented [41, 43]. These calculations are time-consuming, since the steady-state solution of the laser can be found only by a lengthy simulation of the pulse dynamics until the steady state is reached. It is then very difficult to map out the complete parameter range of the laser system. Therefore, for an initial computational resonator design and KLM optimization, the *ABCD*- or *q*-parameter analysis along the lines of [53,54,55] already results in valuable design guidelines.

Two approaches can be taken to achieve a saturable absorber effect due to self-focusing in a laser crystal. The first is called hard-aperture KLM. Figure 15 shows the beam radius as a function of intracavity distance for an asymmetrical z-folded resonator for CW-operation $P = 0$, and an intracavity power of $P = 1.5\,\mathrm{MW}$. An aperture placed in the long arm of the resonator

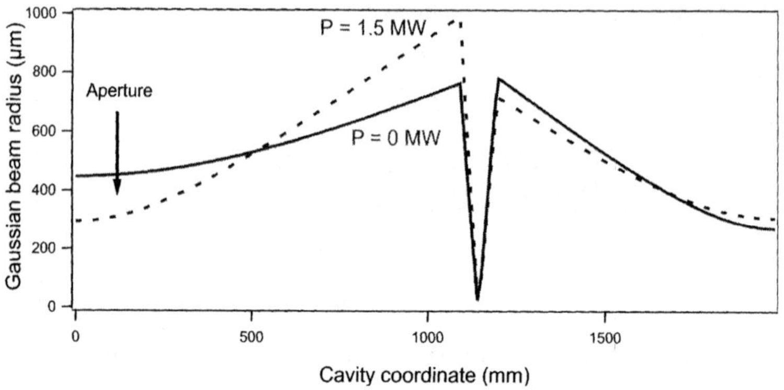

Fig. 15. Example of hard-aperture KLM. Using the *ABCD* model from [53], the intracavity mode of an asymmetrical z-folded resonator (see, e.g., [32]) is calculated for two different pulse peak powers $P = 0$ and 1.5 MW (*dashed*). An aperture placed in the long arm of the cavity would favor the high-power beam

results in an intensity-dependent intracavity loss modulation. The variation in beam waist can become arbitrarily large, depending on how close to its stability limit the cavity is operated.

The second approach is called soft-aperture KLM. In this case, the change in beam waist at the position of the gain medium is exploited to enhance the mode overlap with the pump laser in the laser crystal and, therefore, to enhance the round-trip gain of the laser mode (Fig. 16). The subject of current investigations is the influence of cavity misalignment. In the low-power case (CW or in the wings of the pulse), the intracavity diffraction losses are high due to misalignment. With the increasing power and development of the Kerr lens, the mode shape changes, and diffraction losses are reduced. As a consequence, in all cases due to the lower loss/higher gain in the pulse peak, the pulse shortens in each round-trip.

KLM benefits greatly from the ultrafast response time of the nonresonant Kerr effect, which is of the order of a few femtoseconds, allowing the index to follow a femtosecond laser pulse instantaneously [57]. This supports the classification of KLM as a fast artificial saturable absorber.

The simulation results shown in Fig. 15 and Fig. 16 indicate that considerable waist changes can be achieved by the Kerr-lens effect, and artificial saturable absorbers modulating intracavity gain and loss by several tens of percent can be achieved. A large absorber modulation is important for two reasons. First, in the fast saturable absorber model, the achievable pulsewidth is inversely proportional to the square root of the modulation depth (10). Secondly, nonideal behavior of some laser optical components that impose loss on the pulse, in contrast to a CW-solution may be overcome by strong saturable absorber action. This includes nonperfect dispersion compensation, residual GTI effects from a nonperfect dispersion compensating mirror, additional fil-

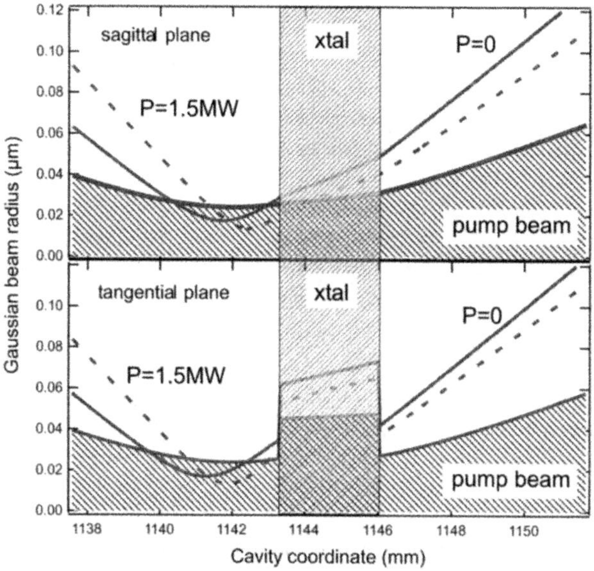

Fig. 16. Example of soft-aperture KLM. The intracavity mode of the resonator from Fig. 15 is shown for two different pulse peak powers $P = 0$ and 1.5 MW (*dashed*), in the *top* for the sagittal plane and in the *bottom* plot for the tangential plane of the Brewster oriented crystal. With increasing power, the overlap of the laser mode with the pump mode in the crystal increases (*shaded area*)

ter effects by roll-off of the output coupler, and other effects that may limit short-pulse generation.

For example, Fig. 17 shows the required KLM modulation depth to overcome the etalon effects introduced into a laser system by double-chirped mirrors with nonperfect AR coatings. The AR coating is characterized by a residual amplitude reflectivity r or power reflectivity r^2 (see Sect. 5.2). The simulations are based on a completely temporal model that is analogous to the simulations presented in Sect. 3.2. The KLM is modeled by a fast saturable absorber with a maximum KLM modulation depth q_0. The nonperfect AR coating generates pre- and post-pulses for each bounce on one of the DCMs. If several bounces on the mirrors occur before the pulse passes the nonlinear gain medium again, which changes the pulse phase with respect to the background, and KLM action is reducing the background, these reflections add up coherently. If the KLM action is not strong enough to reduce this background faster than it is fed by the residual reflections on the mirror, a stable periodic pulse drain cannot emerge. Figure 17 shows the tolerance of the laser against DCM mirror quality, which is quantified by the residual reflectivity of the front AR coating r^2 as a function of the KLM modulation depth q_0.

It is not surprising, and supported by the experimental results presented

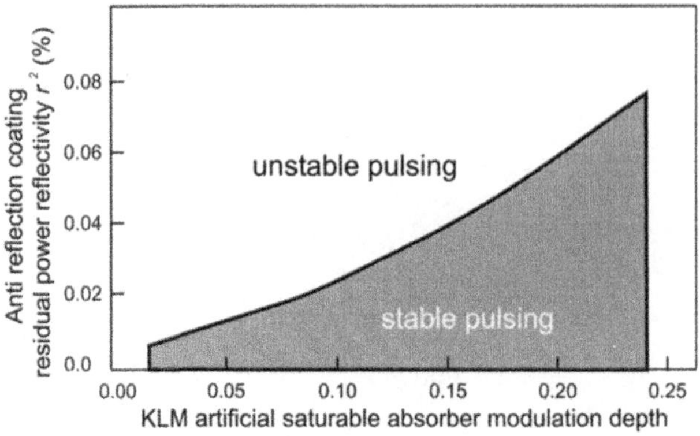

Fig. 17. Tolerance of the laser against DCM mirror quality quantified by the residual reflectivity of the front AR coating r as a function of the KLM modulation depth

in Sect. 6, that stable mode-locked operation with high bandwidth mirrors and, thus with high r^2, can be achieved only with a strong KLM modulation depth.

Though the purely temporal dynamics of a KLM laser is accessible to simulation if the KLM is modeled as a simple saturable absorber, estimation and optimization of the actual intracavity gain/loss modulation depth for a given system are challenging, because they rely on the spatiotemporal complex dynamics of the KLM laser. Full spatiotemporal computer simulations are too time-consuming to allow optimization of laser performance, [41, 43]. Since it appears that dispersion-managed soliton formation is the most important pulseshaping process in today's ultrashort-pulse KLM lasers, it is interesting to extend this model to the spatial domain [58, 59]. Spatiotemporal solitary steady-state solutions can be found taking only dispersion, self-phase modulation, diffraction, and self-focusing into account. As for dispersion-managed solitons, such solutions can be found in a Gaussian or fully numerical approximation. For the temporal and spectral pulseshape, the Gaussian approximation proved appropriate for pulse durations as short as approximately 10 fs [58, 59].

5.2 Dispersion Compensating Mirrors

The generation of few-cycle pulses by external compression [60, 61, 62], as well as direct generation from Kerr-lens mode-locked lasers [28, 30], relies heavily on the availability of chirped mirrors [27, 29, 63] for dispersion compensation. There are two reasons to employ chirped mirrors. First, the high-reflectivity bandwidth, Δf, of a standard dielectric Bragg mirror is determined by the Fresnel reflectivity r_{B} of the high, n_{H}, and low, n_{L}, index materials used for the dielectric mirror

$$r_{\mathrm{B}} = \frac{\Delta f}{f_{\mathrm{c}}} = \frac{n_{\mathrm{H}} - n_{\mathrm{L}}}{n_{\mathrm{H}} + n_{\mathrm{L}}}, \tag{98}$$

where f_c is again the center frequency of the mirror. In general, metal mirrors are too lossy, especially when used as intracavity laser mirrors. For material systems typically used for broadband optical coatings, such as silicon dioxide and titanium dioxide with $n_{\mathrm{SiO_2}} = 1.48$ and $n_{\mathrm{TiO_2}} = 2.4$, (these indexes might vary depending, on the deposition technique used), a fractional bandwidth $\Delta f / f_c = 0.23$ can be covered. This fractional bandwidth is only about one-third of an octave-spanning mirror $\Delta f / f_c = 2/3$. Furthermore, the variation in group delay of a Bragg mirror already impacts pulses that fill half the spectral range $\Delta f = 0.23 f_c$. A solution to this dilemma was found by introducing chirped mirrors [27], the equivalent of chirped fiber Bragg gratings, which were already well-developed components in fiber optics [64]. When the Bragg wavelength of the mirror stack is varied slowly enough, and no limitation on the number of layer pairs exists, an arbitrarily high reflectivity range of the mirror can be engineered. The second reason for using chirped mirrors is based on their dispersive properties due to the wavelength-dependent penetration depth of the light reflected from different positions inside the chirped multilayer structure. Mirrors are filters, and, in the design of any filter, the control of group delay and group delay dispersion is difficult. This problem increases when the design must operate over wavelength ranges up to an octave or more.

5.2.1 The Matching Problem

Several designs for ultrabroadband dispersion-compensating mirrors have been developed over the last few years. For dispersion-compensating mirrors that do not extend the high-reflectivity range far beyond that of a Bragg mirror employing the same materials can already achieve, a multicavity filter design can be used to approximate the desired phase and amplitude properties [65,66]. For dispersion-compensating mirrors that cover a high-reflectivity range of up to $\Delta f / f_c = 0.4$, the concept of double-chirped mirrors (DCMs) has been developed [29,67]. It is based on the following observations. A simple chirped mirror provides high reflectivity over an arbitrary wavelength range and, within certain limits, a custom, designable average group delay achieved through its wavelength-dependent penetration depth [68] (Fig. 18a,b). However, the group delay as a function of frequency shows periodic variations due to the impedance mismatch between the ambient medium and the mirror stack, as well as within the stack (Fig. 18b and Fig. 19). A structure that mitigates these mismatches and gives better control of the group delay dispersion (GDD) is the double-chirped mirror (DCM) (Fig. 18c), in a way similar to that of an apodized fiber Bragg grating [69].

Figure 19 shows the reflectivity and group delay of several Bragg and chirped mirrors composed of 25 index steps, with $n_{\mathrm{H}} = 2.5$ and $n_{\mathrm{L}} = 1.5$,

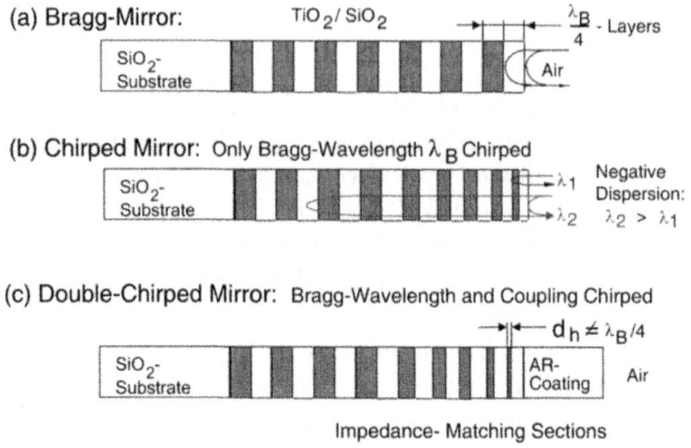

Fig. 18. (**a**) Standard Bragg mirror. (**b**) Simple chirped mirror. (**c**) Double-chirped mirror with matching sections to avoid residual reflections that causes undesired oscillations in the GD and GDD of the mirror

similar to the refractive indexes of TiO_2 and SiO_2, which result in a Fresnel reflectivity of $r_B = 0.25$. The Bragg mirror can be decomposed in symmetrical index steps [29]. The Bragg wave number is defined as $k_B = \pi/(n_L d_L + n_H d_H)$, where d_L and d_H are the thicknesses of the low and high index layer, respectively. The Bragg wave number describes the center wave number of a Bragg mirror composed of equal index steps. In the first case, (Fig. 19, dash-dotted line) only the Bragg wave number is linearly chirped from $6.8\,\mu m^{-1} < k_B < 11\,\mu m^{-1}$ over the first 20 index steps and is held constant over the last 5 index steps. The reflectivity of the structure is computed by assuming the structure imbedded in the low-index medium. The large oscillations in group delay are caused by the different impedances of the chirped grating and the surrounding low-index material causing a strong reflection at the interface of the low-index material and the grating stack. By adiabatic matching of the grating impedance to the low index material, this reflection can be avoided. This is demonstrated in Fig. 19 by the dashed and solid curves, which corresponds to additional chirping of the high index layer over the first 12 steps according to the law $d_H = (m/12)^\alpha \lambda_{B,12}/(4n_H)$ with $\alpha = 1$ and 2 for linear and quadratic adiabatic matching. The argument m denotes the mth index step, and $\lambda_{B,12} = 0.740\,\mu m$.

The strong reduction of the oscillations in the group delay by the double-chirp technique is clearly visible. Quadratic tapering of the high index layer and, therefore, of the grating, already eliminates completely the oscillations in the group delay, which can also be shown analytically by coupled mode analysis [67]. Because of the double chirp, a high transmission window opens at the short wavelength end of the mirror, which is ideally suited for pumping Ti:sapphire lasers. So far, the double-chirped mirror is only matched to the low index material of the mirror. Ideally, the matching can be extended to any

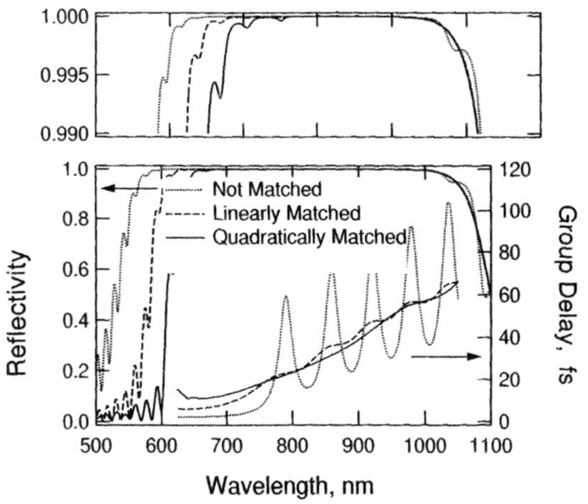

Fig. 19. Comparison of the reflectivity and group delay of chirped mirrors with 25 layer pairs and refractive indexes $n_H = 2.5$ and $n_L = 1.5$. The *upper portion* shows the enlarged top percent of the reflectivity. The *dotted curves* show the result for a simply chirped mirror. The *dashed* and *solid curves* show the result for double-chirped mirrors, where, in addition to the chirp in the Bragg wave number k_B, the thickness of the high index layers is also chirped over the first 12 layer pairs from zero to its maximum value for a linear chirp, i.e., $\alpha = 1$ (*dashed curves*) and for a quadratic chirp, i.e., $\alpha = 2$ (*solid curves*) [29]

other ambient medium by a properly designed AR coating. However, this AR coating must be of very high quality. It must have very low residual reflectivity and ideally, a power reflectivity of 10^{-4}, i.e., an amplitude reflectivity of $r = 10^{-2}$ is required. The quality of the AR coating can be relaxed, if the residual reflection is directed out of the beam path. This is achieved in so-called tilted front-side or back-side coated mirrors [70,71] (Fig. 20a,b). In the back-side coated mirror, the ideal DCM structure, which is matched to the low index material of the mirror, is deposited on the back of the substrate, which is made of the same or at least very similar low index material. The AR coating is deposited on the front of the slightly wedged substrate, so that the residual reflection is directed out of the beam and does not deteriorate the dispersion properties. Thus, the goal of the AR coating is to reduce the Fresnel losses of the mirror at the air–substrate interface. Therefore, it is sufficient for some applications, if the residual reflection at this interface is on the order of 0.5%. However, the substrate must be very thin to keep the overall mirror dispersion negative, typically of the order of 200 μm to 500 μm. Laser-grade optics are difficult to make on such thin substrates, and the stress induced by the coating leads to an undesired deformation of the substrates.

The front-side coated mirror overcomes this shortcoming by depositing an ideal DCM structure that is matched to the index of the wedge material on a regular laser-grade substrate. A 100 μm to 200 μm thin wedge is bonded on top of the mirror, and the AR coating is then deposited on this

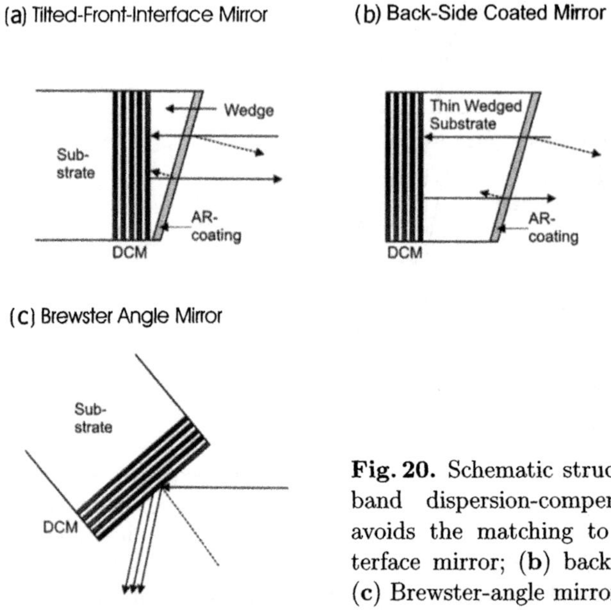

Fig. 20. Schematic structure of proposed broadband dispersion-compensating mirror system avoids the matching to air: (**a**) tilted-front-interface mirror; (**b**) back-side coated mirror and (**c**) Brewster-angle mirror

wedge. This results in stable and octave-spanning mirrors, which have been successfully used in external compression experiments [72]. Both structures have limitations. First, they introduce a wedge into the beam, which leads to an undesired angular dispersion of the beam. This can be partially compensated for by using these mirrors in pairs with oppositely oriented wedges. The second drawback is that it seems to be impossible to make high-quality AR coatings over one, or more than one, octave of bandwidth that have $< 0.5\%$ residual reflectivity [73]. On one reflection such a mirror has at least 1% loss, and, therefore, such mirrors insert high losses inside a laser. For external compression, these losses are acceptable. A third possibility for overcoming the AR-coating problem is to use the ideal DCM under Brewster-angle incidence (Fig. 20) [74]. In that case, the low index layer is automatically matched to ambient air. However, under p-polarized incidence, the index contrast or Fresnel reflectivity of a layer pair is reduced, and more layer pairs are necessary to achieve high reflectivity. The penetration depth into the mirror also increases, so that scattering and other losses in the layers become more pronounced. On the other hand, such a mirror can generate more dispersion per bounce due to the higher penetration depth. For external compression, such mirrors might have advantages because they can cover bandwidths much wider than one octave. This concept is difficult to apply to the fabrication of curved mirrors. There is also a spatial chirp of the reflected beam, which may become sizable for large penetration depth and must be removed by backreflection or an additional bounce on another Brewster-angle mirror that recombines the beam. For intracavity mirrors, a solution is found by using mirror pairs, which cancel the spurious reflections due to an imperfect AR coating and the matching structure in the chirped mirror [33]. This design also has drawbacks and limitations. It requires high-precision fabrication and, depending on the bandwidth of the mirrors, it may be possible to use them only for a restricted range of angles of incidence.

5.2.2 Double-Chirped Mirror Pairs

There have been several proposals to increase the bandwidth of laser mirrors by mutual compensation of GDD oscillations [75, 76, 77] that use computer optimization. These early investigations resulted in a rather low reflectivity of $< 95\%$ over almost half of the bandwidth considered. The ideas leading to the DCMs help to show analytically that a design of DCM pairs covering one octave of bandwidth, i.e., 600 nm to 1200 nm, with high reflectivity and improved dispersion characteristics is possible [33]. The use of these mirror pairs in a Ti:sapphire laser system has resulted in 5 fs pulses with octave-spanning spectra directly from the laser [34]. Yet, the potential of these pairs has not been fully exploited.

A DCM (see Fig. 21a mirror M1) consists of an AR coating and a low-index matched double-chirped back-mirror MB with given wavelength-dependent penetration depth and suppressed spurious reflections. The high reflectivity range of the back-mirror can be easily extended to one octave by simply chirping slowly enough and providing a sufficient number of layer pairs. However, the smoothness of the resulting GDD strongly depends on the quality of matching provided by the AR coating and the double-chirped section. Figure 22 indicates the influence of the AR coating on the GDD of the total DCM structure. The AR coating is represented as a two-port with two incoming waves a_1, b_2 and two outgoing waves a_2, b_1. The connection between the waves at the left port and the right port is described by the transfer matrix

$$\begin{pmatrix} a_1 \\ b_1 \end{pmatrix} = T_{\mathrm{AR}} \begin{pmatrix} a_2 \\ b_2 \end{pmatrix} \quad \text{with} \quad T_{\mathrm{AR}} = \begin{pmatrix} \frac{1}{t} & \frac{r^*}{t^*} \\ \frac{r}{t} & \frac{1}{t^*} \end{pmatrix}, \tag{99}$$

where we assume that the multilayer AR coating is lossless. Here, r and t are the complex coefficients for reflection and transmission at port 1, assuming reflection-free termination of port 2. The back-mirror MB is assumed to be perfectly matched to the first layer in the AR coating, and it has full reflection over the total bandwidth under consideration. Thus, its complex reflectivity in the range of interest is given by

$$\rho_{\mathrm{b}} = \mathrm{e}^{\mathrm{i}\phi_{\mathrm{b}}(\omega)} . \tag{100}$$

The phase $\phi_b(\omega)$ is determined by the desired group delay dispersion

$$\mathrm{GDD_b} = \frac{-\,\mathrm{d}^2\phi_{\mathrm{b}}(\omega)}{\mathrm{d}\omega^2} \tag{101}$$

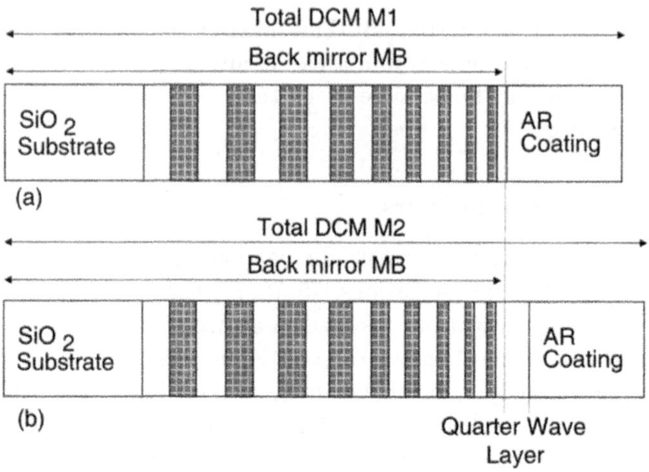

Fig. 21. DCM-Pair M1 (**a**) and M2 (**b**). The DCM M1 can be decomposed into a double-chirped back-mirror MB matched to a medium with the index of the topmost layer. In M2, a layer with a quarter-wave thickness at the center frequency of the mirror and an index equivalent to the topmost layer of the back-mirror MB is inserted between the back-mirror and the AR coating. The new back-mirror comprising the quarter-wave layer can be reoptimized to achieve the same phase as that of MB with an additional π-phase shift over the whole octave of bandwidth

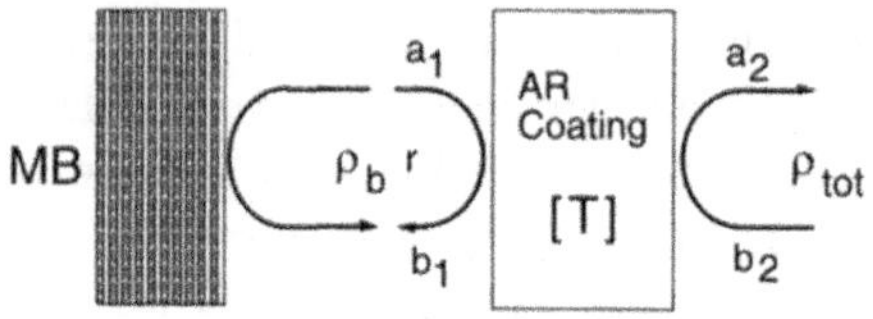

Fig. 22. Decomposition of a DCM into a double-chirped back-mirror MB and an AR coating

up to an undetermined constant phase, ϕ_{b0}, and group delay, $T_{gb0} = \frac{-\mathrm{d}\phi_{\mathrm{b}}(\omega_{\mathrm{c}})}{\mathrm{d}\omega_{\mathrm{c}}}$, at the center frequency of the mirror, ω_{c}. All higher order derivatives of the phase are determined by the desired dispersion of the mirror. Analytic formulas for the design of DCMs that show custom-designed dispersion properties without considering the problem of matching to the ambient air, can be found in [68]. The resulting total mirror reflectivity, including the AR coating, follows from (99):

$$\rho_{\mathrm{tot}} = \frac{t}{t^*}\rho_{\mathrm{b}}\frac{1 - r^*/\rho_{\mathrm{b}}}{1 - r\rho_{\mathrm{b}}} . \tag{102}$$

For the special case of a perfectly reflecting back-mirror according to (100), we obtain

$$\rho_{\mathrm{tot}} = \frac{t}{t^*} \mathrm{e}^{\mathrm{i}\phi_b(\omega)} \frac{1 - z^*}{1 - z}, \quad \text{with} \quad z = r\mathrm{e}^{\mathrm{i}\phi_b(\omega)} . \tag{103}$$

The new reflectivity is again unity, but new contributions in the phase of the resulting reflectivity appear due to the nonperfect transmission properties of the AR coating. With the transmission coefficient of the AR coating

$$t = |t|\mathrm{e}^{\mathrm{i}\phi_{\mathrm{t}}} , \tag{104}$$

the total phase of the reflection coefficient becomes

$$\phi_{\mathrm{tot}} = 2\phi_{\mathrm{t}} + \phi_{\mathrm{b}}(\omega) + \phi_{\mathrm{GTI}} \tag{105}$$

with

$$\phi_{\mathrm{GTI}} = 2 \arctan \left[\frac{\mathrm{Im}(z)}{1 + \mathrm{Re}(z)} \right] . \tag{106}$$

Here, ϕ_{t} is the phase of the transmission coefficient, and ϕ_{GTI} is the phase due to the Gire–Tournois interferometer created by the nonperfect AR coating, i.e., $r \neq 0$, and the back-mirror MB (Fig. 22). The phase ϕ_{t} of a good AR coating, i.e., $|r| < 0.1$, is linear and, therefore, does not introduce undesired oscillations into the GD and GDD. However, the phase ϕ_{GTI} rapidly varies, since $\phi_b(\omega)$ varies by several 2π over the frequency range of interest, due to the monotonous group delay of the back-mirror. The size of these oscillations scale with the quality of the AR coating, i.e., with $|r|$. Thus, the GDD oscillations are reduced with the smaller residual reflectivity of the AR coating. Assuming, that the reflectivity r is real and ≤ 0.1, the oscillations in the group delay and group delay dispersion are easily estimated by

$$T_{\mathrm{g,\,GTI}} = \frac{\mathrm{d}\phi_{\mathrm{GTI}}}{\mathrm{d}\omega} \approx -rT_{\mathrm{gb}}(\omega) \cos[\phi_{\mathrm{b}}(\omega)] \tag{107}$$

with

$$T_{\mathrm{gb}}(\omega) = \frac{-\,\mathrm{d}\phi_{\mathrm{b}}(\omega)}{\mathrm{d}\omega},$$

$$\mathrm{GDD}_{\mathrm{GTI}} = \frac{\mathrm{d}^2\phi_{\mathrm{GTI}}}{\mathrm{d}\omega^2} \approx r \left\{ T_{\mathrm{gb}}^2(\omega) \sin[\phi_{\mathrm{b}}(\omega)] - \mathrm{GDD}_{\mathrm{b}} \cos[\phi_{\mathrm{b}}(\omega)] \right\} . \tag{108}$$

The GTI reflections sum coherently when multiple reflections on chirped mirrors occur inside the laser over one round-trip, thus leading to pre- and postpulses if the mode-locking mechanism is not strong enough to suppress them sufficiently. Experimental results indicate that a residual reflection in the AR coating of $r < 0.01$ and smaller, depending on the number of reflections per round-trip, is required so that the pre- and postpulses are sufficiently suppressed. This corresponds to an AR coating with $< 10^{-4}$ residual power reflectivity, which can be achieved only over a very limited range, as discussed above.

Over a limited wavelength range of 350 nm centered around 800 nm, low residual power reflectivities as small as 10^{-4} have been achieved effectively after reoptimization of the AR coating section and the double-chirped section to form a combined matching section of higher matching quality. For even a larger bandwidth approaching an octave, a residual power reflectivity of 10^{-4} is no longer possible [73]. A solution is offered by the observation that a coherent substraction of the pre- and postpulses to first order in r is possible by reflections on a mirror pair M1 and M2 (Fig. 21a,b). A series of two reflections on a mirror with reflectivity (103) and on a similar mirror with an additional phase shift of π between the AR coating and the back-mirror, having a reflectivity (103) where z is replaced by $-z$, leads to a coherent substraction of the first-order GTI effects. The resulting total reflectivity of the two reflections is given by the product of the individual complex reflectivities, assuming the same AR coating:

$$\rho_{\mathrm{tot},2} = -\left(\frac{t}{t^*}\right)^2 \mathrm{e}^{\mathrm{i}2\phi_b(\omega)} \frac{1 - z^{*2}}{1 - z^2} \,. \tag{109}$$

Now, the GTI effects scale similarly to that of the power reflectivity of the AR coating, r^2, instead of the amplitude reflectivity r This is a tremendous improvement, since it is possible to design AR coatings to the low-index material SiO_2 of the mirror with a residual power reflectivity between 0.001 and 0.01 while covering one octave of bandwidth [73]. However, a single physical layer does not exist that generates a phase shift of $\pi/2$ during one passage for all frequency components contained in an octave. Still, a layer with a quarter-wave thickness at the center frequency is a good initial design. The back-mirror MB in the mirror M2 can then be reoptimized to manage the deviation from a quarter-wave thickness further away from the center frequency because the back-mirror acts as a highly dispersive medium where the phase or group delay can be designed independently.

The top graph in Fig. 23, shows the designed reflectivity of both mirrors of the pair in high resolution, thereby taking into account absorption in the layers. The graph below shows the reflectivity of the mirror, which has, in addition, high transmission between 510 nm to 550 nm for pumping the Ti:sapphire crystal. Each mirror consists of 40 layer pairs of SiO_2 and TiO_2 fabricated by using ion-beam sputtering [78, 79]. Both mirror reflectivities cover more than one octave of bandwidth from 580 nm to 1200 nm or 250 THz to 517 THz, with an average reflectivity of approximately 99.9%, including absorption in the layers. In addition, the mirror dispersion corrects for the second- and higher order dispersion of all intracavity elements, such as the Ti:sapphire crystal and the thin small angle BaF_2 wedges, for fine adjustment of the dispersion from 650 nm to 1200 nm within the 12 bounces occurring in one round-trip. The choice for the lower wavelength boundary

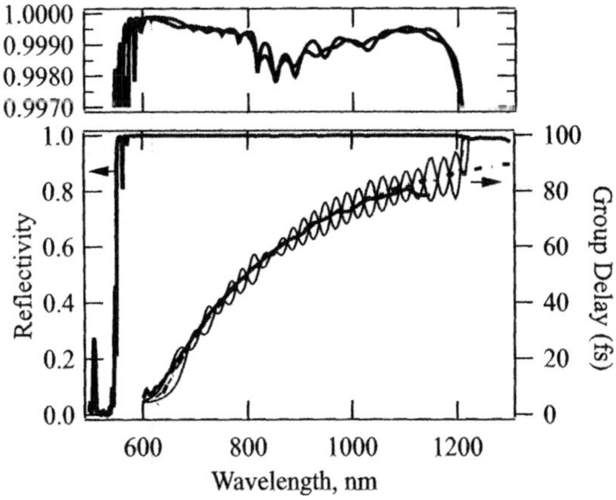

Fig. 23. Reflectivity of a mirror with pump window shown as a *thick solid line* with the *scale to the left.* The group delay design goal for perfect dispersion compensation of a prismless Ti:sapphire laser is shown as a *thick dash-dotted line* with the *scale to the right.* The individual group delay of the designed mirrors is shown as a *thin line* and its average as a *dashed line*, which is almost identical with the design goal over the wavelength range from 650 nm to 1200 nm. The measured group delay, using white light interferometry, is shown as the *thick solid line* from 600 nm to 1100 nm. Beyond 1100 nm, the sensitivity of the Si-detector prevented further measurements

in dispersion compensation is determined and limited by the pump window of Ti:sapphire. The many bounces allow for building of a compact all-mirror, dispersion-controlled, octave-spanning Ti:sapphire laser, even at moderate repetition rates of around 100 MHz. BaF_2 was chosen because it has the lowest ratio between third- and second-order dispersion in this wavelength range of all known materials. This is important, since it appears impossible to design octave-spanning chirped mirrors with large negative third-order dispersion as well as high reflectivity. Another advantage of BaF_2 is that 1 mm BaF_2 has dispersion similar to 2 m of air. Thus, one can replace air by BaF_2 to scale the laser to higher repetition rates, if needed, which is of importance for frequency metrology. The group delays of the individual mirrors of the pair, its average, and the measured average of the mirror pair M1 and M2 are shown beneath the reflectivity trace. The design was realized according to the description given above. The dispersion measurement was performed using white-light interferometry [80], up to approximately 1100 nm, because of the silicon detector roll-off. The oscillations in the group delay of each mirror are about 10 times larger than those of high-quality DCMs that cover 350 nm of bandwidth [32]. However, in the average group delay of both mirrors, the oscillations are ideally suppressed, due to cancellation by more than a factor of 10. Therefore, the effective residual reflectivity of the mirror pair that covers one octave, r^2, is even smaller than that of conventional DCMs. Because of slight fabrication errors, the oscillations in the GD do not cancel precisely, especially close to 900 nm, and 1000 nm deviations from the design goal on the order of 1 fs to 2 fs occur, which will lead to observable spectral features in the spectral output of the laser described below. Note that the fabrication of this new mirror pair has dramatically improved performance, when compared to an earlier version [33]. In addition, earlier pairs were designed for use in combination with prisms, which greatly simplifies the design.

6 Experimental Results on Few-Cycle Infrared Lasers

Because of the limited gain bandwidth of Ti:sapphire of $\sim$ 200 nm, and the limited bandwidth and dispersion control available with conventional laser mirrors, pulses with spectra that cover one octave and pulsewidths shorter than 5 fs (substantially less than two optical cycles at 800 nm) have been possible only by external compression. Impressive results have been achieved using external compression [81, 82]. Spectral broadening by self-phase modulation (SPM) is well known, and it was recently put to spectacular use with photonic bandgap fibers [83, 84] to generate spectra from about 500 nm to 1.5 µm by using the unamplified pulse train of a standard mode-locked Ti:sapphire oscillator. To date the recompression of these impressive spectra has been unsuccessful, and a recent characterization of these wideband spectra has shown that the compression is a significant task [85]. A possible reason for this incompressibility involves the complicated phase structures that emerge from the interplay between SPM and group delay dispersion (GDD) in the fiber, which are highly sensitive to energy fluctuations. The spectral phase can be significantly distorted by modulation instabilities for anomalous GDD. Therefore, a source that directly generates octave-spanning spectra and correspondingly short pulses is highly desirable. Intracavity spectral broadening is advantageous because the spectral phase improves on each round-trip by dispersion-managed mode locking [22] as well as by KLM, thus making the extracavity pulse recompression more successful.

The double-chirped mirror technology described in the previous section leads to laser optics with high reflectivity and the possibility of precise dispersion compensation of up to an octave. In combination with the broadband laser materials discussed in the introduction, a whole class of few-cycle lasers has been developed. In this section, two versions of octave-spanning Ti:sapphire lasers are presented [34, 86] that employ double-chirped mirror pairs (DCMPs) and CaF_2, as well as BaF_2 prisms, for dispersion compensation. Other results obtained with Cr:forsterite and Cr:YAG laser crystals are also briefly described.

The Ti:sapphire laser currently generating the shortest pulses directly from a laser oscillator is shown in Fig. 24. The 2.7 mm thin Ti:sapphire crystal (X) is pumped by a frequency-doubled Nd:YVO_4 laser (Spectra Physics Millennia X). The laser cavity is built up by the double-chirped mirror pairs extending from mirror M_1 to the output coupling mirror OC. The mirrors

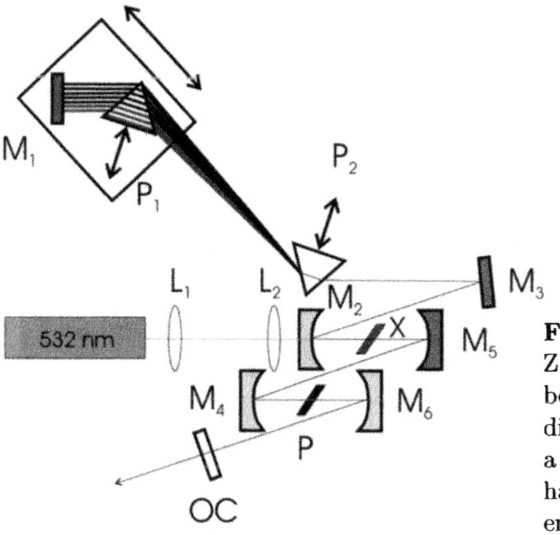

Fig. 24. Setup of the double-Z cavity. Two prisms and eight bounces on DCMPs provide flat dispersion. A second focus in a BK7-plate (P) leads to enhanced SPM, and the laser generates significantly wider spectra

with even subscript are of type 1; the mirrors with odd subscript are of type 2. It is important to note that, ideally, the mirrors must be placed so that the same number of bounces on each mirror type occurs not only within one full round-trip, but also on each side of the laser crystal, since the nonlinearity of the crystal would corrupt the cancellation of the dispersion oscillations. However, in the setup shown in Fig. 24, mirrors M_1 and M_6 are used without a subsequent counterpart. As clarified in the forthcoming paragraph, unbalanced mirror M_6 is a requirement of the dispersion management, and a replacement of mirror M_1 with a protected silver mirror did not increase the overall performance. The curved mirrors M_2 and M_5 focus the beam into the laser crystal, and a second focus is generated in a 2.4 mm thick plate of BK7 by M_4 and M_6. The radius of curvature of all curved mirrors was 100 mm. Two CaF_2 prisms provide positive second-order and higher order dispersion, which, together with the mirror dispersion, compensates for the overall dispersion of the laser crystal and BK7 plate. The prism pairs allow precise fine tuning of the intracavity dispersion. The separation between the prisms is adjustable from 95 cm to 115 cm, thus leading to a repetition rate of $\sim$ 65 MHz. Another important element in an octave-spanning laser is the output coupling mirror. Since a standard Bragg mirror based on TiO_2 and SiO_2 can cover only about 200 nm around a center wavelength of 800 nm, materials with a higher index contrast are needed. The output coupling mirror consists of 5 $ZnSe/MgF_2$ pairs topped with a $\lambda/8$ layer of MgF_2. This mirror has high reflectivity between 700 nm and 1030 nm with transmission of $\sim$ 1% at 800 nm. With absorbed pump power of 4 W, the CW output power was 30 mW. After starting the soft-aperture KLM action by moving prism P_1, the output power increased typically to 120 mW.

The reason for the second focus is twofold. First, in a laser with only

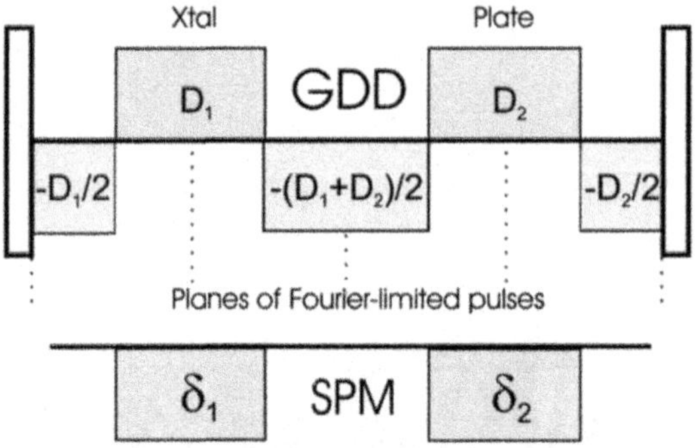

Fig. 25. Second-order dispersion map of the double-Z cavity. The positive dispersion of the laser crystal (D_1) and the glass plate (D_2) is balanced by the DCMPs and the prisms. In the middle of the dispersion cells, the pulse is chirp-free. Additionally, the pulse is modified by SPM in the crystal and the plate with different coefficients δ_1 and δ_2

one focus, where the Ti:sapphire crystal is placed, the pulse energy is limited by the requirement of optimum KLM action. This also limits the amount of SPM and, therefore, the spectral broadening to within one round-trip. To overcome this limitation, we placed a second focus in the cavity. With a glass plate in the second focus, the dispersion map also changed. The resulting dispersion map, which accounted for only second-order dispersion, and the distribution of nonlinearities are shown in Fig. 25.

The second focus produced a larger net self-phase modulation than the first focus, $\delta_2 > \delta_1$, because the plate can be positioned exactly in the focus; whereas, the laser crystal must be shifted slightly off focus for optimum KLM action. Applying even stronger focusing into the glass plate, by using a higher radius of curvature mirrors, leads to even stronger spectral broadening, up to the instability point, where the laser starts to Q-switch. As shown in detail in Sect. 3 of this chapter, the pulse breathes while propagating through the laser cavity. Since we distributed the second-order dispersion symmetrically, ideally, the steady-state pulse was chirp-free and shortest in the center of each dispersion cell. The average dispersion produced by one bounce on a DCM compensated for half the dispersion of the Ti:sapphire crystal. Thus, for optimum spectral broadening, the BK7 plate should produce as much dispersion as the Ti:sapphire crystal. In the current setup, the thickness of the BK7 plate chosen was somewhat less to allow insertion of the prism P_1 far enough into the beam to prevent clipping. Still, the short pulse is transferred and compressed from one focus to the other. Two bounces on M_6 compress the pulse on its way back to the center of the BK7 plate, where considerable additional spectral broadening was provided and more than 100 nm additional bandwidth was produced, when compared to previous results [32]. The influence of the higher order dispersion terms, including the strong partially uncompensated GDD oscillations, as well as gain filtering and KLM action,

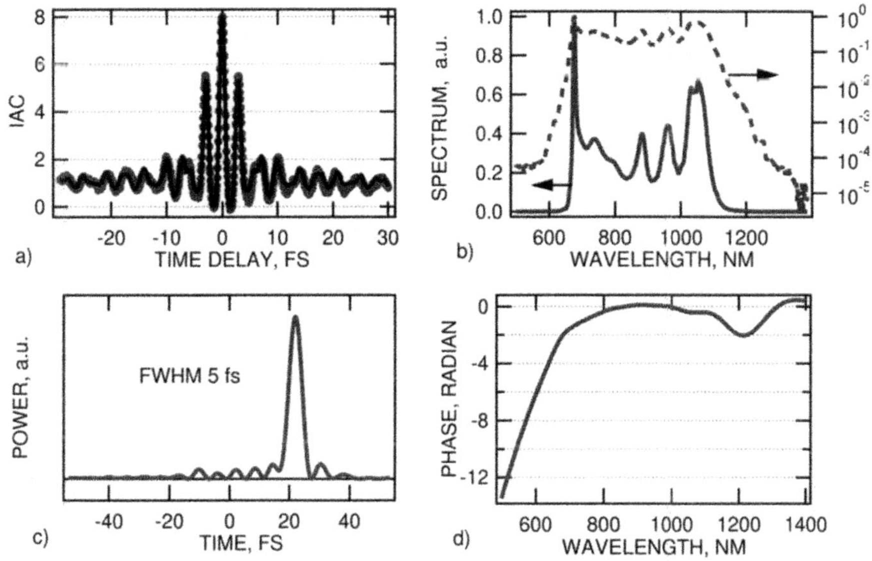

Fig. 26. (**a**) Measured and retrieved interferometric autocorrelation. (**b**) Measured spectrum on linear and logarithmic scales. (**c**) Retrieved intensity envelope. (**d**) Retrieved spectral phase using an adaptive algorithm, as described in the text

can be considered perturbations of the dispersion-managed soliton generated by the processes represented in Fig. 25.

The power spectrum at the laser output is displayed in Fig. 26b on both linear and logarithmic scales. On a logarithmic scale, the spectrum extends from 600nm to 1350 nm above the noise floor. Due to the high transmission of the output coupler below 650 nm and above 1050 nm, the flattop intracavity spectrum of the laser is again significantly emphasized in the wings of the output spectrum, as shown in Fig. 26. This explains the strongly pronounced M-shape of the output spectrum. The FWHM of the corresponding pulse, assuming a flat phase, would be 4.3 fs. The remaining structure in the spectrum is strongly correlated with the oscillations in the measured intracavity GDD. The two peaks at 700 nm and 1050 nm in the spectrum are caused intracavity by the dispersion of the output coupling mirror reflection and are enhanced by the increasing output mirror transmission. Despite the large oscillations in the GDD caused by fabrication tolerances, the spectrum is relatively smooth, due to dispersion-managed soliton formation, as well as enhanced KLM action, by using the second focus. The strong SPM per round-trip leads to an increased nonlinear phase shift. Therefore, subsequent pre- and postpulses caused by GDD oscillations upon reflection of the pulse on the DCMs dephase rapidly, and do not sum coherently over many round-trips. The center wavelength was slightly shifted from the Ti:sapphire gain maximum to 825 nm, which corresponds to an optical period of 2.75 fs.

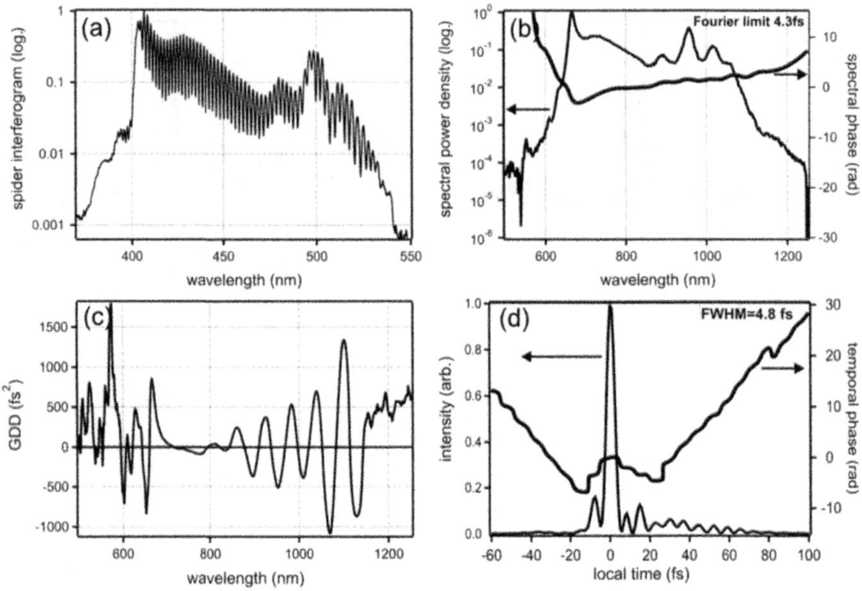

Fig. 27. SPIDER measurement of a 4.8 fs Ti:sapphire laser pulse. (**a**) SPIDER interferogram on a logarithmic scale. (**b**) Spectral power density and spectral phase of the pulse. (**c**) Calculated GDD of the pulse. (**d**) Intensity envelope

The pulses emitted from the laser are recompressed externally by four bounces on DCMPs and a sequence of CaF_2 prisms. The pulses are characterized by an interferometric autocorrelator using thin, metal-coated beam splitters and a thin-wedged KDP doubling crystal (10 μm to 40 μm). The mirror delay was calibrated interferometrically. The interferometric autocorrelation measurement (IAC) is displayed in Fig. 26a. A phase retrieval algorithm [87] was used to reconstruct the actual pulse shape from the measured IAC. The spectral phase was optimized by minimizing the deviation between the measured IAC and the IAC calculated from the measured power spectrum and the assumed spectral phase. The intensity envelope of the reconstructed pulse is shown in Fig. 26c. The intensity envelope leads to a FWHM of 5 fs. The retrieved phase (Fig. 26d) shows that the extracavity dispersion of the shorter wavelengths is not yet optimized. Using a SPIDER setup, as discussed in detail in the Chapter by *Walmsey* et al., the external dispersion compensation was perfected, and 4.8 fs pulses fully characterized with respect to amplitude and phase were obtained [86] (Fig. 27). Figure 27a shows the SPIDER interferogram on a logarithmic scale. Excellent contrast was achieved. Figure 27b shows the measured spectrum and retrieved spectral phase. Figure 27c shows the remaining group delay dispersion of the pulse, which is closely related to the strong dispersion ripple of the double-chirped mirror pairs involved in this experiment [33, 34]. Figure 27d shows the retrieved in-

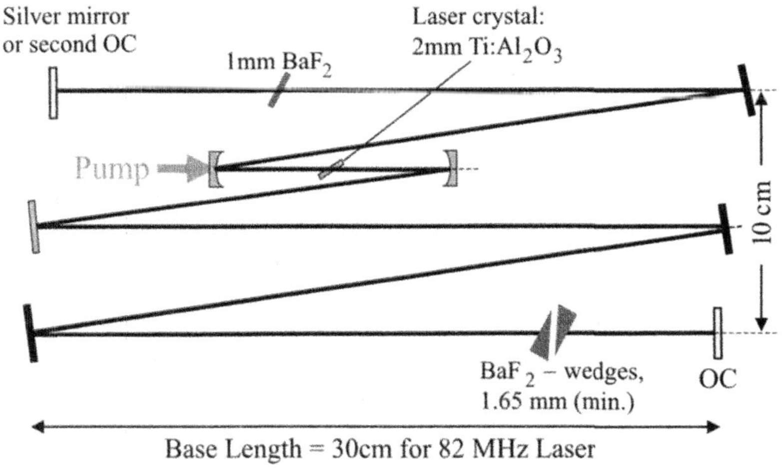

Fig. 28. Schematic diagram of the octave-spanning, prismless Ti:sapphire laser. The Z-fold cavity is astigmatically compensated. Twelve bounces on the DCMPs (six on each mirror type) provide smooth and broadband compensation of the dispersion of the laser crystal, the BaF_2 wedges, and air, which resides inside the laser cavity. The BaF_2 wedges are used to finely adjust the dispersion

tensity profile of the pulse resulting in full-width at half-maximum (FWHM) of 4.8 fs. The octave-spanning spectra of this laser enabled direct detection of the carrier–envelope phase dynamics of the mode comb by applying the 1f–2f self-referencing technique discussed in Chapters by *Udem* et al. and *Jones* et al. in greater detail [88].

For reasons of compactness and long-term stability, a prismless octave-spanning laser is more desirable. The DCM pairs presented in the last section were designed so that the dispersion of all intracavity elements are compensated for, including the thin, small-angle BaF_2 wedges used to fine-tune the dispersion of the resonator. The setup of the Ti:sapphire laser was simplified considerably and is shown in Fig. 28. All cavity mirrors, except the output coupler and the end mirror, are DCMs. The different shadings (gray and black) are used to denote the two different mirrors (M1 and M2) of the pair.

Twelve bounces on these mirrors per round-trip generate the necessary negative dispersion to compensate for the positive second- and third-order dispersion of the laser crystal, the 3.7 m of air inside the 82 MHz laser cavity, and the 11 mm of BaF_2 per round-trip, which is used to fine-tune and balance the dispersion on each side of the laser crystal. A highly doped Ti:sapphire laser crystal with a path length of 2 mm is used. It absorbs approximately 72% of the 532 nm pump light from a frequency-doubled, diode-pumped Nd:YVO_4 laser (Spectra Physics Millennia Xs), which is focused with a 60 mm focal length plano-convex lens into the Ti:sapphire crystal. The curved mirrors shown in Fig. 28 have a radius of curvature of 100 mm.

The twelve bounces on the DCMPs enable a very compact cavity layout. Even for repetition rates as low as 82 MHz, it is possible to fit the resonator on a footprint comparable to a standard sheet of letter-sized paper if an arm ratio of 2 : 1 is used. This compactness, combined with the fact that a slight misalignment of the resonator does not lead to a change in the dispersion, as was the case in a prism-compensated cavity, enables construction of a particularly reliable system for the continuum generation. With the layout shown in Fig. 28, we obtain stable long-term mode locking, even if the laser is optimized for octave-spanning operation, for which precise dispersion compensation over a large bandwidth is crucial [34]. We find that it is possible to operate the laser for several weeks without any realignment. However, to maintain constant output power, the pump power must be adjusted to compensate for the cavity misalignment that naturally occurs over time. Another benefit of the prismless cavity is that fine-tuning of the dispersion can be achieved by simply inserting one of the BaF_2 wedges. This significantly simplifies the handling of the system, since the user has to control only one degree of freedom. This also greatly simplifies the use of such a laser in a synchronized system, such as the one described in the next section. Because, in contrast to a variation of the prism separation, the dispersion in the cavity shown in Fig. 28 is fine-tuned solely by varying the insertion of the BaF_2 wedge, which typically changes the cavity length much less than 100 μm. This cavity length change can be easily corrected by the use of a piezoactuated resonator mirror.

Using the same $ZnSe/MgF_2$ output coupler as before, the laser emits $\sim$ 100 mW of mode-locked power through the 1% output coupler, when pumped with 3.8 W of 532 nm light. The output power is limited by the low output coupling of the broadband output coupler. The efficiency of the laser could be drastically increased by a higher output coupling. On one hand, this spectral shaping can be beneficial to obtain a wider spectrum in the output beam of the laser. On the other hand, it leads to early pulse breaking due to strong spectral filtering. Therefore, it can prohibit further broadening of the intracavity light. The small oscillations in the spectrum originate again from the residual oscillations in the group delay of the DCMPs.

The spectral shape significantly influences the number of optical cycles within the FWHM pulse duration. If we define the center frequency as the weighted average of the power spectrum,

$$f_0 = \frac{\int_0^\infty f\,|E(f)|^2\,\mathrm{d}f}{\int_0^\infty |E(f)|^2\,\mathrm{d}f}\,, \tag{110}$$

where $E(f)$ denotes the Fourier transform of the electric field of the pulse, we obtain the center frequency $f_0 = 365\,\mathrm{THz}$ of the spectrum shown in Fig. 29. Using Fourier transformation, we obtain a FWHM pulse duration of 3.4 fs, assuming a flat spectral phase. However, it might be difficult to fully recompress the pulses emitted by this laser. Therefore, the real pulse duration

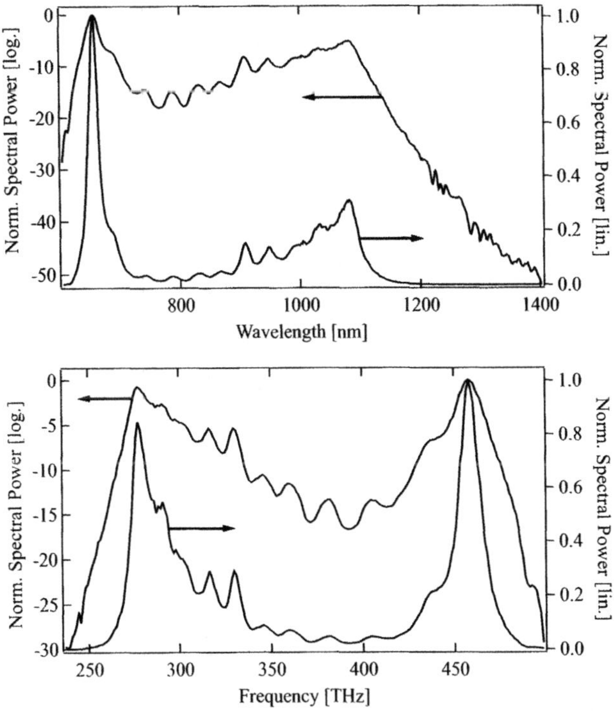

Fig. 29. Typical output spectrum emitted by the laser shown in Fig. 28 on linear (*right*) and logarithmic scales (*left*). The *upper graph* shows the spectral power per $\Delta\lambda$ plotted as a function of wavelength λ. The *lower graph* shows the same spectrum as the power per frequency interval plotted as a function of frequency. The M shape of the spectrum is due to the transmission characteristic of the output coupler. The small oscillations originate from the residual oscillations in the dispersion of the DCMPs. The theoretical FWHM duration of the corresponding pulses, assuming a flat spectral phase is 3.4 fs, which is equivalent to the duration of < 1.3 optical cycles at the weighted center frequency of 365 THz (110)

that can be used for an experiment might be noticeably longer. Finally, 3.4 fs corresponds to a duration of < 1.3 cycles at the weighted center frequency of 365 THz, according to (110). As found earlier [31, 34], the M-shape of the output spectrum has a beneficial influence on the FWHM pulse duration. If we do the same calculation for a rectangularly shaped spectrum with the same FWHM bandwidth from 645 nm to 1100 nm, we obtain a 4.6 fs FWHM pulse duration, again assuming a flat spectral phase. This corresponds to a duration of 1.7 optical cycles at the spectrum's center frequency of 365 THz. Therefore, in an experiment that is sensitive to the electric field rather than to the pulse envelope, one might expect a notable difference between a flattop and an M-shaped spectrum with the same FWHM bandwidth.

Conventional double-chirped mirrors have also been used to extract ultrabroadband spectra from Cr:forsterite [89], Cr:YAG [90], and Cr:LiCAF [91]

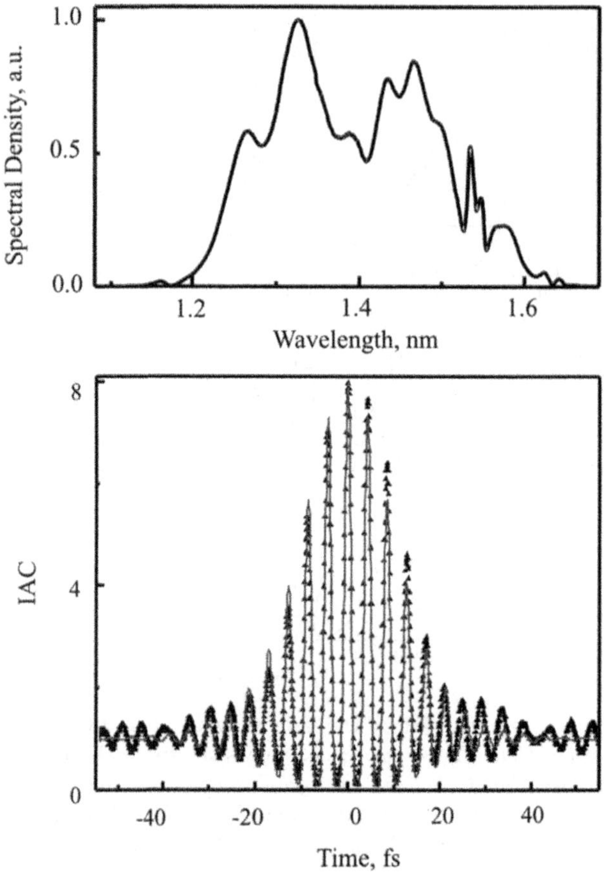

Fig. 30. Optical power spectrum (*top*) and measured interferometric autocorrelation of the Cr:forsterite laser presented in [89]. A sech^2-fit reveals a 14 fs pulse duration

lasers. The Cr:forsterite and Cr:YAG laser systems especially suffer from a large third-order dispersion, since they operate close to the point where the second-order dispersion is zero at 1490 nm for Cr:forsterite and 1580 nm for Cr:YAG. Currently, only chirped mirrors can compensate for the limiting third-order dispersion, in addition to second-order dispersion over a large bandwidth. The spectra and interferometric autocorrelation traces of these laser systems are shown in Fig. 30 and Fig. 31.

The spectra generated from the Ti:sapphire laser, which covers from 600 nm to 1100 nm on a linear scale, and the Cr:forsterite laser, which covers 1100 nm to 1600 nm, have the potential for synthesizing a single-cycle pulse with a spectrum extending over 1.5 octaves, from 600 nm to 1600 nm. A pulse with a rectangularly shaped spectrum spanning this range would result in a

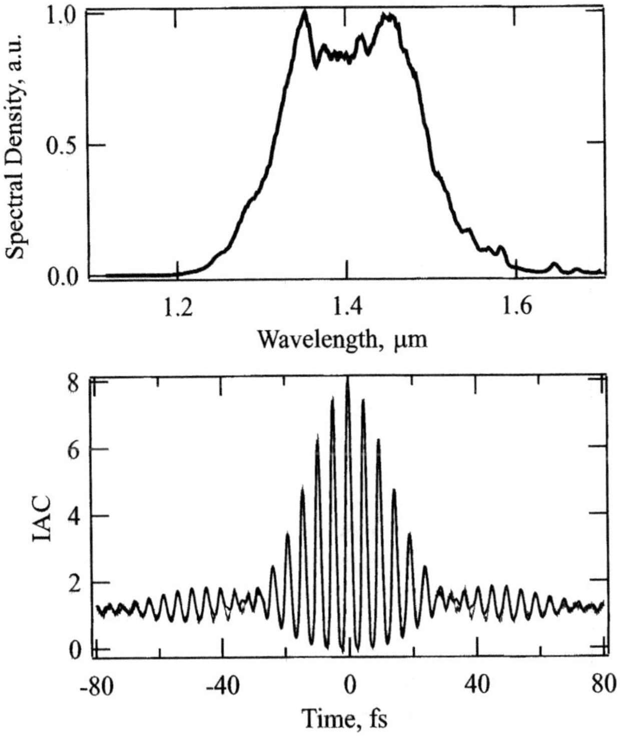

Fig. 31. *Top*: Optical power spectrum of the Cr:YAG laser presented in [90] on a linear scale. *Bottom*: Measured interferometric autocorrelation of the same laser. A phase-retrieval algorithm reveals a 20 fs pulse duration

3 fs pulse; its carrier wavelength would be 875 nm, which corresponds to a carrier oscillation period of 2.9 fs. The challenge involves the phase coherent addition of the output of both lasers. Such a coherent pulse synthesis, which uses longer pulses of 20 fs to 30 fs from two independent Ti:sapphire lasers with slightly shifted center frequencies, has already been demonstrated by *Shelton* et al. [92]. Synchronization of two independent lasers with subcycle precision will be demonstrated in the next section.

7 Toward Single-Cycle-Pulse Generation

As mentioned in the previous section, a single-cycle pulse in the optical domain may be achieved through the phase-coherent superposition of several spectrally overlapped few-cycle lasers. The synchronization of pulse trains from independent mode-locked lasers with subcycle timing fluctuations is the most important and the most challenging step in this synthesis process. Ideally, the relative timing jitter should be less than a tenth of an optical cycle for a high-quality synthesized pulse stream. At a center wavelength of

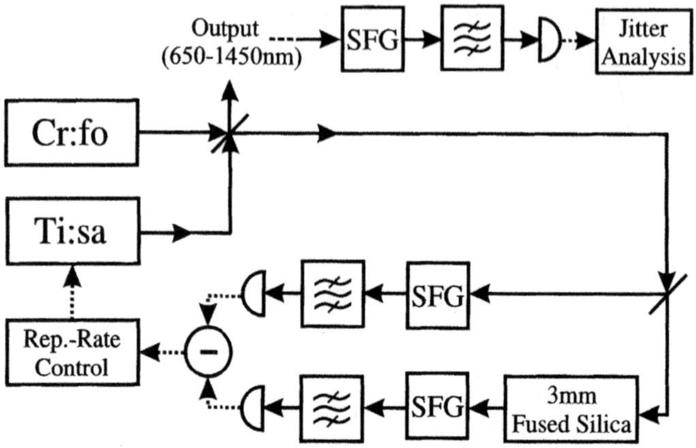

Fig. 32. Experimental setup of synchronized lasers. Cr:fo: passively mode-locked Cr:forsterite laser, Ti:sa: passively mode-locked Ti:sapphire laser; SFG: sum-frequency generation; all band-pass filters transmit only the sum-frequency (1/496 nm = 1/833 nm + 1/1225 nm). The two beam splitters consist of thin-fused silica glass substrates coated with a semitransparent metal film. The third correlator is used to generate the graphs shown in Fig. 35

900 nm, this condition restricts the timing jitter to 300 as or less, measured over the full Nyquist bandwidth, i.e., half the laser repetition rate. Several groups have investigated the possibility of active [92] and/or passive [93, 94] synchronization of multiple lasers. However, a subfemtosecond timing jitter over the Nyquist bandwidth has not been achieved to date.

In this section, a method of synchronization is demonstrated in which the timing jitter between two passively mode-locked lasers is detected by a balanced cross-correlator (see Fig. 32), which is the optical equivalent of a balanced microwave phase detector. The signal is then fed back through an electronic control loop to keep the two lasers synchronized. This method enables drift-free and temperature-independent synchronization between two individual lasers, a task that is difficult to achieve with all-electronic schemes. To ensure the long-term stability of the system against thermal drifts, the two laser beams are combined inside the control loop. To further improve the stability of the system, we used prismless lasers, as discussed in the previous section, to generate the corresponding parts of the continuum. In [95], the generation of ultrabroad spectra with high efficiencies in purely double-chirped-mirror-compensated Ti:sapphire lasers was demonstrated. To improve the long-term stability and mode locking of the Cr:forsterite laser in [89], we used a novel broadband InGaAs saturable absorber on a large area, high-index contrast AlGaAs/Al_xO_y mirror [96]. Due to the high-index contrast between AlGaAs and Al_xO_y, the mirror's reflectivity extended from roughly 1100 nm to 1500 nm, which enabled the generation of sub-30 fs pulses at the 1230 nm wavelength in a self-starting configuration. Even though the spec-

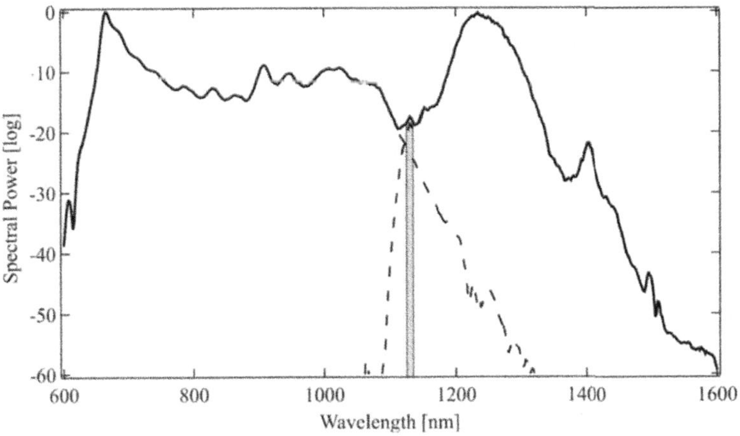

Fig. 33. Optical spectra of mode-locked Ti:sapphire and Cr:forsterite lasers on a logarithmic scale. The *dashed lines* indicate the spectra of the individual lasers in the vicinity of the spectral overlap. The theoretical FWHM duration of the corresponding pulses, assuming a flat spectral phase and perfect phase locking between the two combs, is 3.0 fs, which corresponds to the duration of a single optical cycle at the weighted center frequency (110) of 320 THz

tral width of this laser was significantly narrower than previously reported in [89], the setup did not require purging the cavity with dry nitrogen, and mode locking did not rely on the critical alignment of the cavity. This simplified the operation of this laser. Furthermore, it was possible to operate the laser for several months without any cavity realignment. The output power of the laser was roughly 60 mW of mode-locked power through a 3% output coupler, when pumped with approximately 2 W to 3 W of 1064 nm light, and was limited by the breakup of the pulse into multiple pulses. The tendency for pulse breakup could be reduced by introducing a saturable absorber with more saturable loss, or by the use of additional KLM. To improve the optical-to-optical efficiency, the Cr:forsterite crystal was cooled to −25 °C.

Figure 33 shows the total spectrum of the two lasers (solid line) at the output. The dashed lines indicate the extent of the individual laser spectra in the vicinity of the overlap region. The shaded region indicates the spectral region filtered to record the difference in carrier–envelope offset frequency between the two lasers (see Fig. 33). For phase coherent superposition of the two lasers, the pulse envelopes of the two lasers must be synchronized. In addition, the difference in the carrier–envelope offset frequency between the two lasers must be set to zero with a phase lock. Synchronizing the pulse trains with subcycle precision is the most challenging step in the synthesis process. Ideally, timing accuracy of less than one-tenth of an optical cycle should be achieved. In our case, this required a timing jitter of 300 as or less, measured over the full Nyquist bandwidth. To overcome the typical problems posed by balanced microwave mixers previously used for this task [92], we developed the optical equivalent of such a device: a balanced cross-correlator. As shown in Fig. 32, the outputs of the two lasers were combined on a broadband metallic beam splitter. One part of the combined beam was directed to two nearly identical cross-correlators by using 1 mm-thick LBO crystals that are phase matched for SFG of 833 nm light from the Ti:sapphire laser, and 1225 nm light from the Cr:forsterite laser. The only difference between the two correlators was a 3 mm thick fused silica window in the optical path of one of the correlators. This glass insert a group delay between 833 nm and 1225 nm in order to offset the pulses emitted by the Cr:forsterite and Ti:sapphire lasers by about 45 fs with respect to each other. Due to the low temperature dependence of the chromatic dispersion ($< 1\,\mathrm{as}/^{\circ}\mathrm{C}$), it was legitimate to use this 45 fs delay as a reference for timing offset measurements. For small time differences between the two laser pulses, the difference between the currents of the two photodetectors at the end of each correlator was nearly proportional to the time difference between the two pulses. Furthermore, in the vicinity of zero timing offset, this detector acted like a balanced phase detector operating in the multiple terahertz range, if the signal amplitudes of the two correlators were balanced against each other. At the zero-crossing of the difference of the photocurrents, this detector delivered a perfectly balanced signal. Therefore, the amplitude noise of each laser did not affect the error signal detected. The output of this balanced cross-correlator as a function of the time difference between the Cr:forsterite and the Ti:sapphire pulses is shown in Fig. 34.

The signal from the balanced mixer is used to lock the repetition rates of the two lasers by controlling the cavity length of the Ti:sapphire laser with cavity mirrors mounted on piezoelectric transducers in a manner similar to that discussed in [92]. This finally closed the control loop. The first beam splitter used to combine the two output beams from the lasers was inside this control loop. Since the output beam shown in Fig. 32 originates from this beam splitter, temperature drifts, acoustic noise, or beam fluctuations always affect both laser beams in the same way as they travel along identical paths. Therefore, external noise cannot corrupt the relative jitter, and the output behaves as if it had originated from the same source. Even environmental noise that influences the optical length of the cross-correlators cannot corrupt the timing, since the group delay of the 3 mm fused silica, measured between 833 nm and 1225 nm, is the only timing reference in this system.

Figure 35 shows the resulting timing jitter measurement made with the out-of-loop cross-correlator shown in Fig. 32. The residual timing jitter over the detector's bandwidth of 2.3 MHz is 300 as $\pm$ 100 as. The stated error is determined from the amplitude noise measured at the peak of the cross-correlation. This trace is also shown in Fig. 35. As in most passively mode-locked laser systems, the main contribution to the timing jitter has frequency components equal to a few times the relaxation oscillation frequency of the laser [97]. In the current system, the relaxation oscillation frequencies are

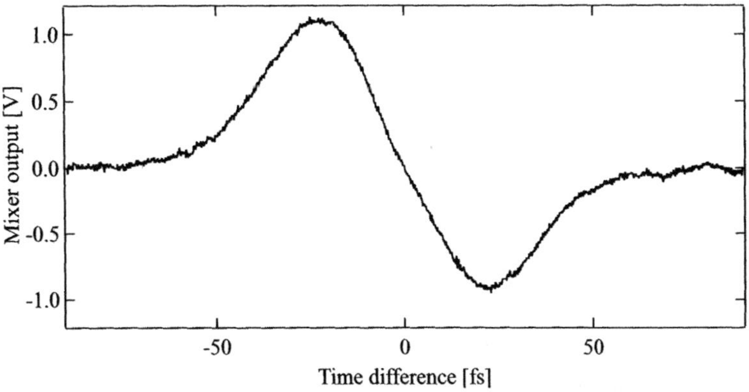

Fig. 34. Output of the balanced cross-correlator as a function of the time difference between the two laser pulses. The trace was taken when the two lasers were unlocked. The *time axis* was calibrated by measuring the frequency difference between the two lasers and by measuring the cavity round-trip time

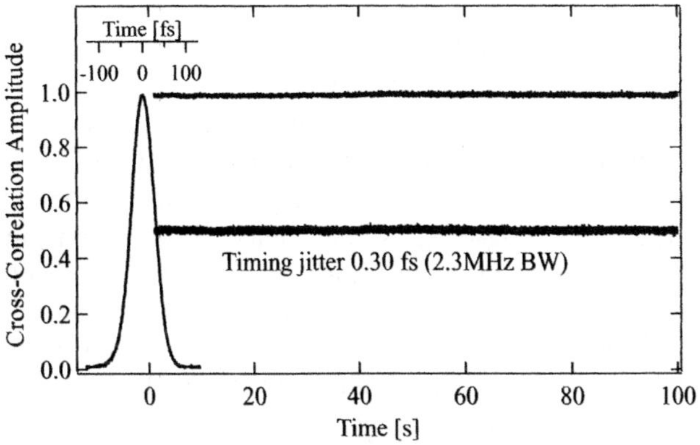

Fig. 35. Timing jitter determined from the amplitude noise of the SFG process of the out-of-loop cross-correlator (Fig. 34). The time delay in the correlator was generated by a dispersive medium in front of the SFG crystal. The rms-jitter measured in a 2.3 MHz BW results in 300 as ± 100 as

$\sim$ 70 kHz for the Ti:sapphire laser and $\sim$ 140 kHz for the Cr:forsterite laser. Therefore, we assume that noise $>$ 2.3 MHz is negligible.

As soon as the two lasers were locked to each other, we observed a strong beat signal in the overlap region of the optical spectrum. The beat signal shown in Fig. 36 was detected with an InGaAs PIN-diode connected to a transimpedance amplifier. To avoid saturation of the detector, only a small part of the optical spectrum was directed to the diode. The maximum transmission of the 10 nm wide band-pass filter is 1130 nm. As described in [98], the beat signal represented the difference in the carrier–envelope offset frequency Δf_{ceo} between the two lasers. In contrast to previous results, it is now

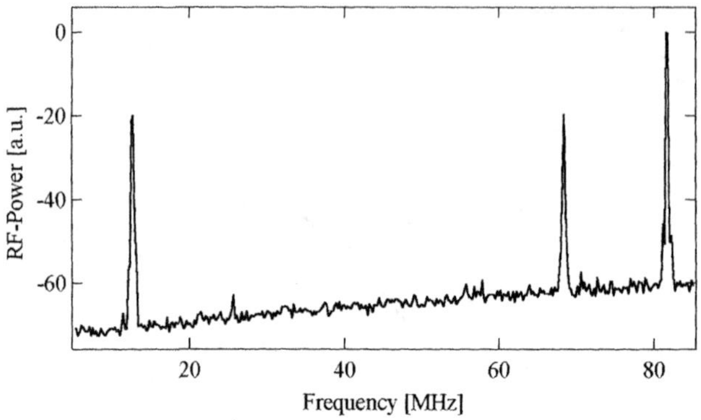

Fig. 36. Heterodyne beat between the Cr:forsterite and the Ti:sapphire lasers obtained behind a 10 nm wide optical band-pass filter centered at 1130 nm. The two beat signals below the repetition rate of 82 MHz represent the difference in the carrier–envelope offset frequency of both lasers. The rf-analyzer filter bandwidth was set to 30 kHz. The noise floor is caused by the uncompensated transimpedance amplifier and posed only a technical limitation

possible to obtain this beat without the use of additional spectral broadening. This helps to provide an exceptionally large signal-to-noise ratio of ~ 50 dB in a 30 kHz bandwidth. To complete the pulse synthesis process, it was necessary to phase-lock this difference offset frequency to dc. The widely used offset-locking technique, which was also employed in [99], could be applied; however, it would not lead to a long-term, stable, phase coherent output, since any change in the path length of the offset lock, would be transferred to the output beam. We are currently investigating a homodyne technique that uses a fraction of each of the two output beams of the first in-loop beamsplitter, i.e., superimposing the Ti:sapphire and Cr:forsterite beams. This would automatically stabilize the phase difference between the two spectral components that constitute the output beam.

8 Conclusion

We present a unified theoretical description of the mode locking of lasers based on the master equation approach that gives insight into the various pulse-generation techniques leading to pulses shorter than two optical cycles. This treatment results in a prediction of the pulseshape and the scaling of the pulsewidth as a function of saturable absorption, gain bandwidth, and the noise and tuning properties of the laser. Soliton-perturbation theory can be used to predict the changes in the phase and group velocities, due to the presence of additional nonlinear terms, as demonstrated by the example of self-steepening. The experimental challenge is in the fabrication of the laser, especially the optical components, in order to meet the theoretical description: maximizing the saturable absorber action of the bandwidth of the mirrors while controlling the intracavity dispersion. The broadband dispersion-compensating mirror technology has been developed to a point where octave-spanning spectra and sub-two-cycle pulses can be generated directly from a Ti:sapphire laser. Few-cycle output has been also achieved from several other infrared laser materials such as Cr:forsterite and Cr:YAG. Together, two of these few-cycle lasers, a Ti:sapphire laser and a Cr:forsterite laser, can potentially span a spectrum from 600 nm to 1600 nm. Tight synchronization between the two lasers is achieved by balanced cross-correlation to one-tenth of an optical cycle. In addition, balanced homodyne detection of the interference beat of both lasers, can be used to stabilize the phase between the two lasers so that true single-cycle optical pulses can be generated in the near future.

Acknowledgements

The authors gratefully acknowledge Volker Scheuer and Gregor Angelow for fabrication of the dispersion-compensating broadband mirrors; Sheila Tandon, Gale Petrich, and Leslie Kolodziejski for fabrication of the broadband saturable absorbers; Max Lederer, Alexander Boiko, and Barry Luther-Davis for fabrication of broadband output couplers; Peter O'Brien for fabrication of the broadband beam splitters; and Alexander Killi for work on the noise and tuning properties of mode-locked lasers. The authors also gratefully acknowledge the work on the broadband Cr:YAG and the Kerr-lens mode-locked Cr:forsterite lasers by Daniel Ripin and Christian Chudoba. We are indebted to Jun Ye, Steven Cundiff, and David Jones for many fruitful discussions. This work was supported in part by MIT Lincoln Laboratory ACC-334, NSF ECS-0119452, AFOSR F49620-01-1-0084, AFOSR F49620-01-1-0186, and ONR-N00014-02-1-0717.

References

[1] P. W. Smith, M. A. Duguay, E. P. Ippen: *Mode-Locking of Lasers*, vol. 3 (Pergamon Press 1974) Part 2
[2] A. J. DeMaria, D. A. Stetser, H. Heynau: Appl. Phys. Lett. **8**, 174–176 (1966)
[3] E. S. L. Shapiro: *Ultrashort Light Pulses*, vol. 18, Top. Appl. Phys. (Springer 1977)
[4] E. P. Ippen, C. V. Shank, A. Dienes: Appl. Phys. Lett. **21**, 348–350 (1972)
[5] C. V. Shank, E. P. Ippen: Appl. Phys. Lett. **24**, 373–375 (1974)
[6] R. L. Fork, C. H. B. Cruz, P. C. Becker, C. V. Shank: Opt. Lett. **12**, 483–485 (1987)
[7] H. A. Haus: Theory of mode-locking with a slow saturable absorber, J. Appl. Phys. **46**, 3049–3058 (1975)
[8] G. H. C. New: Pulse evolution in mode-locked quasicontinuous lasers, IEEE J. Quantum Elect. **10**, 115–124 (1974)
[9] H. A. Haus: Theory of mode-locking with a fast saturable absorber, IEEE J. Quantum Elect. **11**, 736–746 (1975)

[10] O. E. Martinez, R. L. Fork, J. P. Gordon: Opt. Lett. **9**, 156–158 (1984)
[11] L. F. Mollenauer, R. H. Stolen: Opt. Lett. **9**, 13–15 (1984)
[12] J. Mark, L. Y. Liu, K. L. Hall, H. A. Haus, E. P. Ippen: Opt. Lett. **14**, 48–50 (1989)
[13] E. P. Ippen, H. A. Haus, L. Y. Liu: J. Opt. Soc. Am. B **6**, 1736–1745 (1989)
[14] P. N. Kean, X. Zhu, D. W. Crust, R. S. Gant, N. Langford, W. Sibbett: Opt. Lett. **14**, 39–41 (1989)
[15] K. Tamura, E. P. Ippen, H. A. Haus, L. E. Nelson: Opt. Lett. **18**, 1080–1082 (1993)
[16] D. E. Spence, P. N. Kean, W. Sibbett: Opt. Lett. **16**, 42–44 (1991)
[17] D. K. Negus, L. Spinelli, N. Goldblatt, G. Feugnet: Sub-100 femtosecond laser pulse generation by Kerr-lens mode-locking in Ti:Al_2O_3, in G. Dubé, L. Chase (Eds.): OSA Proc. on Advanced Solid-State Lasers (Optical Society of America, Washington 1991) pp. 120–124
[18] F. Salin, J. Squier, M. Piché: Opt. Lett. **16**, 1674–1676 (1991)
[19] M. Piché: Opt. Commun. **86**, 156–160 (1991)
[20] J. H. B. Nijhof, N. J. Doran, W. Forysiak, F. M. Knox: Electron. Lett. **33**, 1726–1727 (1997)
[21] K. Tamura, H. A. Haus, E. P. Ippen: Electron. Lett. **28**, 2226–2228 (1992)
[22] Y. Chen, F. X. Kärtner, U. Morgner, S. H. Cho, H. A. Haus, J. G. Fujimoto, E. P. Ippen: J. Opt. Soc. Am. A **16**, 1999–2004 (1999)
[23] M. T. Asaki, C. P. Huang, D. Garvey, J. Zhou, H. C. Kapteyn, M. N. Murnane: Opt. Lett. **18**, 977–979 (1993)
[24] J. Zhou, G. Taft, C. P. Huang, M. M. Murnane, H. C. Kapteyn, I. P. Christov: Opt. Lett. **19**, 1149–1151 (1994)
[25] C. Spielmann, P. F. Curley, T. Brabec, F. Krausz: IEEE J. Quantum Elect. **30**, 1100–1114 (1994)
[26] A. Stingl, C. Spielmann, F. Krausz: Opt. Lett. **19**, 204–206 (1994)
[27] R. Szipöcs, K. Ferencz, C. Spielmann, F. Krausz: Opt. Lett. **19**, 201–203 (1994)
[28] A. Stingl, M. Lenzner, C. Spielmann, F. Krausz, R. Szipöcs: Opt. Lett. **20**, 602–604 (1995)
[29] F. X. Kärtner, N. Matuschek, T. Schibli, U. Keller, H. A. Haus, C. Heine, R. Morf, V. Scheuer, M. Tilsch, T. Tschudi: Opt. Lett. **22**, 831–833 (1997)
[30] I. D. Jung, F. X. Kärtner, N. Matuschek, D. H. Sutter, F. Morier-Genoud, G. Zhang, U. Keller, V. Scheuer, M. Tilsch, T. Tschudi: Opt. Lett. **22**, 1009–1011 (1997)
[31] D. H. Sutter, G. Steinmeyer, L. Gallmann, N. Matuschek, F. Morier-Genoud, U. Keller, V. Scheuer, G. Angelow, T. Tschudi: Opt. Lett. **24**, 631–633 (1999)
[32] U. Morgner, F. X. Kärtner, S. H. Cho, H. A. Haus, J. G. Fujimoto, E. P. Ippen, V. Scheuer, G. Angelow, T. Tschudi: Opt. Lett. **24**, 411–413 (1999)
[33] F. X. Kärtner, U. Morgner, T. R. Schibli, E. P. Ippen, J. G. Fujimoto, V. Scheuer, G. Angelow, T. Tschudi: J. Opt. Soc. Am. A **18**, 882–885 (2001)
[34] R. Ell, U. Morgner, F. X. Kärtner, J. G. Fujimoto, E. P. Ippen, V. Scheuer, G. Angelow, T. Tschudi: Opt. Lett. **26**, 373–375 (2001)
[35] L. Xu, C. Spielmann, F. Krausz, R. Szipöcs: Opt. Lett. **21**, 1259–1261 (1996)
[36] A. Baltuška, T. Fuji, T. Kobayashi: Phys. Rev. Lett. **88**, 133901 (2002)
[37] R. L. Fork, O. E. Martinez, J. P. Gordon: Opt. Lett. **9**, 150–152 (1984)
[38] S. M. J. Kelly: Electron. Lett. **28**, 806–807 (1992)
[39] K. Tamura, E. P. Ippen, H. A. Haus: Appl. Phys. Lett. **67**, 158–160 (1995)

[40] I. P. Christov, M. M. Murnane, H. C. Kapteyn, J. Zhou, C. P. Huang: Opt. Lett. **19**, 1465–1467 (1994)
[41] I. P. Christov, H. C. Kapteyn, M. M. Murnane, C. P. Huang, J. Zhou: Opt. Lett. **20**, 309–311 (1995)
[42] I. P. Christov, V. D. Stoev, M. M. Murnane, H. C. Kapteyn: Opt. Lett. **21**, 1493–1495 (1996)
[43] I. P. Christov, V. D. Stoev: J. Opt. Soc. Am. B **15**, 1960–1966 (1998)
[44] Y. Chen, H. A. Haus: Dispersion-managed solitons in the net positive dispersion regime, J. Opt. Soc. Am. B **16**, 24–30 (1999)
[45] F. X. Kärtner, J. A. d. Au, U. Keller: IEEE J. Sel. Top. Quant. **4**, 159–168 (1998)
[46] H. A. Haus, K. Tamura, L. E. Nelson, E. P. Ippen: IEEE J. Quantum Elect. **31**, 591–598 (1995)
[47] H. A. Haus, A. Mecozzi: IEEE J. Quantum Elect. **29**, 983–996 (1993)
[48] H. A. Haus, E. P. Ippen: Opt. Lett. **26**, 1654–1656 (2001)
[49] H. A. Haus, Y. Lai: J. Opt. Soc. Am. B **7**, 386–392 (1990)
[50] K. W. Holman, R. J. Jones, A. Marian, S. T. Cundiff, J. Ye: Opt. Lett. **28**, 351–353 (2003)
[51] H. A. Haus, J. G. Fujimoto, E. P. Ippen: IEEE J. Quantum Elect. **28**, 2086–2096 (1992)
[52] G. Cerullo, S. D. Silvestri, V. Magni: Opt. Lett. **19**, 1040–1042 (1994)
[53] A. Penzkofer, M. Wittmann, E. Sigert, S. MacNamara: Opt. Quant. Electron. **28**, 423–442 (1996)
[54] J. Herrmann: J. Opt. Soc. Am. B **11**, 498–512 (1994)
[55] E. J. Grace, A. Ritsataki, P. M. W. French, G. H. C. New: Opt. Commun. **183**, 249–264 (2000)
[56] V. P. Kalosha, M. Müller, J. Herrmann, S. Gatz: J. Opt. Soc. Am. B **15**, 535–550 (1998)
[57] D. Reinisch, G. Vitrant: Prog. Quant. Electron. **18**, 1 (1994)
[58] C. Jirauschek, U. Morgner, F. X. Kaertner: J. Opt. Soc. Am. B **19**, 1716–1721 (2002)
[59] C. Jirauschek, U. Morgner, F. X. Kaertner: J. Opt. Soc. Am. B **20**, 1356–1368 (2003)
[60] A. Baltuška, Z. Wei, M. Pshenichnikov, D. Wiersma: Opt. Lett. **22**, 102 (1997)
[61] A. Baltuska, Z. Wei, M. Pshenichnikov, D. Wiersma, R. Szipöcs: Appl. Phys. B-Lasers O. **65**, 175–188 (1997)
[62] M. Nisoli, S. DeSilvestri, O. Svelto, R. Szipöcs, K. Ferenz, C. Spielmann, S. Sartania, F. Krausz: Opt. Lett. **22**, 522 (1997)
[63] E. J. Mayer, J. Möbius, A. Euteneuer, W. W. Rühle, R. Szipöcs: Opt. Lett. **22**, 528 (1997)
[64] K. O. Hill, F. Bilodeau, B. Malo, T. Kitagawa, S. Theriault, D. C. Johnson, J. Albert, K. Takiguchi: Opt. Lett. **19**, 1314–1316 (1994)
[65] A. V. Tikhonravov, M. K. Trubetskov, A. A. Tikhonravov: in (OSA Topical Meeting on Optical Interference Coatings 1998)
[66] B. Golubovic, R. R. Austin, M. K. Steiner-Shepard, M. K. Reed, S. A. Diddams, D. J. Jones, A. G. V. Engen: Opt. Lett. **25**, 175–278 (2000)
[67] N. Matuschek, F. X. Kärtner, U. Keller: IEEE J. Sel. Top. Quant. **4**, 197–208 (1998)
[68] N. Matuschek, F. X. Kärtner, U. Keller: IEEE J. Quantum Elect. **5**, 129–137 (1999)

[69] M. Matsuhara, K. O. Hill: Appl. Opt. **13**, 2886–2888 (1974)
[70] G. Tempea, V. Yakovlev, B. Bacovic, F. Krausz, K. Ferencz: J. Opt. Soc. Am. B **18**, 1747–1750 (2001)
[71] N. Matuschek, L. Gallmann, D. H. Sutter, G. Steinmeyer, U. Keller: Appl. Phys. B-Lasers O. **71**, 509–522 (2000)
[72] A. Apolonski, A. Poppe, G. Tempea, C. Spielmann, T. Udem, R. Holzwarth, T. Hänsch, F. Krausz: Phys. Rev. Lett. **85**, 740 (2000)
[73] J. A. Dobrowolski, A. V. Tikhonravov, M. K. Trubetskov, B. T. Sullivan, P. G. Verly: Appl. Opt. **35**, 644–658 (1996)
[74] G. Steinmeyer: Brewster-angled chirped mirrors for high-fidelity dispersion compensation and bandwidths exceeding one optical octave, Opt. Express **11**, 2385–2396 (2003)
[75] R. Szipöcs, A. Köhàzi-Kis: Appl. Phys. B-Lasers O. **65**, 115–135 (1997)
[76] V. Laude, P. Tournois: in (CLEO, Baltimore 1999) paper CTuR4
[77] R. Szipöcs, A. Köházi-Kis, S. Lakó, P. Apai, A. P. Kovácz, G. DeBell, L. Mott, A. W. Louderback, A. V. Tikhonravov, M. K. Trubetskov: Appl. Phys. B-Lasers O. **70**, S51–557 (2000)
[78] V. Scheuer, M. Tilsch, T. Tschudi: in (SPIE Conf. Proc. 1994) pp. 445–454
[79] M. Tilsch, V. Scheuer, T. Tschudi: in (SPIE Conf. Proc. 1994) pp. 414–422
[80] K. Naganuma, K. Mogi, H. Yamada: Opt. Lett. **15**, 393 (1990)
[81] M. Pshenichnikov, A. Baltuška, R. Szipöcs, D. Wiersma: *Sub-5 fs Pulses: Generation, Characterization and Experiments* (Springer, Berlin, Heidelberg 1998)
[82] Z. Cheng, G. Tempea, T. Brabec, K. Ferencz, C. Spielmann, F. Krausz: *Generation of Intense Diffraction-Limited White Light and 4 fs Pulses* (Springer, Berlin, Heidelberg 1998)
[83] J. Ranka, R. Windeler, A. Stentz: Opt. Lett. **25**, 25 (2000)
[84] T. Birks, W. Wadsworth, P. Russel: in CLEO (OSA, Washington 2000) p. 59, postdeadline papers
[85] X. Gu, L. Xu, M. Kimmel, E. Zeek, P. O'Shea, A. P. Shreenath, R. Trebino, R. S. Windeler: Opt. Lett. **27**, 1174–1176 (2002)
[86] R. Ell: *Sub-2 cycle Ti:sapphire lasers and phase sensitive nonlinear optics*, Dissertation, University of Karlsruhe (TH), Karlsruhe (2003)
[87] A. Baltuska, A. Pugzlys, M. Pshenichnikov, D. Wiersma, B. Hoenders, H. Ferwerda: in (Ultrafast Optics, Ascona 1999) paper Th10
[88] U. Morgner, R. Ell, G. Metzler, T. R. Schibli, F. X. Kärtner, J. G. Fujimoto, H. A. Haus, E. P. Ippen: Phys. Rev. Lett. **86**, 5462–5465 (2001)
[89] C. Chudoba, J. G. Fujimoto, E. P. Ippen, H. A. Haus, U. Morgner, F. X. Kaertner, V. Scheuer, G. Angelow, T. Tschudi: Opt. Lett. **26**, 292–294 (2001)
[90] D. J. Ripin, C. Chudoba, J. T. Gopinath, J. G. Fujimoto, E. P. Ippen, U. Morgner, F. X. Kaertner, V. Scheuer, G. Angelow, T. Tschudi: Opt. Lett. **27**, 61–63 (2002)
[91] P. C. Wagenblast, U. Morgner, F. Grawert, T. R. Schibli, F. X. Kaertner, V. Scheuer, G. Angelow, M. J. Lederer: Opt. Lett. **27**, 1726–1728 (2002)
[92] R. K. Shelton, S. M. Foreman, L. Ma, J. L. Hall, H. C. Kapteyn, M. M. Murnane, M. Notcutt, J. Ye: Opt. Lett. **27**, 312–314 (2002)
[93] A. Leitensdorfer, C. Fuerst, A. Laubereau: Opt. Lett. **20**, 916–918 (1995)
[94] Z. Wei, Z. Kobayashi, Z. Zhang, K. Torizuka: Opt. Lett. **26**, 1806–1808 (2001)
[95] T. R. Schibli, L. M. Matos, F. J. Grawert, F. X. Kaertner: in (Ultrafast Phenomena, Vancouver 2002)

[96] D.J. Ripin, J.G. Gopinath, H.M. Shen, A.A. Erchak, G.S. Petrich, L.A. Kolodziejski, F.X. Kaertner, E.P. Ippen: Opt. Commun. **214**, 285 (2002)
[97] J. Son, J.V. Rudd, J.F. Whitaker: Opt. Lett. **17**, 733 (1992)
[98] Z. Wei, Y. Kobayashi, K. Torizuka: Opt. Lett. **27**, 2121 (2002)
[99] R.K. Shelton, L.S. Ma, H.C. Kapteyn, M.M. Murnane, J.L. Hall, J. Ye: Science **293**, 1286–1289 (2001)

Few-Cycle Pulses by External Compression

Sandro De Silvestri, Mauro Nisoli, Giuseppe Sansone, Salvatore Stagira, and Orazio Svelto

Istituto Nazionale per la Fisica della Materia, Dipartimento di Fisica, Politecnico di Milano, Piazza Leonardo da Vinci 32, 20133 Milano, Italy
sandro.desilvestri@fisi.polimi.it

Abstract. Recent advances in ultrafast laser technology have led to the generation of light pulses comprising few optical cycles employing different compression techniques. In this chapter, the general scheme of pulse compression will be analyzed, and the most interesting results obtained using different compression schemes will be reviewed. In particular, the hollow-fiber technique, which allows generating sub-5 fs pulses in the millijoule energy range, will be analyzed.

During the last few years the race toward the generation of extremely short light pulses has become very exciting with important achievements [1]. A new era in the technology of femtosecond lasers began in 1991 with the first demonstration of the self-mode-locked Ti:sapphire laser [2]. Since then, a dramatic reduction in achievable pulse duration was obtained. Sub-6 fs pulses have been generated directly from Kerr-lens mode-locked Ti:sapphire lasers using chirped mirrors for dispersion control (see the Chapter by *Kärtner* et al. in this volume). Since the period of the optical cycle in the visible and near-infrared is 2 fs to 3 fs, this pulse duration is nearing the physical limit of devices operating in this wavelength range. In parallel with this progress in femtosecond pulse generation, the introduction, in 1985, of the technique of chirped-pulse amplification (CPA) [3] has made possible the amplification of ultrashort pulses to unprecedented power levels.

Sub-10 fs light pulses can also be generated by external compression. In 1981, *Nakatsuka* et al. [4] introduced a method for optical pulse compression based on the interplay between self-phase modulation (SPM) and group velocity dispersion (GVD) that arises during the propagation of short light pulses in single-mode optical fibers. Using this technique, pulses as short as 6 fs at 620 nm were obtained in 1987 [5] and, employing improved ultrabroad-band dispersion compensation, pulses as short as 4.5 fs at 800 nm were generated in 1997 [6]. The use of single-mode optical fibers limits the pulse energy to a few nanojoules. In 1996, a powerful pulse compression technique based on SPM-induced spectral broadening in a hollow fiber filled with noble gases demonstrated the capability of handling high-energy pulses (submillijoule range) [7]. The implementation of the hollow-fiber technique using 20 fs seed pulses from a Ti:sapphire system and a high-throughput broadband compressor has led to the generation of pulses with duration down to 4.5 fs [8] and energy up to 0.55 mJ [9]. More recently, pulses as short as 3.8 fs have been

F. X. Kärtner (Ed.): Few-Cycle Laser Pulse Generation and Its Applications,
Topics Appl. Phys. **95**, 137–177 (2004)

generated by adaptive compression of a supercontinuum produced in two gas-filled hollow fibers [10]. Using a different compression technique based on phase modulation induced by impulsively excited molecular vibration, pulses as short as 4 fs with an energy of 1.5 μJ have been generated in 2002 [11].

The chapter is organized as follows. In Sect. 1, the general scheme of pulse compression is presented in terms of phase modulation and subsequent ultrabroadband dispersion compensation. Different techniques to achieve ultrabroadband dispersion compensation are presented in Sect. 2. In Sect. 3, the hollow-fiber compression technique is described, a numerical model for nonlinear pulse propagation is presented, and some guidelines for scaling this compression technique are discussed. Experimental results on high-energy pulse compression using the hollow-fiber technique will be presented and compared to numerical predictions. Section 4 reviews experimental results about the compression of ultraviolet pulses in gas-filled hollow fibers. In Sect. 5, alternative schemes for phase modulation and compression of ultrafast light pulses are discussed. Applications and perspectives of the compression techniques are discussed in the last section.

1 General Scheme of Pulse Compression

The general scheme of light pulse compression is the following. The input pulse is first injected into a phase modulator, which broadens the pulse spectrum imposing a frequency sweep (in time). The spectrally broadened and chirped pulse is sent in a dispersive delay line, which rephases all the new frequency components generated by the phase modulation. Ideally, the dispersive delay line would introduce the opposite chirp on the pulse, thus resulting in the compression of the pulse to its minimum width, $\sim 1/\Delta\omega$, where $\Delta\omega$ is the frequency sweep imposed on the pulse in the first step. This general scheme for compression of optical pulses was proposed independently by *Gires* and *Tournois* in 1964 [12] and *Giordmaine* et al. in 1968 [13]. To compress femtosecond pulses, an ultrafast phase modulator has to be used. In 1969, *Fisher* et al. [14] proposed that picosecond pulses could be compressed to femtosecond duration using the large positive chirp produced around the peak of a short pulse by SPM in an optical Kerr liquid. In the same year, *Laubereau* [15] used several cells containing CS_2 and a pair of diffraction gratings to compress, by approximately 10 times, 20 ps pulses generated by a mode-locked Nd:glass laser.

The optical Kerr effect in a medium gives rise to an intensity-dependent change in the refractive index $\Delta n = n_2 I(t)$, where n_2 is the nonlinear index coefficient and $I(t)$ is the optical intensity. The self-induced intensity-dependent nonlinear phase shift experienced by an optical field during its propagation in a Kerr medium of length ℓ is given by $\Delta\phi(t) = (\omega_0/c)n_2 I(t)\ell$ where ω_0 is the pulse central carrier frequency. The induced frequency sweep can be calculated as follows: $\Delta\omega = \mathrm{d}\Delta\phi/\mathrm{d}t$. Around the central part of the pulse, where most of the energy is concentrated, the chirp is approximately linear. The region with linear chirp can be enlarged in the presence of dispersion in the Kerr medium of the same sign as $\Delta\phi$. To compress the spectrally broadened and chirped pulse, a dispersive delay line has to be used, characterized by a nearly linear group delay $T_g(\omega)$.

A fundamental requirement for pulse compression is that the Kerr effect is provided by a guiding nonlinear medium. The spatial intensity profile of the light pulse propagating along a nonguiding medium leads to spatially nonuniform SPM, so that different frequency chirps are generated across the transverse beam distribution. In addition, the radial intensity variation leads to a spatially varying $\Delta\phi$, which gives rise to self-focusing (for positive n_2) and small-scale instabilities [16]. Spatially uniform spectral broadening can be obtained using a guiding nonlinear medium. In 1974, *Ippen* et al. reported the first measurement of SPM in the absence of self-trapping and self-focusing by using a guiding multimode optical fiber filled with liquid CS_2 [17]. In 1978 *Stolen* and *Lin* reported measurements of SPM in single-mode silica core fibers [18]. The important advantage of the single-mode fiber is that the phase modulation can be imposed over the entire transverse profile of the beam, thus removing the problem of unmodulated light in the wings of the beam [17]. In 1981, *Nakatsuka* et al. [4] performed the first pulse compression experiment using fibers as a Kerr medium in the positive dispersion region.

Considering propagation along the z-direction of an optical fiber, the electric field for the fiber modes can be written as follows [19]:

$$E(\boldsymbol{r},\omega) = A(z,\omega)F(x,y)\exp[\mathrm{i}\beta(\omega)z]\,, \tag{1}$$

where $A(z,\omega)$ is the mode amplitude, $F(x,y)$ is the mode-transverse distribution, and $\beta(\omega)$ is the mode-propagation constant. The propagation equation for the guided field splits in two equations for $A(z,\omega)$ and $F(x,y)$. In first-order perturbation theory, a perturbation $\Delta n = \bar{n}_2|E|^2$ of the refractive index does not change the modal distribution $F(x,y)$, and the mode-propagation constant $\bar{\beta}(\omega)$ can be written as $\bar{\beta}(\omega) = \beta(\omega) + \Delta\beta$, where the perturbation $\Delta\beta$ is given by

$$\Delta\beta = \frac{(\omega_0/c)\iint \Delta n|F(x,y)|^2\,\mathrm{d}x\,\mathrm{d}y}{\iint |F(x,y)|^2\,\mathrm{d}x\,\mathrm{d}y}\,. \tag{2}$$

As shown by (2), the perturbation $\Delta\beta$, which includes the effect of fiber nonlinearity, is related to a spatial average on the fiber transverse section of the perturbation Δn. In this way, spatially uniform SPM and related spectral broadening can be realized.

In the last two decades, the general scheme of pulse compression described above has been implemented in different ways. Using a single-mode optical fiber as an ultrafast phase modulator and a prism-grating compressor, pulses as short as 6 fs at 620 nm were obtained in 1987 from 50 fs pulses generated by a colliding-pulse, mode-locking dye laser [5]. More recently, 13 fs pulses from a cavity-dumped Ti:sapphire laser were compressed to 4.5 fs with the same technique using a compressor consisting of a quartz 45° prism pair, a broadband chirped mirror, and a thin-film Gires–Tournois dielectric interferometer compressor [6, 20]. However, the use of a single-mode optical fiber limits the pulse energy to a few nanojoules. In 1996, using a phase modulator consisting of a hollow fiber filled with noble gases, a powerful pulse compression technique was introduced, which demonstrated the capability of handling high-energy pulses [7]. The implementation of the hollow-fiber technique using 20 fs seed pulses from a Ti:sapphire system and a high-throughput broadband prism chirped mirror or chirped mirror only dispersive delay line has led to the generation of pulses with duration down to 4.5 fs [8] and energy up to 0.55 mJ [9]. This technique presents the advantages of a guiding element with a large-diameter mode and of a fast nonlinear medium with high damage threshold and will be discussed extensively in Sect. 3.

The possibility of taking advantage of the ultrabroadband spectrum which can be generated by the phase modulation process, is strictly related to the development of dispersive delay lines capable of controlling the frequency-dependent group delay over such a bandwidth. The next section will overview the main methods for broadband dispersion compensation.

2 Methods for Broadband Dispersion Compensation

The pulse frequency sweep (chirp) imposed by phase modulation is approximately linear near the peak of the pulse, where most of the energy is concentrated. In the presence of dispersion in the phase modulator, the chirp becomes linear over almost the whole pulse. Therefore, optimum temporal compression requires a group delay, $T_g(\omega) = \partial\phi/\partial\omega$, characterized by a nearly linear dependence on frequency in the dispersive delay line. Since the nonlinear index n_2 in SPM is generally positive far from resonance, a negative group delay dispersion (GDD $= \partial T_g/\partial\omega$) is required in the compressor. To generate the shortest pulses, the pulse group delay after the phase modulator and the compressor must be nearly frequency independent. $T_g(\omega)$ can be expanded into a Taylor series around the central frequency ω_0:

$$T_g(\omega) = \phi'(\omega_0) + \phi''(\omega_0)\Delta\omega + \frac{1}{2}\phi'''(\omega_0)\Delta\omega^2 + \frac{1}{3!}\phi''''(\omega_0)\Delta\omega^3 + \cdots, \quad (3)$$

where $\Delta\omega = \omega - \omega_0$, and $\phi''(\omega_0)$, $\phi'''(\omega_0)$, and $\phi''''(\omega_0)$ are the second-, the third-, and the fourth-order-dispersion terms, respectively. Critical values of these dispersion terms above which dispersion causes a significant change in the pulse are given by a simple scaling expression: $\phi^{(n)} = \tau_p^n$, where $\phi^{(n)}$ is the nth-order dispersion term and τ_p is the pulse duration. For example, a second-order dispersion $\phi'' = \tau_p^2$ results in pulse broadening by more than a factor of 2. Therefore dispersion-induced pulse broadening and distortion become increasingly important for decreasing pulse durations. Equation (3) shows that to compress a pulse to near the transform limit, one should eliminate these high-order dispersion terms. For instance, assuming a transform-limited input pulse to the phase modulator, the condition for third-order-dispersion-compensated compression is the following:

$$\begin{aligned}\phi''(\omega_0) &= \phi''_{\text{modulator}} + \phi''_{\text{compresssor}} = 0\,,\\ \phi'''(\omega_0) &= \phi'''_{\text{modulator}} + \phi'''_{\text{compresssor}} = 0\,.\end{aligned} \tag{4}$$

Several compressor schemes have been developed so far that included such components as diffraction gratings, Brewster-cut prism pairs, combination of gratings and prisms, thin prisms and chirped mirrors, and chirped mirrors. In the following, we will briefly outline the main characteristics of these compressor schemes.

2.1 Dispersion Compensation by Grating and Prism Pairs

In 1968, *Treacy* demonstrated, for the first time, the use of a pair of diffraction gratings to achieve negative GDD [21]. In 1984, *Fork* et al. obtained a negative GDD using pairs of Brewster-angled prisms [22], which have been widely used for dispersion control inside laser oscillators since their discovery. In both optical systems, the origin of the adjustable dispersion is the angular dispersion that arises from diffraction and refraction, respectively. The dispersion introduced by these systems can be easily calculated by calculating the phase accumulated between the input and output reference planes [23]. To understand the main properties of such systems, we will refer to Fig. 1. The first element scatters the input beam with wave vector $\boldsymbol{k}_{\text{in}}$ and input path vector $\boldsymbol{l}$ in the direction $\boldsymbol{k}_{\text{out}}$. The beam passes between the first and the second elements and is scattered back in its original direction. The phase accumulated by the scattered beam with respect to the (unscattered) reference is $\phi(\omega) = \boldsymbol{k}_{\text{out}}(\omega) \cdot \boldsymbol{l}$. Considering free-space propagation between the two elements, $|\boldsymbol{k}_{\text{out}}| = \omega/c$, and the accumulated phase can be written as follows:

$$\phi(\omega) = \frac{\omega}{c}|\boldsymbol{l}|\cos[\gamma - \alpha(\omega)] = \frac{\omega}{c}\frac{D}{\cos(\gamma)}\cos[\gamma - \alpha(\omega)]\,, \tag{5}$$

where γ is the angle between the incident wave vector and the normal to the first element; α is the angle of the outgoing wave vector, which is a function of frequency; and D is the spacing between the scattering elements in a direction parallel to their normal. For a grating pair, the frequency dependence of the diffraction angle α is governed by the grating law, that in first-order diffraction is given by

$$\frac{2\pi c}{\omega} = d[\sin\alpha(\omega) - \sin\gamma]\,, \tag{6}$$

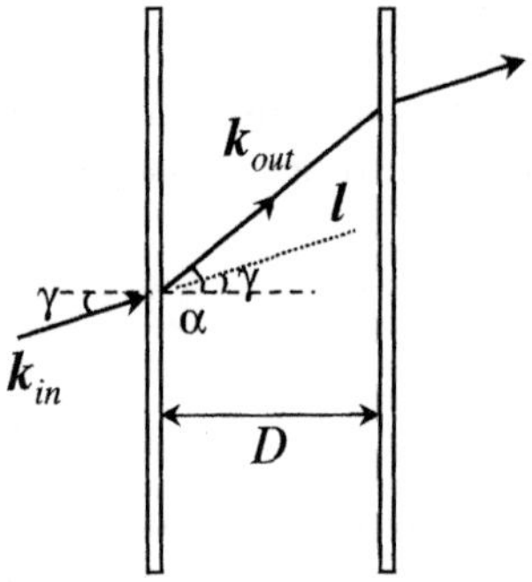

Fig. 1. General scheme of the optical path in a two-element dispersive delay line

where d is the groove spacing of the grating. Using (5) and (6), it is possible to obtain analytic expressions for the GDD and higher order dispersion terms (for single pass):

$$\phi''(\omega) = -\frac{4\pi^2 cD}{\omega^3 d^2 \cos^3 \alpha(\omega)} , \tag{7}$$

$$\phi'''(\omega) = \frac{12\pi^2 cD}{\omega^4 d^2 \cos^3 \alpha(\omega)} \left[1 + \frac{2\pi c \sin \alpha(\omega)}{\omega d \cos^2 \alpha(\omega)} \right] . \tag{8}$$

As evident from (7), where D is the distance between the gratings, the grating pair gives negative dispersion. A disadvantage of the grating pair is the diffraction losses that result in an energy reduction of about a factor of 4, typically, in the double-pass configuration. Moreover, the bandwidth for efficient diffraction is limited.

For a Brewster-angled prism pair, (5) reduces to the following expression (single pass):

$$\phi(\omega) = \frac{\omega}{c} \ell_{\mathrm{p}} \cos \beta(\omega) , \tag{9}$$

where ℓ_{p} is the distance between prism apexes and $\beta(\omega)$ is the angle between the refracted ray at frequency ω and the line joining the two apexes. The second- and third-order dispersions can be expressed in terms of the optical path $P(\lambda) = \ell_{\mathrm{p}} \cos \beta(\lambda)$:

$$\phi''(\omega) = \frac{\lambda^3}{2\pi c^2} \frac{\mathrm{d}^2 P}{\mathrm{d}\lambda^2} , \tag{10}$$

$$\phi'''(\omega) = -\frac{\lambda^4}{4\pi^2 c^3} \left(3\frac{\mathrm{d}^2 P}{\mathrm{d}\lambda^2} + \lambda \frac{\mathrm{d}^3 P}{\mathrm{d}\lambda^3} \right) , \tag{11}$$

with the following derivatives of the optical path with respect to wavelength evaluated at the Brewster wavelength:

$$\frac{\mathrm{d}^2P}{\mathrm{d}\lambda^2} \equiv P'' = 2\left[n'' + (2n - n^{-3})(n')^2\right]\ell_{\mathrm{p}}\sin\beta - 4(n')^2\ell_{\mathrm{p}}\cos\beta\,, \qquad (12)$$

$$\frac{\mathrm{d}^3P}{\mathrm{d}\lambda^3} \equiv P''' = \left[6(n')^3(n^{-6} + n^{-4} - 2n^{-2} + 4n^2) + 12n'n''(2n - n^{-3}) + 2n'''\right]\ell_{\mathrm{p}}\sin\beta + 12\left[(n^{-3} - 2n)(n')^3 - n'n''\right]\ell_{\mathrm{p}}\cos\beta\,, \qquad (13)$$

where n is the refractive index of the prism material; and n', n'', and n''' are, respectively, the first-, second- and third-order derivatives of n, with respect to wavelength. The Brewster-angled-based compressor offers the advantage of reduced losses. Using only fused silica prisms for dispersion compensation, sub-10 fs light pulses were generated directly from an oscillator in 1994 [24]. In 1996, pulses with tens of microjoules energy, spectrally broadened in a gas-filled hollow fiber were compressed down to 10 fs using a prism compressor [7]. In both grating and prism pairs, negative GDD is associated with a significant amount of higher order dispersion, which cannot be lowered or adjusted independently of the desired GDD, thus limiting the bandwidth over which correct dispersion control can be obtained. This drawback has been only partially overcome by combining prism and grating pairs with third-order dispersion of opposite sign. In this way, pulses as short as 6 fs were generated in 1987 [5], and less than 5 fs in 1997 [6], by external compression. This combination cannot be used for few-optical-cycle pulse generation either in laser oscillators, due to the high diffraction losses of the gratings, or in external compressors at high power level, due to the onset of unwanted nonlinearities in the prisms.

2.2 Dispersion Compensation by Chirped Mirrors

More recently, advances in the design of chirped multilayer coatings [25] led to the demonstration of chirped mirrors providing ultrabroadband dispersion control with low losses and the capability of handling high intensities [26,27]. Chirped-mirror-only dispersion compensation offers the additional advantage of greatly simplifying the system design, allowing for compactness, reproducibility, and insensitivity to misalignment, which are of paramount importance in practical application. Chirped mirrors introduce a frequency-dependent group delay by reflecting different spectral components of the incident radiation at different positions in the multilayer structure. The often required linear variation of the group delay with frequency cannot be produced by linearly chirping the multilayer period because this would introduce oscillations in the group delay for longer wavelengths, which penetrate more deeply into the mirror. Such oscillations are due to interferences between slight reflections in the front section and strong reflection from the back, as in a Gires–Tournois interferometer. Replacement of the design technique based exclusively on computer optimization with analytic design methods for the reproducible fabrication of chirped mirrors [27, 28] has achieved simultaneously high reflectivity and dispersion compensation over bandwidths exceeding 160 THz at 800 nm. Chirped mirrors are now key components in broadband delay lines used for generation of sub-10 fs pulses directly from laser oscillators, as reported in the Chapter by *Kärtner* et al. Chirped mirrors are used for the compression of high-energy pulses because they provide high dispersion with little material in the beam path, thus avoiding nonlinear effects in the compressor.

2.3 Methods for Active Dispersion Compensation

In recent years, various schemes for active pulse compression have been reported, based on the use of liquid-crystal modulators (LCM), acousto-optic modulators (AOM), and mechanically deformable mirrors.

2.3.1 Dispersion Compensation Using Liquid Crystal Modulators

A pulseshaping technique [29] based on the use of a LCM for pulse compression offers the advantage of a large bandwidth (300 nm to 1500 nm) and in situ adaptive phase control. In 1997, *Yelin* et al. [30] demonstrated an adaptive method for femtosecond pulse compression based on LCM. Strongly chirped 80 fs pulses generated by an oscillator were sent into a $4f$ pulseshaper composed of a pair of thin holographic transmission gratings. A programmable one-dimensional LCM, placed in the Fourier plane of the shaper, was used as an updatable filter for pulse spectral manipulation. Pulses as short as 11 fs (transform-limited duration: 9 fs) have been obtained, employing an optimization algorithm for adaptive compression based on a search in the two-dimensional space of second- and third-order dispersion coefficients. In 2001, *Karasawa* et al. [31] demonstrated pulse compression, down to 5 fs, of broadband pulses from an argon-filled hollow fiber, using only a LCM for phase compensation. More recently, pulses as short as 3.8 fs have been achieved through a closed-loop combination of a liquid-crystal spatial light modulator for adaptive pulse compression and spectral-phase interferometry for direct electric-field reconstruction (SPIDER) [32] measurements as feedback signal [10]. The experimental apparatus and the results will be discussed in Sect. 3.4.

One problem of the method is pixelization in the Fourier plane owing to the technology of the liquid-crystal active matrix. Diffraction on pixel edges and absorption by the black matrix introduce parasitic effects. The requirement that the actual spectral modulation should approximate a smooth function despite the fixed, finite size of the individual modulator elements, limits the temporal range over which pulse compression can be achieved [33]. Other problems are related to the optical damage of the LCM, which limits the maximum pulse energy, and to the high losses introduced by the device.

Various nonpixelated devices have been proposed. *Dorrer* et al. reported on an optically addressed LCM (liquid crystal light valve) [34]. The light valve consists of two continuous transparent electrodes and continuous layers of a nematic twisted liquid crystal and of photoconductive $Bi_{12}SiO_{20}$ (BSO). A local variation of illumination of the BSO layer (in the blue green spectral region) induces a change in conductivity. When a voltage is applied between the two electrodes, the variation of the BSO conductivity results in a change in the voltage drop across the liquid crystal layer. As the birefringence of the liquid crystal is voltage dependent, a local variation of the refractive index is created, which translates into a variation of the optical phase of the local spectral component. The light valve is addressed by using a display device. Pixelation effects are avoided because the light valve itself is a continuous device. Control of the light valve is more complicated than for the electrically addressed LCM. Moreover, due to its limited spatial frequency response, spectral resolution is limited.

2.3.2 Dispersion Compensation Using Acousto-Optic Modulators

Acousto-optic modulators (AOM) provide useful pulse compression and shaping [35]. The AOM crystal is located in the Fourier plane of a zero-order dispersion line and is driven by a radio-frequency (rf) voltage signal, which is converted into a traveling wave by a piezoelectric transducer. The acoustic wave traveling in the modulator induces a refractive index grating through a photoelastic effect. Such a grating can be phase, amplitude, or frequency modulated by using a suitably modulated rf signal. The optical spectrum of the input light pulse is modified according to the grating spatial modulation function.

In 1997, *Tournois* [36] proposed an acousto-optic programmable dispersive filter (AOPDF), providing for large dispersion-compensation ranges, which does not have to be positioned in the Fourier plane of a dispersion line. The device is based on a collinear acousto-optic interaction in a birefringent uniaxial crystal. The acoustic frequency is a variable function of time and provides control over the group delay of the diffracted optical pulse. At the same time, the spectral amplitude of the diffracted pulse is driven by the intensity of the acoustic signal. As demonstrated in [36], the optical output $E_{\text{out}}(t)$ of the AOPDF is proportional to the convolution of the optical input, $E_{\text{in}}(t)$, and the scaled acoustic signal:

$$E_{\text{out}}(t) \propto E_{\text{in}}(t) \otimes S(t/\alpha)\,, \tag{14}$$

where the scaling factor $\alpha = \Delta n(V/c)$ is the ratio of the speed of sound to the speed of light times the index difference between the ordinary and the extraordinary waves. Therefore, by generating the proper function $S(t)$, it is possible to generate any arbitrary convolution with a temporal resolution given by the inverse of the filter bandwidth. Such a device has been used in a kilohertz chirped-pulse amplification laser chain: compensating for gain narrowing and residual phase errors with the AOPDF, resulted in the generation of 17 fs transform-limited pulses [37]. The total throughput is 30% to 50%, depending on the bandwidth of the device.

2.3.3 Dispersion Compensation Using Deformable Mirrors

Mechanically deformable mirrors can be used for active dispersion control, as proposed by *Heritage* et al. [38]. More recently, pulse compression has been achieved using an electrostatically deformable, gold-coated, silicon nitride membrane mirror, placed in the Fourier plane of a $4f$ zero-dispersion stretcher [39]. The membrane was suspended over an array of 39 actuator electrodes. The potential applied to each actuator generates an electrostatic attraction between the membrane and the electrode, thus inducing a deformation of the mirror surface, which translates into modulation of the phase of the spectral components of the input pulse. The total phase difference is $\phi = 2(2\pi)\Delta z/\lambda$, where Δz is the deflection of the mirror. The minimum radius of curvature of the mirror membrane is given by $R = T/P$, where T is the membrane tension and P is the maximum electrostatic pressure. This limitation of the membrane curvature restricts the possibility of the mirror correction of higher order phases. The main advantages of this method are the following: the phase modulation is smoothly varying, reduced losses due to the high reflectivity (97%) of the mirror, and relatively high actuator density. Experiments have been performed with a mode-locked Ti:sapphire laser, where the deformable mirror recompressed a 15 fs pulse, previously stretched to 90 fs by dispersion in glass, back to approximately the bandwidth limit [39].

Recently, dispersion control over a bandwidth of $\sim$ 220 THz has been demonstrated by *Baltuška* et al. [40] using a compressor consisting of a pair of chirped mirrors and a grating dispersion line with a computer-controlled flexible mirror positioned in the focal plane. The total throughput of the pulseshaper was less than 12% because of the low diffraction efficiency of the grating. Using this compressor, the visible–near-IR pulses, generated by optical parametric amplification, were compressed to 4 fs duration.

3 Hollow-Fiber Compression Technique

In the preceding two sections, we discussed a general method for the compression of femtosecond light pulses. In this section, we will concentrate on a particular compression technique, introduced in 1996 [7], which allows generating few-optical-cycle light pulses with energy in the millijoule range. The technique is based on propagation of laser pulses in a hollow fiber filled with noble gases (the hollow-fiber compression technique). This section is organized as follows: a detailed discussion of the propagation modes in a hollow fiber will be developed in the first part; some basic aspects such as laser-capillary coupling efficiency, fiber dispersion, and radiative losses will be analyzed. In the second part, we will consider nonlinear propagation of laser pulses with an arbitrary polarization state in a gas-filled capillary; the propagation equations will be developed in the framework of the envelope approximation; and some particular cases (linear, circular, and elliptical polarization) will be discussed. Numerical simulation of the spectral broadening of ultrashort pulses achieved by the hollow-fiber technique will be presented, and the compression of such pulses by dispersion compensation will be considered in the general case of an arbitrary polarization state. Experimental results will be briefly reviewed.

3.1 Propagation Modes in Hollow Fibers

The propagation of light in a hollow waveguide is a well-studied problem [41]. Such fibers were considered for a long-distance communication channel when optical fibers were still not reliable. Electromagnetic radiation propagates in hollow fibers by grazing incidence reflections at the dielectric inner surface; only leaky modes are supported because of power losses through the fiber walls. The modes of the hollow waveguide are calculated using the propagation equation for a homogeneous dielectric medium, which is solved separately in the two regions with constant refractive index, the hollow core and the dielectric cladding:

$$\nabla^2 \boldsymbol{E}^{(j)} - \frac{\epsilon_j}{c^2}\frac{\partial^2 \boldsymbol{E}^{(j)}}{\partial t^2} = 0\,, \tag{15}$$

where $\boldsymbol{E}^{(j)}$ is the electric field in the jth region of the waveguide section (core or cladding), ϵ_j the corresponding dielectric constant, and c is the speed of light. Adjacent dielectric regions are connected using boundary conditions. Owing to the cylindrical symmetry of the capillary, it is convenient to project the propagation mode onto a cylindrical basis. In such a case, the electric field of the mode is given by

$$\boldsymbol{E}^{(j)} = E_z^{(j)}\boldsymbol{u_z} + E_r^{(j)}\boldsymbol{u_r} + E_\theta^{(j)}\boldsymbol{u_\theta}\,, \tag{16}$$

where $\boldsymbol{u_z}$ is a unity vector oriented in the propagation direction z; $\boldsymbol{u_r}$ is an outgoing radial unity vector, perpendicular to the z-axis; and $\boldsymbol{u_\theta}$ is a tangential unity vector clockwise oriented with respect to the propagation direction. An approach to solving (15) is to write the electric field as

$$E_{l,\nu}^{(j)}(\boldsymbol{r},t) = \mathrm{Re}\left\{\frac{\exp[\mathrm{i}(\omega t - \beta z)]}{2} \cdot \left[\widetilde{E}_{l,\nu}^{(j)}(r)\exp(\mathrm{i}\nu\theta) + \widetilde{E}_{l,-\nu}^{(j)}(r)\exp(-\mathrm{i}\nu\theta)\right]\right\}, \tag{17}$$

where $l = \{z, r, \theta\}$, ω is the light frequency, β is the propagation constant of the considered mode, and $\nu \geq 0$ is an integer. With these assumptions, the vectorial propagation equation within each of the homogeneous dielectric regions of the waveguide breaks in three equations for the longitudinal, radial, and tangential components of the field:

$$\nabla_r^2 \widetilde{E}_z^{(j)} + \left(\frac{\omega^2}{c^2}\epsilon_j - \beta^2 - \frac{\nu^2}{r^2}\right)\widetilde{E}_z^{(j)} = 0\,, \tag{18}$$

$$\nabla_r^2 \widetilde{E}_r^{(j)} - \frac{2\mathrm{i}\nu}{r^2}\widetilde{E}_\theta^{(j)} + \left(\frac{\omega^2}{c^2}\epsilon_j - \beta^2 - \frac{\nu^2+1}{r^2}\right)\widetilde{E}_r^{(j)} = 0\,, \tag{19}$$

$$\nabla_r^2 \widetilde{E}_\theta^{(j)} + \frac{2\mathrm{i}\nu}{r^2}\widetilde{E}_r^{(j)} + \left(\frac{\omega^2}{c^2}\epsilon_j - \beta^2 - \frac{\nu^2+1}{r^2}\right)\widetilde{E}_\theta^{(j)} = 0\,, \tag{20}$$

where $\nabla_r^2 = \partial^2/\partial r^2 + (1/r)\partial/\partial r$ and the index ν has been omitted for brevity. Equations (18–20) must be solved with boundary conditions derived from the Maxwell equations, which can be written according to the following relations:

$$\widetilde{E}_\theta^{(j)} = \widetilde{E}_\theta^{(j+1)}\,, \tag{21}$$

$$\widetilde{E}_z^{(j)} = \widetilde{E}_z^{(j+1)}\,, \tag{22}$$

$$\epsilon_j \widetilde{E}_r^{(j)} = \epsilon_{j+1}\widetilde{E}_r^{(j+1)}\,, \tag{23}$$

$$\frac{\partial \widetilde{E}_\theta^{(j+1)}}{\partial r} - \frac{\partial \widetilde{E}_\theta^{(j)}}{\partial r} = \frac{\mathrm{i}\nu}{a_j}\left(\widetilde{E}_r^{(j+1)} - \widetilde{E}_r^{(j)}\right), \tag{24}$$

$$\frac{\partial \widetilde{E}_z^{(j+1)}}{\partial r} - \frac{\partial \widetilde{E}_z^{(j)}}{\partial r} = -\mathrm{i}\beta\left(\widetilde{E}_r^{(j+1)} - \widetilde{E}_r^{(j)}\right), \tag{25}$$

$$\frac{\partial \widetilde{E}_r^{(j+1)}}{\partial r} - \frac{\partial \widetilde{E}_r^{(j)}}{\partial r} = -\frac{1}{a_j}\left(\widetilde{E}_r^{(j+1)} - \widetilde{E}_r^{(j)}\right), \tag{26}$$

where a_j is the radius of the cylindrical surface which separates the jth and the $(j+1)$th dielectric regions and the previous relations must be evaluated on that surface.

On the basis of (18–20), three kinds of propagation modes can be obtained, according to the following considerations:

1. For $\nu = 0$, (18–20) decouple, and two solutions can be given:
   (a) If $\widetilde{E}_z = 0$ everywhere, we can argue on the basis of the boundary conditions (23) and (25) that also $\widetilde{E}_r = 0$; the electric field has only a tangential component. Using (20), the mode can be expressed by

$$\widetilde{E}_\theta^{(j)}(r) = A^{(j)} Z_1(\gamma_j r)\,, \tag{27}$$

   where $Z_1(r)$ is one of the Bessel functions of the first order and $\gamma_j = \sqrt{\omega^2\epsilon_j/c^2 - \beta^2}$; the result is a transverse circular electric (TE_{0m}, $m = 0, 1, 2, \ldots$) mode, in which the electric field lines are transverse concentric circles centered on the propagation axis;
   (b) If $\widetilde{E}_\theta = 0$ everywhere, the other two components are present and given by the relations

$$\widetilde{E}_r^{(j)}(r) = B^{(j)} Z_1(\gamma_j r)\,, \tag{28}$$

$$\widetilde{E}_z^{(j)}(r) = C^{(j)} Z_0(\gamma_j r)\,. \tag{29}$$

It can be shown from the Maxwell equations that in this case a transverse circular magnetic (TM_{0m}) mode is obtained.

2. For $\nu \neq 0$, (18–20) are coupled; all field components are present in this case, giving rise to hybrid modes, which are indicated as $EH_{\nu m}$. It can be shown that axial components in $EH_{\nu m}$ are small, thus allowing this kind of mode to be thought of as transverse. In such a case, the general solution to (18–20) can be written as

$$\widetilde{E}^{(j)}_{\theta,\nu}(r) = A^{(j)} Z_{\nu+1}(\gamma_j r) + B^{(j)} Z_{\nu-1}(\gamma_j r)\,, \tag{30}$$

$$\widetilde{E}^{(j)}_{r,\nu}(r) = \mathrm{i}\left[A^{(j)} Z_{\nu+1}(\gamma_j r) - B^{(j)} Z_{\nu-1}(\gamma_j r)\right]\,, \tag{31}$$

$$\widetilde{E}^{(j)}_{z,\nu}(r) = C^{(j)} Z_{\nu}(\gamma_j r)\,. \tag{32}$$

Determination of coefficients $A^{(j)}, B^{(j)}, C^{(j)}$ and of the propagation constant β is obtained from the boundary conditions. It is worth pointing out that two additional boundary conditions must be added to the problem: 1. a finite value of the electric field must be imposed on the axis of the waveguide (so that in the core of the waveguide, only linear combinations of Bessel functions of the first kind $J_\nu(r)$ are allowed); 2. the energy flow of the leaky mode in the cladding of the capillary must be directed outward. The last condition is respected if the mode solution in the cladding is assumed to be a linear combination of outgoing Hankel functions $H^{(2)}_\nu(r)$. Here, we are mainly interested in those capillary modes that can be efficiently coupled to a laser beam. Such modes must show quasi-linear polarization, which corresponds to the condition $\nu = 1$, thus restricting our analysis to EH_{1m} modes. In this case, the (complex) values of the propagation constant for the EH_{1m} modes can be achieved by looking for the roots of the following determinant, which derives from (21–26):

$$\begin{bmatrix} -J_0 & -J_2 & H_0 & H_2 \\ \mathrm{i}J_0\epsilon_1 & -\mathrm{i}J_2\epsilon_1 & -\mathrm{i}H_0\epsilon_2 & \mathrm{i}H_2\epsilon_2 \\ -(\partial J_0 - J_0/a) & -(\partial J_2 + J_2/a) & (\partial H_0 - H_0/a) & (\partial H_2 + H_2/a) \\ \mathrm{i}(\partial J_0 + J_0/a) & -\mathrm{i}(\partial J_2 + J_2/a) & -\mathrm{i}(\partial H_0 + H_0/a) & \mathrm{i}(\partial H_2 + H_2/a) \end{bmatrix},$$

where $\epsilon_1 = n^2_{\mathrm{core}}$ is the relative dielectric constant in the (hollow) core, with refractive index n_{core}; $\epsilon_2 = n^2_{\mathrm{clad}}$ is the relative dielectric constant in the cladding, with refractive index n_{clad}; a is the inner capillary radius; $\partial \equiv \partial/\partial r$; $J_0 = J_0(\gamma_1 r)$, $J_2 = J_2(\gamma_1 r)$, $H_0 = H^{(2)}_0(\gamma_2 r)$, $H_2 = H^{(2)}_2(\gamma_2 r)$, and all functions are evaluated or derived at $r = a$.

For fiber diameters sufficiently larger than the optical wavelength, an approximate expression for the electric field components of the EH_{1m} modes and for the corresponding propagation constant can be derived analytically [41].

The complex propagation constant $\beta(\omega)$ of an EH_{1m} mode is given by

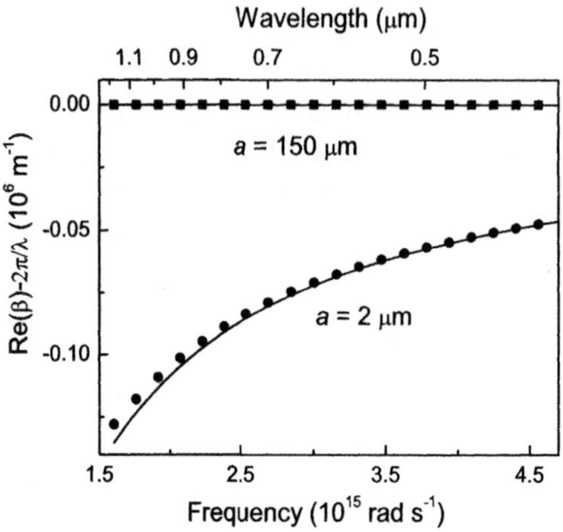

Fig. 2. Difference between the real part of the propagation constant β of the EH_{11} mode and the free propagation constant $2\pi/\lambda$ as a function of frequency for $a = 2\,\mu m$ (*solid dots*) and for $a = 150\,\mu m$ (*solid squares*). *Solid lines*: analytical approximation of the same quantity. In both cases, it was assumed that $n_{\text{core}} = 1$, $n_{\text{clad}} = 1.45$

$$\beta(\omega) = \frac{\omega n_{\text{core}}(\omega)}{c}\left[1 - \frac{1}{2}\left(\frac{u_m c}{\omega n_{\text{core}}(\omega) a}\right)^2\right] + \frac{\mathrm{i}}{a^3}\left(\frac{u_m c}{\omega n_{\text{core}}(\omega)}\right)^2 \frac{n^2(\omega)+1}{\sqrt{n^2(\omega)-1}}, \quad (33)$$

where $n(\omega)$ is the ratio between the refractive indexes of the external and internal media and u_m is the mth zero of $J_0(r)$. Among the hybrid modes of the first order, EH_{11} has the lower leakage; for this reason, we will call it the fundamental mode. The very good approximation given by (33) is shown in Fig. 2, where the difference between the real part of the propagation constant β of the EH_{11} mode and the free propagation constant $2\pi/\lambda$ is reported as a function of frequency; it was assumed in the calculations that $n_{\text{core}} = 1$ and $n_{\text{clad}} = 1.45$. The difference between the exact values of $\text{Re}(\beta)$ (solid symbols) and the approximated ones (solid lines) is negligible for a $150\,\mu m$ fiber radius and appears very low even for a $2\,\mu m$ capillary. It is worth noting that, for fiber radii in the range of a few hundreds of microns, the propagation constant in the visible spectral region is similar to the free propagation value. Strong differences are observed for radii in the range of a few tens of microns. Similar considerations can be done by comparing the exact and the approximated values of $\text{Im}(\beta)$, which are shown in Fig. 3 for the same values of refractive index and fiber radii reported in the previous figure; the analytical approximation of the propagation constant is also very good in this case, even for very small capillaries. It must be recalled here that the attenuation coefficient for guided radiation is proportional to $\text{Im}(\beta)$; as can

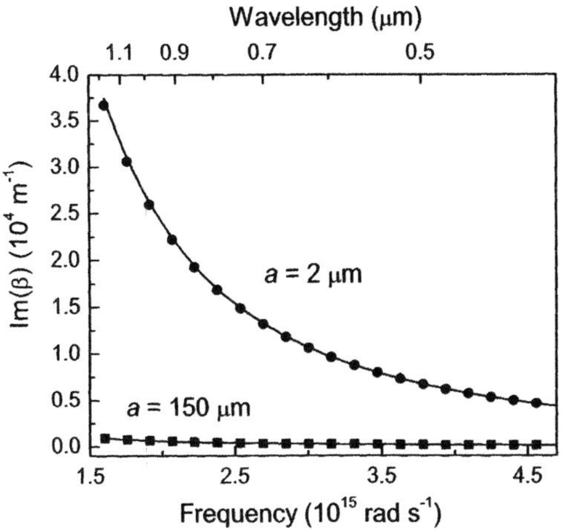

Fig. 3. Imaginary part of the propagation constant β of the EH_{11} mode as a function of frequency for $a = 2\,\mu m$ (*solid dots*) and for $a = 150\,\mu m$ (*solid squares*). *Solid lines*: analytical approximation of the same quantity. In both cases, it was assumed that $n_{core} = 1$, $n_{clad} = 1.45$

be seen from the figure, the attenuation changes with frequency, especially for small fibers. For a capillary radius of the order of a hundred microns, the spectral dispersion in the attenuation cannot be neglected because it is responsible for spectral reshaping of the input field. An ultrashort pulse at the input of the hollow fiber would suffer spectral narrowing and consequently temporal broadening.

Once the propagation constant has been determined, the field pattern in the waveguide can be obtained from (30–32) by using the boundary conditions previously mentioned. Although EH_{11} should be a quasi-linearly polarized mode, departures from linear polarization can be observed for very small capillaries. This behavior is ascribed to the refraction of field lines across the fiber walls due to the discontinuity of the radial electric field component. Figure 4 shows the electric field lines inside a hollow fiber with (a) a $150\,\mu m$ and (b) a $2\,\mu m$ core radius. As can be seen from the plot, the departure from linear polarization is not negligible only for fibers with a radius of the order of a few microns; in this case, the difference in linear polarization is more pronounced on the sides than in the central part of the mode. Figure 5 shows the normalized intensity of the EH_{11} mode as a function of position across the hollow fiber for an inner radius $a = 2\,\mu m$. The mode presents a bell-shaped central part corresponding to the portion guided by grazing incidence reflections on the glass walls; the oscillatory pattern around the peak is due to the propagating wave across the cladding which is related to mode leakage. Owing to the very small inner radius, strong losses are experienced in this case, so that the outgoing light wave is clearly visible. The mode shape is substantially unchanged for larger fibers; nevertheless,

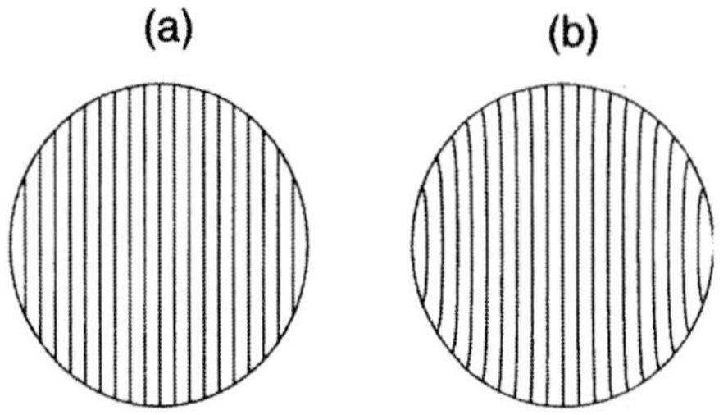

Fig. 4. Field lines of the EH_{11} mode calculated for a capillary with inner radius of (**a**) 150 μm and (**b**) 2 μm. In the calculations, it was assumed that $n_{core} = 1$, $n_{clad} = 1.45$

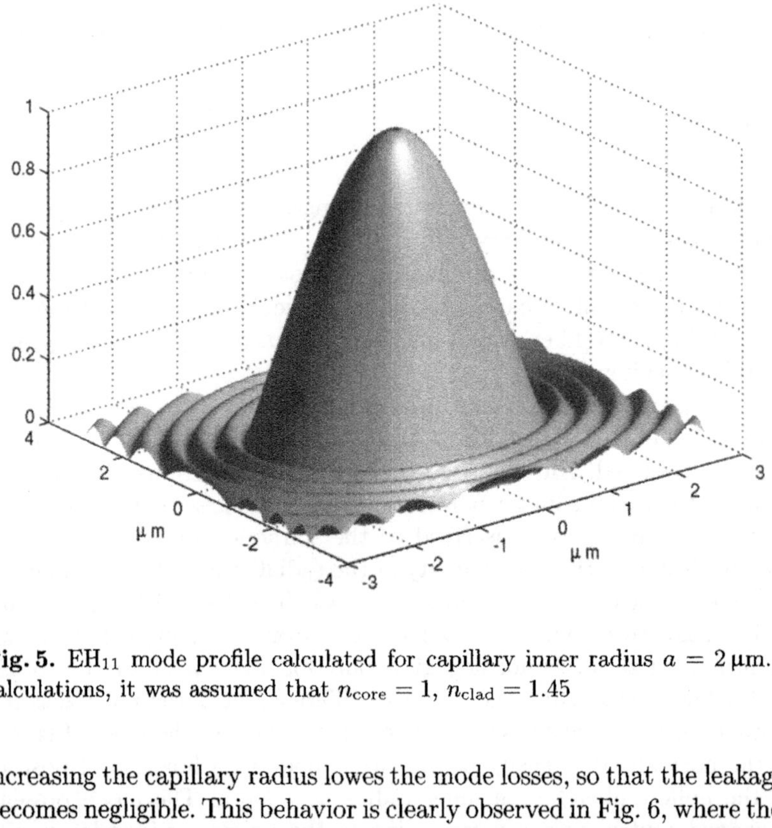

Fig. 5. EH_{11} mode profile calculated for capillary inner radius $a = 2$ μm. In the calculations, it was assumed that $n_{core} = 1$, $n_{clad} = 1.45$

increasing the capillary radius lowes the mode losses, so that the leakage wave becomes negligible. This behavior is clearly observed in Fig. 6, where the EH_{11} mode of a 150 μm capillary is shown. In the limit of large fiber diameters, the mode profile inside the capillary can be approximated analytically. In such a case, the intensity of EH_{1m} modes is given by $I_c(r) = I_{c0} J_0^2(u_m r/a)$, where I_{c0} is the peak intensity.

When the laser beam is injected into the capillary, many modes can be excited. Nevertheless, single-mode operation is generally required for pulse

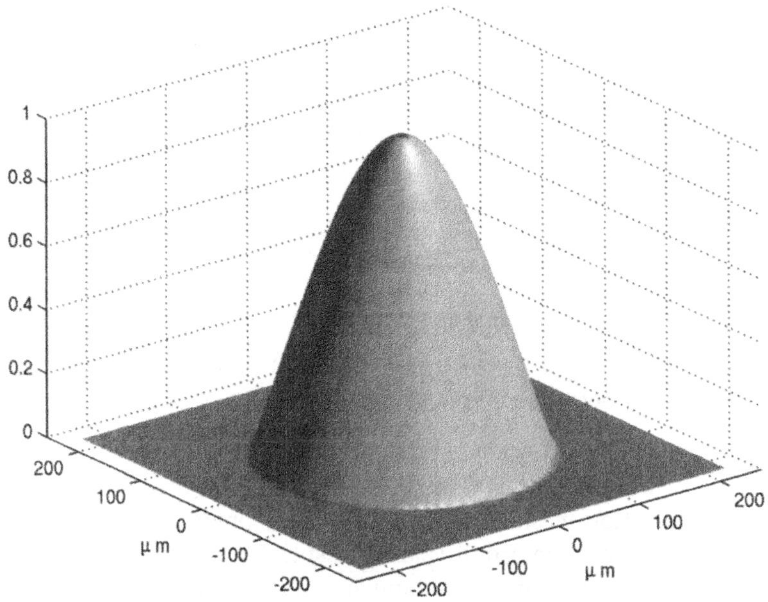

Fig. 6. EH_{11} mode profile calculated for a capillary inner radius $a = 150\,\mu m$. In the calculations, it was assumed that $n_{core} = 1$, $n_{clad} = 1.45$

compression, thus mode discrimination must be actuated. This goal can be easily achieved by optimal coupling between the input laser beam and the fundamental fiber mode EH_{11}. Assuming a Gaussian linearly polarized input beam with an intensity profile $I_l(r) = I_{l0} \exp(-r^2/w_l^2)$, it is possible to determine the expression for the coupling efficiency between the input beam and the capillary modes, which is given by [42]

$$\eta = \frac{4\left[\int rJ_0(u_{1m}r/a)\exp(-r^2/w_l^2)\,\mathrm{d}r\right]^2}{w_l^2\int rJ_0^2(u_{1m}r/a)\,\mathrm{d}r}. \tag{34}$$

As can be seen from Fig. 7, for an optimum value of $w_l/a = 0.65$, the coupling efficiency of the EH_{11} mode with the input Gaussian beam is $\sim 98\%$; higher modes show a value lower than 0.5%. It is worth pointing out that, even if higher order modes were excited, mode discrimination would be achieved anyway, owing to the higher loss rate of EH_{1m} with respect to the fundamental mode [41]. Mode discrimination in the capillary also allows spatial filtering of the input beam [43], which, it was recently demonstrated, is of fundamental importance in some nonlinear phenomena, like high-order harmonic generation [44].

An important issue in nonlinear pulse compression is the dispersion experienced by light pulses during propagation in a gas-filled capillary. As already seen in Fig. 2, the propagation constant departs from free-propagation behav-

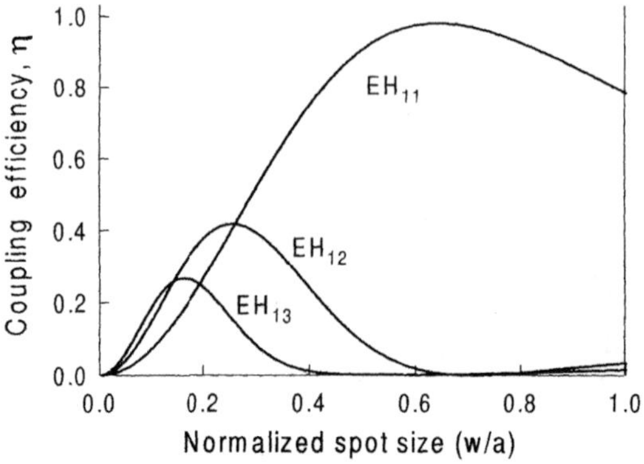

Fig. 7. Coupling efficiency η for the hybrid modes EH_{1m} excited by a Gaussian beam as a function of normalized input spot size w/a

ior as the fiber radius lessens; we can thus expect higher dispersion induced by guided propagation in smaller capillaries. This prediction is confirmed by Fig. 8, which shows the group velocity dispersion GVD $= \partial^2\beta/\partial\omega^2$ as a function of frequency for different capillary radii. Negligible dispersion is achieved for a fiber diameter in the range between 300 μm and 500 μm. In the following, we will show how optical compression can be obtained with dispersion compensation of the order of tens of square femtoseconds; such values are comparable to those induced by guided propagation in small capillaries (with diameters between 100 μm and 200 μm); in these cases, the physics of nonlinear propagation of a femtosecond pulse in a gas-filled capillary can differ substantially with respect to larger fiber diameters.

3.2 Nonlinear Pulse Propagation in Hollow Fibers

The propagation of ultrashort pulses in nonlinear media is often treated by considering the evolution of the pulse envelope alone. Such an approach is valid down to single-cycle optical pulses [45], provided that the slowly evolving wave approximation (SEWA) is applicable. The SEWA has two requirements: 1. The pulse envelope undergoes "small" changes during propagation, and 2. the carrier–envelope phase does not change significantly across a distance equal to a wavelength. Both conditions are met if nonlinear propagation in noble gases is considered. In the following, we will develop a numerical model for nonlinear propagation of ultrashort laser pulses with an arbitrary polarization state in the framework of the SEWA. Let us consider an input pulse, whose electric field in a hollow fiber is given in the temporal domain by [46]

$$\boldsymbol{E}(t) = [\boldsymbol{\sigma}_+ A_+(t) + \boldsymbol{\sigma}_- A_-(t)]\,\mathrm{e}^{-\mathrm{i}\omega_0 t} + \mathrm{c.c.}\,, \tag{35}$$

where the quantities,

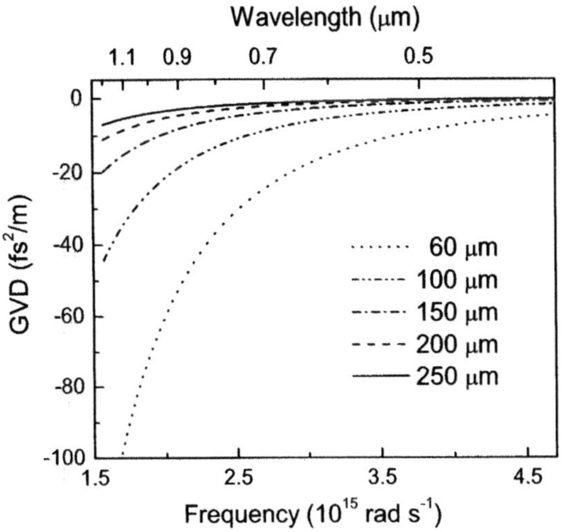

Fig. 8. Group velocity dispersion, GVD, of the EH_{11} mode as a function of frequency for various fiber radii

$$\boldsymbol{\sigma}_\pm = (\boldsymbol{u}_x \pm \mathrm{i}\boldsymbol{u}_y)/\sqrt{2}\,, \tag{36}$$

represent two circularly polarized waves with counterclockwise ($\boldsymbol{\sigma}_+$) and clockwise ($\boldsymbol{\sigma}_-$) helicity, respectively; and ω_0 is the optical carrier frequency. With these assumptions, the electric field of an ultrashort pulse propagating inside a hollow waveguide can be expressed as the superposition of two components:

$$E_\pm(r,z,t) = F(r)\,A_\pm(z,t)\,\mathrm{e}^{\mathrm{i}(\beta_0 z - \omega_0 t)}\,, \tag{37}$$

where z is the propagation coordinate along the capillary axis; $F(r)$ is the modal field distribution along the section, which is supposed to be independent of laser frequency and polarization; and $\beta_0 = \beta(\omega_0)$ is the propagation constant of the capillary mode excited by the laser input beam at the center frequency ω_0. It is worth noting that the mode distribution and the corresponding propagation constant are independent of the wave polarization as long as the waveguide has axial symmetry. Modal birefringence would be introduced in a nonsymmetrical capillary, thus requiring a more involved theoretical model. Moreover, we are assuming that the nonlinear effects are sufficiently weak, so that we are allowed to discard changes in mode properties during propagation along the waveguide. Introducing a moving reference frame with the new coordinates $T = t - z/v_\mathrm{g}$ and $Z = z$, where $1/v_\mathrm{g} = \mathrm{Re}\left[(\partial\beta/\partial\omega)_{\omega_0}\right]$, the evolution of the $A_\pm(Z,T)$ envelopes obeys the following relations:

$$\frac{\partial A_\pm}{\partial Z} = \mathrm{i}\hat{\mathcal{D}}A_\pm + \mathrm{i}\gamma\left(1+\frac{\mathrm{i}}{\omega_0}\frac{\partial}{\partial T}\right)(A_\pm \Delta n_\pm)\,, \tag{38}$$

$$\Delta n_\pm = \frac{2n_2}{3}\left(|A_\pm|^2 + 2|A_\mp|^2\right)\,, \tag{39}$$

where $\hat{\mathcal{D}}$ is a differential operator accounting for radiation losses and dispersion, $\gamma = \omega_0/(cS_{\mathrm{eff}})$ with $S_{\mathrm{eff}} = 0.48\pi a^2$ the effective area of a hollow fiber [19], n_2 is the nonlinear index coefficient of the medium (calculated for linear polarization). The nonlinear coefficient of the gas is proportional to the gas pressure p according to the relation $n_2 = p\kappa_2$, where κ_2 is the nonlinear refractive index per unit pressure, which can be determined from tabulated values reported in the literature [47]. One of the most used solving procedures for (38) is the so called split-step Fourier method [19], in which the action of the dispersive operator on the pulse is calculated in the spectral domain, assuming that the Fourier representation of $\hat{\mathcal{D}}$ is

$$\hat{\mathcal{D}}(\omega) = \beta(\omega+\omega_0) - \mathrm{Re}[\beta(\omega_0)] - \frac{\omega}{v_{\mathrm{g}}}\,. \tag{40}$$

It must be pointed out that (40) takes into account the overall dispersion in a hollow fiber. The expression for $\beta(\omega)$ reported in (33) considers both the dispersion of the noble gas and that of the guiding structure, the refractive index $n_{\mathrm{core}}(\omega)$ of the gas inside the capillary can be calculated at standard conditions (gas pressure $p = 1$ atm, temperature $T_0 = 273.15\,\mathrm{K}$) by tabulated dispersion relations [48], and the actual refractive index can then be easily determined under the operating conditions used for pulse compression [47].

In the general case, (38) is coupled. The interplay between cross- and self-phase modulation can induce different spectral evolutions of the two pulse components, which are responsible for sensible changes in the polarization state of the pulse. Two special cases can be individuated for which (38) reduce to one equation, for linear and for perfect circular polarization of pulses. In the first case, (38) becomes identical and reduces to the ordinary propagation equation for a linearly polarized pulse [19]; in the second case, only one component is present, and its propagation equation is formally identical to that for linear polarization except for the effective value of the a nonlinear coefficient. For this reason, circular polarization is kept during propagation in nonlinear medium.

The evolution of the pulse during nonlinear propagation can be analyzed in a concise way with the help of two parameters [49], the ellipticity,

$$\epsilon(\omega) = \frac{|\widetilde{A}_+(\omega)| - |\widetilde{A}_-(\omega)|}{|\widetilde{A}_+(\omega)| + |\widetilde{A}_-(\omega)|}, \tag{41}$$

and the azimuth of the ellipse in the xy-plane,

$$\psi(\omega) = \frac{1}{2}\arg\left[\frac{\widetilde{A}_+(\omega)}{\widetilde{A}_-(\omega)}\right], \tag{42}$$

which are defined for every spectral component of a pulse. Similar parameters could also be defined in the temporal domain; nevertheless, "temporal ellipticity" could be meaningless for very short pulses, because $\epsilon(t)$ can be defined only over a reasonable number of optical cycles. It is worth pointing out that, for linearly polarized pulses ($\epsilon = 0$), the azimuth ψ corresponds to the angle that the electric field direction forms with the x-axis.

3.3 Compression of High-Energy Pulses with an Arbitrary Polarization State: Numerical Simulations

The formalism introduced in the previous sections allows us to study the nonlinear propagation and the compression of high-energy pulses with an arbitrary polarization state. The more interesting cases that can be analyzed with our model are those of an elliptically or a circularly polarized pulse; nevertheless, more general situations can be considered. Numerical simulations were performed for a 0.5 mJ input pulse with hyperbolic secant envelopes of 25 fs duration with elliptical polarization ellipticity $\epsilon(\omega) = 0.35$, and ellipse azimuth $\psi(\omega) = \pi/4$ over the whole pulse spectrum. Under these input conditions, the A_+ component carries 80% of the overall pulse energy. Figure 9 shows the normalized intensity of the two components of a pulse at the output of a nonlinear waveguide. Both components show a duration of ~ 30 fs; the A_- envelope shows pronounced asymmetry due to nonlinear steepening. The corresponding spectral intensities are shown in Fig. 10. The different spectral broadening achieved by the two components of a pulse is due to unbalancing between the self-phase and the cross-phase modulation terms. It is interesting to note that the weaker component (corresponding to the dashed lines in Fig. 9 and Fig. 10) undergoes stronger broadening and steepening effects, which are essentially due to the cross-phase modulation term in (38). A factor 2 appears in this term, thus accounting for such a strong effect. On the other hand, steepening and spectral broadening of the more intense component of the pulse are essentially due to self-phase modulation, which is less effective with respect to cross-phase modulation. On the basis of the theoretical results, we have calculated the spectral ellipticity and ellipse azimuth of pulses at the output of a hollow waveguide. Figure 11a shows the pulse ellipticity at the input (dashed line) and at the output (solid line) of a hollow fiber. As can be seen from the figure, the ellipticity does not remain constant during propagation in a capillary. Such an effect can be also noted in the spectral behavior of the ellipse azimuth ψ, as shown in Fig. 11b. The nonlinear propagation results in rotation of the ellipse from the initial azimuth of $\pi/4$ to more than $\pi/2$ (on average). The ellipse rotation induced by the nonlinear interaction between the two components of a pulse is a well-known effect that can also be observed at low energies and long pulse durations [50].

The compression of pulses with an arbitrary polarization state is more difficult to achieve with respect to linearly polarized ones because the phase compensation needed for one component of a pulse could be unsuitable for

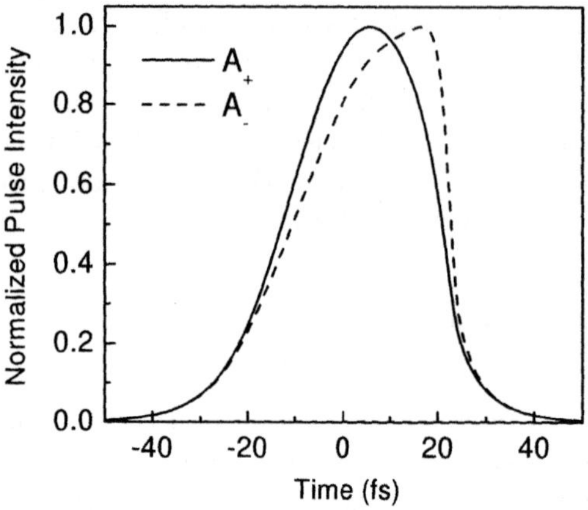

Fig. 9. Normalized intensity of the pulse components at the output of the hollow fiber for an elliptically polarized pulse with input energy of 0.5 mJ, ellipticity $\epsilon = 0.35$, and ellipse azimuth $\psi = \pi/4$

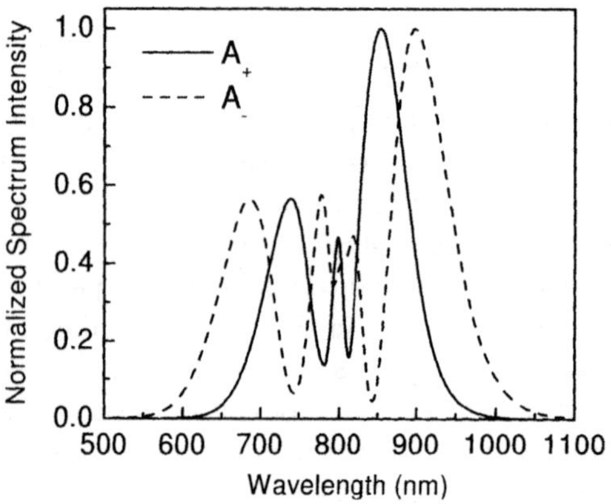

Fig. 10. Normalized power spectra of pulse components at the output of a hollow fiber for an elliptically polarized pulse with input energy of 0.5 mJ, ellipticity $\epsilon = 0.35$, and ellipse azimuth $\psi = \pi/4$

the other. A partial solution to this problem is to compress the more intense component. In the following, we will investigate the phase compensation of the elliptically polarized pulses previously considered, whose counterclockwise polarized component carries the larger energy fraction. After solution of (38) by the split-step Fourier method, the optical compression was simulated by applying a phase filter to the whole electric field; the filter was determined to obtain ideal compression for the counterclockwise component. The resulting electric field reported in Fig. 12, shows the temporal behavior of the pulse in a tridimensional plot. A pulse ellipse appears to form a $\pi/2$ angle around the peak region with respect to the horizontal direction; a complex evolution is observed in the wings, which appears to be completed in about 20 fs from the peak; the central portion of the pulse lasts about 2.5 optical cycles. One may wonder if the result obtained with the ideal compression of one component is the best one can obtain because the potentially shorter duration of the weak clockwise envelope, which has a larger spectral extension, cannot be reached in this way. For this reason we repeated the calculation for ideal compression

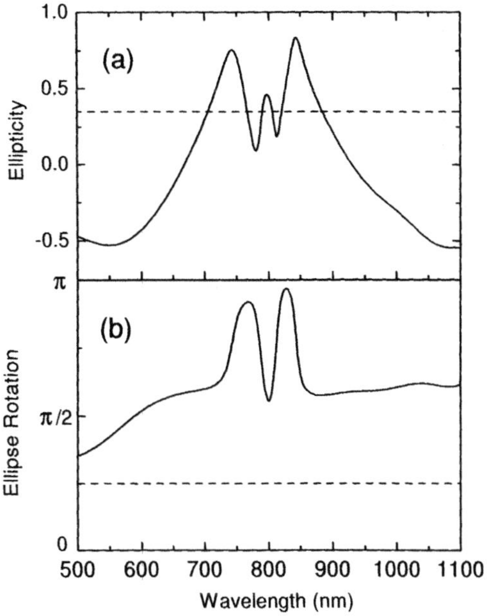

Fig. 11. (**a**) Spectral pulse ellipticity ϵ at the input (*dashed line*) and at the output (*solid line*) of a hollow fiber for the elliptically polarized pulse considered in the text. (**b**) Spectral ellipse azimuth ψ at the input (*dashed line*) and at the output (*solid line*) of a hollow fiber

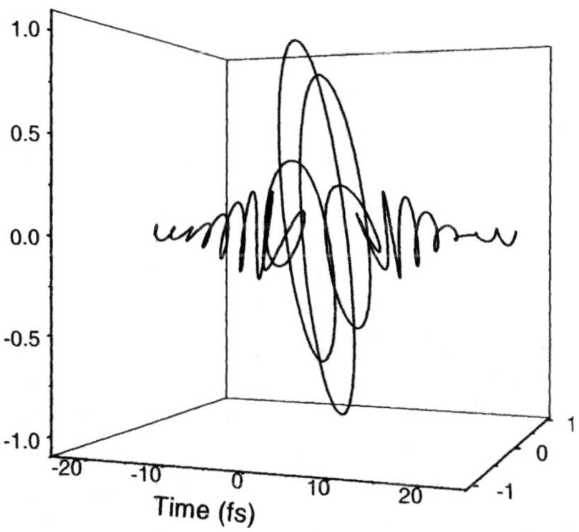

Fig. 12. Tridimensional view of the electric field of an elliptically polarized compressed pulse, obtained from numerical simulations for optimal compression of the counterclockwise component (see text)

of both components of the pulse; it is worth noting that the result cannot be generated using standard procedures because the two components should be separated, compressed in different ways, and then recomposed. Nevertheless the calculations show a similar pulse duration with respect to the compression of a single component.

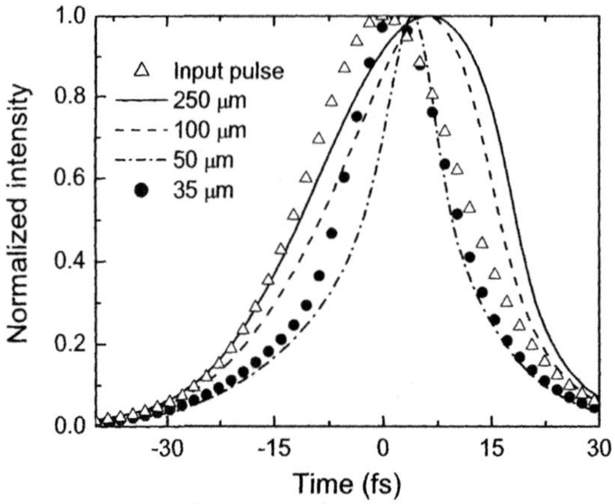

Fig. 13. Normalized intensity of the output pulse after propagation in an argon-filled capillary as a function of fiber radius. Input pulse: 800 nm central wavelength, 25 fs duration, 10^{13} W $\cdot$ cm^{-2} intensity; fiber length: 50 cm; gas pressure: 0.7 bar

Numerical calculations have been performed of the temporal intensity profile of a linearly polarized light pulse at the output of a hollow fiber before the compression stage. The results of the simulations are shown in Fig. 13, which displays the normalized intensity of the output pulse after propagation in an argon-filled fiber for different values of the fiber bore radius, ranging from 250 μm to 35 μm. In the calculations, the input pulse intensity was kept constant and the gas pressure constant, so that similar cases could be compared. Moreover, the input intensity chosen was a value well below the threshold for significant ionization of the gas. From Fig. 13, it is evident that for the largest fiber, with a 250 μm inner radius, the spectrally broadened output pulse (solid line) appears longer than the input pulse (empty triangles) because the negative dispersion of the waveguide cannot compensate for the positive dispersion of the gas; the signature of self-steepening is also clearly visible. It must be pointed out that the (theoretical) transmission for this fiber is near 98%. If we consider smaller capillaries, we observe a shortening of the output pulse; the duration reaches its minimum for a 50 μm radius, where good balancing among losses, dispersion, and spectral broadening is obtained. This trend is inverted when smaller capillaries are considered, as can be seen for a 35 μm fiber (filled dots); this behavior is related to many factors: the strong dispersion in the attenuation coefficient, responsible for spectral narrowing of the pulse; the very low transmission experienced by the pulse (less than 0.2% in this case) which strongly reduces spectral broadening; and the excess in negative dispersion induced by the guide. From our investigation, we conclude that pulse compression can be achieved directly in a capillary at the expense of greatly reduced power transmission; this method is not suitable for compression of high-energy laser pulses.

3.4 Compression of High-Energy Pulses with Linear and Circular Polarization: Experimental Results

In 1996, the hollow-fiber compression technique was first demonstrated [7] using 140 fs, 0.66 mJ pulses from a Ti:sapphire laser system. Spectral broadening was obtained in a 70 cm long, 140 μm diameter hollow fiber filled with krypton at a pressure of 2 bar. Pulse compression was achieved using Brewster-cut prism pairs. Pulses as short as 10 fs with energy up to 0.24 mJ were generated. A second set of experiments was performed using 20 fs input pulses with energy of 40 μJ and a 60 cm long, 160 μm diameter hollow fiber filled with krypton [8]. A typical broadened spectrum for krypton pressure $p = 2.1$ bar is shown in Fig. 14a. The frequency-broadened pulses emerging from the hollow fiber were collimated and propagated through a dispersive delay line based on chirped mirrors and two pairs of fused silica prisms of small apex angle (20°). The compressed pulses were measured by interferometric second-harmonic autocorrelation. The second-order interferometric autocorrelation signal is generated in a 15 μm thick BBO frequency-doubling crystal and recorded at a scan rate of 1 Hz. To evaluate the pulse duration, we took the inverse Fourier transform of the spectrum and assumed, as a free parameter, some residual cubic phase distortion. By best compression of the pulse whose spectrum is shown in Fig. 14a, we measured the interferometric second-harmonic autocorrelation trace of Fig. 14b. From this trace, a pulse duration of 4.5 fs (FWHM) was determined. Pulse energy after compression was 20 μJ.

Scaling the input energy to higher values, the generation of sub-10 fs pulses at a subterawatt peak power level has been demonstrated [9]. Input pulses of 20 fs with energy of 1 mJ were coupled in a 60 cm long, 500 μm diameter hollow fiber filled with argon. After recollimation by a silver mirror, the beam is directed toward the dispersive delay line. At the millijoule level, the prism-chirped-mirror compressor previously described cannot be employed because of self-focusing in the prisms. In this case, only ultrabroadband chirped dielectric mirrors were used. These chirped mirrors introduce a nearly constant negative group-delay dispersion over a spectral range as broad as 150 THz (650 nm to 950 nm) and exhibit high reflectivity over the wavelength range of 600 nm to 1000 nm. The overall transmissivity of the compressor, including the recollimating and steering optics, is $\sim$ 80%. By best compression, the interferometric autocorrelation trace of Fig. 15 has been measured, which corresponds to a pulse duration of 5 fs. Pulse energy after compression was 0.55 mJ, which corresponds to a peak intensity of 0.11 TW.

The extension of the hollow-fiber technique to the sub-4 fs regime requires two main achievements: 1. the possibility of extending the induced spectral broadening preserving a single-mode operation, which is essential for subsequent compression; and 2. the development of ultrabroadband dispersive delay lines for dispersion compensation. In 2002, a novel spectral broadening technique, based on hollow-fiber cascading, was reported, which allows gen-

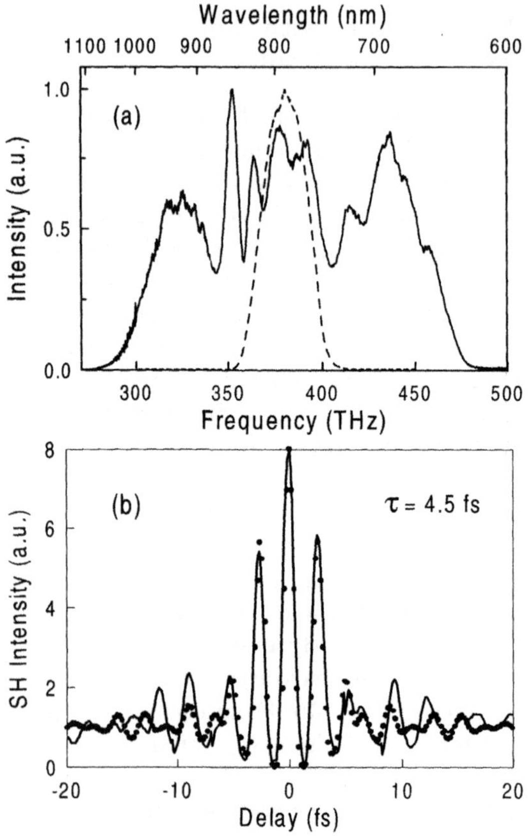

Fig. 14. (**a**) Spectral broadening in krypton (pressure $p = 2.1$ bar) of 20 fs, 40 μJ pulses. The spectrum of the input pulses is shown as a *dashed line*. (**b**) Measured (*solid line*) and calculated (*dots*) interferometric autocorrelation trace of the compressed pulses; an evaluation of the pulse duration (FWHM) is also given

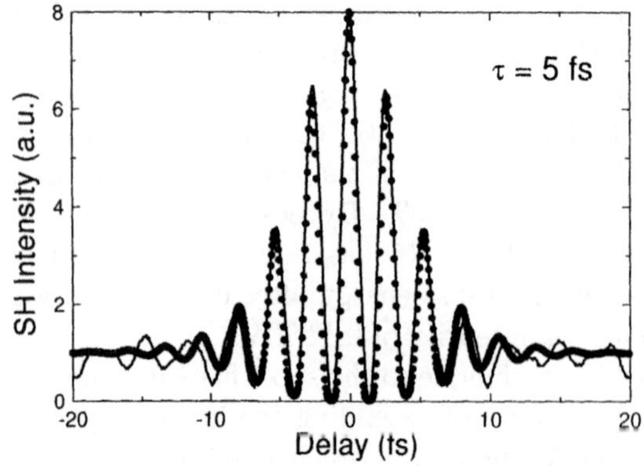

Fig. 15. Measured (*solid line*) and calculated (*dots*) interferometric autocorrelation trace of compressed pulses with 0.11 TW peak power; an evaluation of pulse duration (FWHM) is also given

eration of a supercontinuum extending on a bandwidth exceeding 510 THz with excellent spatial beam quality [51]. Pulses of 25 fs at 0.5 mJ pulses were coupled in an argon-filled (gas pressure 0.2 bar), 60 cm long hollow fiber, with an inner radius of 0.25 mm. Gas pressure was chosen to obtain pulses with duration of about 10 fs after compression, using broadband chirped mirrors with completely negligible wings.

Such pulses were then injected into a second argon-filled (gas pressure less then 0.1 bar), 60 cm long hollow fiber with an inner radius of 0.15 mm. The pulse spectrum at the output of the second fiber extends from $\sim$ 400 nm to $>$ 1000 nm. The output beam presents excellent spatial characteristics (single-mode operation), and it is diffraction limited. Output pulse energy is 0.1 mJ. Dispersion compensation over such a spectral bandwidth is challenging. In 2003, ultrabroadband dispersion compensation was achieved using a liquid-crystal spatial light modulator (SLM). The beam at the output of a hollow-fiber cascade was collimated and sent into a pulseshaper consisting of a 640 pixel liquid-crystal SLM, two 300 line/mm grating, and two 300 mm focal-length spherical mirrors ($4f$ setup). The pulse energy at the entrance of the pulseshaper was 0.1 mJ, and it was 15 μJ at the exit. The pulse was characterized using the SPIDER technique, optimized for sub-10 fs pulses. The measured spectral phase was used to compress the pulse iteratively. Compression was started with an initially flat phase written on the liquid-crystal mask. Then, the measured spectral phase was inverted and added to the phase applied to the SLM. Typically, five iterations were necessary to yield the shortest pulse. The spectral phase of the supercontinuum was very stable and remarkably constant for several hours. Figure 16 shows the temporal profile of the shortest pulse. The full-width-at-half-maximum (FWHM) was measured at (3.77 ± 0.11) fs with a pulse energy of 15 μJ. The spectrum spans a bandwidth of about 270 THz (see Fig. 17) and has been cut at 566 THz by inserting a knife-edge into the spectrally dispersed beam inside the pulseshaping apparatus. The reason for this is twofold: Because of the lack of a strong enough upconversion signal from the shorter wavelength region, the spectral phase remained undetermined in the cut region. The second reason is the strong scattered fundamental light which has to be kept from interfering with the SPIDER signal in the spectrometer and therefore from decreasing the modulation depth. Figure 17 shows that the spectral phase is essentially flat over the entire bandwidth, except for a nearly 2π-step at 711 nm (422 THz). The phase step in conjunction with the strongly modulated spectrum accounts for the pre- and afterpulses in the temporal domain (Fig. 16).

As discussed in the preceding subsections, the hollow-fiber compression technique allows generating few-optical-cycle light pulses with circular polarization [46]. Laser pulses of 25 fs, 0.5 mJ at 800 nm were sent through a zero-order quarter-wave plate (QWP) designed for femtosecond pulses, to induce circular polarization. The laser beam was then coupled to a hollow fiber with

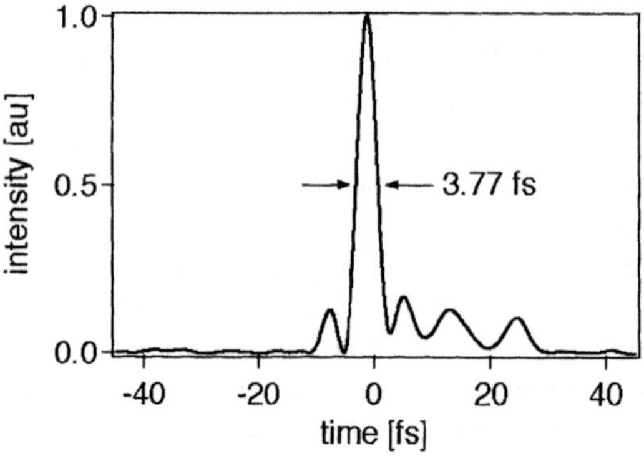

Fig. 16. Reconstructed temporal pulse profile showing a FWHM of 3.77 fs

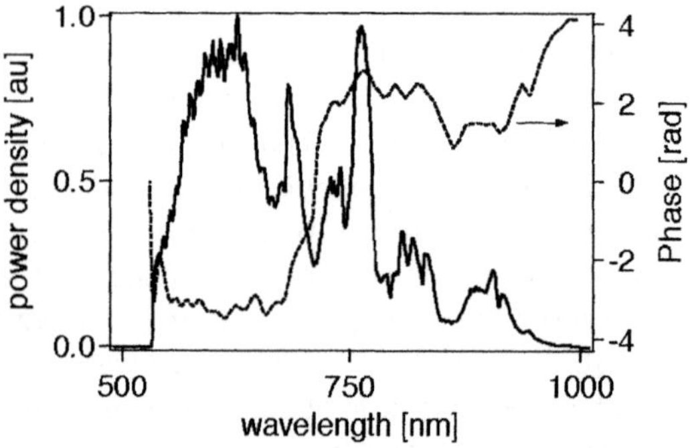

Fig. 17. The measured pulse spectrum is shown in *solid*. The spectrum has been cut at wavelengths below 530 nm (above 566 THz) and spans a bandwidth of 270 THz. The reconstructed spectral phase is shown *dashed*

a 250 μm inner radius and 65 cm length, filled with argon at a pressure of 0.8 bar. The spectrally broadened pulses emerging from the fiber were then sent into a chirped mirror compressor; the beam direction was kept as near to normal incidence as possible to preserve the circular polarization during compression and to induce the same dispersion compensation for the vertical and horizontal components of the electric field. The temporal behavior of the compressed pulses was determined by second-order autocorrelation measurements. Two distinct autocorrelations were recorded for the horizontal and vertical components of the electric field of the pulse. Figure 18 shows the autocorrelation measured for the vertical component (solid line); a similar trace (not shown) was obtained for the horizontal component of the optical field. A duration of $\sim$ 6 fs was estimated from the measured autocorrelations. We compared the experimental results to a numerical simulation of the nonlinear pulse propagation and compression. To consider the influence of the QWP on the input pulse, we took into account the dispersion and the group velocity

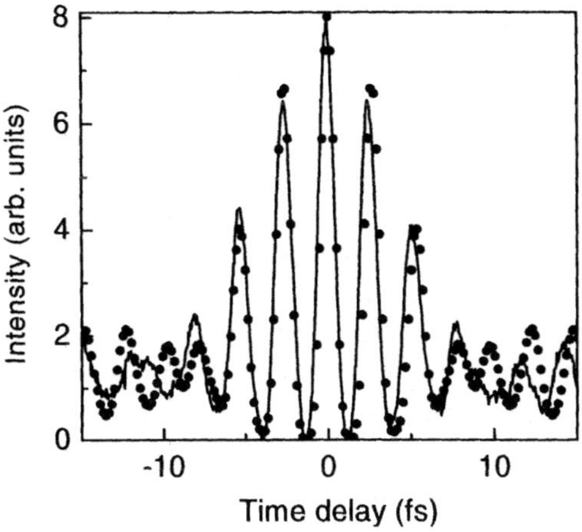

Fig. 18. *Solid line:* experimental autocorrelation of circularly polarized compressed pulses. *Dots:* calculated autocorrelation of the compressed pulses according to the theoretical model described in the text

mismatch in our birefringent plate. The calculated autocorrelation, taking into account the effects of the chirped mirror compressor, is shown in Fig. 18 as solid dots; good agreement between the experimental and the theoretical traces is observed. The pulse duration obtained from the theoretical model is 6.4 fs; the transform-limited duration corresponds to 5.4 fs. The calculated compressed pulses showed negligible differences from those obtained in an ideal QWP.

3.5 Design Criteria for the Hollow-Fiber Compression Technique

Equation (38) can be generalized to a wider class of a nonlinear effects, such as the noninstantaneous response of nonlinear medium or the Raman effect [19]. Both phenomena can take place in molecular gases such as N_2, H_2, and so on. The noninstantaneous response is related to the alignment of molecules in the field direction; its signature is the redshift of the broadened spectrum with respect to the initial center frequency of the pulse [43]. High-energy pulse compression techniques based on the Raman effect have also been recently proposed, as will be discussed in Sect. 5. In compression techniques based on the instantaneous Kerr effect, the retarded response of the nonlinearity or the Raman effect must be avoided; this is the case for the hollow-fiber compression technique, in which noble gases are usually employed. A further reason for this choice is the high damage threshold, which ranges from $\sim 10^{14}\,\mathrm{W \cdot cm^{-2}}$ (in Xe) to $\sim 10^{15}\,\mathrm{W \cdot cm^{-2}}$ (in He). Gas ionization at high intensities must be avoided because it can be a source of unwanted nonlinear behavior of the pulse; moreover, ionization at a capillary entrance can be detrimental for optimal coupling with a laser beam. As a consequence, ionization imposes an upper limit on the input peak intensity of the pulse; for an assigned input pulse energy, this restriction corresponds to a fiber radius larger than a certain value $a_{\min}$. Once the fiber radius has been chosen, one can play with two free parameters to achieve the desired spectral broadening, the gas pressure and the fiber length. Nevertheless, the gas pressure must be kept lower than a critical value p_{crit}, above which self-focusing and filamentation of the beam take place. This critical pressure is given, for a free propagating beam, by $p_{\mathrm{crit}} = \lambda^2/2\pi\kappa_2 P_0$, where P_0 is the peak power of the input pulse and λ the laser wavelength. The guiding structure contrasts this phenomenon, so that the critical pressure for self-focusing is substantially higher with respect to free propagation [52]. It is worth pointing out that self-focusing in a hollow fiber gives rise to power transfer from the fundamental toward high-order propagation modes, thus introducing intermodal dispersion; moreover, the spectral broadening of the pulse becomes an uniform across the beam section. Once gas type and pressure p are chosen, the fiber length must be selected to achieve the desired spectral broadening. Assuming an input pulse with duration τ_0 (FWHM) and intensity I, the fiber length l that must be used to compress the pulse ideally down to duration τ_{c} is

$$l = 0.154 \frac{\tau_0}{\tau_{\mathrm{c}}} \frac{\lambda}{p I \kappa_2}, \qquad (43)$$

where a Gaussian spectral shape was assumed. These scaling criteria must be considered as general guidelines that should be followed in the design of the compression setup; the degrees of freedom in the choice of fiber characteristics, gas type, and pressure are sufficient to adapt the technique to a large variety of laser sources and desired compression performance.

4 Pulse Compression in the Ultraviolet

Using cross-phase modulation in gas-filled hollow fibers, the generation of sub-10 fs light pulses in the ultraviolet was demonstrated in 1999 by *Durfee* et al. [53]. The fundamental and the second harmonic of a femtosecond laser were injected into a gas-filled hollow fiber. Pulses at the third-harmonic were generated through parametric frequency mixing ($3\omega = 2\omega + 2\omega - \omega$) in a phase-matching condition, with about 30% to 40% conversion efficiency from the pump light at 2ω [54]. The combined action of self- and cross-phase modulation induces significant spectral broadening of the idler pulse at 3ω, which can be subsequently compressed. The key element of this method is the use of guided-wave, phase-matched optical parametric generation in gases that preserves high conversion efficiency and good output beam quality while maintaining a short pulsewidth. The propagation constant for a hollow fiber filled with a homogeneous medium is given by (33), that, neglecting the contribution to dispersion from free electrons, can be approximated in the following way [55]:

$$\mathrm{Re}(\beta) \equiv k \approx \frac{2\pi}{\lambda} + \frac{2\pi P \delta(\lambda)}{\lambda} - \frac{u_{nm}^2 \lambda}{4\pi a^2}, \qquad (44)$$

where P is the gas pressure and $\delta(\lambda)$ is related to the index of refraction of the gas by $n = 1 + P\delta(\lambda)$. The difference-frequency mixing between two beams propagating in a hollow fiber at central frequencies ω_1 and ω_2, gives rise to a signal beam at frequency $\omega_3 = N\omega_2 - M\omega_1$, where M and N are the numbers of photons involved in the nonlinear process. The phase mismatch is $\Delta k = Nk_2 - Mk_1 - k_3$ and, using (44), can be written as

$$\begin{aligned}\Delta k &= 2\pi P\left(\frac{N\delta_2}{\lambda_2} - \frac{M\delta_1}{\lambda_1} - \frac{\delta_3}{\lambda_3}\right) - \frac{1}{4\pi a^2}(u_3^2\lambda_3 + Mu_1^2\lambda_1 - Nu_2^2\lambda_2) \\ &= \Delta k_{\mathrm{mat}} - \Delta k_{\mathrm{mode}}\,, \end{aligned} \tag{45}$$

where u_1, u_2, and u_3 are the modal constants for the three interacting beams. The material phase mismatch, Δk_{mat}, is positive for normal dispersion. When $M = 1$, $N = 2$, $\lambda_1 = 2\lambda_2$, and $u_1 = u_2$, the modal phase mismatch, Δk_{mode}, is positive for any signal mode. In this situation, it is possible to achieve phase matching ($\Delta k = 0$) by adjusting the gas pressure properly.

The experiments [53] were performed using 35 fs idler pulses at 800 nm. The pump pulses at 400 nm were obtained by second-harmonic generation in a BBO crystal. Pump and idler beams were subsequently coupled in an argon-filled hollow fiber. The argon pressure was adjusted to maximize the phase-matched UV signal. The UV pulses were directed to a grating compressor with a single-pass configuration. Pulses as short as 8 fs were measured by self-diffraction frequency-resolved optical gating. The compressed pulse energy was greater than 1 μJ. The spectral width of the generated signal pulse, centered at 270 nm, was about 16 nm. Such spectral broadening can be explained in terms of cross-phase modulation. Since the 800 nm idler pulse in the experiments is the most intense, it is the dominant source of nonlinear spectral broadening. Neglecting pump depletion and walk-off, the phase modulation of the pump is $\phi_{\mathrm{p}}(t) = [n_2 I_{\mathrm{p}}(t) + 2n_2 I_{\mathrm{id}}(t)]k_{\mathrm{p}} z$, where I_{p}, and I_{id} are the pump and idler intensities, respectively, n_2 is the nonlinear index of refraction; and k_{p} is the pump wave vector. The first term in the expression of ϕ_{p} is due to pump self-phase modulation; the second is due to cross-phase modulation. Integrating the equation which describes the evolution of the signal field, it is possible to calculate the phase modulation induced on the signal:

$$\phi_{\mathrm{sig}}(t) \approx n_2\left[2k_{\mathrm{p}} I_{\mathrm{p}}(t) + \frac{11}{2} k_{\mathrm{id}} I_{\mathrm{id}}(t)\right] z\,. \tag{46}$$

The signal pulse frequency shift, $\Delta\omega = -\partial\phi_{\mathrm{sig}}/\partial t$, gives spectral broadening in very good agreement with the experimental measurement.

5 Phase Modulation Based on Molecular Motion

Alternative methods for phase modulation of short pulses, not based on SPM, have been proposed and experimentally demonstrated in the last few years.

In the following, three methods for phase modulation of short pulses based on the use of molecular motion will be described: one method uses two narrowband lasers, slightly detuned from the Raman resonance, to generate a spectrum of Raman sidebands, whose Fourier transform is a periodic train of subfemtosecond pulses [56]. A second technique is based on impulsive excitation of molecular motion in Raman-active gases [57]; another method, closely related to the previous one, uses impulsively excited molecular rotational wave packets [58].

5.1 Raman Sideband Generation by Quasi-CW Lasers

The potential of stimulated Raman scattering for broadband generation has been demonstrated by several groups. In 2000, using molecular deuterium (D_2), *Sokolov* et al. [59] demonstrated collinear generation of a Raman spectrum extending over $50\,000\,\mathrm{cm}^{-1}$. The spectrum consists of seventeen mutually coherent sidebands, spaced by the fundamental vibrational frequency of D_2 and ranging from 2.94 μm to 195 nm. The basic idea of the method is the use of a Raman transition with a sufficiently large coherence, so that the generation length and the phase-slip length are of the same order. This coherence is established by driving the molecular transition using two narrowband lasers, chosen so that the (tunable) frequency difference between the two beams is approximately equal to the fundamental vibrational frequency in deuterium. The molecular motion induced modulates the driving laser frequencies, giving rise to the collinear generation of a very broad spectrum. The spectrum generated has approximately Bessel function sideband amplitudes and in the temporal domain, corresponds to a periodic beat of two frequency-modulated signals with center frequencies corresponding to the driving frequencies. Numerical simulation predicts that the group velocity dispersion of the Raman medium itself leads to pulse compression, giving rise to the generation of a train of subfemtosecond pulses [59, 60].

5.2 Phase Modulation by Impulsively Excited Molecular Vibrations

Another method for molecular phase modulation is based on impulsive excitation of molecular motion in Raman-active gases [57,61]. In this case, a first high-intensity light pulse, with a duration shorter than the molecular vibrational period of a molecular gas, impulsively excites a coherent molecular motion (vibration or rotation) in the gas. A second, relatively weak, delayed pulse propagating in the excited medium is strongly phase-modulated due to the modulation of the gas refractive index induced by the molecular motion. This gives rise to the generation of Stokes and anti-Stokes sideband frequencies. The condition for impulsive excitation is easily met, since the vibrational frequencies, Ω_{v}, of molecular motion in Raman-active gases and liquids are typically of the order of several hundreds of reciprocal centimeters, so that for exciting pulse durations below 100 fs, the molecular vibrational period turns out to be longer than the laser pulse duration. Therefore, in the following, we will assume a driving pulse duration, τ_p, much shorter than the molecular vibrational period $T_v = 2\pi/\Omega_v$. To gain physical insight into this phase-modulation method, we will briefly outline the main conclusions of the theoretical model [57, 62, 63].

When the driving pulse in the impulsive regime is turned off, the molecules normal vibrational coordinate Q exhibits freely dumped oscillations:

$$Q(t) \approx Q_0 \exp(-t/T_2) \sin\left(\widetilde{\Omega}_v t\right) , \tag{47}$$

where T_2 is the dephasing time ($T_2 \gg T_v$), $\widetilde{\Omega}_v^2 = \Omega_v^2 - (1/T_2)^2$, and the initial amplitude Q_0 is proportional to the pump energy. $Q(z,t)$ is the waveform of the vibrational excitation, which moves with the pump pulse group velocity. A second relatively weak delayed pulse, which does not change the state of free molecular vibrational motion, will propagate in the excited medium in a linear scattering regime. Neglecting the relaxation of $Q(z,t)$ on the timescale of the injection pulse duration, τ_i, the solution of the wave equation for the delayed pulse field can be written as follows:

$$E_i(z,\tau) = E_i(0,\tau) \exp\left[-iG_i \frac{\sin(\Delta K z)}{\Delta K} \sin(\Omega_v \tau + \Delta K z)\right] , \tag{48}$$

where $E_i(0,\tau)$ is the input amplitude of the injection pulse; G_i, proportional to the amplitude Q_0 of the molecular vibrations, has the sense of the Raman gain for the injection field; $\Delta K = \Omega_v \left[(v_i)^{-1} - (v_p)^{-1}\right]/2$ is the group velocity mismatch between pump and injected pulses (v_p and v_i are the group velocities of the pump pulse and of the injected pulse, respectively). The solution (48) shows that the field amplitude of the injected pulse does not change during propagation, whereas the phase is periodically modulated with the molecular vibrational period. The result of this modulation depends on the injected pulse duration, τ_i. If $\tau_i > T_v$ and the pulse covers more than one period of molecular vibration, its power spectrum can be calculated as follows:

$$|F_i(\Delta\omega)|^2 = \sum_{n=-\infty}^{\infty} |f_i(\Delta\omega - n\Omega_v)|^2 J_n^2(\xi) , \tag{49}$$

where $f_i(\Delta\omega) = f_i(\omega - \omega_i)$ is the Fourier transform of the input injection pulse electric field $E_i(0,\tau)$, and $\xi = G_i \sin(\Delta K z)/\Delta K$. The power spectrum (49) shows sidebands at frequencies $\omega_n = \omega_i + n\Omega_v$ ($n = 0, \pm 1, \ldots$). Due to the periodic character of the phase-modulation function, pulse compression can be obtained with positive or negative GVD, thus leading to the generation of periodic pulse sequences [63]. The method has been demonstrated, using SF_6 as a Raman medium, by the generation of a regular sequence of 5.8 fs pulses at 400 nm, with a period equal to the period of the excited vibrational mode of SF_6 [64].

When the probe pulse duration is shorter than the vibrational period, the delayed pulse covers only part of the sinusoidal modulation function. In this case, the induced time-dependent phase, frequencies, and frequency chirp of the injected pulse will depend strongly on which part of the modulation function is imposed on the pulse; this is determined by the relative delay between pump and probe pulses within the molecular vibrational period. The experiments were performed in a 1 m long, 128 μm diameter, hollow fiber filled with SF_6 gas [11]. A Ti:sapplire laser system generates 0.8 mJ, 30 fs pulses at 800 nm. About 0.5 mJ was used to impulsively excite the gas inside the hollow fiber. The remaining part of the beam was used to generate the probe pulse at 400 nm in a BBO crystal to separate the pump and probe beams easily at the output of the fiber. Typical probe energy before injection into the hollow fiber was about 10 μJ. The SF_6 molecule has a strong symmetrical vibrational mode A_{1g} with a period $T_v = 43$ fs. The electronic absorption band of SF_6 is far above the pump and probe wavelengths, so that only the A_{1g}-Raman mode is efficiently impulsively excited by the 30 fs pump pulses. To probe only a portion of the sinusoidal phase-modulation function significantly shorter than the modulation period T_v, a probe pulse with a duration of 15 fs was used. Such pulses were generated using self-phase modulation in a second gas-filled hollow fiber. Since, during the propagation along the hollow fiber, the probe pulse should interact with the same portion of the molecular modulation function spread with the group velocity of the pump pulse, the pressure of the Raman gas inside the fiber was adjusted to obtain velocity matching between the 800 nm pump and the 400 nm probe pulse [65]. To adjust pulse compression, a broadband chirped mirror with negative GDD was used. Under optimum conditions, a quasi-continuous spectrum with a width of almost 100 nm was generated, for a pump–probe delay of 242 fs. Such delay corresponds to the best phase compensation for the dispersion of the whole system. The self-diffraction autocorrelation of the compressed pulses was measured, that corresponded to a deconvoluted pulsewidth of less than 4 fs $\big((3.8 \pm 0.15)\,\text{fs}\big)$. Output pulse energy was 1.5 μJ.

5.3 Phase Modulation Using Molecular Rotational Wave Packets

A variant of the previous method is based on time-dependent phase modulation induced by molecular rotational wave packets [58]. In this case, a short pump pulse impulsively excites a rotational wave packet in CO_2 gas that undergoes periodic full revivals at time delays of $T_r = h/2B$, where h is the Planck constant and B is the rotational constant of the molecule. A suitably temporally delayed probe pulse, propagating in the excited medium, is phase-modulated by the time-dependent index of refraction during the rotational revival of a wave packet.

The dephasing and rephasing of the coherent rotational wave packet, induced by the excitation pulse, give rise to periodic modulation of the gas refractive index, which is given by [58]

$$n^2(t) \approx 1 + \frac{N}{\epsilon_0}\left[\alpha_\perp + \Delta\alpha\langle\cos^2\theta(t)\rangle\right], \tag{50}$$

where N is the gas number density; and $\Delta\alpha = \alpha_\parallel - \alpha_\perp$ and $\alpha_\parallel$, $\alpha_\perp$ are the components of the anisotropic polarizability for fields, respectively, parallel and perpendicular to the molecular axis. The alignment cosine, $\langle\cos^2\theta(t)\rangle$, is the thermally averaged expectation value of $\cos^2\theta(t)$, where θ is the angle between the applied electric field and the internuclear axis. The refractive index presents periodic peaks due to rotational revivals, where the wave packets rephase completely, separated by additional peaks where partial rephasings occur. In the case of CO_2, full revivals occur every $T_r \sim 42.8$ ps, and partial rephasings occur at intervals of $T_{par} = T_r/4$. The delay of the probe pulse can be adjusted to match either a minimum or a maximum of the refractive index in correspondence with either a complete or a partial revival of the rotational wave packets. In this way, the probe pulse experiences predominately positive or negative quadratic temporal phase modulation, thus leading to an almost linear chirp, either positive or negative. Moreover, the induced spectral broadening of the probe pulse can be maximized by adjusting the duration of the probe to match the time separation of the inflection points of $n(t)$.

The experiments were performed using 20 fs, $\sim$ 1 mJ energy pump pulses at 800 nm [58]. A fraction of the pump beam ($\sim$ 30%), is frequency doubled in a 25 mm KDP crystal to generate a 270 fs probe pulse at 400 nm. A long probe pulse was used to increase the spectral broadening, as discussed above. Pump and probe pulses were coupled with adjustable delay into a 60 cm long, 450 μm diameter, hollow fiber filled with CO_2 gas at a pressure of 433 Pa. The probe delay was adjusted to match the first complete rotational revival to achieve the maximum spectral broadening in a region of negative chirp. Pulse compression was achieved by propagating the probe pulse in a 25.4 mm thick fused silica glass. Output pulse duration, measured by self-diffraction, frequency-resolved optical gating, was 30 fs, corresponding to a compression ratio of 9. Further improvements of the technique are expected to allow for the generation of shorter pulses.

6 Applications and Perspectives

Many fundamental processes of light–matter interaction take place in nature on extremely short timescales, typically below 1 ps. As an example, electronic motion and dephasing in molecules and solids occurs in the 10 fs time domain, nuclear motion on the 100 fs timescale, and vital functions in living system such as energy relaxation, energy transfer, or isomerization occur within a few picosecond. Spectroscopy is the natural tool for getting information on the physical mechanisms underlying these phenomena, provided that techniques with suitable time resolution are available. The development of laser sources delivering sub-10 fs light pulses, either directly from an oscillator or by using external compression techniques, allows pushing to an extreme performance in the traditional fields of time-resolved spectroscopy. In particular, the generation of powerful light pulses in the 5 fs regime by the hollow-fiber compression technique has opened up new frontiers for experimental physics. One of these, more appealing for future investigation of molecular or solid-state physics, is extreme nonlinear optics [1], i.e., the wealth of phenomena taking place when ultrashort pulses are focused on small spots, thus reaching unprecedented peak intensities so that the electric field of the pulse, rather than the intensity, is relevant. The spatial extension of a 5 fs pulse in the propagation direction is limited to a few times the wavelength of the radiation ($\approx 0.5\,\mu$m to $1\,\mu$m in the visible and near-infrared spectral regions). Since such pulses are delivered in a nearly diffraction-limited beam, they can be focused on a spot size comparable to the wavelength, thus leading to a concentration of light in a volume of a few cubic micrometers. This extreme temporal and spatial confinement achieves peak intensities higher than $10^{15}\,\mathrm{W/cm^2}$ with pulse energies in the microjoule range. At these intensity levels, the amplitude of the electric field approaches $10^9\,\mathrm{V/cm}$, which exceeds the static Coulomb field experienced by outer shell electrons in atoms, thus leading to optical-field ionization, and the light–matter interactions take place in the strong-field regime. Using few-optical-cycle pulses, the optical-field ionization rate becomes comparable to the laser field oscillation frequency, and the electron is set free near the peak of the pulse. As evidenced in [1], in the visible/near-infrared spectral regions, multiphoton preionization becomes negligible only in the 10 fs regime, and optical-field ionization takes over fully. The shorter the driving pulse, the stronger the laser field that an electron experiences at the instant it is detached from an atom. This has opened the way to important applications of intense sub-10 fs pulses in the field of high-order harmonic generation in noble gases. Since the ground-state depletion is shifted to higher intensities, the cutoff is shifted to higher harmonics. It has been demonstrated that sub-10 fs pulses can drive the emission of coherent radiation down to the soft X-ray region of the electromagnetic spectrum [66]. Moreover, as shown in [44], the use of fundamental beams with a truncated-Bessel intensity profile allows significantly improved spatial quality in the harmonic beam. Therefore, the hollow-fiber compression technique, which allows one to obtain such an intensity profile directly, provides an efficient tool for optimizing the temporal and spatial characteristics of the high-order harmonic generation process. This is of special concern in the application of XUV emission in various experiments, where extreme temporal resolution and high brightness are required.

Another extremely important advantage offered by the use of intense few-cycle pulses for high-order harmonic generation is that the emission of photons with energy in the cutoff region of the harmonic spectrum can be confined to one half-oscillation period near the peak of the pulse. Using 40 fs pulses, *Paul* et al. experimentally demonstrated that a group of harmonics (ranging from the 11th to the 19th) generated in a jet of argon are locked in phase and form a train of 250 as pulses, separated by one half-cycle of the laser period and extending over about 10 fs [67]. The most promising way to produce single attosecond pulses is to use intense few-optical-cycle driving pulses. The cutoff harmonics generated by such pulses, it was predicted, would merge into a continuum, thus generating a single attosecond pulse, as recently experimentally demonstrated [68], using sub-10 fs pulses generated using the hollow-fiber compression technique. This important achievement paves the way to the extension of time-resolved spectroscopy into the attosecond domain.

Moreover, one of the most intriguing issues in ultrafast laser physics is the role of the carrier–envelope (absolute) phase of a few-cycle laser pulse in laser–matter interaction. Since intense laser–matter interactions depend on the electric field of the pulse, the absolute phase is important for a number of nonlinear processes. Recently the role of the absolute phase in strong-field photoionization was experimentally demonstrated [69] using circularly polarized, 6 fs pulses generated by the hollow-fiber compression technique. For few-cycle pulses, depending on the absolute phase, the generation of photoelectrons violates inversion symmetry. The absolute phase is thus expected to cause an anticorrelation in the number of electrons escaping in directions opposite to the direction of propagation of the laser beam. Such anticorrelation has been experimentally measured using a correlation technique that does not require phase stabilization [69]. This is a promising step toward the measurement of the absolute phase and the control of the electric field of a few-cycle laser pulse.

Perspectives in the use of the hollow-fiber technique are related to its energy scalability. The technique can be easily employed up to millijoule-level laser pulses; the upgrade toward higher energies (tens of mJ) is hindered by fiber damage and gas ionization, thus appearing to be problematic. It must be mentioned that the laser beam profile should be controlled to keep good coupling with the capillary and avoid damage of the fiber entrance. Novel compression techniques for very high pulse energies, based on nonlinear laser–plasma interaction, have been proposed [70], but are still far from operation.

References

[1] T. Brabec, F. Krausz: Intense few-cycle laser fields: Frontiers of nonlinear optics, Rev. Mod. Phys. **72**, 545–591 (2000)

[2] D. E. Spence, P. N. Kean, W. Sibbett: 60 femtosecond pulse generation from a self-mode-locked ti:sapphire laser, Opt. Lett. **16**, 42–44 (1991)

[3] D. Strickland, G. Mourou: Compression of amplified chirped optical pulses, Opt. Commun. **56**, 219–221 (1985)
[4] H. Nakatsuka, D. Grischkowsky, A. C. Balant: Nonlinear picosecond-pulse propagation through optical fibers with positive group velocity dispersion, Phys. Rev. Lett. **47**, 910–913 (1981)
[5] R. L. Fork, C. H. B. Cruz, P. C. Becker, C. V. Shank: Compression of optical pulses to six femtoseconds by using cubic phase compensation, Opt. Lett. **12**, 483–485 (1987)
[6] A. Baltuška, Z. Wei, M. S. Pshenichnikov, D. A. Wiersma, R. Szipöcs: Optical pulse compression to 5 fs at a 1-MHz repetition rate, Opt. Lett. **22**, 102–104 (1997)
[7] M. Nisoli, S. D. Silvestri, O. Svelto: Generation of high energy 10 fs pulses by a new pulse compression technique, Appl. Phys. Lett. **68**, 2793–2795 (1996)
[8] M. Nisoli, S. D. Silvestri, O. Svelto, R. Szipöcs, K. Ferencz, C. Spielmann, S. Sartania, F. Krausz: Compression of high-energy laser pulses below 5 fs, Opt. Lett. **22**, 522–524 (1997)
[9] G. Cerullo, S. D. Silvestri, M. Nisoli, S. Sartania, S. Stagira, O. Svelto: Few-optical-cycle laser pulses: from high peak power to frequency tunability, IEEE J. Sel. Top. Quant. **6**, 948–958 (2000)
[10] B. Schenkel, J. Biegert, U. Keller, C. Vozzi, M. Nisoli, G. Sansone, S. Stagira, S. D. Silvestri, O. Svelto: Generation of 3.8 fs pulses from adaptive compression of a cascaded hollow fiber supercontinuum, Opt. Lett. **28**, 1987–1989 (2003)
[11] N. Zhakoronkov, G. Korn: Generation of single intense short optical pulses by ultrafast molecular phase modulation, Phys. Rev. Lett. **88**, 203901–1–4 (2002)
[12] F. Gires, P. Tournois: Interféromètre utilisable pour la compression d'impulsions lumineuses modulées en fréquence, C. R. Hebd. Acad. Sci. **258**, 6112 (1964)
[13] J. A. Giordmaine, M. A. Duguaym, J. W. Hansen: Compression of optical pulse, IEEE J. Quantum Elect. **4**, 252–255 (1968)
[14] R. A. Fisher, P. L. Kelly, T. K. Gustafson: Subpicosecond pulse generation using the optical Kerr effect, Appl. Phys. Lett. **14**, 140–143 (1969)
[15] A. Laubereau: External frequency and pulses, compression of picosecond, Phys. Lett. **29A**, 539–540 (1969)
[16] A. J. Campillo, S. L. Shapiro, B. R. Suydam: Periodic breakup of optical beams due to self-focusing, Appl. Phys. Lett. **23**, 628–630 (1973)
[17] E. P. Ippen, C. V. Shank, T. K. Gustafson: Self-phase modulation of picosecond pulses in optical fibers, Appl. Phys. Lett. **24**, 190–192 (1974)
[18] R. H. Stolen, C. Lin: Self-phase-modulation in silica optical fibers, Phys. Rev. A **17**, 1448–1453 (1978)
[19] G. P. Agrawal: *Nonlinear Fiber Optics* (Academic, San Diego 1995)
[20] A. Baltuška, Z. Wei, R. Szipöcs, M. S. Pshenichnikov, D. A. Wiersma: All solid-state cavity-dumped sub-5 fs laser, Appl. Phys. B-Lasers O. **65**, 175–188 (1997)
[21] E. B. Treacy: Compression of picosecond light pulses, Phys. Lett. **28A**, 34–35 (1968)
[22] R. L. Fork, O. E. Martínez, J. P. Gordon: Negative dispersion using pairs of prisms, Opt. Lett. **9**, 150–152 (1984)
[23] I. Walmsley, L. Waxer, C. Dorrer: The role of dispersion in ultrafast optics, Rev. Sci. Inst. **72**, 1–29 (2001)

[24] J. Zhou, G. Taft, C. P. Huang, M. M. Murnane, H. C. Kapteyn, I. P. Christov: Pulse evolution in a broad-bandwidth Ti:sapphire laser, Opt. Lett. **19**, 1149–1159 (1994)

[25] R. Szipöcs, K. Ferencz, C. Spielmann, F. Krausz: Chirped multilayer coatings for broadband dispersion control in femtosecond lasers, Opt. Lett. **19**, 201–203 (1994)

[26] N. Matuschek, F. X. Kärtner, U. Keller: Theory of double-chirped mirrors, IEEE J. Sel. Top. Quant. **4**, 197–208 (1998)

[27] G. Tempea, F. Krausz, C. Spielmann, K. Ferencz: Dispersion control over 150 THz with chirped dielectric mirrors, IEEE J. Sel. Top. Quant. **4**, 193–196 (1998)

[28] F. X. Kärtner, N. Matuschek, T. Schibli, U. Keller, H. A. Haus, C. Heine, R. Morf, V. Scheuer, M. Tilsch, T. Tschudi: Design and fabrication of double-chirped mirrors, Opt. Lett. **22**, 831–833 (1997)

[29] A. M. Weiner, D. E. Leaird, J. S. Patel, J. R. Wullert: Programmable femtosecond pulse shaping by use of a multielement liquid-crystal phase modulator, Opt. Lett. **15**, 326–328 (1990)

[30] D. Yelin, D. Meshulach, Y. Silberberg: Adaptive femtosecond pulse compression, Opt. Lett. **22**, 1793–1795 (1997)

[31] N. Karasawa, L. Li, A. Suguro, H. Shigekawa, R. Morita, M. Yamashita: Optical pulse compression to 5.0 fs by use of only a spatial light modulator for phase compensation, J. Opt. Soc. Am. B **18**, 1742–1746 (2001)

[32] C. Iaconis, I. A. Walmsley: Self-referencing spectral interferometry for measuring ultrashort optical pulses, IEEE J. Quantum Elect. **35**, 501–509 (1999)

[33] A. M. Weiner: Femtosecond pulse shaping using spatial light modulators, Rev. Sci. Inst. **71**, 1929–1960 (2000)

[34] C. Dorrer, F. Salin, F. Verluise, J. P. Huignard: Programmable phase control of femtosecond pulses by use of a nonpixelated spatial light modulator, Opt. Lett. **23**, 709–711 (1998)

[35] M. A. Dugan, J. X. Tull, W. S. Warren: High-resolution acousto-optic shaping of unamplified and amplified femtosecond laser pulses, J. Opt. Soc. Am. B **14**, 2348–2358 (1997)

[36] P. Tournois: Acousto-optic programmable dispersive filter for adaptive compensation of group delay time dispersion in laser systems, Opt. Commun. **140**, 245–249 (1997)

[37] F. Verluise, V. Laude, Z. Cheng, C. Spielmann, P. Tournois: Amplitude and phase control of ultrashort pulses by use of an acousto-optic programmable dispersive filter: Pulse compression and shaping, Opt. Lett. **25**, 575–577 (2000)

[38] J. P. Heritage, E. W. Chase, R. N. Thurston, M. Stern: A simple femtosecond optical third-order disperser, in OSA Tech. Dig. Ser. (Opt. Soc. Am., Washington 1991) paper CTuB3

[39] E. Zeek, K. Maginnis, S. Backus, U. Russek, M. M. Murnane, G. Mourou, H. Kapteyn, G. Vdovin: Pulse compression by use of deformable mirrors, Opt. Lett. **24**, 493–495 (2000)

[40] A. Baltuška, T. Fuji, T. Kobayashi: Visible pulse compression to 4 fs by optical parametric amplification and programmable dispersion control, Opt. Lett. **27**, 306–308 (2002)

[41] E. A. Marcatili, R. A. Schmeltzer: Hollow metallic and dielectric waveguide for long distance optical transmission and laser, AT&T Tech. J. **43**, 1783–1809 (1964)

[42] E. G. Neumann: *Single-Mode Fibers* (Springer, Berlin, Heidelberg 1988)
[43] M. Nisoli, S. Stagira, S. D. Silvestri, O. Svelto, S. Sartania, Z. Cheng, M. Lenzner, C. Spielmann, F. Krausz: A novel high energy pulse compression system: Generation of multigigawatt sub-5 fs pulses, Appl. Phys. B-Lasers O. **65**, 189–196 (1997)
[44] M. Nisoli, E. Priori, G. Sansone, S. Stagira, G. Cerullo, S. D. Silvestri, C. Altucci, R. Bruzzese, C. de Lisio, P. Villoresi, L. Poletto, M. Pascolini, G. Tondello: High-brightness high-order harmonic generation by truncated bessel beams in the sub-10 fs regime, Phys. Rev. Lett. **88**, 33902–1–4 (2002)
[45] T. Brabec, F. Krausz: Nonlinear optical pulse propagation in the single-cycle regime, Phys. Rev. Lett. **78**, 3282–3285 (1997)
[46] S. Stagira, E. Priori, G. Sansone, M. Nisoli, S. D. Silvestri, C. Gadermaier: Nonlinear guided propagation of few-optical-cycle laser pulses with arbitrary polarization state, Phys. Rev. A **66**, 33810 (2002)
[47] H. J. Lehmeier, W. Leupacher, A. Penzkofer: Nonresonant third order hyperpolarizability of rare gases and N_2 determined by third order harmonic generation, Opt. Commun. **56**, 67–72 (1985)
[48] A. Dalgarno, A. E. Kingston: The refractive indices and verdet constants of the inert gases, Proc. R. Soc. Lon. Ser.-A **259**, 424–429 (1966)
[49] R. M. A. Azzam, N. M. Bashara: *Ellipsometry and Polarized Light* (North-Holland, Amsterdam 1977)
[50] R. W. Boyd: *Nonlinear Optics* (Academic, San Diego 1992)
[51] M. Nisoli, G. Sansone, S. Stagira, C. Vozzi, S. D. Silvestri, O. Svelto: Ultrabroadband continuum generation by hollow fiber cascading, Appl. Phys. B-Lasers O. **75**, 601–604 (2002)
[52] G. Tempea, T. Brabec: Theory of self-focusing in a hollow waveguide, Opt. Lett. **23**, 762–764 (1998)
[53] C. G. Durfee, III, S. Backus, H. C. Kapteyn, M. Murnane: Intense 8 fs pulse generation in the deep ultraviolet, Opt. Lett. **24**, 697–699 (1999)
[54] L. Misoguti, S. Backus, C. G. Durfee, R. Bartels, M. M. Murnane, H. C. Kapteyn: Generation of broadband VUV light using third-order cascaded processes, Phys. Rev. Lett. **87**, 23601–1–4 (2001)
[55] C. G. Durfee, III, S. Backus, M. Murnane, H. C. Kapteyn: Ultrabroadband phase-matched optical parametric generation in the ultraviolet by use of guided waves, Opt. Lett. **22**, 1565–1567 (1997)
[56] S. E. Harris, A. V. Sokolov: Subfemtosecond pulse generation by molecular modulation, Phys. Rev. Lett. **81**, 2894–2897 (1998)
[57] A. Nazarkin, G. Korn, M. Wittmann, T. Elsaesser: Generation of multiple phase-locked Stokes and anti-Stokes components in an impulsively excited Raman medium, Phys. Rev. Lett. **83**, 2560–2563 (1999)
[58] R. A. Bartels, T. C. Weinacht, N. Wagner, M. Baertschy, C. H. Greene, M. M. Murnane, H. C. Kapteyn: Phase modulation of ultrashort light pulses using molecular rotational wave packets, Phys. Rev. Lett. **88**, 13903–1–4 (2002)
[59] A. V. Sokolov, D. R. Walker, D. D. Yavuz, G. Y. Yin, S. E. Harris: Raman generation by phased and antiphased molecular states, Phys. Rev. Lett. **85**, 562–565 (2000)
[60] A. V. Sokolov, D. D. Yavuz, S. E. Harris: Subfemtosecond pulse generation by rotational molecular modulation, Opt. Lett. **24**, 557–559 (1999)

[61] V. P. Kalosha, J. Herrmann: Pulse compression without chirp control and frequency detuning by high-order coherent Raman scattering in impulsively excited media, Opt. Lett. **26**, 456–458 (2001)

[62] A. Nazarkin, G. Korn: Pulse self-compression in the subcarrier cycle regime, Phys. Rev. Lett. **83**, 4748–4751 (1999)

[63] M. Wittmann, A. Nazarkin, G. Korn: Fs-pulse synthesis using phase modulation by impulsively excited molecular vibrations, Phys. Rev. Lett. **84**, 5508–5511 (2000)

[64] M. Wittmann, A. Nazarkin, G. Korn: Synthesis of periodic femtosecond pulse trains in the ultraviolet by phase-locked Raman sideband generation, Opt. Lett. **26**, 298–300 (2001)

[65] A. Nazarkin, M. Wittmann, G. Korn: Group velocity-matched interactions in fs-nonlinear optics: Enhanced high-order Raman scattering by impulsively excited vibrations, in OSA Tech. Dig. Ser. (Opt. Soc. Am., Washington, DC 2001) p. 226

[66] C. Spielmann, N. H. Burnett, S. Sartania, R. Koppitsch, M. Schnürer, C. Kan, M. Lenzner, P. Wobrauschek, F. Krausz: Generation of coherent X-rays in the water window using 5 fs laser pulses, Science **278**, 661–663 (1997)

[67] P. M. Paul, E. S. Toma, P. Breger, G. Mullot, F. Augé, P. Balcou, H. G. Muller, P. Agostini: Observation of a train of attosecond pulses from high harmonic generation, Science **292**, 1689–1692 (2001)

[68] M. Hentschel, R. Kienberger, C. Spielmann, G. A. Reider, N. Milosevic, T. Brabec, P. Corkum, U. Heinzmann, M. Drescher, F. Krausz: Attosecond metrology, Nature **414**, 509 (2001)

[69] G. G. Paulus, F. Grasbon, H. Walther, P. Villoresi, M. Nisoli, S. Stagira, E. Priori, S. D. Silvestri: Absolute-phase phenomena in photoionization with few-cycle laser pulses, Nature **414**, 182–184 (2001)

[70] V. M. Malkin, G. Shvets, N. J. Fisch: Fast compression of laser beams to highly overcritical powers, Phys. Rev. Lett. **82**, 4448–4451 (1999)

Parametric Amplification and Phase Control of Few-Cycle Light Pulses

Andrius Baltuška[1] and Takayoshi Kobayashi[2]

[1] Photonics Institute, Vienna University of Technology, Gusshausstr. 27, 1040 Vienna, Austria
[2] Department of Physics, Faculty of Science, University of Tokyo, 7-3-1 Hongo, Tokyo 113-0033, Japan
kobayashi@femto.phys.s.u-tokyo.ac.jp

Abstract. We have investigated several different regimes of seeded parametric amplification. The choice of the seed configuration determines whether the CEP of the idler pulses becomes fixed or, alternatively, it drifts while the idler and signal pulses remain phase-locked. In our opinion, an OPA of the former kind (phase-stabilizing) can be effectively employed in a sequence of femtosecond laser amplifiers for production of intense CEP-locked pulses, thus dispensing with intricate schemes of active CEP control. On the other hand, phase-repeating OPAs offer a relatively easy way to synthesize single-cycle pulses and, additionally, can be used for accurate determination of the phase offset for each individual laser shot. We believe that this work provides a new insight into the experimental capabilities of various OPA systems and opens attractive possibilities for their use in field-sensitive applications of nonlinear optics.

1 Generation and Shaping of 4 fs Pulses

Parametric generation and amplification of light [1, 2, 3] play a key role in extending the versatility of laser applications. Several unique properties give optical parametric amplifiers (OPA) a favorable edge over conventional laser amplifiers based on media with population inversion. Just to name a few distinct features, OPAs emit coherent wavelength-tunable radiation, supply high gain, offer exceptional pulse quality, and support a broad bandwidth of simultaneous amplification. The broadband operation makes OPAs, especially those employing noncollinear phase matching, ideally suited for the generation of the shortest optical pulses. Moreover, unique phase properties of OPAs provide a direct way to control the oscillations of the light field with respect to the envelope of a few-cycle pulse. The first part of this chapter will be devoted to the topics of ultrabroad parametric amplification and subsequent pulse compression to 4 fs in the visible optical range, whereas phase-control of OPA pulses will be investigated in part two.

1.1 Broadband Parametric Amplification

Ultrashort laser pulses, which consist of just several oscillations of an electromagnetic field, are required to probe ultrafast processes in physics and trace

F. X. Kärtner (Ed.): Few-Cycle Laser Pulse Generation and Its Applications,
Topics Appl. Phys. **95**, 179–227 (2004)

primary reaction steps in photochemistry and photobiology. The extremely high temporal resolution of femtosecond lasers has enabled numerous groundbreaking studies in nonlinear spectroscopy [4,5] and coherent control [6,7,8,9]. Compact tabletop sources of few-cycle terawatt pulses are used for high harmonic generation and production of coherent soft X-rays [10, 11, 12, 13, 14]. A broad spectral width of ultrashort pulses is also essential in applications that require the shortest possible length of coherence, such as optical coherent tomography [15].

As the utility of different femtosecond laser systems experiences an explosive growth in both the number of users and variety of applications, the corresponding pulse duration continues to shrink. Optical pulses carrying just a few optical cycles were first achieved at a fixed wavelength around 620 nm in the 1980s [16] by use of an amplified dye laser, the output of which was recompressed following a self-phase modulation (SPM) action in a quartz fiber. With the proliferation of diverse types of rugged Ti:sapphire lasers in the 1990s, sub-5 fs pulses became routinely available around 800 nm by employing SPM spectral broadening in both quartz [17, 18, 19] and hollow waveguides [20, 21, 22] in combination with novel pulse compression techniques, such as chirped mirrors [23, 24, 25].

Despite achieving record-short 4 fs pulse durations [14,26], the SPM-based techniques have two important drawbacks. First, they do not offer tunable pulses. Limited tunability can be achieved only at the expense of reduced bandwidth by filtering out various components of an SPM-broadened spectrum. The second, more fundamental drawback comes from the spectral modulation inherent in SPM action. On the one hand, the interplay between material dispersion and SPM can substantially improve the smoothness of the broadened spectra [27]. On the other hand, pulse lengthening, which is due to dispersion, lowers the peak intensity and makes it impossible to sustain strong SPM action over a great propagation length, thus limiting the amount of achievable spectral broadening. It is worth noting that the recent invention of a microstructure fiber [28,29] has led to generation of extremely broadband white-light continua with spectra that support 2 fs pulses, assuming perfect compression. This remarkable achievement is a consequence of a dramatic lowering of fiber dispersion. However, the very same feature, which has led to such an unprecedented spectral broadening, is also responsible for a great amount of spectral modulation. The latter makes it nearly impossible to use this type of pulse in many demanding applications of nonlinear spectroscopy that are sensitive to both the time and frequency distribution of the electric field.

Unlike the above-mentioned schemes that involve cumbersome fiber chirping stages, the closing years of the last decade also saw remarkable progress in record-short pulse generation directly from Ti:saphhire oscillators. Two- to three-cycle pulses at nanojoule energy level have been reported by several research groups [30, 31, 32, 33]. While offering, in general, greater simplicity compared with the elaborate extracavity compression arrangement [18,21,26], these 5 fs oscillators share with them the same spectral range and the lack of wavelength tunability, unless the latter is implemented at the expense of elongated pulse duration. An entirely new chapter in broadband mode locking and a promising route toward generation of even shorter pulses directly from laser cavities has been opened recently by demonstrating the possibility of phase-locking two (formerly) independent Ti:sapphire oscillators [34, 35] and two oscillators based on different gain media [36]. This approach potentially provides the most straightforward way to yield an ultimate single-cycle optical pulse by coherently combining the spectra of several lasers employing various gain media [14].

Both the intra- and extracavity methods of pulse shortening, reviewed above, imply generation of broad laser spectra at the fundamental frequency or around that of its harmonics, making continuous wavelength tuning impossible in the shortest pulse regime. Both difficulties, however, can be readily overcome in parametric pulse generation and amplification. It has been recognized that the group velocity mismatch between the pump, signal, and idler waves in a parametric amplifier can be effectively compensated for by employing noncollinear beam geometry [37,38]. In this configuration, the group velocity of a faster traveling wave (the idler) is projected at an angle in the direction of a slower wave (the signal), thus improving the temporal overlap of the two and achieving broadband phase matching. Furthermore, parametric conversion in the visible and near-infrared has been radically changed by the discovery of the unique phase-matching conditions in a Type I BBO crystal pumped by the second harmonic (SH) of the Ti:sapphire laser [39,40,41]. As a result, widely tunable sub-20 fs noncollinear OPAs (NOPAs) have become routine [42,43]. Tunable operation in the 10 fs regime in both the visible and infrared has also been demonstrated [44, 45, 46, 47, 48]. Following the development of sophisticated pulse compressors, phase correction over the entire parametrically amplified bandwidth became possible, which resulted in the generation of sub-5 fs pulses [49, 50, 51, 52].

The great potential of the sub-10 fs NOPAs has been revealed by several spectacular applications of nonlinear spectroscopy, such as time-domain studies of ultrafast molecular dynamics in the condensed phase [51, 53, 54, 55, 56, 57, 58, 59, 60, 61, 62]. Despite the fact that other methods of ultrashort-pulse generation surpass NOPAs in both the shortness of the pulse and spectral extent [26, 63], parametric amplifiers produce noticeably smoother spectra, which plays a vital role in high-resolution time- and frequency-resolved spectroscopy. Therefore, the demand for continual improvement of NOPA pulse quality becomes evident.

Though the concepts of spectral broadening, discussed above, are fundamental for pulse shortening, no less important are the methods of phase control that lead to improved pulse compression and/or facilitate dispersion manipulation in phase-sensitive applications of nonlinear spectroscopy. Several dispersion control techniques and their combinations have been employed to obtain the shortest pulses from NOPAs. For instance, the sub-5 fs regime has been achieved by using both a combination of a 45° angled-prism compressor and a set of ultrabroad chirped mirrors (UBCM) [49], and custom-designed UBCMs alone [52]. It is feasible to fabricate a set of "ideal" UBCMs, but the fixed dispersion of such multilayer dielectric structures makes it impossible to introduce fine wavelength-selective control of group delay, which is required in the daily optimization of the pulse compressor. Similarly, typical elements of tunable negative dispersion, such as prism and grating pairs [64, 65, 66], are rendered virtually nonadjustable because of the large spectral bandwidth of a sub-5 fs pulse.

Therefore, it is evident that further improvement of pulse quality cannot be achieved without implementing flexible phase correction. The power of the latter approach, applied to a NOPA, has been demonstrated in computer-controlled shaping of tunable 16 fs pulses using a liquid-crystal mask [67]. Recently, the invention of micromachined flexible mirrors with electrostatic actuators has provided a new exciting possibility for adaptive pulseshaping. An inexpensive commercially available membrane of this type was first applied to obtain almost ideally spectrum-limited pulses of 15 fs duration from a Ti:sapphire laser [68, 69, 70] and subsequently employed for shaped-pulse optimization of coherent emission of soft X-rays produced by high-harmonic generation [71]. By implementing this method in a compressor of a NOPA signal wave, *Miller* et al. [72] produced bandwidth-limited 7 fs pulses in the visible. Successful optimization of a compression stage in a similar 10 fs tunable NOPA has been reported by *Riedle* et al. [73]. They inserted a micromachined flexible mirror in a prism compressor thus drastically improving the pulse quality and cutting down the time needed for readjusting the parametric amplifier after selecting a new wavelength of operation.

Here, we report on further progress in generation of short visible pulses using a NOPA. The first aspect of this study deals with additional enhancement of the phase-matching bandwidth, which allowed us to produce wider and smoother parametrically amplified spectra. The second subject investigated in this work concerns improvement of pulse compression and adaptive control of the pulse quality. To this end, we employed a micromachined flexible mirror identical to those used by other groups of researchers [68,69,70,71,72,73]. We demonstrate how computer-controlled optimization combined with SHG FROG pulse diagnostic [74,75], allows us to reduce the output pulse duration down to about 4 fs at the central wavelength around 600 nm [76,77].

1.2 Noncollinear Parametric Amplifier

The NOPA, the design of which is described in this work, consists of three major parts. The first element is the parametric amplifier itself that has a gain bandwidth sufficient to support sub-4 fs operation. The second component is

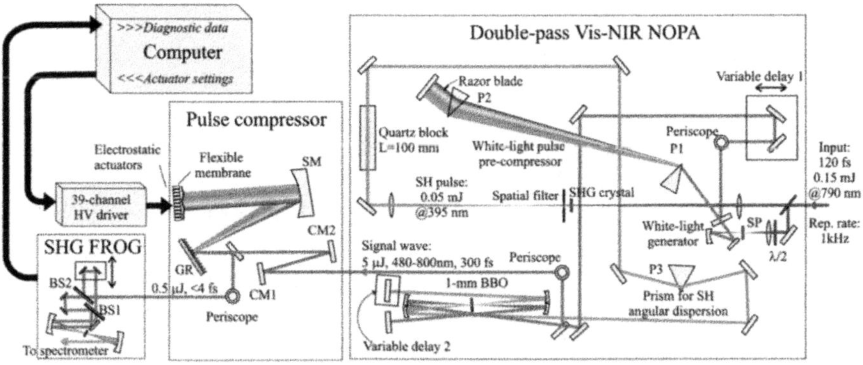

Fig. 1. Schematic of experimental setup. $\lambda/2$: 800 nm wave-plate; SP: 2 mm sapphire plate; P1,2: 45° quartz prisms; P3: 69° quartz prism, the distance from P3 to the NOPA crystal is 80 cm; CM1,2: ultrabroadband chirped mirrors (Hamamatsu Photonics); GR: 300 lines per millimeter ruled diffraction grating (Jobin Yvon); SM: spherical mirror, $R = -400$ mm; BS1,2: chromium-coated $d = 0.5$ mm quartz beam splitters. SHG crystal: 0.4 mm $\theta = 29°$ BBO (EKSMA); NOPA crystal: 1 mm $\theta = 31.5°$ BBO (Casix); SHG FROG crystal: $\theta = 29°$ BBO wedge plate $d = 5\,\mu$m to 20 μm (EKSMA). Spherical mirrors around the NOPA crystal have $R = -200$ mm; *Thick arrows on the left* indicate the data flow from the pulse diagnostic setup (SHG FROG) and the feedback to the flexible mirror

a grating-chirped-mirror compressor, used for rough group delay compensation, with a flexible mirror that serves for fine spectral phase adjustment. The third part comprises a pulse diagnostic setup based on SHG FROG with feedback to a personal computer that also controls the actuators of the flexible mirror. An overview of the setup is presented in Fig.1. The system is pumped by a 120 fs, 1 kHz repetition rate regenerative amplifier CPA1000 (MXR Clark) that is seeded by a Femtolite fiber oscillator (IMRA). Below, we address separately the design and functions of each stage of the setup.

1.3 Extension of Phase-Matching Bandwidth

The discovery of "magic" phase-matching conditions in a Type I BBO crystal pumped by a 400 nm light [39, 40] opened the way to producing amplified visible pulses that have bandwidths of nearly 200 THz [44, 49]. The uniqueness of this pumping arrangement also comes from the fact that the pump inclination to the direction of the seed (ca. 3.7°) matches almost perfectly the angle of birefringent walk-off between the o (ordinary) and e (extraordinary) waves inside the crystal. As a result, BBO pieces as long as 1 mm to 2 mm can be employed even in 5 fs NOPAs. The implications of noncollinear phase matching are well understood and discussed in numerous papers [38, 39, 40, 41, 42, 43, 44, 45, 46, 47, 48, 49, 78, 79, 80, 81, 82, 83, 84, 85, 86].

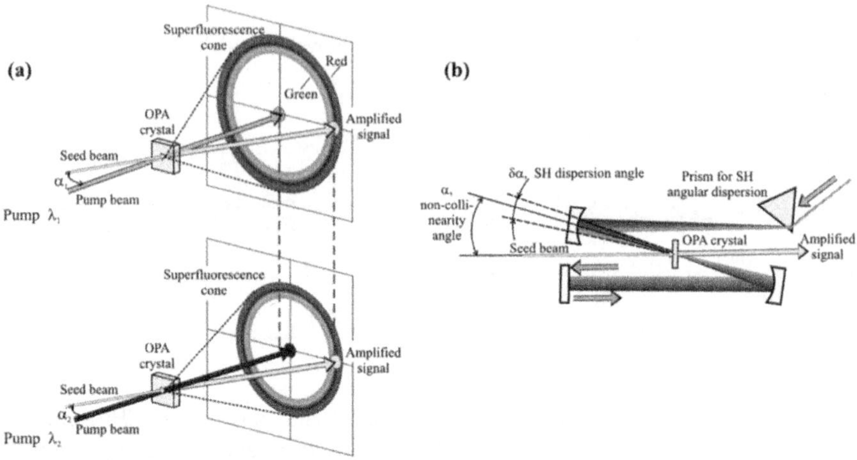

Fig. 2. Extension of parametric bandwidth by adjusting the incidence angle of individual pump wavelengths on a NOPA crystal. (**a**) Schematic representation of noncollinear pump geometry. (**b**) SH dispersion adjustment based on the use of a prism and a focusing optic

Further attention to the subtleties of parametric amplification in a noncollinear configuration has been paid by *Shirakawa* et al. [48,49], who considered the effect of pulse front tilt of the signal wave on the ability to compress it into a sub-5 fs pulse. To prevent tilting of the signal pulse in space, an effect that also results in angular dispersion of the amplified pulse, these researchers proposed to use a pump beam with a tilted wavefront. This configuration, named pulse-front matching, was implemented by sending the pump beam through a prism and adjusting the pulse tilt with a telescope consisting of two convex lenses [86].

Angular dispersion of the pump beam in the sub-5 fs NOPA is also important in enhancing of the phase-matching bandwidth. To understand the mechanism that is responsible for this broadening, we recall the fact that the pump is not monochromatic and its spectral extent is determined by the duration of the input fundamental pulses and conditions of frequency-doubling. Even with relatively thick SH crystals (1 mm to 2 mm LBO or BBO) and comparatively long (120 fs to 150 fs) pulses from standard regenerative amplifiers employed to pump NOPAs [44, 49], the resulting SH radiation has a bandwidth of several nanometers. The concept of how this can be used to improve the phase-matching conditions is schematically depicted in Fig. 2. We notice first, that a certain angle of noncollinearity, α, results in broadband emission of superfluorescence within the narrowest possible solid angle.

Wide-bandwidth parametric amplification can subsequently be achieved by pointing the seed beam in that direction. In other words, the optimal value of the angle α is individual for each pump wavelength. The adjustment of the pump beam dispersion, required for phase-matching optimization, is obtained by selecting the apex angle of a prism in the SH pathway (Fig. 2b) and the distance from the prism to the focusing optic. The required SH dispersion can be evaluated using the following simplified equation:

$$\delta\alpha(\lambda) = \arctan\left\{\left[\tan\left[\gamma(\lambda_0) - \gamma(\lambda)\right]\frac{l}{f}\right]\right\}, \tag{1}$$

where λ is pump (SH) wavelength; λ_0 denotes the central component of the SH spectrum; l and f are the distance from the prism to the focusing optic and the focal distance, respectively; and $\gamma(\lambda)$ is the exit angle of the SH beam behind the prism:

$$\gamma(\lambda) = \arcsin\left\{n(\lambda)\sin\left[\beta - \arcsin\left(\frac{\sin\gamma_0}{n(\lambda)}\right)\right]\right\}. \tag{2}$$

In (2), γ_0 is the angle between the SH beam and the normal to the input face of the prism (incidence angle onto the prism), β stands for the apex angle of the prism, and $n(\lambda)$ is the refractive index of glass.

In our setup, the distance between the Brewster-angled prism for SH dispersion and a $R = -200$ mm focusing mirror (Fig. 1) is 80 cm. The effective increase of the overall bandwidth of parametric amplification is achieved by staggering phase-matching curves that correspond to individual pump wavelengths with specifically chosen angles of incidence. Phase-matching conditions representing our experimental situation are depicted in Fig. 3.

Standard formalism for Type I phase matching in crystals [87] was used in this calculation. We assumed that the efficiency of the parametric frequency conversion is low enough to disregard pulse-reshaping mechanisms, which can lead to effective reduction of the group-velocity mismatch between the pump and signal waves. In other words, the situation shown in Fig. 3 represents the low estimate of the signal bandwidth that can be simultaneously phase-matched in the NOPA. Dark-shaded contours in Fig. 3 represent individual pump wavelengths. The thickness of each dark stripe gives the FWHM of the angular phase matching [87] and is determined by the nonlinear interaction length, assumed equal to the crystal thickness (1 mm). The whole shaded area in Fig. 3 corresponds to the combined bandwidth produced by the entire SH spectrum, shown as a shaded contour in the inset in Fig. 3. To maximize the effect by employing the maximum number of pump frequencies, we used a 0.4 mm BBO thinner than in previous research, for the frequency doubling of the fundamental 120 fs pulses. This crystal offers an $\sim$ 30% conversion efficiency as opposed to $\sim$ 50% in a 1 mm BBO, but it ensures a larger spectral width of the SH. The corresponding frequency conversion efficiency is given in the inset in Fig. 3 by a solid curve and indicates that this crystal causes no significant reduction of the resulting SH bandwidth.

Finally, we point out that analogous ideas about bandwidth extension have been applied in other parametric schemes. Angular dispersion of the

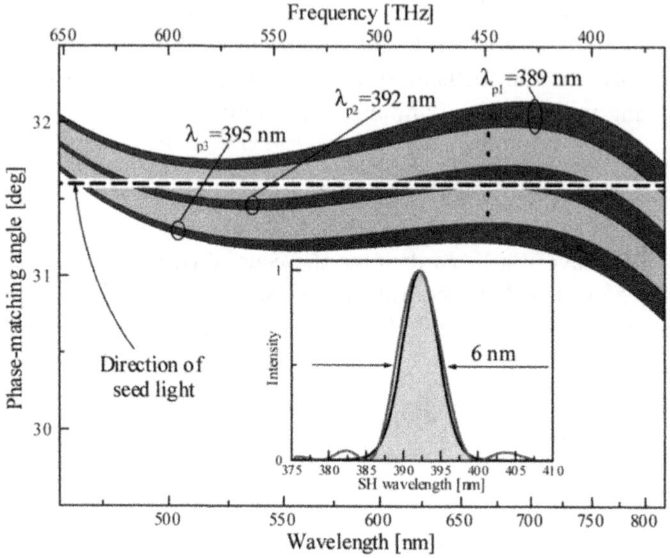

Fig. 3. Phase matching in a 1 mm Type I BBO crystal, $\theta = 31.5°$. *Dark shaded* contours indicate FWHM of the angular phase matching for individual monochromatic pump wavelengths after optimization of their incidence angles. The *dashed line* denotes the direction of the seed beam (see text for details). The *inset* depicts an experimentally measured SH spectrum used to pump the NOPA (*shaded contour*) and the calculated conversion efficiency of a 0.4 mm Type I BBO SHG crystal

frequencies in the fundamental beam has been used to obtain achromatic phase matching for SHG [88, 89], whereas a multipass OPA with slightly different beam directions in every pass [90] dramatically broadens the resultant bandwidth of the amplified pulse. Recently, a single-stage OPA pumped by two noncollinear beams has also been demonstrated as an efficient means of boosting the spectral width [91].

In summary to this section, the phase-matching bandwidth in a NOPA can be extended by increasing the spectral width of the pump and adjusting frequency-dependent pump angles.

1.4 Time Window of Broadband Parametric Amplification

In addition the need to satisfy phase-matching conditions, addressed above, we turn our attention to considerations about time-domain shaping of the amplified radiation. Despite the extremely broad bandwidth of parametric gain, only the part of the spectral components of the seed, which overlaps the pump pulse in time, can be amplified. The problem of the time window of amplification can be approached in two ways, by elongating the pump pulse and by compressing the seed. Both methods are employed in our system.

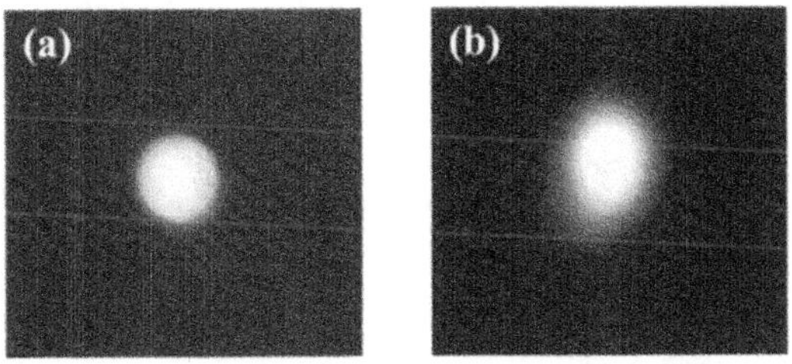

Fig. 4. Far-field beam profile images of a NOPA, operating in the broadest spectrum regime, recorded with a CCD camera. (**a**) After the first pass through the crystal (pulse energy 1.4 μJ). (**b**) After the second pass (pulse energy 5 μJ)

To stretch the pump pulse, we installed a 10 cm block of quartz in the pathway of the SH beam, as is shown in Fig. 1. According to our estimate, this lengthens the SH pulse of the bandwidth given in the inset in Fig. 3 to $\sim$ 500 fs. Despite a significant reduction in the peak intensity of the blue pump, the implementation of the stretching scheme is justified in view of the high frequency-conversion efficiency of the SHG crystal. Additionally, this reduces the intensity of the pump below the damage threshold of the BBO crystal used in the NOPA. The same idea of amplifying stretched pulses, which allows us to maintain a safe level of the peak intensity, is also behind the concept of chirped parametric pulse amplification [92, 93, 94, 95].

As the result of SH pulse stretching, we were able to employ a confocal focusing arrangement of the pump beam (Fig. 1), which significantly cleaned the structure of the output beam mode in comparison with the previously used nonconfocal pump arrangement [48, 49]. The mode pattern after the first and the second pass through the NOPA crystal in the temporal shaping of the pump pulse is shown in Fig. 4a and Fig. 4b, respectively. Though the mode shape after a single pass is nearly perfectly centrosymmetric, the onset of a crescent-like distortion becomes visible after an additional pass. This phenomenon is explained by preferential amplification along the superfluorescence cone rather than in a lateral direction. Therefore, as the parametric amplification approaches its saturation in the center of the beam, the output mode of the NOPA begins to spread out of the pump–seed plane in the direction of the highest parametric gain.

It has also been pointed out that the use of a glass block to stretch the pump pulse helps to reduce the sensitivity to the drift of the pump–seed delay [96], thus improving the robustness of the parametric amplifier to environmental disturbances.

Beside SH stretching in the quartz block, some further pulse broadening occurs in the prism employed for radial dispersion of the pump beam (*vide supra*). Note that the use of such a prism with a large insertion depth in the previous NOPA scheme [48, 49] also contributed to improving the temporal

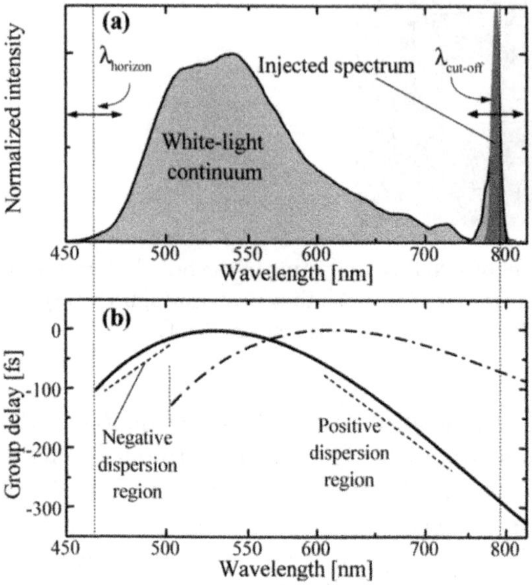

Fig. 5. Overview of spectral and temporal parameters of the seed pulse. (**a**) Optimized spectrum of the white-light seed (*shaded contour*) and injected fundamental spectrum (*dark contour*). (**b**) Various settings of the group delay of the seed following its precompression by a prism pair (see text for details). *Horizontal arrows* indicate adjustable parameters of the prism precompressor

overlap between the pump and seed pulses and resulted in the enhancement of the simultaneously phase-matched bandwidth of the NOPA.

We now turn our attention to temporal shaping of the seed pulse. As follows from the considerations presented above, the white-light seed does not require meticulous precompression to achieve the maximum bandwidth of simultaneous phase matching. In our NOPA design, the white-light continuum produced in a 2 mm sapphire window runs through a pair of 45° quartz prisms separated by $\sim$ 50 cm (Fig. 1). The prism precompressor plays several distinct roles in the system. First of all, by placing an adjustable razor blade behind the inner prism, we can eliminate the intense frequency components of the fundamental pulse injected into the sapphire plate and prevent their undesirable amplification in the NOPA crystal. Second, the insertion depth of the inner prism determines the so-called horizon wavelength, i.e., the minimum wavelength transmitted through the prism sequence. Finally, by balancing the dispersion of the precompressor, we can optimize the temporal window of parametric amplification and the output spectral shape.

The function of the precompressor is schematically presented in Fig. 5. The fundamental spectrum injected into the sapphire window and the resultant white-light continuum are depicted in Fig. 5a.

Also shown are the flexible parameters of the prism sequence, allowing us to select a spectral window of amplification. Figure 5b presents results of a computer simulation of the seed group delay at the input of the NOPA crystal. The solid line indicates the conditions for the broadest spectral range of operation, whereas the shaded area corresponds to the optimal temporal overlap of the pump pulse. As can be immediately noticed from the shape of the solid curve in Fig. 5b, the combination of white-light generation in the sapphire window, prisms, and propagation through air leads to a predominantly positive dispersion. This is consistent with our considerations about the frequency sequencing of the pump and seed pulses, presented above. The exception is the short-wavelength region, which exhibits negative dispersion as a result of a rapid change in the refractive index of the prism material (quartz).

It can be further shown that the reduction of the insertion depth of the inner prism can dramatically reduce the required time overlap with the pump pulse. An example of this is given in Fig. 5b (dash-dotted curve). However, the accompanying shift of the horizon wavelength toward the red decreases the usable bandwidth of amplification.

We further examine the results of the spectral shaping of the NOPA output, achieved by manipulating the dispersion of the precompressor and varying the independent pump–seed delays of the two passes through the NOPA crystal. The latter are labeled Variable delay 1 and 2 in Fig. 1. Three characteristic cases of output spectra are presented in Fig. 6a–c. Figure 6a corresponds to the redshifted horizon wavelength, shown as a dash-dotted curve in Fig. 5b. A nearly Lorentzian spectral shape in Fig. 6a was obtained by advancing the pump pulse, in both of the passes, ahead of the seed pulse, which results in predominant amplification of the spectral components that are around the extremum of the group delay (dash-dotted curve in Fig. 5b). Next, study the maximum bandwidth operation that corresponds to the group delay in Fig. 5b, given by a solid curve. The spectrum presented in Fig. 6b was recorded with identical pump–seed delays in each pass, whereas the data in Fig. 6c were obtained by balancing both delays to enhance the redshifted and blue-blueshifted amplification in respective NOPA passes. Temporal intensities, computed under the assumption of perfect pulse compression, are given in Fig. 6d–f. It is evident that the spectrum in Fig. 6c has the largest width and corresponds to the narrowest pulsewidth measured in terms of FWHM (Fig. 6f). However, the steep frequency cutoffs, especially in the NIR edge of the spectrum, make the wing structure of the pulse more pronounced.

In brief, in this section we have shown that a high level of seed precompression is not required for broadband amplification. On the other hand, flexible spectral shaping of the amplified signal is attained by balancing pump–seed delays for each of the two passes through the NOPA crystal.

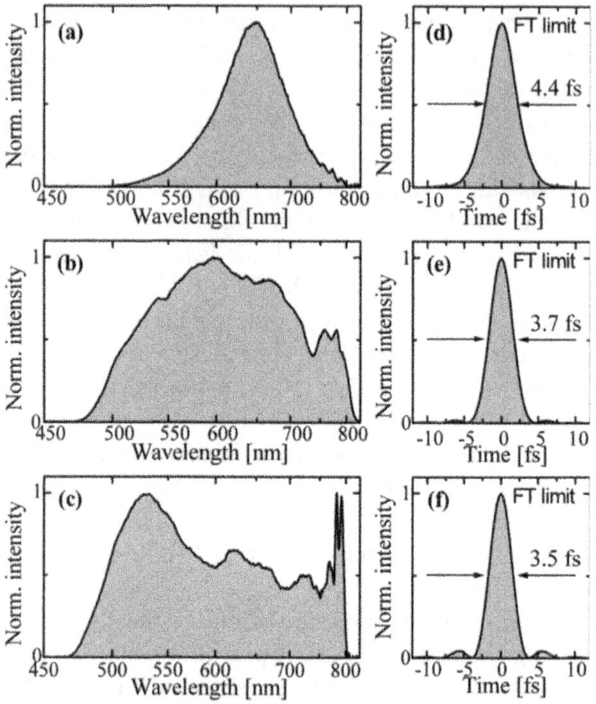

Fig. 6. Spectral shaping of the amplified signal pulses after two NOPA passes. (**a**) A "shallow" group delay of the seed pulse (corresponding to the *dash-dotted curve* in Fig. 5b). (**b**) Higher precompressor dispersion and wider seed spectrum (*solid curve* in Fig. 5b). (**c**) Same as (**b**) with the pump–seed delays optimized to enhance the blue and the red wings of the spectrum in the first and the second NOPA pass, respectively. (**d–f**) Corresponding temporal pulse profiles calculated assuming ideal pulse compression

1.5 Amplitude-Phase Characterization of Amplified Pulses

Accurate phase measurement of the NOPA pulses plays several key roles in this research. First of all, it is required to characterize the phase of the chirped amplified signal pulses and design an adequate pulse compressor. Second, phase measurement is needed for controlling the quality of adaptive correction of the residual phase distortion. Assuming that the pulse compressor can potentially provide perfect compensation, the resultant spectrum-limited pulses should be tested in the end.

To tackle these challenges, we chose the technique of SHG FROG because of the lowest possible order of nonlinearity involved, the simplicity of the experimental setup (only two input beams), and an identical beam arrangement with an excite–probe scheme. The latter feature is particularly valuable considering the applications of NOPA pulses in nonlinear spectroscopy [51, 53, 56, 57, 58, 59], since broadband NOPA pulses can be easily reshaped even as they propagate through air.

Ideally, single-shot, amplitude-phase characterization is preferred for use with an adaptive pulseshaper to reduce measurement time. However, a single-shot variety of SHG FROG is not applicable in our case as it requires transverse beam expansion, which considerably lowers the SH signal and employs a large intersection angle between the autocorrelator beams [97, 98]. Moreover, the configuration of the apparatus for pulse diagnostics can no longer be used for excite–probe spectroscopic applications.

Another pulse characterization technique based on the second-order nonlinearity, SPIDER [99, 100], has recommended itself in the measurement of few-cycle pulses [101]. Nevertheless, this method also requires a separate setup dissimilar to the scheme of an excite–probe experiment. Therefore, the use of SPIDER faces incompatibility comparable to that in single-shot FROG.

As is apparent from the scope of problems for amplitude-phase characterization, outlined above, it can be divided into the measurement of strongly chirped pulses and of (nearly) compressed pulses. In the first case, a long delay scan is required in SHG FROG due to the duration of the chirped pulse, and the low intensity of the SH signal may pose a problem. The peak signal intensity is considerably higher in the second case, whereas the precision of time-delay calibration, rather than the scan range, becomes essential for correct FROG measurement.

The schematic of our convertible excite–probe setup, extended for SHG FROG detection, is presented in Fig. 7. To facilitate both the wide- and narrow-range scans and ensure adequate calibration accuracy, we employed a nanomotion translation stage (with a minimal delay increment of 10 nm) equipped with a linear motor as well as with a piezoactuator. Additionally, this delay line features a built-in interferometer and active position stabilization. To balance the dispersion in both beams of the FROG apparatus, two identical Cr-coated, 0.5 mm thick quartz beam splitters were used. Unlike their dielectric counterparts, these beam splitters exhibit virtually flat reflectivity across the visible and NIR. However, about 30% of the input light is absorbed by the coating in our case. The beams are focused on the SHG crystal and collimated behind it by $R = -250$ mm spherical mirrors.

The greatest obstacle facing SHG FROG measurement of ultrabroadband pulses is the limited phase-matching bandwidth of the SHG crystal. A very thin crystal, required for broadband SHG, reduces the amount of useful SH signal that it can yield. On the other hand, a thick crystal supports a narrow SHG bandwidth and leads to an effect known as spectral filtering [102], whereupon vital information is irrecoverably lost from the measured FROG trace.

Recently, a technique of angle dithering of the SHG crystal has been proposed as a method to overcome the limited phase-matching bandwidth [103].

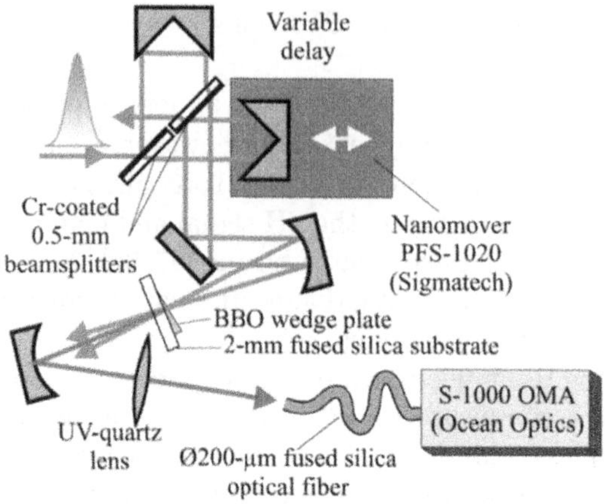

Fig. 7. Schematic of an excite–probe setup adapted for SHG FROG measurement

A relatively thick crystal in this case is rapidly tilted back and forth in the intersection plane of the autocorrelator beams, thus leading to a periodic change in the phase-matching angle. Whereas this technique has been successfully applied to measure picosecond white-light continua stretching over hundreds of nanometers, the use of this method for the measurement of a compressed few-cycle pulse has not been attempted so far. A severe potential limitation in this case is the concern for temporal filtering that may occur as a result of the loss or shift of the geometric overlap. The most important problem here is the demand for highly accurate positioning, within several microns, of the rotational axis of the crystal at the intersection of the autocorrelator beams. This restriction, however, applies only to the measurement of the compressed pulses.

A possible solution of the problem of insufficient phase-matching bandwidth can be achieved in principle by constructing a multilayer crystal [104]. In this configuration, two or three thin crystals, each about 20 μm thick and cut for different wavelengths of phase matching, are to be bonded together by optical contact. Compared with a single crystal, the resultant SHG bandwidth of the layered structure could then be substantially broadened. To the best of our knowledge, this design has not been carried out to date because of manufacturing difficulties.

Instead, to improve the situation with the thin BBO wedge available for our measurements, we chose to compensate for the spectral filtering in the crystal by applying a numerical correction to the recorded FROG trace. It has been shown before that the shape of the spectral filter can be estimated reasonably well if the crystal thickness is known and a precise angular reference for the crystal orientation is available [105]. However, the use of a wedge with a variable thickness precludes such a calculation. Additionally, an angular tilt of the wedge leads to uncertainty about the shape of the spectral filter. To overcome this difficulty in this research, we adopted the method of correction by spectral marginals, which was first suggested and implemented by *Taft* et al. [106]. The marginals of a FROG trace can be computed by integrating the FROG trace along the frequency and time-delay axes, yielding time and frequency marginals, respectively. The latter marginal can be expressed in the following form [98, 107]:

$$M(\Omega) = \int I_{\mathrm{FROG}}(\Omega, \tau)\, \mathrm{d}\tau\,. \tag{3}$$

As determined by the mathematical form of the SHG FROG signal, the shape of an ideal frequency marginal is identical to the autoconvolution of the spectral intensity, i.e.,

$$M_{\mathrm{ideal}}(\Omega) \equiv \int I(\Omega - \omega) I(\omega)\, \mathrm{d}\omega\,. \tag{4}$$

The spectral correction curve for a detected SHG FROG trace can then be obtained by calculating the ratio between the autoconvolution contour and the frequency marginal computed from the experimental data, $M_{\mathrm{ideal}}(\Omega)/M(\Omega)$. The correction curve, obtained in this way, automatically accounts for the spectral modulation of the measured SH signal in the transport optics and spectrometer and includes the spectral sensitivity of the detector. We employ this method of postexperimental validation of the FROG measurement to characterize both the chirped and compressed pulses from the NOPA. Subsequently, the spectrally corrected FROG traces were processed using a standard commercially available FROG retrieval program, purchased from Femtosoft Ltd.

1.6 Design of Pulse Compressor

The group delay of chirped NOPA pulses, characterized by SHG FROG, is shown in Fig. 8 by filled dots. This measurement provides the target for the design of the pulse compressor. As we argued above, appropriate chirped dielectric reflectors can in principle be created to compensate for the target group delay. However, implementation of adaptive dispersion control based on a micromachined flexible mirror requires spatial dispersion of the laser beam to actuate, independently, the path length of all frequency components across the NOPA spectrum. Therefore, the pulse compressor should inevitably incorporate a dispersion line that includes a focusing element and thereby provides a Fourier plane for spatial modulation of the dispersed laser beam. The main challenge here is the selection of a suitable configuration that introduces a considerable amount of negative dispersion required for pulse shortening and simultaneously removes the spatial chirp on the output beam.

Our experimental arrangement, shown in Fig. 1, is based on a 300 line per millimeter ruled diffraction grating and an $R = -400$ mm spherical reflector

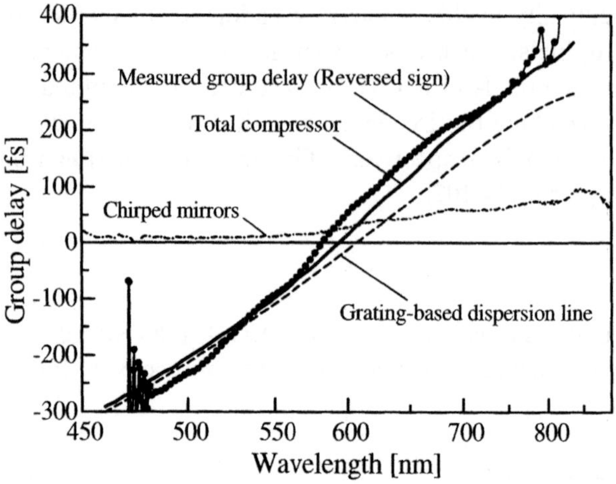

Fig. 8. Overview of compressor dispersion. *Solid circles* show the target group delay, corresponding to time-inverted data obtained by SHG FROG. The *solid curve* gives the overall group delay of the compressor excluding the contribution of the flexible mirror. The *dashed curve* depicts the estimated group delay of the telescopic grating-based disperser, whereas the *dash-dotted curve* represents the contribution of dielectric chirped mirrors measured by white-light interferometry

with the flexible mirror positioned in the focal plane. This scheme operates in the negative dispersion regime when the -1 diffraction order of the grating is selected. The latter fact is of crucial importance because of the low, 40 fs range of the group delay correction provided by the membrane deflection of the micromachined mirror. Therefore, the main correction of the spectral phase should be carried out by other elements of the pulse compressor, whereas the computer-controlled flexible mirror can be employed only for fine group delay tuning.

The overall dispersive properties of the designed compressor are presented in Fig. 8. The main part of the required negative dispersion is contributed by the telescopic grating disperser (dashed curve). The dispersion from positive to negative can be tuned by selecting the angle of reflection off the spherical mirror and the distance between it and the grating. However, such manipulations can easily result in spatial chirp of the output beam. We performed a careful dispersive ray-tracing analysis of the system to find the optimal conditions satisfying both the requirements of the group delay and spatial beam quality. To introduce additional negative dispersion that cannot be provided by the grating–mirror arrangement, we also employed two dielectric ultrabroadband chirped mirrors (UBCM). The UBCM coatings were specially designed and manufactured by Hamamatsu Photonics for pulse compression in visible–near-IR NOPAs. The reflectance bands of these mirrors fully support the spectrum of the signal wave. The group delay of UBCMs (dash-dotted curve in Fig. 8) was measured by using a white-light interferometer [108]. In addition to contributing approximately one-third of the negative dispersion required for pulse compression, these mirrors supply higher order phase correction in the spectral wings. The total calculated group delay, excluding the adaptive flexible mirror, is given in Fig. 8 by the solid curve. The remaining estimated deviation from the target, shown as solid dots, is within the correction range of the flexible mirror and can be eliminated by changing the shape of the reflective surface.

The micromachined device employed in our work is identical to those used in [68, 69, 70]. The membrane is controlled by 39 actuators that are lined up in three rows and provides a clear aperture of $\sim 30\,\mathrm{mm} \times 7\,\mathrm{mm}$. Potentially, such a modulator can nearly perfectly eliminate smooth variations of the spectral phase that can be described by a cubic spline drawn through 13 nodes, distributed, in this particular case, over a 300 nm wide spectral range. To reduce the number of controlled parameters from 39 to 13, we lock together the actuator voltages in the vertical direction, i.e., perpendicularly to the dispersed beam. This simplification is justified because the surface of the flexible mirror is located in the focal plane of the spherical mirror, which results in a negligibly low sensitivity of the output beam to actuation in the vertical plane. The channels of the flexible mirror are connected to 8-bit high-voltage drivers, controlled by a personal computer (Fig. 1), which allows changing the setting of each actuator in the range of 0 V to 280 V. The stiffness of the 0.5 μm silicon-nitride membrane depends on the material and number of layers of the deposited optical coating. An Au coating with a Cr bonding layer underneath, used in our case, enables a maximum deflection in excess of 6 μm in the center of the mirror.

The components of the pulse compressor are placed on the optical bench according to the ray-tracing calculation. The distance between the diffraction grating and the spherical mirror is subsequently adjusted within a ± 1 mm range to optimize the SH signal from a thin doubling crystal. The last step of the pulse compression – adaptive phase correction using the flexible mirror – will be described in detail in the following section.

1.7 Adaptive Pulse Shaping

Two different strategies of final adaptive phase control are feasible; one-step (also known as open-loop) and iterative search (a feedback based on a learning algorithm). The differences between the two approaches were reviewed in detail by *Weiner* [102]. Below, we briefly discuss the merits and drawbacks of both methods in the specific case of a shaper based on a flexible mirror and motivate our choice in application to the compression of NOPA pulses.

As has been pointed out in the previous section, phase characterization of an uncompressed (or partially compressed) NOPA pulse provides full information about the required dispersion adjustment. Therefore, if these data could be programmed directly on the flexible mirror, the task of producing spectrum-limited pulses would be solved straightforwardly. However, such a move requires several types of calibration of the flexible membrane. The first, and the easiest, is the measurement of the wavelength distribution vs. spatial coordinate and actuator position. Whereas such a procedure is identical to that in a flat-field polychromator, there might be a need to carry it out after every realignment of the NOPA, which is likely to shift the laser beam position relevant to the coordinates of the mirror. Another, and by far a more challenging problem, is calibration of the membrane response as a function of the voltage applied to the electrostatic actuators. Since the elasticity of the membrane is considerably lower at its edges that are contained by the rigid perimeter of the mirror, the displacement of the reflective surface, exerted by an identical actuator voltage, differs dramatically across the working area. Also, the complexity of the calibration task is increased by the fact that the amount of deflection at given coordinates of the micromachined reflector depends on the pull from several actuators at the same time. Therefore, a multidimensional database would be required to account for all possible combinations of the actuator voltages, which is clearly highly impractical. An alternative and more viable approach is to apply a mathematical model to describe the shape of the membrane.[1] Direct one-step phase compensation using a micromachined mirror was demonstrated by *Zeek* et al. [68]. The resultant pulse compression was incomplete and was subsequently improved by iterative optimization based on a generic algorithm.

In contrast to single-step adaptive correction, the use of iterative search does not require a priori calculations and/or an extensive knowledge of the response functions of the pulseshaper. The power of this approach is illustrated by the fact that it even permits finding a solution to an unknown pulse profile that is optimal for a complex light–matter interaction. An early attempt to employ a model nonlinear process as a reference for pulse optimization simply relied on the enhancement of SH intensity through adaptive phase control [109]. Very recently, coherent wave-packet control in mixed alkali-metal clusters has been performed using adaptively phase-shaped pulses [110]. Remarkably, the femtosecond laser in such experiments is "learning" different pathways of a chemical reaction as the feedback algorithm automatically searches for an appropriate solution for the pulse properties. In this case, rather than producing a certain predetermined pulse, the shaper actually advises the user on the best pulseshape judging by the emission efficiency of the target product, e.g., a chemical compound. Thus, the feedback scheme provides undeniable benefits for optimization of complex experimental systems. Obviously, iterative adaptive control requires a longer time to converge to an appropriate solution in comparison with the one-step strategy. Therefore, the efforts should be concentrated on devising an informative feedback and a robust searching algorithm. Whereas the use of SH intensity as a merit function for adaptive optimization proved effective for sub-20 fs pulse compres-

[1] The system of Poisson equations and sample program codes is available at the manufacturer's website `http://www.okotech.com`

sion [67, 109], nevertheless, this quantity is clearly inadequate to reflect the whole pulse complexity. Real-time autocorrelation trace measurements have also been employed as a feedback parameter [111, 112], though they contain relatively poor information on the actual spectral phase of the pulse. Above all, especially in very broadband pulses, both feedback varieties are irrecoverably affected by spectral filtering in the SHG crystal. Exhaustive feedback schemes were implemented by using FROG pulse characterization [68,69,113] or TADPOLE [113]. Another single-shot pulse diagnostic technique is SPIDER, which potentially is suitable for adaptive feedback but, as pointed above, is incompatible with the excite–probe geometry of NOPA applications. Also, if the reference for SPIDER characterization is derived from the fundamental pulse of the Ti:sapphire laser, a time jitter could appear between the replicas of the amplified signal pulses and the reference pulse as a consequence of air turbulence that causes a slight drift of the membrane in the flexible mirror.

Unfortunately, as outlined in Sect. 1.5, single-shot FROG acquisition and/or rapid FROG scans could not be performed in our system, whereas SPIDER requires a separate setup. Therefore, in the choice of the feedback, we opted for a compromise between complete pulse measurement at the end of optimization and fast partial characterization during the adaptive search.

First of all, we note that the measurement of the SH spectrum, which corresponds to the SHG FROG signal at zero time delay, is highly informative and obviates the need to perform time-consuming delay scans:

$$S(\Omega) = \left| \int A(\Omega - \omega) A(\omega) \exp\left\{ \mathrm{i} \left[\phi(\Omega - \omega) + \phi(\omega) \right] \right\} \mathrm{d}\omega \right|^2 , \tag{5}$$

where $A(\omega) = \sqrt{I(\omega)}$ is the spectral amplitude and $\phi(\omega)$ is the spectral phase of the pulse. Unlike the time-domain autocorrelation signal, which is centrosymmetric by definition, (5) features an autoconvolution. Whereas a straightforward complex deconvolution is impossible because only the modulus squared of the signal is measured, nevertheless, with a narrow class of exceptions, both the shape and the magnitude of $S(\Omega)$ are highly sensitive to changes in $\phi(\omega)$. Figure 9 provides examples of different distortions of the spectral phase and their effect on the resultant shape of the SH spectrum. We chose a super-Gaussian model spectrum (shaded contour in Fig. 9a), which corresponds to a 4 fs spectrum-limited pulse. The latter is shown in Fig. 9c as a solid curve. Our example considers three phase shapes of the second, third, and fourth orders, depicted in Fig. 9a as dashed, dash-dotted, and dotted curves, respectively. The amount of phase distortion was selected such that the resultant pulses have equal peak intensities in all three cases and the FWHM of the pulse is broadened by only 10% to 12% compared with the spectrum-limited duration, as shown in Fig. 9c. The ideal SH spectra, corresponding to these three phase shapes, as well as a flat spectral phase are given in Fig. 9b. Whereas the pulse intensities, given by dashed curves in

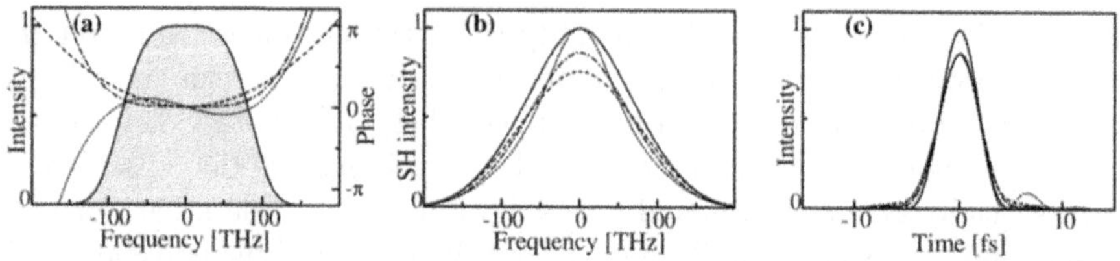

Fig. 9. Numerical simulation of the influence of the spectral phase on the SH spectrum. *Dashed*, *dotted*, and *dash-dotted* curves represent second, third, and fourth order phase distortions, respectively. (**a**) shows the spectrum of a 4 fs pulse (*shaded contour*) and the three phase patterns used in the simulations. (**b**) depicts corresponding SH spectra. The *solid curve* presents the spectrum-limited pulse (i.e., a flat spectral phase). (**c**) gives temporal intensities. The *solid curve* shows an ideally compressed pulse

Fig. 9c are similar with the exception of the minute wing structure, the SH spectra in Fig. 9b exhibit substantial phase-dependent differences. Therefore, even when the pulse is compressed to within $\sim 10\%$ of its spectrum-limited duration, the SH spectrum is sensitive enough to differentiate among various types of residual phase distortion. An additional advantage of this simple measurement is that it can be employed for adaptive feedback, even in the presence of appreciable spectral filtering in the SHG crystal, provided there is at least some SH intensity at the wavelength of interest.

Having demonstrated the sensitive link between the SH spectrum and the spectral phase of the pulse, we next consider how to employ this type of measurement to obtain the flattest possible phase by feedback optimization. As can be seen from Fig. 9b, the SH intensity detected at any wavelength is highest for the spectrum-limited pulse (shown as a solid curve). The assertion that the global maximum of the SH intensity at all wavelengths corresponds to optimal pulse compression is intuitively clear. Assuming low depletion of the fundamental and instantaneous nonlinearity in the SHG crystal, the strongest SH signal corresponds to the pulse with the highest peak intensity. The link between the global minimum of pulse duration and the global maximum of pulse intensity is also easy to see as it immediately follows from Parseval's theorem. According to this theorem, the total power of a pulse in the time domain and in the frequency domain is the same. Therefore, two pulses, which differ from each other only by their spectral phase, carry identical total power, i.e., the area underneath the intensity envelope of the pulse. Consequently, the "narrowest" pulse necessarily corresponds to the "tallest" pulse.

However, the considerations stated above do not constitute a rigorous proof and do not explain why the SH intensity is maximized across the whole spectrum if the pulse is perfectly compressed. A stricter proof is supplied in Appendix A.

Now, we turn our attention to the algorithm that uses the measurement of the SH spectrum as input and produces a set of actuator voltages at the output. The latter variables constitute a multidimensional parameter space, in which the problem of function maximization (minimization) can be solved in a number of ways. The popular approaches include generic algorithms, simulated annealing, and several types of gradient descent in a multidimensional space [102, 114]. One of the chief concerns in the selection of a suitable algorithm is its cost-effectiveness with respect to the time required for achieving the global extremum. This requirement renders some techniques of multidimensional search impractical. For instance, the convergence of the simulated annealing method is assured only if the adjustment of all variables proceeds in very small steps, thus making the algorithm intrinsically slow. In comparison, gradient search methods allow extensive parameter adjustment in a single iteration, but they frequently lose the speed advantage because of the need to estimate a set of derivatives. It must be pointed out that the gradient approach is very effective for solving problems in which the derivatives can be computed (especially from an analytical expression). Nevertheless, in our pulse compression experiment, finding derivatives only helps to reduce the number of iterations but still requires the same number of functional evaluations (i.e., changes of the flexible mirror settings and SH spectrum measurements) as simpler algorithms.

Several additional difficulties associated with the implementation of adaptive control deserve mentioning. Notably, from the mathematical standpoint, actuator voltages do not constitute a set of orthogonal parameters, which would considerably simplify the problem of finding the global extremum. As pointed out earlier, these voltages depend heavily on each other as a result of their combined action on the membrane of the flexible mirror. To complicate the situation further, the adjustment of an individual actuator does not produce the desired local effect on the SH spectrum (or the full FROG trace) confined to a narrow wavelength region, but rather manifests itself throughout the entire SH bandwidth. The latter is a consequence of sum frequency generation, which is present in SHG in combination with direct frequency doubling. Curiously, the use of a cross-correlation with a narrowband pulse does not lessen the problem of the local spectral response but instead it rather lowers the overall sensitivity to the duration of the would-be compressed pulse. Last, but not least, the feedback algorithm, based on a one-dimensional spectral scan, should be reasonably robust with respect to the intensity noise of the laser, which causes misguiding fluctuations of the SH signal.

In our experiment, we tried out several different algorithms, among them the multidimensional search, using first-order derivatives and hybrid variations of this method. Despite its seeming simplicity, Brent's method of parabolic interpolation performed best against laser noise. Since this algorithm contains only a one-dimensional extremum search, we looped it to process the settings of each of the 13 actuators independently the mutual parameter dependence was accounted for in the merit function. The choice of the latter is as important as the selection of an appropriate search algorithm. The most straightforward parameter to evaluate is the area of the SH spectrum, i.e., $\int S(\Omega)\,\mathrm{d}\Omega$. However, as is easy to notice from the SH spectra presented in Fig. 9b, the area does not change appreciably after the pulse duration is confined to within $\sim 10\%$ of its spectral limit. As a result, direct area maximization is almost tantamount to direct SH intensity maximization [67, 109]. Also, from the point of view of the algorithm practicality, global minimization (of residuals) is preferred to maximization. However, the difficulty of applying the former strategy in our case is the lack of a target (unnormalized) SH spectrum with respect to which the fine residual structure can be computed. We circumvent this problem by keeping track of the all-time highest signals at every SH wavelength, which are stored in a "maximum envelope" array, $S_{\mathrm{max}}(\Omega)$. The merit function that has to be minimized is then computed in the following way:

$$Z(U_i) = \int |S_{\mathrm{max}}(\Omega) - S(U_i, \Omega)|\,\mathrm{d}\Omega\,, \tag{6}$$

where U_i stands for the voltage of the ith actuator. The flowchart of the whole minimization procedure is presented in Fig. 10.

Before the start of this routine, $S_{\mathrm{max}}(\Omega)$ is reset, and it is further automatically updated as the algorithm learns combinations of actuator settings that increase S_{max} at various wavelengths. An apparent contradiction with the stated goal of the minimization of the merit function comes from the fact that $Z = 0$ in the very first iteration, and it increases instead of decreasing in the following iteration. However, in trying to find the downhill direction of the $Z(U_i)$, Brent's method attempts both a higher and a lower value of U_i, which immediately results in modification of S_{max}, unless U_i is already optimal. Therefore, only the initial iteration is partially incorrect as it might potentially result in an uphill rather than a downhill step. The sudden variations of S_{max} in subsequent iterations might belie the steepness of the $Z(U_i)$ slope but retain the general direction toward the minimum. The lack of a steady reduction of area residuals, expressed by Z, makes the algorithm more agile and prevents stagnation. The same ability that allows the program to adjust itself to a sudden increase in S_{max} also makes our search method less sensitive to the perturbations introduced by laser noise.

The algorithm is incorporated in our FROG measurement software, which serves to check the pulse quality before and after adaptive phase correction. The time required for satisfactory convergence depends mostly on the integration constant of the optical multichannel analyzer (OMA) employed for the measurement of the SH signal. With a 300 ms time constant, the adaptive optimization requires just 1–5 minutes and less than a minute if the routine is initialized from the configuration used a day earlier.

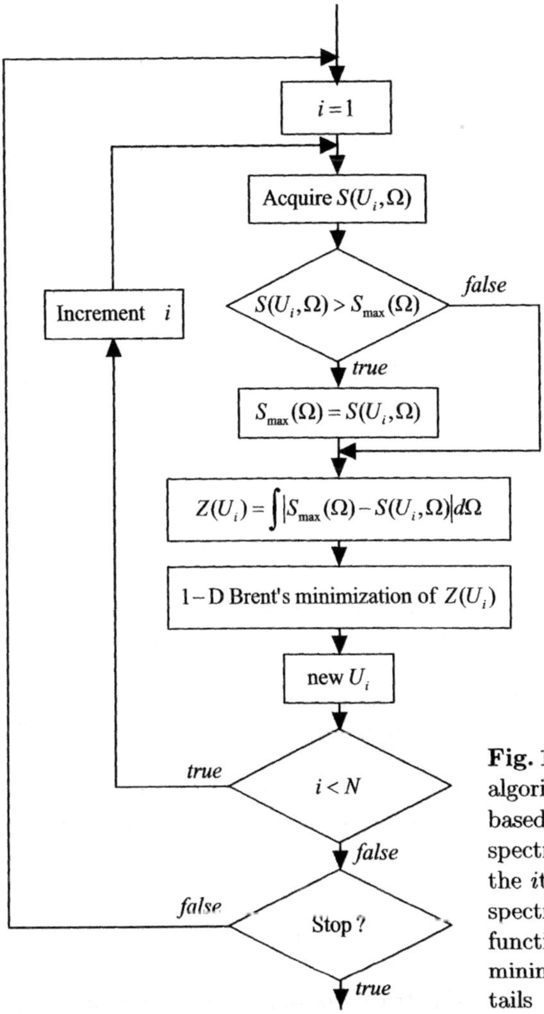

Fig. 10. Flowchart of the feedback algorithm for adaptive pulseshaping based on measurement of the SH spectrum. U_i is the voltage setting of the ith actuator; $S(U_i, \Omega)$ is the SH spectrum; $Z(U_i)$ denotes the merit function used for multidimensional minimization. See text for further details

The results obtained by applying the approach developed to the final phase correction of NOPA pulses are summarized in Fig. 11. Figure 10a,c shows the measured and retrieved SHG FROG traces before adaptive phase correction and correspond to all values of U_i set to zero. This should not be confused with a case of a flat reflector in the focal plane of the spherical mirror, as the membrane shape remains deformed and introduces a mostly parabolic group delay, i.e., third-order phase correction. The resultant duration of this partially compressed pulse is 5.3 fs and its intensity is shown in Fig. 11f as a solid curve. Whereas a time cut across the FROG trace in its middle (around 300 nm) corresponds to a reasonably narrow "pseudo-autocorrelation" trace for that wavelength region, an apparent wing structure around the spectral edges of the FROG trace exposes the eventual pulse lengthening. The wings and sidelobes of the FROG pattern are clearly trace-

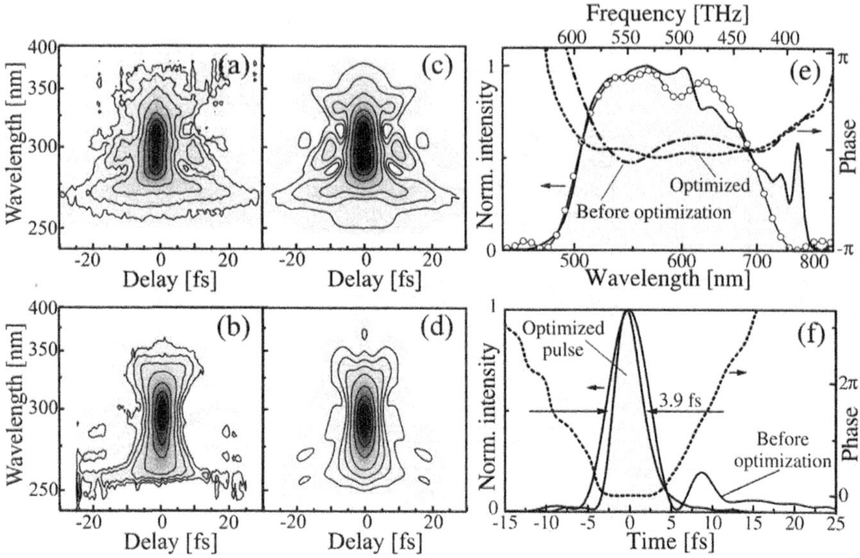

Fig. 11. Overview of pulse-shaping results. (**a**) and (**b**) depict measured SHG FROG traces before and after adaptive phase correction, respectively. Corresponding retrieved traces are displayed in (**c**) and (**d**). (**e**) shows the fundamental spectrum (*shaded contour*) measured at the crystal location in the FROG apparatus and the spectrum recovered by the FROG retrieval algorithm (*open circles*). The *dash-dotted curve* represents the spectral phase prior to adaptive shaping, whereas the *dashed curve* shows the optimized phase. (**f**) Initial (*solid curve*) and optimized (*shaded contour*) temporal intensity profiles. The *dashed curve* depicts the temporal phase of the optimized pulse

able to the shape of the recovered distortion of the spectral phase, shown in Fig. 11e as a dash-dotted curve.

The measured and retrieved FROG traces, obtained upon automated optimization of the membrane configuration, are presented in Fig. 11b,d, respectively, and show a significantly cleaner wing structure. The deviation of the spectral phase is noticeably smaller compared with the case registered before optimization and is confined to a $< \pi/8$ fluctuation range within the phase-matching spectral region of the SHG crystal. Phase improvement outside this region, where FROG detection is enabled as a result of spectral correction by the marginal method, is less pronounced. The recovered intensity profile of the adaptively compressed pulse is depicted as a shaded contour in Fig. 11f. The pulse duration, deduced by the FROG inversion algorithm, is 3.9 fs. Nevertheless, given the problem with insufficient SHG phase matching, which is apparent from a comparison of the measured and retrieved fundamental spectra (Fig. 11e), an $\sim 10\%$ error margin should be assigned to the pulse duration obtained. The accuracy is estimated from numerical simulations [105] and the dispersion of FROG inversion results of the experimental data.

In brief, the use of a robust adaptive algorithm based on a simple spectral measurement permits rapid, improvement of the pulse quality, within several minutes. The inability to remove the remaining discrepancy of the spectral phase is caused only by the limited SHG bandwidth.

1.8 Conclusions and Outlook

Adaptive phase correction is indispensable to high-quality compression of extremely broadband laser pulses owing both to the complexity of the phase distortion and the lack of a broad range, as well as wavelength-selective tunability of traditional pulse compressors.

In this work, we present the design of a 4 fs visible–near-IR parametric amplifier with adaptive pulse tailoring, which permits a factor-of-30 compression of pump pulses from a standard Ti:sapphire chirped-pulse-amplification system. In comparison with the NOPAs reported earlier, we have performed several modifications that allowed us to increase the output bandwidth and furnished new means of spectral shaping. First, we demonstrated, both numerically and experimentally, that angular dispersion of pump wavelengths can be employed for broadening the parametric gain bandwidth of a noncollinearly pumped BBO crystal. Next, we studied the aspects of time ordering of the spectral component in the pump pulse and the white-light continuum seed and established that broadband amplification can be achieved without high-accuracy chirp precompensation of the seed. The introduction of a flexible precompressor based on a prism pair provides an easy tool for wavelength separation of the seed spectrum and balancing the shape of the amplified signal spectrum. Whereas smooth parametric spectra were generated that correspond to 3.5 fs pulses, if ideally compressed, we managed to obtain $\sim$ 4 fs pulses after passing the signal beam through a hybrid compressor based on a diffraction-grating telescopic dispersion line, specially designed multilayer dielectric chirped mirrors, and a computer-controlled flexible mirror. A robust adaptive control algorithm was developed in this work, which is aimed at practical compatibility between the pulse diagnostic/compression and standard applications of ultrashort, time-resolved, nonlinear spectroscopy. Automated optimization permits rapid adjustment of the pulse parameters with no human interference and simultaneously ensures that the amplitude-phase properties of the measured compressed pulse exactly match those used in a subsequent spectroscopic experiment.

We foresee several technical issues that are important for further development of the NOPA technology.

As shown in this work, efficient use of all pump frequencies leads to broadening of the amplification band. With sub-50 fs amplified Ti:sapphire systems now becoming standard, commercially available equipment, the parametric bandwidth can easily be expanded even further by employing wider SH pump spectra. However, a potential drawback lies in the derivation of an appropriate seed spectrum, as the fundamental radiation around 800 nm will have a large overlap with the NOPA gain band. To avoid undesirable amplification of intense fundamental spectral components, they have to be attenuated in the seed spectrum, or alternatively, a gas-filled capillary [115] may be used for white-light generation instead of bulk chirping in a sapphire or glass window. Despite this obvious difficulty, NOPA configurations with a shorter pulse pump laser are definitely worth exploring due to the advantage of increased bandwidth.

Another immediate challenge is improvement of the throughput of the adaptive pulse compressor. The efficiency of a grating dispersion line is necessarily low due to the need to support an extraordinarily broad input bandwidth with minimal possible spectral narrowing or reshaping. A prism disperser, based on refraction in a transparent material, offers both high throughput and virtually no spectral distortion. Additionally, unlike that of a diffraction grating, the angular dependence of a dispersed beam is almost linear on the frequency scale. The latter is essential for adequate sampling of broadband laser spectra on an adaptive shaper. Nevertheless, a prism-based line with a Fourier plane cannot operate in the negative-dispersion regime. Therefore, implementation of this design requires simultaneous addition of broadband optical elements with powerful negative dispersion and a highly smooth group delay. We believe that fabrication of gradient-index chirped dielectric reflectors [116] might enable the use of low-loss refractive dispersion lines in the near future.

One of the fundamental issues facing further progress in the development of a NOPA with an even shorter pulse duration is the difficulty with pulse diagnostics. In a typical time-resolved excite–probe experiment, it is mandatory to carry out precise and unambiguous pulse characterization exactly at the sample position. Among different techniques of pulse measurement reported to date, only FROG provides this capability. Whereas Type I phase-matching SHG FROG, employed in this work, is stretched to the limit due to the finite phase-matching bandwidth of the SHG crystal, implementation of a heterodyned third-order FROG variety, based on an automatically phase-matched, frequency-resolved, pump probe in a transparent medium [117], may provide an excellent alternative. In contrast to a broadband operation and high wavelength-specific sensitivity to the distortion of the spectral phase, this method relies on measuring the signal in the spectral region of the fundamental radiation. Therefore, the use of this technique involves no setup modifications that are required otherwise for the detection of a noncollinear SH beam. However, the challenging problem is the engineering of rapid feedback to the adaptive optics, which would obviate the need to collect the entire FROG trace in each iteration of the optimization loop.

Speaking about further development of parametric devices in a broader perspective, several exciting trends that emerged recently are worth mentioning. One area of intense research concentrates on the design of midinfrared parametric oscillators [118] and amplifiers [119] producing four to five cycle pulses. Such sources are ideal for probing vibrational dynamics of molecules, for example, the stretching of OH groups [120, 121]. Also, several interesting new approaches to parametric amplification have been reported. One of them is chirp reversal in the seed shaper of a NOPA. By using a mirror with specially designed geometry, it was possible to compensate for higher order spectral phase distortions. As a result, negatively chirped visible–near-IR pulses could be amplified to 40 μJ and compressed below 20 fs in a block of quartz [122]. Whereas this method drastically simplifies the NOPA design, its limitation lies in the level of amplification. Upscaling the pulse energy would require beam expansion to avoid SPM in the bulk material used for chirping. Undoubtedly, the route toward high-energy few-cycle parametric pulses lies in chirped parametric pulse amplification [84, 94, 95] and its hybridization with laser amplification techniques [123].

Finally, we mention yet another fascinating use of the NOPA technique. Remarkably, parametric amplifiers provide an alternative approach to the generation of few-cycle optical pulses by using multiple source spectra rather than a single one [124, 125]. This becomes possible as a result of unique phase relations among the signal, pump, and idler waves in a white-light seeded NOPA [126]. Feasible suggested routes toward monocycle pulses include merging of the signal and idler spectra [50, 51] and the spectra of the signal and the fundamental broadened around 800 nm [127] into a single coherent bandwidth. In our opinion, these developments open new possibilities of using NOPAs in nonlinear optical spectroscopy as well as in high-intensity field-sensitive applications of nonlinear optics.

2 Parametric Schemes for Phase-Controlled Pulse Generation

Our investigation of few-cycle OPAs has been limited so far to the problem of efficient, broadband amplification and subsequent pulse compression. In other words, we have considered the temporal intensity profile without addressing the distribution of light-field oscillations underneath the envelope of a few-cycle pulse. Below, we discuss optical field reproducibility in several schemes of parametric amplification.

2.1 Carrier–Envelope Dynamics in Parametric Amplifiers

As shown by *Xu* et al. [128], the relative position of a carrier wave with respect to the pulse peak in a mode-locked laser varies from shot to shot. Such an offset between the most intense field oscillation and the highest point of the envelope is described by the so-called carrier–envelope phase (CEP) [14, 128, 129]. The value of the CEP, which is directly linked to the peak strength of the electric field of an intense few-cycle pulse, critically determines the outcome of many challenging applications of nonlinear optics, such as high-order harmonic generation and production of X-ray pulses with attosecond duration [10,13,14,130]. Recently, a number of methods to measure both the shot-to-shot variation of CEP [131,132,133,134,135,136] and its precise value [14,129,137,138] have been proposed and some of them have been demonstrated experimentally [131, 132, 133, 134]. Moreover, the first direct evidence of the decisive influence of CEP in a few-cycle light–matter interaction emerged from a photoionization experiment in which the direction of the emitted photoelectron is predetermined by the CEP of the laser pulse [139]. It has also been suggested that the duration of the recently observed attosecond spikes produced by superposing high harmonics [13] can be compressed to 10 by proper phasing of all the harmonics, which requires maintaining the CEP value of the driving laser pulse constant. To facilitate further exploration of these and similar fascinating phenomena, several groups of researchers are currently pursuing the goal of constructing CEP-stabilized sources of high-intensity femtosecond pulses. Whereas high-precision CEP control of megahertz and gigahertz oscillators has been achieved by locking the laser cavity to a servoloop feedback based on a phase-tracking system [130,131,132], the same method of active stabilization is difficult to apply in power amplification at kilohertz repetition rates.

In this chapter, we explore an alternative approach to the control of carrier–envelope dynamics, which is based on *all-optical* phase stabilization in OPAs. We show that the use of fundamental physical properties of the parametric interaction of light enables locking the CEP of the idler pulse to a fixed value regardless of the carrier slippage in the driving laser source. The concept presented and experimentally verified in this research can be implemented for generation of exactly reproducible high-intensity few-cycle laser fields that are indispensable to the study of XUV and attosecond physics. Next, we show that a small change in OPA configuration permits straightforward and accurate estimation of the CEP pulse-to-pulse drift of both input and output pulses. The latter method can be used as a complementary phase diagnostic of intense pulses from random-CEP laser amplifiers in single-shot experiments designed to probe extreme nonlinear processes.

Though determination of the CEP value is beyond the scope of this work, we investigate the shot-to-shot variation of CEP that we term ψ_j, with $j = 0, 1, 2, \ldots$ indicating the pulse number. For convenience, we choose ψ_0 equal to zero and assume that, with the exception of the CEP drift, all pulses in the sequence are identical. The pulse-to-pulse variation of ψ_j in a femtosecond oscillator originates from the frequency fluctuations of phase-locked modes that form the laser spectrum [140]. Since the frequency comb of an ultrashort pulse consists of an extremely high number of densely spaced modes, for convenience, we assume that the laser spectrum, $I(\omega)$, is continuously filled with frequencies and all frequencies are positive [141,142]. In this treatment, the role of ψ_j may be seen as a frequency-independent phase offset of the spectral phase, which is evident from the following Fourier-transform link between the time- and frequency-domain electric field:

$$E_j(t) \equiv E_0(t)\exp(\mathrm{i}\psi_j) = \frac{1}{2\pi}\int \sqrt{I(\omega)}\exp\{\mathrm{i}\,[\phi(\omega)+\psi_j-\omega t]\}\,\mathrm{d}\omega\,. \tag{7}$$

In (7), $\phi(\omega)$ denotes the relative spectral phase that determines the envelope of the pulse, i.e., $|E_0(t)|^2$. The *spectral* interpretation of ψ_j makes it easier to understand four- and three-wave mixing phenomena taking place in an OPA and explains the manifestation of the CEP in the measurements of frequency-domain interference addressed below. At this point, it is worth noting that our simplification cannot be used in applications of frequency metrology and synthesis, which rely on counting individual frequency modes of a mode-locked spectrum [140] $f_n = f_{\mathrm{rep}}(n + \psi/2\pi)$, where f_{rep} is the repetition frequency and $n = 0, 1, 2, \ldots$ is the mode number. Nevertheless, (7) adequately describes the field dynamics of an individual pulse and enables interpretation of CEP phenomena observed in single-shot experiments.

Our line of reasoning about the mechanism of CEP stabilization in a parametrically amplified wave relies on three key observations. First, we notice that a white-light continuum pulse produced by SPM retains the same value of ψ_j as that of the input pulse. This property is used in CEP measurement schemes that draw on white-light generation with octave-spanning bandwidths [130, 131, 132, 133]. Consider ω_4 as a new frequency component added by SPM. The result of the underlying four-wave mixing can be expressed as follows:

$$\begin{aligned}\omega_4 &= \omega_1 + \omega_2 - \omega_3\,,\\ \psi_{\omega 4} &= \pi/2 + \psi_{\omega 1} + \psi_{\omega 2} - \psi_{\omega 3}\,,\end{aligned} \tag{8}$$

where $\omega_j,\ j = 1, 2, 3$ are an arbitrary triplet of frequencies within the laser spectrum. The second line in (8) describes the buildup of the frequency-independent part of the pulse phase, which readily explains why the newly added frequency components of the broadened spectrum inherit the original value of the pulse-to-pulse phase drift. The complete pulse propagation equation describing SPM is treated in detail in Appendix B.

Second, we investigate the phase relationship between the seed and amplified waves in a parametric interaction. In general, it is impossible to determine beforehand the value of the phase offset, ψ_{S}, carried by the amplified signal pulse. In optical parametric oscillators and in superfluorescence-seeded OPAs, ψ_{S} is likely to be random because parametric amplification is initiated by quantum noise. The situation is different, however, when an externally produced seed pulse is supplied. The latter corresponds to OPAs built on amplification of sections of white-light continua. In this configuration, ψ_{S} is determined by the seed pulse and preserved intact in the amplified signal pulse.

Third, we recall that the phases of all three pulses that take part in a parametric interaction are linked. The simplified frequency and phase equations of the relevant three-wave mixing can be given as

$$\begin{aligned} \omega_{\mathrm{I}} &= \omega_{\mathrm{P}} - \omega_{\mathrm{S}} \,, \\ \psi_{\mathrm{I}} &= -\pi/2 + \psi_{\mathrm{P}} - \psi_{\mathrm{S}} \,, \end{aligned} \tag{9}$$

where indexes I, S, and P stand for the idler, signal, and pump waves, respectively.

Similarly to parametric frequency conversion, summarized by (9), the buildup of the frequency-independent phase shift in second harmonic generation (SHG) is expressed by

$$\psi_{\mathrm{SH}} = \pi/2 + 2\psi \,. \tag{10}$$

The condition given by (10) is widely used in measurements of CEP drift, which are based on beating overlapping frequencies of the second-harmonic (SH) and fundamental spectra [130, 131, 132, 133, 135] to obtain the so called *f-to-2f* interference signal [143].

In the more generalized case of three-wave parametric interaction, (9) implies that the phase of the idler pulse accommodates the phase difference between the other two pulses. In an experimental demonstration of this concept, a single signal pulse was generated by employing three noncollinear pump beams, resulting in three independent idler pulses [144].

We now notice that if the values of ψ_{S} and ψ_{P} are identical, then the value of ψ_{I} is constant regardless of any pulse-to-pulse fluctuations in ψ_{S} (or ψ_{P}). A similar effect is encountered in terahertz pulse generation by optical pulse rectification [145], which may be viewed as a special case of three-wave parametric mixing expressed by (9).

Having stated the conditions necessary to maintain a constant phase offset in one of the parametric waves, we further discuss possible practical schemes. Table 1 summarizes three different configurations of white-light-seeded OPAs.

Table 1. Summary of phase properties of various OPA designs

OPA configuration	A	B	C
Pump frequency ω_{P}	ω_0	$2\omega_0$	$2\omega_0$
Central frequency of white-light continuum	ω_0	ω_0	$2\omega_0$
Phase offset of pump ψ_{P}	ψ	$2\psi + \pi/2$	$2\psi + \pi/2$
Phase offset of signal ψ_{S}	$\psi + \pi/2$	$\psi + \pi/2$	$2\psi + \pi$
Phase offset of idler ψ_{I}	$-\pi$	$\psi - \pi/2$	$-\pi$
Self-stabilization of ψ_{I}	yes	no	yes

Here, we denote the central frequency of fundamental light as ω_0 and the offset of the fundamental phase as ψ. For simplicity, in the ψ_{I} values listed in

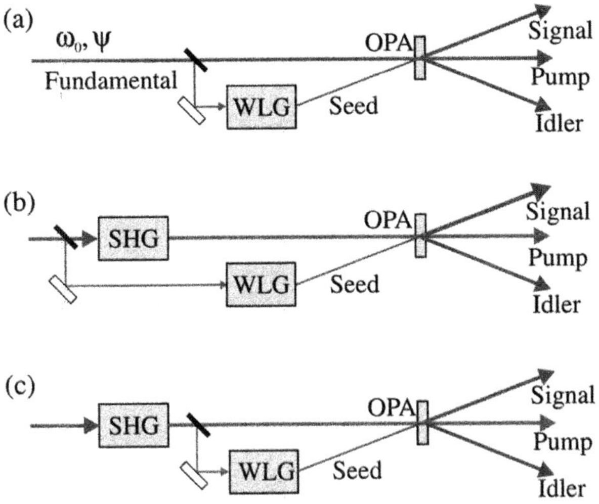

Fig. 12. Schematics of three OPA configurations corresponding to those in Table 1. WLG denotes white-light generator; SHG stands for second-harmonic generator; OPA indicates the crystal in which the seed pulse is amplified. For clarity, the seed and pump beams shown intersect in the crystal (noncollinear phase matching). However, conclusions presented in Table 1 are also valid for a collinear beam arrangement (collinear phase matching)

Table 1, we omitted the additional phase shift that accumulates due to propagation of pump and seed pulses prior to interaction in a parametric crystal. OPA schematics, corresponding to the configurations listed in Table 1, are presented in Fig. 12. According to Table 1, automatic CEP control is achieved when the OPA is seeded with an SPM-broadened replica of the pump pulse in Configurations A and C. We refer to these schemes as *CEP-stabilizing OPAs*, the working of which will be described in the following section. Unlike the other two designs presented in Table 1, Configuration B does not exhibit a CEP-stabilizing property. However, the latter scheme reproduces the CEP drift of the fundamental light in both the signal and idler waves. This *phase-repeating OPA* will be discussed in Sect. 2.3.

2.2 All-Optical Phase Stabilization

Next, we experimentally verify the idea of CEP locking by employing a Configuration C OPA that is based on a noncollinearly pumped BBO crystal (Fig. 13a). As follows from (10), cancellation of the pulse-to-pulse CEP drift of the idler implies that frequency-doubled idler pulses (idler-SH) should also experience CEP stabilization. Therefore, the presence of phase control in an OPA can be ascertained by observing the interference between the idler and idler-SH, provided the two have some spectral overlap. In a noncollinear

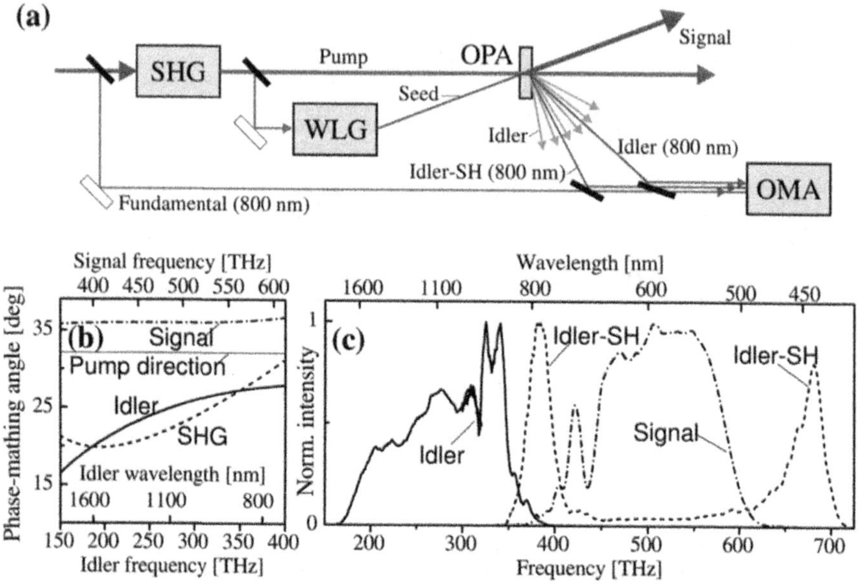

Fig. 13. Schematic of experimental setup for CEP self-stabilization measurement. (**a**) OPA and three-beam interferometer. OMA stands for a spectrometer with an optical multichannel analyzer. (**b**) Phase-matching configuration of 31.5° Type I BBO showing the direction of various parametric waves inside the crystal. Note that the generation of the frequency-doubled idler at $\sim$ 790 nm is phase-matched. (**c**) Output spectra of the parametric amplifier. Note the spectral overlap between the idler and its second harmonic around 790 nm

OPA, an octave-spanning idler spectrum can be generated, directly obviating the need for additional spectral broadening that was required in the CEP drift measurements reported previously [131, 132, 133]. We also took advantage of the fact that frequency doubling of the idler can be attained in the same OPA crystal. Indeed, as can be seen from the phase-matching curves in Fig. 13b, the idler SHG is nearly phase-matched in a broad spectral range. The resultant signal, idler, and idler-SH spectra, produced in a single BBO crystal, are presented in Fig. 13c. Assuming ideal compression, the spectra of both the idler and the signal support sub-5 fs pulses, which corresponds to less than two optical cycles in the case of the idler. Whereas the signal spectrum does not cover a full octave, this becomes possible for the idler. The latter has the same bandwidth as the signal but lies at a lower frequency, thus fulfilling the condition for *f-to-2f* interference. To pump the OPA, we frequency-doubled the output of a 120 fs Ti:sapphire laser amplifier operating at the wavelength of 790 nm. To generate the white-light seed, we focused a fraction of the 395 nm pulses into a 2 mm thick CaF_2 window. The energy of the 395 nm pump pulses sent to the OPA crystal was 9 μJ, whereas the output energy from a 1 mm BBO was 1.6 μJ and 1.1 μJ in the signal and idler waves, respectively. Noted that the pump energy was chosen to provide the necessary amount of idler-SH, whereas efficient parametric amplification and idler emission were obtained at much lower pump levels.

In addition to providing the spectral overlap between the idler and idler-SH, seen in Fig. 5c, the experimental scheme chosen furnishes one more important means of examining the CEP-conserving properties of the OPA through interference between the idler and the fundamental light of the Ti:sapphire laser. The latter experiment is necessary to estimate the amount of CEP drift carried by the pulses that drive the OPA.

Because of the radial dispersion of the idler beam, which is the result of noncollinear phase matching (Fig. 13b), the 790 nm components of the idler and idler-SH are emitted from the BBO at different angles. These beams are subsequently interferometrically recombined by use of two beam splitters, as shown in Fig. 13a. Additionally, a double pass through a quarter-wave plate (not shown) rotates by the idler-SH polarization 90° to match it with that of the idler. A fraction of the residual fundamental light is superposed on the idler–idler-SH beam, upon which all three beams are transmitted through a 50 μm pinhole, placed between a confocal lens pair, to ensure spatial coherence.

The results of the CEP pulse-to-pulse drift measurement are summarized in Fig. 14. The idler–fundamental and idler–idler-SH interferences were examined separately. The spectra carried by each individual beam are depicted in Fig. 14a and Fig.14b; Fig. 14c,d shows corresponding single-shot interferograms (dotted curves) and their average intensities (solid curves). The fringe pattern observed in the phase-locked idler–idler-SH interference (Fig. 14d) is remarkably stable, whereas it is absent in the average idler–fundamental intensity (Fig. 14c, solid curve) due to CEP fluctuations in the input pulse. To determine further the shot-to-shot phase drift from individual interferograms (Fig. 14c,d, dotted curves), we employed a Fourier phase retrieval algorithm [146] well known in spectral interferometry that involves Fourier transformation of the data, filtering, and back transformation. The results of this procedure are presented in Fig. 14e,f. The rapid phase fluctuation in Fig. 14e is indicative of the random CEP value of the pulses emitted by the Ti:sapphire laser, whereas the CEP of the idler stays locked within a $\pm\pi/10$ range (Fig. 14f). The rms residual phase drift is 0.18 rad and is attributed to the intensity instability of the pump laser.

Several remarks about the operation of the phase-stabilizing OPA are due at this point. First of all, we pay attention to the fact that the phase-matching configuration of a white-light-seeded OPA makes it possible to amplify the superfluorescence induced by the pump beam. (A schematic representation of the corresponding radiation cone emitted in the phase-matching direction is given in Fig. 2a.) In a multipass OPA driven by a sufficiently intense pump, efficient parametric amplification of superfluorescence takes place even in the absence of a seed beam. As discussed in Sect. 2.1, such a regime does not ren-

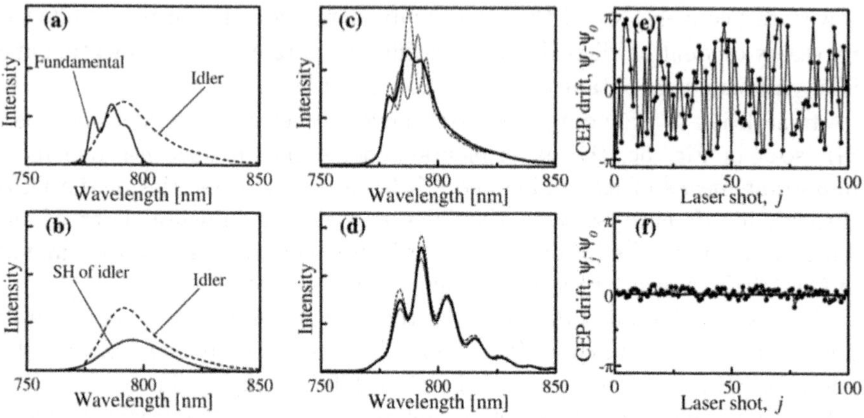

Fig. 14. Experimental results of CEP self-stabilization measurement. (**a,b**) Overlapped spectral fragments of the residual fundamental, idler, and idler-SH beams. (**c,d**) The *solid curves* show the interference pattern averaged over 1000 shots, whereas *dotted curves* represent single-shot interferorgams. (**e,f**) *Dots* show relative CEP jumps wrapped on a $\pm\pi$ interval. Note that the stable phase pattern obtained from the interference of the idler and its second harmonic is direct proof of CEP self-stabilization

der CEP-stable pulses. This mechanism can compete with the amplification of a white-light seed causing extra CEP jitter and even excessive amplitude noise. The origin of the latter amplitude instability becomes apparent when one considers that the combined seed is a product of interference between a SPM-produced pulse and the superfluorescence that carries the (nearly) random phase. Therefore, to suppress the amplification of superfluorescence, the pump energy of the first pass through the OPA should be kept down. As a consequence, the initial parametric gain should be limited to 10^4 to 10^5 to avoid the loss of phase stability. In subsequent passes through the OPA, the pump energy can be increased considerably since the intensity contrast between the seed pulse and pump-induced superfluorescence is already dramatically improved. The recommendation on the gain limitation in the first pass is applicable to white-light seeded amplifiers in general, since it also helps to improve their amplitude noise performance.

The next comment is mostly relevant to the specific OPA scheme used in our experiments to demonstrate parametric CEP stabilization. It concerns the difficulty in finding the appropriate material for white-light generation in Configuration C. Unlike in 800 nm driving pulses, it is impossible to obtain sufficient SPM-induced spectral broadening of the SH pulses in sapphire (Sect. 1.2) maintaining a single-filament regime. On the other hand, optical materials with a narrower transparency range in the UV exhibit deterioration through the formation of color centers. The latter most likely are due to multiphoton absorption. In our experiments, we employed excimer-laser tested

CaF_2 as a medium for SPM. Nevertheless, the onset of optical degradation in this material was already noticeable after about a 40 minute continuous exposure to submicrojoule, 395 nm pulses at a 1 kHz repetition rate. Unfortunately, we are not able to offer an easy practical solution to the problem of optical degradation. Spinning a CaF_2 disk has been demonstrated as a way of increasing its lifetime limitlessly as an SPM medium [147] used to seed an OPA. In the context of CEP stabilization, however, such a method does not seem applicable because of the excessive demands for the flatness and homogeneity of the CaF_2 window. A potential way out might be continuous heating of CaF_2 to $\sim$ 1500 K to prevent the formation of color centers. Fortunately, the selection of stable SPM materials for use at higher wavelengths (Configurations A, B if pumped by a Ti:sapphire laser) is much broader.

Another remark is due about control over the long-term CEP drift in a phase-stabilizing OPA and the way of changing the value of the CEP of the output pulses. In principle, any phase shift accumulated between the pump and the seed will automatically result in a change in the CEP of the output idler pulses by the same amount. In this respect, the seed and pump pathways are like the arms of an interferometer. Therefore, the longer the path from the beam splitter to the OPA crystal, the harder it is to maintain the required interferometric stability. By the same token, the output CEP value of the idler can be adjusted simply by changing the time delay between the pump and the seed. Since only minute path length variations (within one full wavelength) are required to scan the phase in a 2π range, this does not affect the intensity of the idler and the signal. By contrast, whereas a linear phase (i.e., time delay) variation is required in an OPA to adjust the CEP, control of the latter in a mode-locked laser oscillator [128, 140] involves changing the dispersion of the cavity and/or manipulating the phase offset through the strength of the Kerr lens. In principle, it should be feasible to make even large-arm OPAs suitable for phase stabilization by furnishing them with simple path length control loops that are standard in Fourier-transform IR spectrometers.

A curious feature of the OPA presented in this section (Fig. 13) is that the spectrum of a CEP-stabilized idler overlaps the spectrum of an idler-SH, which is also CEP-stabilized. As a result, the continuous bandwidth of the CEP-controlled output spans over 500 THz. One possible application of this OPA, in which the idler alone spans the whole octave, can be the study of photocurrent in semiconductor switches [148, 149]. The direction and the magnitude of the photocurrent are expected to vary, depending on the CEP of such a pulse as a result of the interplay between the one- and two-photon resonant absorption.

Next, we outline the prospects for practical implementation of parametric CEP stabilization for production of high-intensity pulses. Several routes for further pulse amplification are envisaged. First, the energy of CEP-locked idler pulses can be boosted in an *all-parametric* scheme by using chirped pulse parametric amplification [92,95], which, it was recently shown, produces terawatt femtosecond pulses [94]. Two-cycle optical pulses are feasible along this road. To simplify the chain of OPAs, the idler produced by a CEP-stabilizing stage can be seeded into subsequent OPAs and amplified as a signal without destroying the CEP stability. This approach would permit the use of broadband noncollinear phase matching and, at the same time, circumvent the difficulty with the collimation of a radially dispersed idler beam. Notably, the choice of phase matching (Type I or Type II, collinear or noncollinear pump) determines the bandwidth and central frequencies of the amplified waves but does not affect the OPA phase performance outlined in Sect. 2.2.

Another route to increasing the intensity of parametrically CEP-stabilized pulses is to inject the OPA output into widespread chirped-pulse amplifiers [150]. For the most commonly used Ti:sapphire-based systems, a stabilized pulse at 800 nm can be derived from two parametric sources. One of them is the frequency-doubled idler obtained from an OPA device in Configuration A, many of which are commercially available. Another applicable scheme of parametric downconversion, also described by (9), is difference-frequency generation between selected portions of a white-light spectrum produced by 400 nm pulses [151]. A drawback in the latter case is the lower efficiency of parametric conversion compared to an OPA.

2.3 Phase Self-Referencing

In this section, we address the working of a phase-repeating OPA. Unlike the other two designs presented in Table 1, Configuration B does not possess a CEP-stabilizing property. Nevertheless, it corresponds to the popular broadband noncollinear phase matching in a BBO crystal for the pump by frequency-doubled pulses from a Ti:sapphire laser. It corresponds to the two-cycle OPA discussed earlier in Sect. 1.2–Sect. 1.4. In this design, both the signal and the idler carry the same phase offset, the fluctuation of which is determined by the laser source. Interestingly, this opens an intriguing possibility of combining the white-light spectrum, produced by fundamental pulses, with signal (visible) and idler (near infrared) spectra into a single bandwidth with the prospect of obtaining high-contrast monocycle laser fields. Synchronous fluctuation of ψ in all three pulses reduces the task of a seamless "stitching" of the three overlapping spectra to mechanical stabilization of the path lengths that the pulses travel before recombination. However, rather than considering further ways of decreasing the pulse duration, the following discussion will focus on using this OPA as a tool to estimate CEP drift.

As mentioned in the previous section where we introduced the concept of *f-to-2f* interference, pulse-to-pulse fluctuations of the CEP can be determined through interference between different orders of harmonics [135]. Typically, the *f-to-2f* scheme involves generation of an octave-wide supercontinuum with a subsequent frequency-doubling of the redshifted portion of the

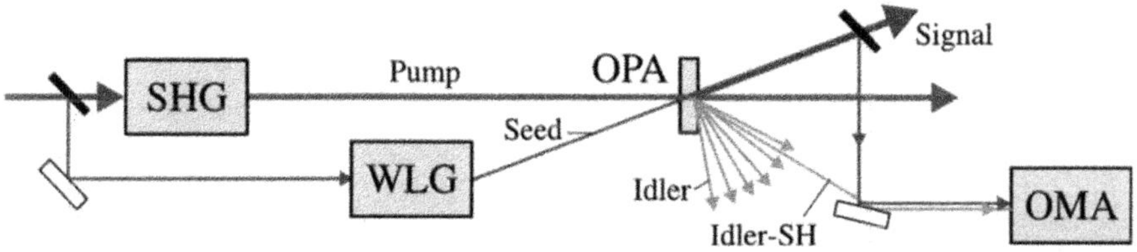

Fig. 15. Schematic of experimental setup for the measurement of CEP shot-to-shot drift

resultant spectrum. This method has been employed to stabilize megahertz-repetition-rate Ti:sapphire oscillators through an active feedback loop [130, 132] and measure CEP drift from a chirped-pulse-amplification system [133]. The phase relation among subharmonics in a synchronously pumped optical parametric oscillator has also been measured [134] and used to phase-lock the output parametric waves [125].

Similarly to the *f-to-2f* scheme, using an OPA in Configuration B, ψ can be extracted from the interference between the signal and SH of the idler. Remarkably, the white-light seed, required to produce the appropriate signal and idler frequencies (i.e., $\omega_S = 2\omega_I$), has to span much less than an octave since the spectral shift between the signal and the pump is much smaller (9).

In the previous section, we already explained the efficient frequency doubling of the idler, which takes place in BBO as a byproduct of parametric amplification (Fig. 13b,c). There, we used the lower frequency part of the two-pronged idler-SH spectrum (dashed curve in Fig. 13c). In the currently discussed application, we will use the high-frequency part of the idler-SH that has a spectral overlap with the signal. One of the important advantages of such SH generation in comparison with an additional doubling crystal is the fact that the phase-matching bandwidth is not limited by the thickness of the BBO.

To use the phase-matching properties of BBO outlined above, we constructed a simple one-pass OPA depicted in Fig. 15. The radial dispersion of the idler-SH beam is compensated for by the use of a diffraction grating and two collimating spherical mirrors. To match the polarization of the two waves, a periscope was introduced in the pathway of the signal beam. For measurement of the CEP drift, we diverted a fraction of the signal beam by reflecting it from a 0.5 mm thick quartz plate. The remaining signal intensity was used for sub-6 fs pulse compression and pulse-envelope measurement by frequency-resolved optical gating [75] (cf. Sect. 1.5). The idler SH pulses were downchirped by passing them through a 5 mm plate of BK7 glass. The latter pulse reshaping was required to minimize the phase difference with respect to the signal wave that carries a mirrored chirp sign [152]. To improve the fringe contrast of the measured interferograms, we employed a confocal detection arrangement, consisting of a 50 μm pinhole inserted between a pair of $f = 50$ mm lenses. Further details of the apparatus can be found in [153].

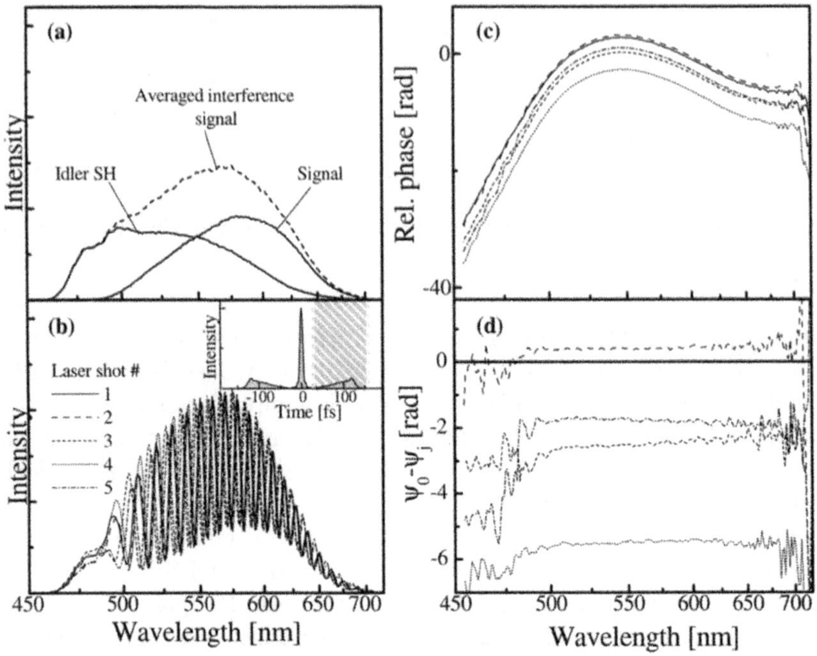

Fig. 16. Summary of experimental results of CEP drift measurement. (**a**) Spectra of signal and frequency-doubled idler waves transmitted through the pinhole, and mean intensity of the interference averaged over 300 laser shots. (**b**) Single-shot interferograms for five successive laser pulses. The *inset* shows the power spectrum of the frequency-to-time Fourier transform, where the *dashed band* indicates the filter used for phase recovery. (**c**) Relative phase between signal and SH idler pulses recovered for the five spectrograms shown in (**b**). A linear phase slope corresponding to a 104 fs delay between the signal and idler pulses at $\lambda = 550$ nm is subtracted from all curves. (**d**) Phase drift relative to the initial pulse

The results of the CEP pulse-to-pulse drift measurement are summarized in Fig. 16. As can be seen in Fig. 16a, the signal-wave and idler-SH spectra have a sizable overlap in the 480 nm to 680 nm range. A sequence of single-shot interferograms is shown in Fig. 16b.

The relative signal–idler-SH phase can be recovered from this measurement directly by employing a Fourier-transform algorithm [146] that we also used in Sect. 2.2. The band-pass filter, used by us to isolate the relevant phase information, is depicted in the inset in Fig. 16b. The phases obtained by the backtransformation are shown in Fig. 16c and are offset from each other by nearly frequency-independent shifts, as can be judged from relative differences with respect to the initial pulse (Fig. 16d). The CEP drift data shown in Fig. 16d exhibit very small (within 0.2 rad) excursions from a constant value within a 200 nm spectral range, which indicates the high fidelity of the measurement. Interestingly, one of the reasons why the trustworthiness

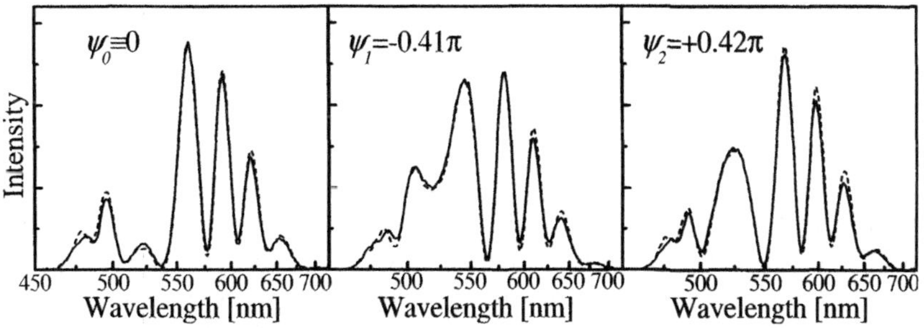

Fig. 17. Interferograms recorded for three successive pulses. The delay between the signal and idler-SH pulses is now shorter in comparison with the situation in Fig. 16. *Solid curves* show experimental data. *Dashed lines* correspond to a single-parameter fit described in the text. Note the exceptionally high fringe visibility achieved because of the use of confocal detection

of the relative phase recovery between the signal and idler-SH is boosted so dramatically is the thickness of the BBO crystal, 0.5 mm, which ensures that the contribution of sum frequency components in the resulting idler-SH spectrum is negligible. The results presented in Fig. 16 are in favorable contrast with previously reported data of CEP slip determination by *f-to-2f* interference [133]. Additionally, no assumptions about the duration of the interfering pulses have to be made in our scheme.

To demonstrate further the precision of the described measurement, we analyze interferograms consisting of just several fringes. The Fourier method for phase recovery cannot be applied in this situation because of the temporal overlap between the signal and idler-SH pulses, which makes it impossible to use time-domain filtering similar to that shown in the inset in Fig. 16b. Instead, we apply a one-parameter fit directly to the spectrograms shown in Fig. 17 by solid curves. The fits, given on the same plots by dashed curves, were obtained from the signal and idler-SH spectra, presented in Fig. 16a, and the relative phase depicted in Fig. 16c by the solid curve. The only degree of freedom was the value of ψ that was estimated by a least-squares fitting routine. The high quality of the fits shown in Fig. 17 is a further illustration of the reliability of the measuring technique used.

We next recall that the CEP fluctuation of the signal wave in the parametric scheme discussed is identical to that of the laser pulse emitted by the Ti:sapphire amplifier that drives the OPA. Therefore, phase characterization in CPA lasers by use of an OPA could be a method of choice with clear advantages over employing a fiber, required for production of an octave-wide supercontinuum, and an extra SHG crystal [133]. In comparison, the OPA scheme demands a much narrower spectral width of the white-light seed, which can be readily obtained from a thin sapphire plate. An additional advantage of the method discussed in this section is its superior reliability, which results from a very broad, ~200 THz, spectral overlap of the interfering waves and the configuration of frequency doubling. Finally, we point out that, unlike the *f-to-2f* scheme based on synthesizing harmonics from *peripheral* frequencies of a white-light spectrum, the OPA method uses *central* spectral components of the parametric waves. This reduces the sensitivity of the technique described to the input intensity noise.

In a brief summary, in this section we have characterized the pulse-to-pulse CEP drift directly in an optical parametric amplifier without employing any additional nonlinear optical elements. Despite the fact that neither CEP-stabilization nor the estimation of the complete CEP value can be achieved with this method, the approach discussed here is promising for field-sensitive applications of nonlinear optics. For example, the shot-to-shot statistics of the CEP drift can be directly correlated with that of the direction of electron photoemission in an ionization experiment [139] or single-shot high-harmonic emission spectra [154, 155]. In an experiment designed in this way, the OPA-based diagnostic would provide the necessary information on the CEP drift of the driving laser pulse, whereas distinct features of the high-order-nonlinear response, e.g., the shape of cutoff harmonic spectra [154], can be used to calibrate the actual value of CEP.

2.4 Conclusions

We have investigated several different regimes of seeded parametric amplification. The choice of the seed configuration determines whether the CEP of the idler pulses becomes fixed or, alternatively, it drifts while the idler and signal pulses remain phase-locked. In our opinion, an OPA of the former kind (phase-stabilizing) can be effectively employed in a sequence of femtosecond laser amplifiers for production of intense CEP-locked pulses, thus dispensing with intricate schemes of active CEP control. On the other hand, phase-repeating OPAs offer a relatively easy way to synthesize single-cycle pulses and, additionally, can be used for accurate determination of the phase offset for each individual laser shot. We believe that this work provides a new insight into the experimental capabilities of various OPA systems and opens attractive possibilities for their use in field-sensitive applications of nonlinear optics.

A Global Spectral Maximum of Second Harmonic

Here, we obtain a proof of the assertion about the global maximum of SH intensity at all wavelengths in a flat spectral phase. Since laser pulses can be described by bandwidth-limited functions, we discretize (5) in the following form:

$$\widehat{S}_k = \left| \sum_{j=0}^{k} A_j A_{k-j} \exp\left[\mathrm{i} \left(\phi_j + \phi_{k-j} \right) \right] \right|^2 , \tag{A.1}$$

where A_j is a real nonnegative spectral amplitude and $j = 0$ is the first nonzero element on a finite support selected arbitrarily along the frequency axis. In a spectrum-limited pulse ($\phi_j = \text{const}$), (A.1) becomes

$$S_k = \left(\sum_{j=0}^{k} A_j A_{k-j} \right)^2 . \tag{A.2}$$

Next, we employ the method of mathematical induction to prove that

$$S_k \geq \widehat{S}_k . \tag{A.3}$$

If the assertion (A.3) is valid for $k = 0$ and, assuming that it also holds for $k = m$, one can prove that relation (A.3) is true for $k = m + 1$; then it is also valid for any integer $k > 0$.

Evidently, the statement holds for $k = 0$, since $S_0 = \widehat{S}_0$.

Now, we assume that (A.3) is valid for $k = m$ and introduce the following notations:

$$P_j \equiv A_j A_{k-j} , \tag{A.4}$$

$$\psi_j \equiv \phi_j + \phi_{k-j} , \tag{A.5}$$

$$M_m \equiv \sum_{j=0}^{m} P_j , \tag{A.6}$$

$$\widehat{M}_m e^{i\Phi_m} \equiv \sum_{j=0}^{m} P_j e^{i\psi_j} , \tag{A.7}$$

where $\widehat{M}_m$ and Φ_m represent the modulus and the phase of the complex sum, respectively.

From the assumption $S_m \geq \widehat{S}_m$, it follows that the moduli of the sums are similarly related:

$$M_m \geq \widehat{M}_m , \tag{A.8}$$

Next, we examine the case of $k = m + 1$ and obtain

$$S_{m+1} = M_m^2 + P_{m+1}^2 + 2 M_m P_{m+1} \tag{A.9}$$

and

$$\widehat{S}_{m+1} = \widehat{M}_m^2 + P_{m+1}^2 + 2 \widehat{M}_m P_{m+1} \cos\left(\Phi_m - \psi_{m+1}\right) . \tag{A.10}$$

Recalling relation (A.8) and noticing that $\cos\left(\Phi_m - \psi_{m+1}\right) \leq 1$, we conclude that $S_{m+1} \geq \widehat{S}_{m+1}$. Therefore, (A.3) is now proved for all $k \geq 0$, i.e., for any SH frequency.

B Phase Coherence Between the Input Pulse and Its SPM-Broadened Replica

The propagation equation in the perturbative regime of nonlinear optics [14] can be derived directly from Maxwell's equation [14, 27, 156, 157] and written in the frequency domain as

$$\left[\nabla^2 + k^2(\omega_4)\right] E(\boldsymbol{r}, \omega_4) = \frac{\omega_4^2}{\varepsilon_0 c^2} P^{\mathrm{nl}}(\boldsymbol{r}, \omega_4)\,, \tag{B.1}$$

where $E(\boldsymbol{r}, \omega)$ and $P^{\mathrm{nl}}(\boldsymbol{r}, \omega)$ denote the electric field and generated nonlinear polarization, respectively. The particular form of $P^{\mathrm{nl}}(\boldsymbol{r}, \omega)$, which describes both the delayed Raman response (slow nonlinearity) and parametric four-wave mixing governing the SPM process [27, 156, 157] (instantaneous nonlinearity) can be given by

$$\begin{aligned} P^{\mathrm{nl}}(\boldsymbol{r}, \omega_4) = \iint \mathrm{d}\omega_1\, \mathrm{d}\omega_2 E(\boldsymbol{r}, \omega_1) E(\boldsymbol{r}, \omega_2) E^*(\boldsymbol{r}, \omega_1 + \omega_2 - \omega_4) \\ \cdot \chi^{\langle 3 \rangle}(\omega_4 : \omega_1, \omega_2, \omega_1 + \omega_2 - \omega_4)\,, \end{aligned} \tag{B.2}$$

where $\chi^{\langle 3 \rangle}$ represents the effective third-order susceptibility and conjugation marks emission [158]. The appearance of the conjugation operator in the convolution integral in (B.2) accounts for the preservation of the initial phase *fluctuation* [cf. (8)] in the output pulse. The fluctuation should not be confused with the phase *offset* accumulated by the pulse as a result of linear and nonlinear propagation [14]. Unlike the fluctuation, the offset remains the same for every successive pulse, except for some special cases. One such exclusion is a nonnegligible third-harmonic generation (THG) occurring simultaneously with SPM, which requires the addition of another term in (B.2) with a EEE product in the convolution integral. Curiously, the inclusion of resonant or nonresonant fifth-order parametric wave mixing (EE^*EE^*E) [159] during SPM does not influence the conclusion about the preservation of the phase fluctuation in the output pulse.

It is important to note that that the $\pi/2$ phase lag in simplified (8) corresponds to the solution for the special case of instantaneous nonlinearity in a transparent medium under the assumption of a slowly evolving wave approximation [14, 141]. The application of the exact propagation (B.1) to all frequency-conversion processes in a white-light-seeded OPA, with the exception of nonnegligible THG, confirms the conclusion about the parametric stabilization of CEP in a white-light-seeded OPA.

Also, it must be pointed out that the intensity fluctuation and/or beam pointing instability of the laser source driving the SPM may result in additional phase noise. This means that the above-mentioned phase offset that accumulates during the nonlinear propagation will differ from pulse to pulse and effectively result in phase jitter between the injected pulse and that behind the SPM medium. The phase stability (or, in the context of CEP stabilization, the quality of CEP drift replication in white light) can be simply verified by observing single-shot, frequency-domain interference between the injected pulse and its spectrally broadened replica. Such an experiment is trivial for a laser amplifier delivering pulses with micro- to millijoule energies. For nanojoule-energy oscillators, however, the study of SPM-caused phase jitter is far more complicated. Recently, the measurement of phase coherence of oscillator pulses broadened in a microstructure fiber [28] was carried out by referencing one *f-to-2f* interferometer to another *f-to-2f* loop while actively locking the CEP of the laser [143,160,161]. Complementing the efforts of experimentalists, *Haus* and *Ippen* in a recent theoretical paper [162] derived an expression for the intensity-induced phase shift of a soliton traveling through a Kerr medium.

Acknowledgements

The authors are grateful to Y. Ohbayashi, H. Suzuki, and T. Hara for designing and manufacturing the chirped dielectric optics used in this work. We are indebted to T. Fuji, H. Kano, A. Yabushita, T. Saito, A. Shirakawa, A. Sugita, Y. Kobayashi, M. Kakehata, G. Steinmayer, R. Danielius, A. P. Piskarskas, E. Riedle, S. De Silvestri, J. A. Gruetzmacher, K. Osvay, A. E. Kaplan, M. S. Pshenichnikov, and F. Krausz for stimulating discussions. M. Matsuo is acknowledged for assistance in constructing the flexible-mirror setup.

References

[1] J. A. Armstrong, N. Bloembergen, J. Ducuing, P. S. Pershan: Interactions between light waves in a nonlinear dielectric, Phys. Rev. **127**, 1918 (1962)
[2] R. A. Baumgartner, R. L. Byer: Optical parametric amplification, IEEE J. Quantum Elect. **15**, 432 (1979)
[3] A. P. Piskarskas: Optical parametric generators, tunable, powerful & ultrashort, Opt. Photonics News **8**, 24 (1997)
[4] A. H. Zewail (Ed.): *Femtochemistry* (World Scientific, Singapore 1994)
[5] A. H. Zewail: Femtochemistry-advances over a decade, in M. Chergui (Ed.): *Femtochemistry* (World Scientific, Singapore 1997)
[6] C. J. Bardeen, Q. Wang, C. V. Shank: Selective excitation of vibrational wavepacket motion using chirped pulses, Phys. Rev. Lett. **75**, 3410 (1995)
[7] B. Kohler, V. V. Yakovlev, J. Che, J. L. Krause, M. Messina, K. R. Wilson, N. Schwentner, R. M. Whitnell, Y. Yan: Quantum control of wave packet evolution with tailored femtosecond pulses, Phys. Rev. Lett. **74**, 3360 (1995)
[8] V. V. Yakovlev, C. J. Bardeen, J. Che, J. Cao, K. R. Wilson: Chirped pulse enhancement of multiphoton absorption in molecular iodine, J. Comp. Phys. **108**, 2309 (1998)
[9] C. J. Bardeen, J. Che, K. R. Wilson, V. V. Yakovlev, P. Cong, B. Kohler, J. L. Krause, M. Messina: Quantum control of NaI photodissociation reaction product states by ultrafast tailored light pulses, J. Phys. Chem. **101**, 3815 (1997)

[10] M. Drescher, M. Hentschel, R. Kienberger, G. Tempea, C. Spielmann, G. Reider, P. Corkum, F. Krausz: X-ray pulses approaching the attosecond frontier, Science **291**, 1923 (2001)

[11] Z. Chang, A. Rundquist, H. Wang, M. M. Murnane, H. C. Kapteyn: Generation of coherent soft X-rays at 2.7 nm using high harmonics, Phys. Rev. Lett. **79**, 2967 (1997)

[12] Z. Chang, A. Rundquist, H. Wang, I. Christov, M. M. Murnane, H. C. Kapteyn: Generation of coherent, femtosecond, X-ray pulses in the "water window", IEEE J. Sel. Top. Quant. **4**, 266 (1998)

[13] P. M. Paul, E. S. Toma, P. Breger, G. Mullot, F. Augé, P. Balcou, G. H. Muller, P. Agostini: Observation of a train of attosecond pulses from high harmonics generation, Science **292**, 1689 (2001)

[14] T. Brabec, F. Krausz: Intense few-cycle laser fields: frontiers of nonlinear optics, Rev. Mod. Phys. **72**, 545 (2000)

[15] W. Drexler, U. Morgner, F. X. Kärtner, C. Pitris, S. A. Boppard, X. D. Li, E. P. Ippen, J. G. Fujimoto: In vivo ultrahigh-resolution optical coherence tomography, Opt. Lett. **24**, 1221 (1999)

[16] R. L. Fork, C. H. B. Cruz, P. C. Becker, C. V. Shank: Compression of optical pulses to six femtosecond by using cubic phase compensation, Opt. Lett. **12**, 483 (1987)

[17] A. Baltuška, Z. Wei, M. S. Pshenichnikov, D. A. Wiersma: Optical pulse compression to 5 fs at a 1-MHz repetition rate, Opt. Lett. **22**, 102 (1997)

[18] A. Baltuška, Z. Wei, M. S. Pshenichnikov, D. A.Wiersma, R. Szipöcs: All-solid-state cavity-dumped sub-5-fs laser, Appl. Phys. B-Lasers O. **65**, 175 (1997)

[19] A. Baltuška, M. S. Pshenichnikov, D. A. Wiersma: Amplitude and phase characterization of 4.5-fs pulses by frequency-resolved optical gating, Opt. Lett. **23**, 1474 (1998)

[20] M. Nisoli, S. D. Silvestri, O. Svelto: Generation of high energy 10 fs pulses by a new pulse compression technique, Appl. Phys. Lett. **68**, 2793 (1996)

[21] M. Nisoli, S. D. Silvestri, R. Szipöcs, K. Ferencz, C. Spielmann, S. Sartania, F. Krausz: Compression of high-energy laser pulses below 5 fs, Opt. Lett. **22**, 522 (1997)

[22] M. Nisoli, S. Stagira, S. D. Silvestri, O. Svelto, S. Sartania, Z. Cheng, M. Lenzner, C. Spielmann, F. Krausz: A novel high-energy pulse compression system: generation of multigigawatt sub-5-fs pulses, Appl. Phys. B-Lasers O. **65**, 189 (1997)

[23] R. Szipöcs, K. Ferencz, C. Spielmann, F. Krausz: Chirped multilayer coatings for broadband dispersion control in femtosecond lasers, Opt. Lett. **19**, 201 (1994)

[24] R. Szipöcs, A. Köházi-Kis: Theory and design of chirped dielectric mirrors, Appl. Phys. B-Lasers O. **65**, 115 (1997)

[25] R. Szipöcs, F. Krausz: *Dispersive Dielectric Mirror* (1998) USA patent #5724503

[26] Z. Cheng, G. Tempea, T. Brabec, K. Ferencz, C. Spielmann, F. Krausz: Generation of intense diffraction-limited white light and 4-fs pulses, in Ultrafast Phenomena XI (Springer, Berlin, Heidelberg 1998) p. 8

[27] G. P. Agrawal: *Nonlinear Fiber Optics*, 3rd ed. (Academic, San Diego 2001)

[28] J. K. Ranka, R. S. Windeler, A. J. Stentz: Visible continuum generation in air-silica microstructure optical fibers with anomalous dispersion at 800 nm, Opt. Lett. **25**, 25 (2000)
[29] J. K. Ranka, R. S. Windeler, A. J. Stentz: Optical properties of high-delta air-silica microstructure fibers, Opt. Lett. **25**, 796 (2000)
[30] I. D. Jung, F. X. Kärtner, N. Matuschek, D. H. Sutter, F. Morier-Genoud, G. Hang, U. Keller, V. Scheuer, M. Tilsch, T. Schudi: Self-starting 6.5-fs pulses from a ti:sapphire laser, Opt. Lett. **22**, 1009 (1997)
[31] Y. Chen, F. X. Kaertner, U. Morgner, S. H. Cho, H. A. Haus, E. P. Ippen, J. G. Fujimoto: J. Opt. Soc. Am. B **16**, 1999 (1999)
[32] U. Morgner, F. X. Kärtner, S. H. Cho, Y. Chen, H. A. Haus, J. G. Fujimoto, E. P. Ippen, V. Scheuer, G. Angelow, T. Tschudi: **24**, 411 (1999)
[33] G. Steinmeyer, D. Sutter, L. Gallmann, N. Matuschek, U. Keller: Science **286**, 1507 (1999)
[34] R. K. Shelton, L. S. Ma, H. C. Kapteyn, M. M. Murnane, J. L. Hall, J. Ye: Science **293**, 1286 (2001)
[35] R. K. Shelton, S. M. Foreman, L. S. Ma, J. L. Hall, H. C. Kapteyn, M. M. Murnane, M. Notcutt, J. Ye: Opt. Lett. **27**, 312 (2002)
[36] Z. Y. Wei, Z. G. Z. Y. Kobayashi, K. Torizuka: Opt. Lett. **26**, 1806 (2001)
[37] G. Banfi, P. D. Trapani, R. Danielius, A. Piskarskas, R. Righini, I. Santa: Opt. Lett. **18**, 1547 (1993)
[38] P. D. Trapani, A. Andreoni, C. Solcia, P. Foggi, R. Danielis, A. Dubietis, A. Piskarskas: J. Opt. Soc. Am. B **12**, 2237 (1995)
[39] T. J. Driscoll, G. M. Gale, F. Hache: Opt. Commun. **110**, 638 (1994)
[40] G. M. Gale, M. Cavallari, T. J. Driscoll, F. Hache: Opt. Lett. **20**, 1562 (1995)
[41] G. M. Gale, F. Hache, M. Cavallari: **4**, 224 (1998)
[42] T. Wilhelm, J. Piel, E. Riedle: Opt. Lett. **22**, 1494 (1997)
[43] E. Riedle, M. Beutter, S. Lochbrunner, J. Piel, S. Schenkl, S. Spoerlein, W. Zinth: Appl. Phys. B-Lasers O. **71**, 457 (2000)
[44] G. Cerullo, M. Nisoli, S. Stagira, S. D. Silvestri: Opt. Lett. **23**, 1283 (1998)
[45] G. Cerullo, M. Nisoli, S. Stagira, S. D. Silvestri, G. Tempea, F. Krausz, K. Ferencz: Opt. Lett. **24**, 1529 (1999)
[46] G. Cerullo, N. M., S. S., S. D. Silvestri, G. Tempea, F. Krausz, K. Ferencz: Appl. Phys. B-Lasers O. **S70**, S253 (2000)
[47] A. Shirakawa, T. Kobayashi: Appl. Phys. Lett. **72**, 147 (1998)
[48] A. Shirakawa, I. Sakane, T. Kobayashi: Opt. Lett. **23**, 1292 (1998)
[49] A. Shirakawa, I. Sakane, M. Takasaka, T. Kobayashi: Appl. Phys. Lett. **74**, 2268 (1999)
[50] T. Kobayashi, A. Shirakawa: Appl. Phys. B-Lasers O. **70**, S239 (2000)
[51] T. Kobayashi, A. Shirakawa, T. Fuji: **7**, 525 (2001)
[52] M. Zavelani-Rossi, G. Cerullo, S. D. Silvestri, L. Gallmann, N. Matuschek, G. Steinmeyer, U. Keller, G. Angelow, V. Scheuer, T. Tschudi: Opt. Lett. **26**, 1155 (2001)
[53] G. Cerullo, G. Lanzani, M. Muccini, C. Taliani, S. D. Silvestri: Phys. Rev. Lett. **83**, 231 (1999)
[54] R. A. G. Cinelli, V. Tozzini, V. Pellegrini, F. Beltram, G. Cerullo, M. Zavelani-Rossi, S. D. Silvestri, M. Tyagi, M. Giacca: Phys. Rev. Lett. **86**, 3439 (2001)
[55] G. Lanzani, G. Cerullo, M. Zavelani-Rossi, S. D. Silvestri, D. Comoretto, G. Musso, G. Dellepiane: Phys. Rev. Lett. **87**, 187402 (2001)

[56] A. Sugita, T. Saito, H. Kano, M. Yamashita, T. Kobayashi: Phys. Rev. Lett. **86**, 2158 (2001)
[57] T. Fuji, T. Saito, T. Kobayashi: Chem. Phys. Lett. **332**, 324 (2000)
[58] T. Kobayshi, T. Saito, H. Ohtani: Nature **414**, 531 (2001)
[59] T. Kobayashi, A. Shirakawa, H. Matsuzawa, H. Nakanishi: Chem. Phys. Lett. **321**, 385 (2000)
[60] A. J. Wurzer, T. Wilhelm, J. Piel, E. Riedle: Chem. Phys. Lett. **299**, 296 (1999)
[61] S. Lochbrunner, A. J. Wurzer, E. Riedle: J. Comp. Phys. **112**, 10699 (2000)
[62] K. Stock, T. Bizjak, S. Lochbrunner: Chem. Phys. Lett. **354**, 409 (2002)
[63] M. Wittman, A. Nazarkin, G. Korn: Opt. Lett. **26**, 298 (2001)
[64] E. Treacy: IEEE J. Quantum Elect. **5**, 454 (1969)
[65] R. L. Fork, O. E. Martinez, J. P. Gordon: Opt. Lett. **9**, 150 (1984)
[66] Z. Bor, B. Racz: Opt. Commun. **54**, 165 (1985)
[67] D. Zeidler, T. Hornung, D. Proch, M. Motzkus: Appl. Phys. B-Lasers O. **70**, S125 (2000)
[68] E. Zeek, K. Maginnis, S. Backus, U. Russek, M. Murnane, G. R. Mourou, H. Kapteyn, G. Vdovin: Opt. Lett. **24**, 493 (1999)
[69] E. Zeek, R. Bartels, M. Murnane, H. Kapteyn, S. Backus: Opt. Lett. **25**, 587 (2000)
[70] E. Zeek: *Pulse Shaping for High Harmonic Generation*, Ph.D. thesis, Univ. of Michigan (2000)
[71] R. Bartels, S. Backus, E. Zeek, L. Misoguti, G. Vdovin, I. P. Christov, M. M. Murnane, H. C. Kapteyn: Nature **406**, 164 (2000)
[72] M. Armstrong, P. Plachta, E. Ponomarev, R. J. D. Miller: Opt. Lett. **26**, 1152 (2001)
[73] P. Baum, S. Lochbrunner, L. Gallmann, G. Steinmeyer, U. Keller, E. Riedle: Appl. Phys. B-Lasers O. **74**, S219 (2002)
[74] R. Trebino, D. J. Kane: J. Opt. Soc. Am. A **10**, 1101 (1993)
[75] R. Trebino, K. DeLong, D. Fittinghoff, J. Sweetser, M. A. Krumbügel, B. Richman, D. Kane: Rev. Sci. Inst. **68**, 3277 (1997)
[76] A. Baltuška, T. Fuji, T. Kobayashi: Opt. Lett. **27**, 306 (2002)
[77] A. Baltuška, T. Kobayashi: Appl. Phys. B-Lasers O. **75**, 427 (2002)
[78] A. Andreoni, R. Danielius, P. D. Trapani, A. Dubietis, P. Foggi, A. Piskarskas, C. Solcia: J. Opt. Soc. Am. B **12**, 2237 (1995)
[79] A. Andreoni, G. P. Banfi, R. Danielius, P. D. Trapani, M. Monguzzi, A. Piskarskas, C. Solcia, C. Sozzi: Phys. Rev. A **51**, 3164 (1995)
[80] J. Wang, M. H. Dunn, C. F. Rae: Opt. Lett. **22**, 763 (1997)
[81] L. A. W. Gloster, I. T. McKinnie, T. A. King: Opt. Commun. **112**, 328 (1994)
[82] G. Cerullo, M. Nisoli, S. D. Silvestri: Appl. Phys. Lett. **71**, 3613 (1997)
[83] S. Lochbrunner, T. Wilhelm, J. Piel, S. Spörlein, E. Riedle: in *Advanced Solid State Lasers*, vol. 26, TOPS (Opt. Soc. Am. 1999) p. 366
[84] J. Piel, M. Beutter, E. Riedle: Opt. Lett. **25**, 180 (2000)
[85] I. N. Ross, P. Matousek, G. H. New, K. Osvay: J. Opt. Soc. Am. B **19**, 2945 (2002)
[86] R. Danielius, A. Piskarskas, P. D. Trapani, A. Andreoni, C. Solcia, P. Foggi: Opt. Lett. **21**, 973 (1996)
[87] V. G. Dmitriev, G. G. Gurzadyan, D. N. Nikogosyan: *Handbook of Non-Linear Optical Crystals* (Springer, Berlin, Heidelberg 1991)

[88] B. A. Richman, S. E. Bisson, R. Trebino, M. G. Mitchell, E. Sidick, A. Jacobson: Opt. Lett. **22**, 1223 (1997)
[89] B. A. Richman, S. E. Bisson, R. Trebino, E. Sidick, A. Jacobson: Opt. Lett. **23**, 497 (1998)
[90] T. Sosnowski, P. B. Stephens, T. B. Norris: Opt. Lett. **21**, 140 (1996)
[91] E. Žeromskis, A. Dubietis, T. G., A. Piskarskas: Opt. Commun. **203**, 435 (2002)
[92] A. Dubietis, G. Jonušaskas, A. Piskarskas: Opt. Commun. **88**, 437 (1992)
[93] A. Galvanauskas, A. Hariharan, D. Harter, M. A. Arbore, M. M. Fejer: Opt. Lett. **23**, 210 (1998)
[94] I. N. Ross, J. L. Collier, P. Matousek, C. N. Danson, D. Neely, R. M. Allot, D. A. Pepler, C. Hernandez-Gomez, K. Osvay: Appl. Opt. **39**, 2422 (2000)
[95] I. N. Ross, P. Matousek, M. Towrie, A. J. Langley, J. L. Collier: Opt. Commun. **144**, 125 (1997)
[96] D. Zeidler, T. Witte, D. Proch, M. Motzkus: Opt. Lett. **26**, 1921 (2001)
[97] G. R. Fleming: *Chemical Applications of Ultrafast Spectroscopy* (Oxford Univ. Press, New York 1986)
[98] K. W. DeLong, D. N. Fittinghoff, R. Trebino: IEEE J. Quantum Elect. **32**, 1253 (1996)
[99] C. Iaconis, I. A. Walmsley: Opt. Lett. **23**, 792 (1998)
[100] C. Iaconis, I. A. Walmsley: IEEE J. Quantum Elect. **35**, 501 (1999)
[101] L. Gallmann, D. H. Sutter, N. Matuschek, G. Steinmeyer, U. Keller: Appl. Phys. B-Lasers O. **70**, S67 (2000)
[102] A. Weiner: Rev. Sci. Inst. **71**, 1929 (2000)
[103] P. O'Shea, M. Kimmel, X. Gu, R. Trebino: Opt. Express **7**, 342 (2000)
[104] H. C. Kapteyn, J. Zhou: (2001), personal communication
[105] A. Baltuška, M. S. Pshenichnikov, D. A. Wiersma: IEEE J. Quantum Elect. **35**, 459 (199)
[106] G. Taft, A. Rundquist, M. M. Murnane, I. P. Christov, H. C. Kapteyn, K. W. DeLong, D. N. Fittinghoff, M. A. Krumbügel, J. Sweetser, R. Trebino: IEEE J. Sel. Top. Quant. **2**, 575 (1996)
[107] K. W. DeLong, R. Trebino, J. Hunter, W. E.White: J. Opt. Soc. Am. B **11**, 2206 (1994)
[108] A. P. Kovacs, K. Osvay, Z. Bor, R. Szipöcs: Opt. Lett. **20**, 788 (1995)
[109] D. Yelin, D. Meshulach, Y. Silberberg: Opt. Lett. **22**, 1793 (1997)
[110] A. Bartelt, S. Minemoto, C. Lupulescu, S. Vajda, L. Wöste: Eur. Phys. J. D **16**, 127 (2001)
[111] T. Baumert, T. Brixner, A. Seyfried, M. Strehle, G. Gerber: Appl. Phys. B-Lasers O. **65**, 779 (1997)
[112] T. Brixner, M. Strehle, G. Gerber: Appl. Phys. B-Lasers O. **68**, 281 (1999)
[113] T. Brixner, A. Oerlein, M. Strehle, G. Gerber: Appl. Phys. B-Lasers O. **70**, S119 (2000)
[114] W. H. Press, S. A. Teukolsky, W. T. Vetterling, B. P. Flannery: *Numerical Recipes in C* (Cambridge Univ. Press, New York 2000)
[115] M. Nisoli, S. D. Silvestri, O. Svelto: Appl. Phys. Lett. **68**, 2793 (1996)
[116] M. Namikawa, N. Nakamura, T. Kawano, K. Itoh, T. Fuji, A. Baltuska, T. Kobayashi: upublished
[117] S. Yeremenko, A. Baltuška, F. de Haan, M. S. Pshenichnikov, D. A. Wiersma: Opt. Lett. **27**, 1171 (2002)

[118] T. Beddard, M. Ebrahimzadeh, T. D. Reid, W. Sibbett: Opt. Lett. **25**, 1052 (2000)

[119] J. A. Gruetzmacher, N. F. Scherer: Rev. Sci. Inst. **73**, 2227 (2002)

[120] S. Woutersen, H. J. Bakker: Nature **402**, 507 (1999)

[121] J. Stenger, D. Madsen, P. Hamm, E. T. J. Nibbering, T. Elsaesser: Phys. Rev. Lett. **87**, 027401 (2001)

[122] R. Danielius: (2002), personal communication

[123] I. Jovanovic, C. A. Ebbers, C. P. J. Barty: Opt. Lett. **27**, 1622 (2002)

[124] O. Albert, G. Mourou: Appl. Phys. B-Lasers O. **69**, 207 (1999)

[125] Y. Kobayashi, K. Torizuka: Opt. Lett. **26**, 1295 (2001)

[126] A. Baltuška, T. Fuji, T. Kobayashi: Phys. Rev. Lett. **88**, 133901 (2002)

[127] M. S. Pshenichnikov: (2001), personal communication

[128] L. Xu, C. Spielmann, A. Poppe, T. Brabec, F. Krausz: Opt. Lett. **21**, 2008 (1996)

[129] P. Dietrich, F. Krausz, P. B. Corkum: Opt. Lett. **25**, 16 (2000)

[130] A. Apolonski, A. Poppe, G. Tempea, C. Spielmann, T. Udem, R. Holzwarth, T. W. Hänsch, F. Krausz: Phys. Rev. Lett. **85**, 740 (2000)

[131] S. A. Diddams, D. J. Jones, J. Ye, T. Cundiff, J. L. Hall: Phys. Rev. Lett. **84**, 5102 (2000)

[132] D. J. Jones, S. A. Diddams, J. K. Ranka, A. Stentz, R. S. Windeler, J. L. Hall, S. T. Cundiff: Science **288**, 635 (2000)

[133] M. Kakehata, H. Takada, Y. Kobayashi, K. Torizuka, Y. Fujihara, T. Homma, H. Takahashi: Opt. Lett. **26**, 1436 (2001)

[134] Y. Kobayashi, K. Torizuka: Opt. Lett. **25**, 856 (2000)

[135] M. Mehendale, S. A. Mitchell, J. P. Likforman, D. M. Villeneuve, P. B. Corkum: Opt. Lett. **25**, 1672 (2000)

[136] H. R. Telle, G. Steinmeyer, A. E. Dunlop, J. Stenger, D. H. Sutter, U. Keller: Appl. Phys. B-Lasers O. **69**, 327 (1999)

[137] I. P. Christov: Opt. Lett. **24**, 1425 (1999)

[138] A. de Bohan, P. Antoine, D. B. Miloševic, B. Piraux: Phys. Rev. Lett. **81**, 1837 (1998)

[139] G. G. Paulus, F. Grabson, H. Walther, P. Villoresti, M. Nisoli, S. Stagira, E. Priori, S. D. Silvestri: Nature **414**, 182 (2001)

[140] S. T. Cundiff, J. Ye, J. L. Hall: Rev. Sci. Inst. **72**, 3746 (2001)

[141] T. Brabec, F. Krausz: Phys. Rev. Lett. **78**, 3282 (1997)

[142] S. M. Gallagher, A. W. Albrecht, J. D. Hybl, B. L. Landin, B. Rajaram, D. M. Jonas: J. Opt. Soc. Am. B **15**, 2338 (1998)

[143] S. T. Cundiff: J. Phys. D: Appl. Phys. **35**, R43 (2002)

[144] A. Dubietis, R. Danielius, G. Tamošauskas, A. Piskarskas: J. Opt. Soc. Am. B **15**, 1135 (1998)

[145] X. C. Zhang, B. Hu, J. Darrow, D. Auston: Appl. Phys. Lett. **56**, 1011 (1990)

[146] M. Takeda, H. Ina, S. Kobayashi: josaa **72**, 156 (1982)

[147] P. Tzankov, I. Buchvarov, T. Fiebig: Opt. Commun. **203**, 107 (2002)

[148] A. Haché, Y. Kostoulas, R. Atanasov, J. L. P. Hughes, J. E. Sipe, H. M. van Driel: Phys. Rev. Lett. **78**, 306 (1997)

[149] A. Haché, J. E. Sipe, H. M. van Driel: IEEE J. Quantum Elect. **34**, 1144 (1998)

[150] M. D. Perry, G. Mourou: Science **264**, 917 (1994)

[151] F. Krausz, H. Kapteyn: (2001), personal communication

[152] R. Danelius, A. Piskarskas, V. Sirutkaitis, A. Stabinis, A. Yankauskas: JETP Lett. **42**, 122 (1985)
[153] A. Baltuška, T. Fuji, T. Kobayashi: Opt. Lett. **27**, 1241 (2002)
[154] G. Tempea, M. Geissler, T. Brabec: J. Opt. Soc. Am. B **16**, 669 (1999)
[155] N. Milosevic, A. Scrinzi, T. Brabec: Phys. Rev. Lett. **88**, 093905 (2002)
[156] P.V. Mamyshev, S.V. Chernikov: Opt. Lett. **15**, 1076 (1990)
[157] J. Botineau, R.H. Stolen: josaa **72**, 1592 (1982)
[158] N. Bloembergen: *Nonlinear Optics*, 4th ed. (World Scientific, Singapore 1996) p. 74
[159] S. Mukamel: *Principles of Nonlinear Optical Spectroscopy* (Oxford Univ. Press, New York 1995)
[160] T.M. Fortier, J. Ye, S.T. Cundiff, R.S. Windeler: Opt. Lett. **27**, 445 (2002)
[161] T.M. Fortier, D.J. Jones, J. Ye, S.T. Cundiff: Opt. Lett. **27**, 1436 (2002)
[162] H.A. Haus, E.P. Ippen: Opt. Lett. **26**, 1654 (2001)

Part II

Characterization of Ultrashort Laser Pulses

Measuring Ultrashort Pulses in the Single-Cycle Regime: Frequency-Resolved Optical Gating

Rick Trebino[1], Andrius Baltuška[2], Maxim S. Pshenichnikov[3], and Douwe A. Wiersma[3]

[1] School of Physics, Georgia Institute of Technology, Atlanta, GA 30332, USA
rick.trebino@physics.gatech.edu
[2] Photonics Institute, Vienna University of Technology, Gusshausstr. 27, 1040 Vienna, Austria
[3] Department of Chemistry, Ultrafast Laser and Spectroscopy Laboratory, University of Groningen, The Netherlands

Abstract. In this chapter, we developed the theory of frequency-resolved optical gating (FROG) in the few-cycle regime and applied it to the SHG FROG measurement of 2.5-optical-cycle pulses with a central wavelength around 800 nm. These are some of the shortest pulses that have been completely characterized to date. We also describe the measurement of uncompressed strongly nontransform-limited weak-intensity continuum pulses. These two key measurements, which are required to design, test, and optimize the pulse compressor, were both performed without a single change in the SHG FROG apparatus. No other pulse measurement technique known allows such simplicity and versatility.

1 Introduction

Extremely short ($\sim$ 10 fs or less) pulses are now available in several labs and have a wide range of applications. It might seem that such extremely short pulses could not possibly have sufficient bandwidth to be distorted and still be so short, but that is not the case. Intensity and phase distortions not only exist in such incredibly short pulses, but they play decisive – and even useful – roles in many phenomena. For instance, pulses with identical spectra but different spectral phases yield wildly different efficiencies in high-harmonic generation processes [1]. The spectral phase also heavily affects wave-packet motion in organic molecules [2, 3], population inversion in liquid [4] and gas [5] phases, and even the direction of a chemical reaction [6]. Moreover, a totally automated search for the best shaped pulse to optimize a preselected reaction channel was recently demonstrated [7, 8, 9]. Measuring the phase and amplitude of the excitation pulses in such experiments then allows backreconstruction of potential surfaces of the parent molecule.

The complete determination of the electric field of femtosecond pulses also uncovers the physics behind their generation, as has been demonstrated for femtosecond Ti:sapphire lasers [10, 11]. Such information is invaluable in

F. X. Kärtner (Ed.): Few-Cycle Laser Pulse Generation and Its Applications, Topics Appl. Phys. **95**, 231–263 (2004)

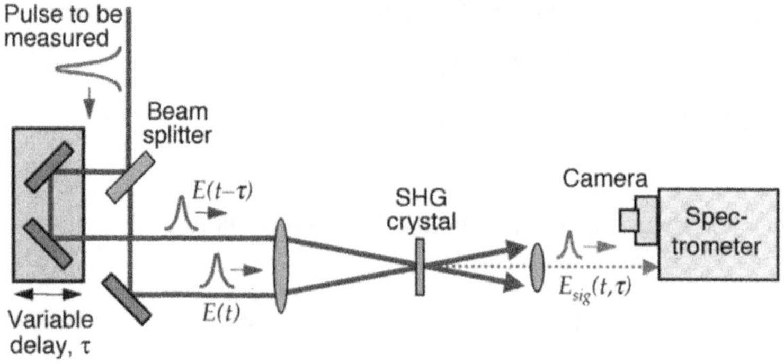

Fig. 1. Second-harmonic-generation FROG. The spectrum of the signal pulse is measured as the relative delay is varied between two replicas of the pulse to be measured. There are many variations and alternatives on the FROG class of techniques. This simple device allows complete determination of a pulse's intensity and phase. Finally, note the similarity of FROG to an excite–probe apparatus. This means that any excite–probe geometry can be easily modified (by replacing the sample with a nonlinear crystal) to yield a FROG measurement of the pulse at the sample medium

determining the methods for their generation and also the ultimate limits to further pulse shortening. Due to the great complexity of the broadband phase correction required to produce transform-limited pulses with duration shorter than 5 fs [12, 13, 14, 15], such measurements are as difficult as they are important. And, because the broader the pulse spectrum, the shorter the pulse can potentially be, it is important to develop methods for measuring very broadband pulses, such as the continuum generated by self-phase modulation and other nonlinear-optical processes at high intensity and long path lengths.

Due to their insensitivity to pulse intensity and phase structure and their high sensitivity to noise, simple techniques based on autocorrelation and spectrum are unreliable, especially in the few-cycle regime. Fortunately, it is now possible to achieve full characterization of ultrashort pulses using FROG [16, 17, 18]. FROG simply involves measuring a spectrally resolved autocorrelation, which yields a spectrogram of the pulse (see Fig. 1). The pulse intensity and phase are then determined from the FROG trace using one of several commercially available phase-retrieval algorithms written specifically for this purpose. The FROG technique is described in great detail in a recent book [18]. Notably, no a priori information about the pulse shape is necessary to reconstruct the pulse from the experimental FROG trace.

In general, FROG is quite accurate and rigorous [19]. The large number of data points in the two-dimensional FROG trace and the resulting overdetermination of the pulse also make it much less sensitive to noise than pulse diagnostics based on one-dimensional measurements, such as ordinary autocorrelation and interferometric methods. Last, but not least, FROG offers data self-consistency checks that are unavailable in other pulse measuring techniques. These feedback mechanisms involve computing the temporal and spectral marginals that are the integrals of the FROG trace along the delay and frequency axes [18]. A comparison of the marginals with the independently measured fundamental spectrum and autocorrelation verifies the validity of the measured FROG trace [11, 18, 20, 21]. To date, FROG methods have been applied to measure a vast variety of pulses with different duration, wavelength, and complexity [22].

A number of outstanding features make FROG especially valuable for the measurement of sub-10 fs pulses.

First, because FROG uses a type of excite–probe geometry, common for most nonlinear optical experiments, it is ideally suited to measure pulses used in spectroscopic experiments. Unlike other pulse diagnostics [23, 24, 25, 26, 27, 28, 29, 30, 31, 32], FROG does not require splitting of auxiliary laser beams and prefabrication of reference pulses. This is a great practical advantage, since the setup complexity in many spectroscopic experiments is already quite high [33, 34, 35, 36, 37, 38, 39] FROG directly offers pulse characterization *precisely at the position of the sample* by simply interchanging the sample with a nonlinear medium for optical gating [18]. The last point is essential for $\sim$ 10 fs pulses [12, 13, 14, 15, 40] currently used in spectroscopy because the dispersive lengthening that such pulses experience, even due to propagation through air, precludes the use of a separate diagnostic device. Although an alternative technique, spectral phase interferometry for direct electric-field reconstruction (SPIDER) [25, 29, 30, 31, 32], can also measure sub-6 fs pulses [28], it requires a cumbersome additional setup, which can lead to errors in measurement. In this respect, the simplicity of FROG and its compatibility with ultrafast spectroscopy provide an undisputed advantage both in saving experimental effort and in ensuring that the "right pulse at the right location" is being measured. Thus, FROG is the ideal way to measure and optimize pulses on a target prior to carrying out a spectroscopic experiment.

Second, it is possible to measure sub-10 fs pulses correctly by FROG *even in presence of massive systematic error*. Several types of such errors will inevitably appear in any measurement (whether by FROG or any other technique) of pulses whose spectra span a hundred nanometers or more. And FROG is unique in its ability to deal with such distortions. For example, a FROG trace affected by wavelength-dependent detector sensitivity and wavelength-dependent frequency conversion efficiency can be corrected and validated via its consistency checks [11, 18]. In contrast, an autocorrelation or SPIDER trace measured under identical conditions may be corrupted irreparably, and it would be difficult to know.

Third, the temporal resolution of a FROG measurement is not limited by the sampling increment in the time domain. Assuming that the whole time-frequency spectrogram of the pulse is properly contained within the measured FROG trace, short-time information is also provided by large-scale variations in frequency. The broadest feature in the frequency domain determines the shortest feature in the time domain. Therefore, no fine temporal pulse structure can be overlooked [22], even if the delay increment used to collect the FROG trace is larger than the duration of such a structure. Thus, the reliability of the FROG data depends simply on the proper delay-axis calibration, rather than on very fine sampling in time, which might otherwise be troublesome, since the pulse itself extends only a few microns in space.

Fourth, pulses with large bandwidths can often be quite complex, and FROG easily measures even extremely complex pulses. It is important to remember that, if a technique *can* measure only simple pulses, it *will* measure only simple pulses. In other words, alternative techniques can measure only simple pulses and hence will erroneously yield a simple pulse even when presented with a complex one. And because they lack feedback, it would be difficult to know that such an error has occurred.

Fifth, recent developments have made FROG even more ideal for measuring near-single-cycle pulses. A crystal-angle-dithering technique yields extremely large phase-matching bandwidths, independent of crystal thickness, and now eliminates the need for extremely thin crystals (required in other techniques). And geometric time-smearing effects, which are generally negligible in FROG, can be made to be zero by using single-shot beam geometry.

As the FROG technique and its many other advantages are well known, and numerous articles and a book [18] exist describing it in detail, we will assume familiarity with FROG's basic ideas and concentrate here on the issues that arise in the measurement of sub-10 fs and other extremely broadband pulses.

There are as many beam geometries for FROG measurements as there are nonlinear-optical effects. As a result, choosing the appropriate FROG geometry [18, 20, 22] for measuring sub-10 fs pulses requires some thought. In particular, in some $\chi^{(3)}$-based techniques (for instance, polarization-gating, transient grating etc.), the finite response time due to the Raman contribution to the nonlinearity can play a role in the measurement of $\sim$ 20 fs pulses [41]. Interestingly, these effects are less important for sub-10 fs pulses, which are shorter than the Raman timescale. Nevertheless, SHG FROG in transparent crystals [42, 43, 44] and third-harmonic-generation FROG near a surface [45], both of which have essentially instantaneous nonlinearities, present the best overall choices for measurement of the shortest pulses available to date.

Another important experimental issue is the sensitivity of the FROG measurement. Among the FROG geometries, SHG FROG is most appropriate for low-energy pulses because it [42] is second order, rather than third order. The signal pulse's different color and polarization allow effective suppression of scattered input light, adding to the suppression provided by the geometry. Low-order nonlinearity, combined with this background suppression, gives

SHG FROG a higher dynamic range than any other FROG geometry – and any other technique.

In general, FROG pulse reconstruction does not depend on pulse duration since the FROG traces simply scale in the time-frequency domain. Because the pulse gates itself, the shorter the pulse to be measured, the shorter the gate pulse that is automatically available to measure it. However, in all pulse-measurement techniques, with the decrease in pulse length and the accompanying increase in bandwidth, the measured trace can deviate from the simplest "ideal" trace due to limitations of the nonlinear medium. Besides the obvious pulse lengthening during its propagation before and inside the nonlinear medium due to group velocity dispersion, there are more fundamental issues – issues relevant to all techniques that may hope to address the few-cycle regime.

For example, the conventional description of nonlinear optics applicable to multicycle pulses becomes questionable for near-single-cycle pulses. Clearly, for such extremely short pulses, the conventionally employed slowly varying envelope approximation [46, 47, 48], which assumes that the pulse amplitude varies negligibly over one optical cycle, fails. Furthermore, the phase-matching bandwidth [49] – which is finite due to dispersion in the nonlinear medium – rapidly becomes problematic as the pulse spectrum broadens [10, 11]. This effect can, however, be essentially eliminated in FROG measurements using crystal angle-dithering. Another issue is the frequency-dependent variation in the reflectivity of the optics and gratings and the sensitivity of the photodetector used to detect the signal light generated in the nonlinear process. These effects all potentially distort the signal pulse spectrum and hence are known as the *spectral-filter* effect [49, 50, 51]. Geometric distortions due to the multishot noncollinear geometry employed in spectroscopic experiments [11,51,52] can also potentially artificially lengthen the measured pulse in any measurement technique [18]. Geometric distortions can be eliminated completely in FROG measurements if a single-shot beam geometry is used. In any case, if a portion of the signal field is accidentally spectrally filtered out or if the signal is artificially "blurred" in time, this could influence the measured trace [53] and subsequently lead to erroneous measurements.

In this chapter, we give a detailed description of FROG's performance for ultrabroadband pulses that would correspond to pulses 3 fs in length, if the spectral phase were flat. Starting with Maxwell's equations, *we derive a complete expression for the SHG FROG signal that is valid even in the single-cycle pulse regime.* It includes phase matching in the crystal, beam-geometry effects, dispersive pulse-broadening inside the crystal, and dispersion of the second-order nonlinearity. Subsequently, we describe a simplified – but still quite accurate – expression that decomposes the SHG FROG signal into a simple product of the ideal SHG FROG trace and a spectral filter. Remarkably, this simplification allows using the same commercial FROG code that retrieves longer pulses even for single-cycle pulses, provided that the trace is first divided by the spectral-filter correction factor. Then we describe numerical simulations, which convincingly show that the approximations made in the derivation of the simplified expression are well justified. We also show that, despite all these potential sources of error, accurate measurements of single-cycle pulses are straightforward. Indeed, this is the case even without the simplifications provided by crystal-angle-dithering and the use of a geometric-distortion-free single-shot beam geometry.

It is important to realize that, although the spectral filter effect is important for accurately retrieving the pulse spectrum, it is less important if only the spectral phase is desired. However, the check of the measurement provided by the FROG retrieval of the pulse spectrum and its comparison with the independently measured spectrum is very helpful, as any measurement of such broadband pulses is subject to error.

We do not discuss the performance of $\chi^{\langle 3\rangle}$-based FROG techniques – transient grating (TG) and self-diffraction (SD) – in the single-cycle regime. They are considered (for longer pulses) in [18].

Next, we discuss the practical implementation of these ideas in the FROG measurement of near-single-cycle and very broadband pulses. Finally, we present experimental results of SHG FROG characterization of a white-light continuum and pulses less than 5 fs long.

2 Theory

The ideal FROG signal is a simple mag-squared Fourier transform of a time- and delay-dependent signal field [22]. This formalism describes an ideal nonlinear medium, which has zero thickness and an infinitely fast nonlinear response. This, of course, can never occur in a real experiment. In this section, we present a comprehensive theoretical analysis in which the formalism describing the FROG signal is thoroughly reexamined. We derive complete expressions valid even for single-cycle-pulse applications. Among other results, we show that the slowly varying envelope approximation, with its rejection of derivatives of the time-domain electric field, is *unnecessary.* We also show that geometric delay smearing does not introduce a significant distortion of the observed traces, even for near-single-cycle pulses, provided that the geometry is carefully optimized.

2.1 SHG FROG in the Single-Cycle Regime

We consider first the case of $\chi^{\langle 2\rangle}$-based FROG and derive a general formalism for pulses of arbitrarily broad bandwidth, allowing us to carry out a scrupulous numerical analysis of the performance of SHG FROG in various experimental situations.

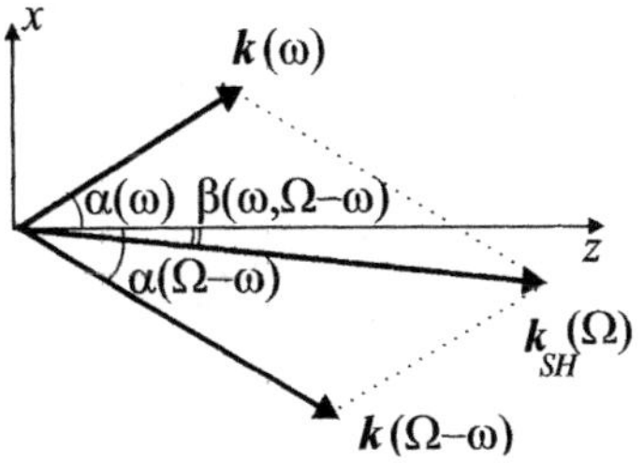

Fig. 2. Noncollinear phase matching for three-wave interaction. $\boldsymbol{k}(\omega)$ and $\boldsymbol{k}(\Omega-\omega)$ are the wave vectors of the fundamental fields that form an angle α with the z-axis. $k_{\mathrm{SH}}(\Omega)$ is the wave vector of the second-harmonic that intersects the z-axis at an angle β

2.1.1 General Formalism

In this section, we derive the complete equation that describes SHG FROG for pulses as short as one optical cycle. We include such effects as the phase-matching condition, time-smearing effects due to noncollinear geometry, spectral filtering of the second harmonic radiation, and dispersion of the second-order nonlinearity.

We consider a noncollinear geometry in which the fundamental beams intersect at a small angle (Fig. 2). As has been pointed out [49], pulse broadening due to crystal bulk dispersion is negligibly small compared to the group-velocity mismatch. This means that the appropriate crystal thickness is determined mostly from the phase-matching condition. For instance, in a 10 µm BBO crystal, the bulk dispersion broadens a single-cycle pulse by only ∼ 0.1 fs, whereas the group-velocity mismatch between the fundamental and second-harmonic pulses is as much as 0.9 fs.

We assume that the confocal parameter and the longitudinal beam overlap of the fundamental beams are considerably longer than the crystal length. For instance, for an ideal Gaussian beam of ∼ 2 mm diameter focused by a 10 cm achromatic lens, the confocal parameter, that is, the longitudinal extent of the focal region, is ∼ 1.2 mm. This is considerably longer than the nonlinear crystal. Under such conditions, all wavefronts inside the crystal are practically flat. Therefore, second-harmonic generation will be a function only of the longitudinal coordinate and will include the transversal coordinates only at the last step to account for the spatial beam profile. Note that the constraint on focusing is not necessarily always fulfilled. For example, very tight focusing using a 1 cm lens in the situation described above reduces the length of the focal region to only 12 µm, and, in this case, it would be impossible to disregard the dependence on transverse coordinates.

We assume that the second-harmonic field is not absorbed in a nonlinear crystal, which is well justified even for single-cycle pulses. Absorption bands of the crystals that are transparent in the visible occur below ∼ 200 nm. If the central wavelength of a single-cycle pulse is located at 800 nm, the field of the SH components around 200 nm will be more than a thousand times smaller than the SH peak at 400 nm and can be safely ignored. We also require that the SHG efficiency be low enough to neglect depletion of the fundamental beams, so the system of two coupled equations describing the nonlinear interaction [54] will simplify to one. The equation that governs propagation of the second-harmonic wave in the $+z$-direction inside the crystal can be obtained directly from Maxwell's equations [55]:

$$\frac{\partial^2}{\partial z^2}\mathcal{E}_{\mathrm{SH}}(z,t) - \varepsilon_0\mu_0\frac{\partial^2}{\partial t^2}\int_{-\infty}^{t}\varepsilon(t-t')\mathcal{E}_{\mathrm{SH}}(z,t')\,\mathrm{d}t' = \mu_0\frac{\partial^2}{\partial t^2}\mathcal{P}^{\langle 2\rangle}(z,t)\,, \quad (1)$$

where $\mathcal{E}_{\mathrm{SH}}(z,t)$ is the second-harmonic field, $\mu_0\varepsilon_0 = 1/c^2$, ε is the relative permittivity, and $\mathcal{P}^{\langle 2\rangle}(z,t)$ is the induced second-order dielectric polarization. We Fourier-transform both $\mathcal{E}_{\mathrm{SH}}(z,t)$ and $\mathcal{P}^{\langle 2\rangle}(z,t)$, obtaining a result equivalent to (1) in the frequency (Ω) domain:

$$\frac{\partial^2}{\partial z^2}\widetilde{\mathcal{E}}_{\mathrm{SH}}(z,\Omega) + k_{\mathrm{SH}}^2(\Omega)\widetilde{\mathcal{E}}_{\mathrm{SH}}(z,\Omega) = -\mu_0\Omega^2\widetilde{\mathcal{P}}^{\langle 2\rangle}(z,\Omega)\,, \quad (2)$$

where $\widetilde{\mathcal{E}}_{\mathrm{SH}}(z,\Omega)$ and $\widetilde{\mathcal{P}}^{\langle 2\rangle}(z,\Omega)$ are Fourier transforms of $\mathcal{E}_{\mathrm{SH}}(z,t)$ and $\widetilde{\mathcal{P}}^{\langle 2\rangle}(z,t)$, respectively; Ω is the frequency; and $k_{\mathrm{SH}}(\Omega)$ is the wave vector of the second-harmonic field: $k_{\mathrm{SH}}^2(\Omega) = \Omega^2\varepsilon_0\mu_0\tilde{\varepsilon}(\Omega)$, where $\tilde{\varepsilon}(\Omega)$ is the Fourier transform of the relative permittivity $\varepsilon(t)$.

To simplify the left part of (2), we write the second-harmonic field as a plane wave propagating along the z-axis (and ignore the complex conjugate):

$$\widetilde{\mathcal{E}}_{\mathrm{SH}}(z,\Omega) = \widetilde{\boldsymbol{E}}_{\mathrm{SH}}(z,\Omega)\mathrm{e}^{\mathrm{i}k_{\mathrm{SH}}(\Omega)z}\,, \quad (3)$$

where $\widetilde{\boldsymbol{E}}_{\mathrm{SH}}(z,\Omega)$ is a spatial envelope function. Substituting (3) in (2) yields

$$2\mathrm{i}k_{\mathrm{SH}}(\Omega)\frac{\partial}{\partial z}\widetilde{\boldsymbol{E}}_{\mathrm{SH}}(z,\Omega) + \frac{\partial^2}{\partial z^2}\widetilde{\boldsymbol{E}}_{\mathrm{SH}}(z,\Omega) = -\mu_0\Omega^2\widetilde{\mathcal{P}}^{\langle 2\rangle}(z,\Omega)\mathrm{e}^{-\mathrm{i}k_{\mathrm{SH}}(\Omega)z}\,. \quad (4)$$

So far, we have made no simplifications concerning pulse duration. Now, we apply the slowly varying amplitude approximation [55], i.e.,

$$\left|\frac{\partial}{\partial z}\widetilde{\boldsymbol{E}}_{\mathrm{SH}}(z,\Omega)\right| \ll \left|2k_{\mathrm{SH}}(\Omega)\widetilde{\boldsymbol{E}}_{\mathrm{SH}}(z,\Omega)\right|\,, \quad (5)$$

to omit the term $\partial^2\widetilde{\boldsymbol{E}}_{\mathrm{SH}}(z,\Omega)/\partial z^2$.

Note that the standard time-domain description of the signal wave propagation results in a second-order differential equation, similar in its structure to our (4). Unlike (4), however, simplification of the time-domain expression requires rejection of the second-order *temporal* derivative of the envelope, i.e.,

$$\left|\frac{\partial^2}{\partial t^2}E(t)\right| \ll \left|\frac{4\pi}{T_{\mathrm{per}}}\frac{\partial}{\partial t}E(t)\right|\,, \quad (6)$$

where T_{per} is the characteristic period of light oscillation. Such an approach implies the assumption of slow envelope variation vs. time. This condition is *not* fulfilled for pulses only a few cycles long, since the change of the envelope within one optical period is comparable to the magnitude of the envelope itself. *Brabec* and *Krausz* [56], who explored the time-domain approach for the propagation of near-single-cycle pulses, found that rejection of the second-order derivative term is warranted when the phase and the group velocities of light are close to each other. To this point, we notice that the application of condition (5) to the frequency domain (4) does not require any assumptions on the change of the temporal envelope altogether. Therefore, inequality (5) is safe to apply even to single-cycle pulses, provided there is no appreciable linear absorption at lengths comparable to the wavelength. The only remaining concern is the lowest frequencies, for which k_{SH} becomes close to zero. However, the amplitude of the low-energy spectral wing is negligible even for single-cycle optical pulses and therefore can be disregarded. Consequently, (4) can be readily solved by integration over the crystal length L:

$$\widetilde{\boldsymbol{E}}(L,\Omega) = \mathrm{i}\frac{c\mu_0\Omega}{2n_{\mathrm{SH}}(\Omega)}\int\limits_0^L \widetilde{\mathcal{P}}^{\langle 2\rangle}(z,\Omega)\mathrm{e}^{-k_{\mathrm{SH}}(\Omega)z}\,\mathrm{d}z\,, \tag{7}$$

where $n_{\mathrm{SH}}(\Omega) = \sqrt{\tilde{\varepsilon}(\Omega)}$ is the refractive index of the second harmonic.

Now, we calculate the second-order polarization $\widetilde{\mathcal{P}}^{\langle 2\rangle}(z,\Omega)$. We assume that two fundamental fields cross in the xz-plane at a small angle θ (Fig. 2). Each beam's angle with the z-axis inside the crystal is then

$$\alpha(\omega) = \arcsin\left[n(\omega)\sin(\theta/2)\right] \approx n(\omega)\theta/2\,.$$

The relative delay between the pulses is τ. An additional delay for off-axis components of the beam due to the geometry can be expressed for a plane wave as

$$\tau'(x) = xn(\omega)\sin\alpha(\omega)/c = x\sin\alpha_0/c \approx x\alpha_0/c$$

for the beam propagating in the $+\alpha$-direction and $\tau'(x) \approx -x\alpha_0/c$ for the beam in the $-\alpha$-direction. The electric fields in the frequency domain are the Fourier transforms,

$$\begin{aligned}\widetilde{E}_1(\omega) &= \widetilde{E}(\omega)\mathrm{e}^{\mathrm{i}\omega\left(\frac{x\alpha_0}{c}\right)}\,,\\ \widetilde{E}_2(\omega) &= \widetilde{E}(\omega)\mathrm{e}^{\mathrm{i}\omega\left(\frac{-x\alpha}{c}-\tau\right)}\,.\end{aligned} \tag{8}$$

To calculate the second-order dielectric polarization induced at frequency Ω by the two fundamental fields, we must sum over all possible permutations of fundamental frequencies:

$$\begin{aligned}\widetilde{\mathcal{P}}^{\langle 2\rangle}(z,\Omega) &= \int \tilde{\chi}^{\langle 2\rangle}\left(\Omega,\omega,\Omega-\omega\right)\widetilde{E}_1(\omega)\widetilde{E}_2(\Omega-\omega)\,\mathrm{d}\omega\\ &= \mathrm{e}^{\mathrm{i}\Omega(\tau+x\alpha_0/c)}\int \tilde{\chi}^{\langle 2\rangle}\left(\Omega,\omega,\Omega-\omega\right)\widetilde{E}(\omega)\widetilde{E}(\Omega-\omega)\\ &\quad\cdot\exp\left\{\mathrm{i}\left[k_z(\omega)z + k_z(\Omega-\omega)z + \omega(\tau+2x\alpha_0/c)\right]\right\}\mathrm{d}\omega\,.\end{aligned} \tag{9}$$

In (9), we have included the frequency dependence of the nonlinear susceptibility $\tilde{\chi}^{\langle 2\rangle}(\Omega, \omega, \Omega-\omega)$, and we have represented the fundamental field analogously to (3). The electric field of the second harmonic therefore becomes

$$\begin{aligned}\widetilde{\boldsymbol{E}}_{\mathrm{SH}}(L,\Omega) = \mathrm{i}\frac{c\mu_0 \Omega L}{2n(\Omega)} \exp\left[\mathrm{i}\Omega(\tau + x\alpha_0/c)\right] \int \tilde{\chi}^{\langle 2\rangle}(\Omega, \omega, \Omega-\omega) \\ \cdot \widetilde{\boldsymbol{E}}(\Omega-\omega)\widetilde{\boldsymbol{E}}(\omega)\exp\left[\mathrm{i}\frac{\Delta k(\omega,\Omega-\omega)L}{2} + \mathrm{i}\omega\left(\tau + \frac{2x\alpha_0}{c}\right)\right] \\ \cdot \operatorname{sinc}\left(\frac{\Delta k(\omega,\Omega-\omega)L}{2}\right)\mathrm{d}\omega\,, \quad (10)\end{aligned}$$

where $\Delta k(\omega, \Omega-\omega)$ is the phase mismatch given by the equation

$$\begin{aligned}\Delta k(\omega,\Omega-\omega) = k(\omega)\cos\left[\alpha_0 n_1(\omega)\right] + k(\Omega-\omega)\cos\left[\alpha_0 n_2(\Omega-\omega)\right] \\ - k_{\mathrm{SH}}(\Omega)\cos\left[\beta(\omega,\Omega-\omega)\right]\,, \quad (11)\end{aligned}$$

where n_1 and n_2 are the refractive indexes of the fundamental waves and $\beta(\omega, \Omega-\omega)$ is the angle between $k_{\mathrm{SH}}(\Omega)$ and the z-axis inside the crystal. The appearance of this angle can be easily understood from Fig. 2. The momentum conservation law determines the direction of the emitted second-harmonic field:

$$\boldsymbol{k}(\omega) + \boldsymbol{k}(\Omega-\omega) = \boldsymbol{k}_{\mathrm{SH}}(\Omega)\,, \quad (12)$$

where $\boldsymbol{k}(\omega)$ and $\boldsymbol{k}(\Omega-\omega)$ are the wave vectors of the incident fundamental waves. For $\boldsymbol{k}(\omega) \neq \boldsymbol{k}(\Omega-\omega)$, β is nonzero and can be found from the following equation

$$\sin\left[\beta(\omega,\Omega-\omega)\right] = \left[\frac{k(\omega)n_1(\omega) - k(\Omega-\omega)n_2(\Omega-\omega)}{k_{\mathrm{SH}}(\Omega)}\right]\sin\alpha_0\,. \quad (13)$$

As β is of the same order as the intersection angle, the correction $\cos[\beta(\omega, \Omega-\omega)]$ is required only in the Δk expression (11). Elsewhere, this correction can be dropped.

The values of the wave vectors and refractive indexes in (11) and (13) depend on the actual polarization of the three interacting waves. For type I, we obtain

$$\begin{aligned}\Delta k(\omega,\Omega-\omega) = k_{\mathrm{o}}(\omega)\cos\left[\alpha_0 n_{\mathrm{o}}(\omega)\right] + k_{\mathrm{o}}(\Omega-\omega)\cos\left[\alpha_0 n_{\mathrm{o}}(\Omega-\omega)\right] \\ - k_{\mathrm{E}}(\Omega)\cos\left[\beta(\omega,\Omega-\omega)\right]\,, \quad (14)\end{aligned}$$

and for type II:

$$\begin{aligned}\Delta k(\omega,\Omega-\omega) = k_{\mathrm{e}}(\omega)\cos\left[\alpha_0 n_{\mathrm{e}}(\omega)\right] + k_{\mathrm{o}}(\Omega-\omega)\cos\left[\alpha_0 n_{\mathrm{o}}(\Omega-\omega)\right] \\ - k_{\mathrm{e}}(\Omega)\cos\left[\beta(\omega,\Omega-\omega)\right]\,. \quad (15)\end{aligned}$$

Here indexes o and e indicate the ordinary and extraordinary waves, respectively.

To calculate the total FROG signal, one should integrate the signal intensity,

$$\widetilde{I}_{\rm SH}(L,\Omega) = \varepsilon_0 \frac{n_{\rm SH}(\Omega)}{c} \left| \widetilde{\boldsymbol{E}}_{\rm SH}(L,\Omega) \right|^2 , \tag{16}$$

over the transverse coordinates x and y. Hence, for the second-harmonic signal detected in FROG, we obtain

$$\begin{aligned} I_{\rm meas}(\Omega,\tau,L) = {} & \frac{\Omega^2 L^2 Q(\Omega)}{2c^3 \varepsilon_0 n_{\rm SH}(\Omega)} (\ln 2)^{3/2} \int \exp\left[-4\ln 2 \left(\frac{x}{d_0}\right)^2\right] \\ & \cdot \left| \int_0^{\Omega} \widetilde{\chi}^{\langle 2\rangle}(\Omega,\omega,\Omega-\omega)\, \widetilde{\boldsymbol{E}}(\Omega-\omega) \widetilde{\boldsymbol{E}}(\omega) \exp\left[\mathrm{i}\frac{\Delta k(\omega,\Omega-\omega)L}{2} \right. \right. \\ & \left. \left. + \mathrm{i}\omega\left(\tau + \frac{2x\alpha_0}{c}\right)\right] \mathrm{sinc}\left(\frac{\Delta k(\omega,\Omega-\omega)L}{2}\right) \mathrm{d}\omega \right|^2 \mathrm{d}x\,. \end{aligned} \tag{17}$$

In (17), $Q(\Omega)$ is the spectral sensitivity of the photodetector, and the transverse beam profile is assumed Gaussian with the FWHM diameter of d_0.

We have limited our discussion to the low-efficiency SHG, i.e., when the depletion of the fundamental waves can be disregarded – and where useful pulse measurements must be performed. In the high conversion efficiency regime, additional effects play an important role, distorting the trace in both type I [57] and type II crystals [58, 59].

Another important example of undesirable distortions in the high-conversion-efficiency regime is nonlinear (two-photon) absorption of the frequency-doubled radiation inside a SHG crystal [60]. Fortunately, at the minuscule crystal thicknesses necessary for measuring such short pulses, these effects are not usually a problem.

To conclude this section, we wish to emphasize the many advantages of the frequency-domain – vs. the time-domain – approach to the wave equation (1) in the single-cycle regime. The frequency representation allows us *automatically* to include dispersive broadening of both fundamental and second-harmonic pulses, as well as their group-velocity mismatch, frequency-dependence of the nonlinear susceptibility, frequency-dependent spatial profiles of the beams, and the blueshift of the second-harmonic spectrum (the analog of self-steepening in fibers [61]). Furthermore, we have had to make only a single approximation (5), which itself is easily avoidable in computer simulations. A similar frequency-domain approach to ultrashort-pulse propagation in optical fibers [62] helped solve a long-standing question of the magnitude of the shock term [61, 63].

2.1.2 Ultimate Temporal Resolution of SHG FROG

For arbitrary pulses, the complete expression for the SHG FROG signal (17) must be computed numerically. However, for linearly chirped Gaussian pulses, (17) can be evaluated analytically.

With such a result, we can study such effects as geometric delay smearing, as discussed in [18]. Recall that this distortion occurs because, with crossed beams, the relative delay varies across the beam. This transverse geometric effect can be quite advantageous: it allows single-shot pulse measurement by mapping the delay onto the transverse position [64]. In addition, in [18], it is shown that a single-shot SHG FROG measurement precisely removes *all* geometric smearing from the FROG trace – even when a pulse has near-single-cycle length. However, as this is not yet well appreciated and multishot measurements of such pulses remain common, we consider here the multishot case, where geometric smearing could in principle be an issue.

As can be seen from (17), the dependence on the transverse coordinate x yields a range of delays across the beam simultaneously, which broadens the FROG trace along its delay axis. As in *Taft* et al. [11], we assume Gaussian-intensity pulses and, under perfect phase-matching conditions, we find that geometric smearing yields a SHG FROG trace that is temporally wider that it should be. We find that the measured pulse duration τ_{meas} is

$$\tau_{\text{meas}}^2 = \tau_{\text{p}}^2 + \Delta\tau^2 , \tag{18}$$

where τ_{p} is the true pulsewidth and $\Delta\tau$ is the delay smear parameter

$$\Delta\tau = \frac{\theta d}{2c} , \tag{19}$$

where d is the beam diameter in the crystal and θ the intersection angle of the fundamental beams.

We consider two input Gaussian beams separated by their input diameter d_{i} on the focusing optic. In this case, the intersection angle $\theta = d_{\text{i}}/f$, and the beam diameter in the focal plane $d = f\lambda/\pi d_{\text{i}}$, where f is the focal length of the focusing optic. Thus, at $\lambda = 800\,\text{nm}$, the resulting delay smearing amounts only to $\Delta\tau = \lambda/2\pi c = 0.4\,\text{fs}$. This value represents the ultimate resolution of a pulse measurement in noncollinear geometry, and it is quite good, allowing accurate measurement of even a single-cycle pulse. Furthermore, this result is independent of the focusing optic, angle, and beam diameters. The temporal resolution deteriorates, however, if the beams are other than Gaussian: for beams with a rectangular spatial profile, the resulting temporal resolution worsens to 0.7 fs. But this still represents minimal distortion, even in a single-cycle pulse measurement. And shorter pulses will have a shorter wavelength, reducing this effect to a correspondingly small value.

Even if a measurement is poorly designed, the temporal resolution can be improved by filtering out signal light of the wrong delay. This can be achieved simply with a narrow slit behind the nonlinear medium, or, better, by imaging the nonlinear medium onto a slit (like the entrance slit of the spectrometer in a FROG), as discussed in [18, 65].

It is also possible to use a collinear geometry to remove any geometric smearing effects [66, 67, 68, 69]. Or even better, as mentioned earlier, a single-shot SHG FROG geometry avoids this effect entirely. Thus, only a truly poorly designed FROG apparatus will suffer from geometric distortions.

2.1.3 Approximate Expression for the SHG FROG Signal

In this section, we obtain a simplified expression for SHG FROG, accurate even for single-cycle pulses. We show that, remarkably, the measured signal can be simply described by an ideal, i.e., perfectly phase-matched, SHG FROG trace multiplied by a spectral filter. In other words, the measured SHG FROG trace can simply be divided by this spectral filter, and the standard SHG FROG algorithm will yield the pulse field.

To simplify (17), we make several reasonable approximations. First, as shown in the previous section, using a carefully chosen beam geometry (with a small crossing angle), the effect of geometric smearing is negligibly small. For instance, it causes only a 10% error in the duration measurement of a 3 fs pulse and hence can be safely neglected for all pulses that have been created at this time. With such an approximation, the integral along x in (17) can be performed analytically. Second, we expand $k_o(\omega)$ and $k_o(\Omega-\omega)$ into a Taylor series around $\omega = \Omega/2$ and keep the terms that are linear with frequency.[1] Hence, for type I phase matching, we write:

$$\begin{aligned}\Delta k\left(\omega, \Omega-\omega\right) &\approx 2k_o\left(\frac{\Omega}{2}\right)\cos\left[\frac{\theta}{2}n_O\left(\frac{\Omega}{2}\right)\right] - k_e(\Omega) \\ &= \Delta k\left(\frac{\Omega}{2}, \frac{\Omega}{2}\right). \end{aligned} \tag{20}$$

Third, we estimate the dispersion of the second harmonic and calculate the second-order susceptibility $\tilde{\chi}^{\langle 2\rangle}(\Omega, \omega, \Omega-\omega)$ using the dispersion of the refractive index. For a classical anharmonic oscillator model [54] $\tilde{\chi}^{\langle 2\rangle}(\Omega, \omega, \Omega-\omega) \propto \tilde{\chi}^{\langle 1\rangle}(\Omega)\tilde{\chi}^{\langle 1\rangle}(\omega)\tilde{\chi}^{\langle 1\rangle}(\Omega-\omega)$, where the first-order susceptibility $\tilde{\chi}^{\langle 1\rangle}(\Omega) = n^2(\Omega) - 1$. Equation (17) can now be decomposed into a product of the spectral filter $R(\Omega)$, which originates from the finite conversion bandwidth of the second-harmonic crystal and varying detector sensitivity, and an ideal FROG signal $I_{\mathrm{FROG}}^{\mathrm{SHG}}(\Omega, \tau)$:

$$I_{\mathrm{meas}}(\Omega, \tau, L) \propto R(\Omega)\, I_{\mathrm{FROG}}^{\mathrm{SHG}}(\Omega, \tau) \tag{21}$$

[1] Alternatively, one can perform a Taylor expansion around the central frequency of the fundamental pulse $\omega = \omega_0$ [24, 49, 70]. However, in this case, the first derivative terms do not cancel each other and must be retained. Our simulations also prove that the expansion around $\omega = \Omega/2$ provides a better approximation for broadband pulses

where

$$I_{\mathrm{FROG}}^{\mathrm{SHG}}(\Omega,\tau)=\left|\int \widetilde{\boldsymbol{E}}(\Omega-\omega)\widetilde{\boldsymbol{E}}(\omega)\exp\left(\mathrm{i}\omega\tau\right)\mathrm{d}\omega\right|^2 \tag{22}$$

and

$$R(\Omega)=Q(\Omega)\frac{\Omega^2}{n_\mathrm{e}(\Omega)}\left\{\left[n_\mathrm{e}^2(\Omega)-1\right]\left[n_\mathrm{o}^2(\Omega/2)-1\right]^2\right\}^2 \cdot \mathrm{sinc}^2\left(\frac{\Delta k(\Omega/2,\Omega/2)L}{2}\right). \tag{23}$$

In (21–23), we have kept only the terms that are Ω-dependent.

The FROG signal given by (22) is the well-known definition of SHG FROG [16,18,20,42] written in the frequency domain [18]. The same formula is also used in existing FROG retrieval algorithms. Note that, alternatively, the complete (17) can be readily implemented in the algorithm based on the method of generalized projections [71]. However, relation (21) is better numerically because the integral (22) takes the form of an autoconvolution in the time domain and can be rapidly computed via a fast Fourier transform [72]. It is also important that the use of (21) permits a direct check of FROG marginals to validate experimental data [18].

The spectral filter $R(\Omega)$, given by (23), is a product of several factors (Fig. 3). The Ω^2 dependence follows directly from Maxwell's equations. The meaning of this factor is that higher frequency components are more efficiently generated than lower frequency ones. It has been shown that in some cases, it is appropriate to include an additional Ω-dependence that accounts for the difference in the efficiency of generation of various SH components as a result of spatial overlap of the fundamental spectral components [51,73]. In other words, this extra Ω-term originates from a frequency-dependent beam waist as a consequence of pulse propagation and focusing on the SH crystal. The combined dependence of the FROG signal on the signal frequency in such a case is Ω^3. This situation is applicable, for instance, for relatively short (several millimeter or less) pieces of a single-mode fiber [51,73]. In other cases, such as the output of a hollow fiber [70, 74] or of a Kerr-lens mode-locked laser [75], where the distribution of the beam waist size of the fundamental beam in the SH crystal does not change appreciably with wavelength, the additional Ω-term can be disregarded. To bring our theoretical considerations into accordance with the experimental results of the fiber-compressed-pulse FROG characterization, which will be presented later, we include the effect of the spatial overlap in our numerical calculations and use the combined Ω^3-dependence. As can be seen from Fig. 3, the Ω^3-term (dotted curve) leads to substantial distortion of the second-harmonic spectrum of ultrabroadband pulses. For instance, due to this factor alone, the upconversion efficiency of a spectral component at 600 nm is 4.5 times higher than that at 1000 nm.

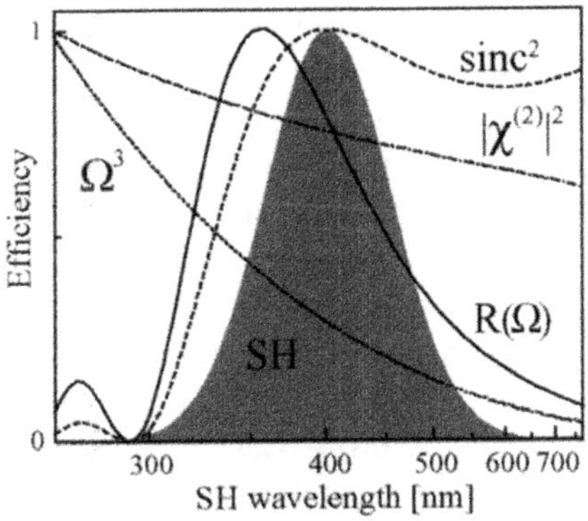

Fig. 3. Constituent terms of the spectral filter $R(\Omega)$, given by (23): the Ω^3-dependence (*dotted line*), estimated squared magnitude of second-order susceptibility $\chi^{\langle 2 \rangle}$ (*dash-dotted line*), the crystal phase-matching curve for a type I 10 µm BBO crystal cut at $\Theta = 29°$ (*dashed line*), and their product (*solid curve*). The second-harmonic spectrum of a 3 fs Gaussian pulse is shown for comparison (*shaded contour*)

The variation of the second-order susceptibility with frequency (dash-dotted curve), expressed in (23) as the dependence on the refractive indexex, plays a much smaller role than the Ω^3-factor (dotted curve). For example, for a BBO crystal, the squared magnitude of $\tilde{\chi}^{\langle 2 \rangle}$ for the 600 nm component of the fundamental wave is only 1.3 times larger than that for the 1000 nm component. Such a virtually flat second-order response over such an immense bandwidth is a good indication of the almost instantaneous nature of $\tilde{\chi}^{\langle 2 \rangle}$ in transparent crystals. Nonetheless, the contribution of the $\tilde{\chi}^{\langle 2 \rangle}$-dispersion would affect the measurement of pulses with spectra that are hundreds of nanometers wide.

The last factor contributing to $R(\Omega)$ is the phase-matching curve of the SHG crystal (Fig. 3, dashed line). The shape and the bandwidth of this curve depend on the thickness, orientation, and type of crystal.

2.2 SHG FROG Characterization of Fiber-Compressed Pulses

In this section, we put to work experimentally the various ideas we have discussed regarding SHG FROG pulse characterization. We demonstrate the practical application of SHG FROG to strongly chirped ultrabroadband pulses and compressed 4.5 fs pulses from a cavity-dumped-laser-based white-light generator. Several practical implications for dealing with these extraordinarily broad bandwidths will be considered here. The basic experimental requirements for a broadband SHG FROG apparatus are adequate phase-matching bandwidth of the SHG crystal and low overall dispersion of the optical elements. The device should also be able to yield two replicas of the pulse that do not differ from each other in their spectral content or phase.

In particular, the practical details of the SHG crystal in a FROG measurement are important. One reason is the usual SHG crystal conflict: a thick crystal is more efficient, but a thin crystal has a greater phase-matching bandwidth. Another issue is the angular orientation of the crystal or the wavelength for which it is cut. The difficulty here comes from the fact that a "red" crystal cut (a crystal cut for preferentially phase matching the red wavelengths of the pulse) typically provides nearly flat frequency conversion efficiency over most of the bandwidth of an ultrabroadband pulse. Such an orientation, however, dispenses with the blueshifted wing of the spectrum where the conversion efficiency falls dramatically. On the other hand, use of a "blue" crystal-cut significantly lowers the frequency-conversion efficiency in the red wing of the spectrum. To find a reasonable balance that satisfies the demands of various pulses, we develop a useful criterion for crystal selection in this section. Finally, we compare the merits of the two most commonly used SHG crystals for FROG measurements, BBO and KDP.

Next, we focus our attention on details of the FROG apparatus and the peculiarities of the measurement by this technique of strongly chirped and nearly fully compressed laser pulses. The spectral phase of a white-light pulse measured before and after a pulse compressor permits good verification of the ray-tracing routine employed to design it. We then present valuable observations on how extra information about the level of pulse compression can be gained from a simple examination of the SHG FROG trace, which is normally considered quite unintuitive.

2.2.1 The Choice of the SHG Crystal

In this section, we provide several guidelines for selecting the correct SHG crystal for a FROG measurement. We use a simple criterion to determine the required crystal thickness: the conversion efficiency calculated according to (23) must be $> 50\%$ of the peak conversion efficiency over the entire FWHM of the FROG spectral marginal. For pulses that are Gaussian in frequency, the ideal spectral marginal (the autoconvolution of the fundamental spectrum) is $\sqrt{2}$ times broader than the pulse bandwidth. Using this criterion, we evaluated BBO and KDP crystals. For each crystal, we consider two different angle cuts one for type I phase matching at the wavelength of 800 nm and the other for 600 nm. Figure 4 depicts the appropriate crystal thickness of BBO (solid curve) and KDP (dashed curve) as a function of the duration of a bandwidth-limited Gaussian pulse.

As can be seen from Fig. 4, an approximately 10 μm BBO should be used to measure 5 fs pulses at 800 nm. The KDP crystal is approximately 2.5 times thicker due to its lower dispersion. However, while clearly providing an advantage in thickness, the KDP crystal has a disadvantage in SHG efficiency. The signal level that can be obtained with a BBO crystal 2.5 times thinner is still approximately a factor of 6 larger than from KDP due to the higher nonlinear coefficient and lower phase-matching angle in the BBO crystal [76].

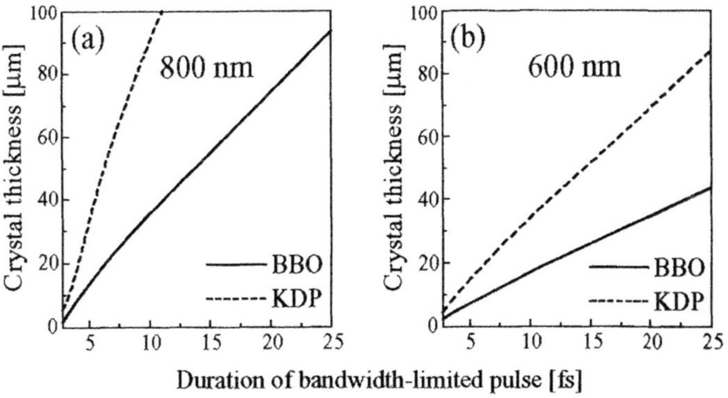

Fig. 4. Type I crystal thickness required for SHG FROG measurement as a function of the pulse duration at the central wavelength of 800 nm (**a**) and 600 nm (**b**). The crystals are cut at $\Theta = 29°$ for BBO (*solid line*) and $\Theta = 44°$ for KDP (*dashed line*)

Therefore, BBO is more suitable for characterization of weak-intensity pulses. For high-intensity pulses, where the low level of the second-harmonic signal is not really the issue, KDP presents a better choice [70].

Once the phase-matching bandwidth is sufficiently flat, the Ω^3-dependence (23) begins to dominate the conversion efficiency. This dependence blueshifts the second-harmonic spectrum. If the phase-matching bandwidth of the SHG crystal exceeds that required by the pulse bandwidth, angle-tuning the crystal to the red can effectively counteract such a blueshift [70]. To illustrate the point, we consider a 10 μm BBO crystal for measuring 8 fs Gaussian pulses at 800 nm. Figure 5a shows the blueshift of the FROG spectral marginal (filled circles) with respect to the autoconvolution (solid curve), if the crystal is perfectly phase-matched at 800 nm, i.e., $\Theta = 29°$. However, after adjusting the phase-matching angle to $\Theta = 24.4°$ that now corresponds to the central wavelength of 970 nm (Fig. 5b), the phase-matching curve of the crystal (dashed curve) nearly perfectly balances the Ω^3-dependence (dotted curve). The overall conversion efficiency becomes almost flat, and no spectral correction of the FROG trace is required. Experimentally, *Taft* et al. [11] demonstrated the effectiveness of the angular adjustment that enabled them to obtain correct FROG data.

The mutual compensation of the Ω^3- and phase-matching terms is only possible for relatively long ($\sim$ 10 fs) pulses. As a thinner crystal is chosen to measure shorter pulses, the high-frequency slope of the phase-matching curve grows steeper than the low-frequency one (Fig. 5c,d). This is to be expected, since crystal dispersion is low in the infrared but is high in the UV, near the UV absorption band. Tuning the central wavelength of the crystal from 800 nm (Fig. 5c) to 970 nm (Fig. 5d) substantially narrows the SH spectrum in the blue due to crystal phase matching. It also becomes difficult to correct the FROG trace for the spectral filter because the conversion efficiency falls

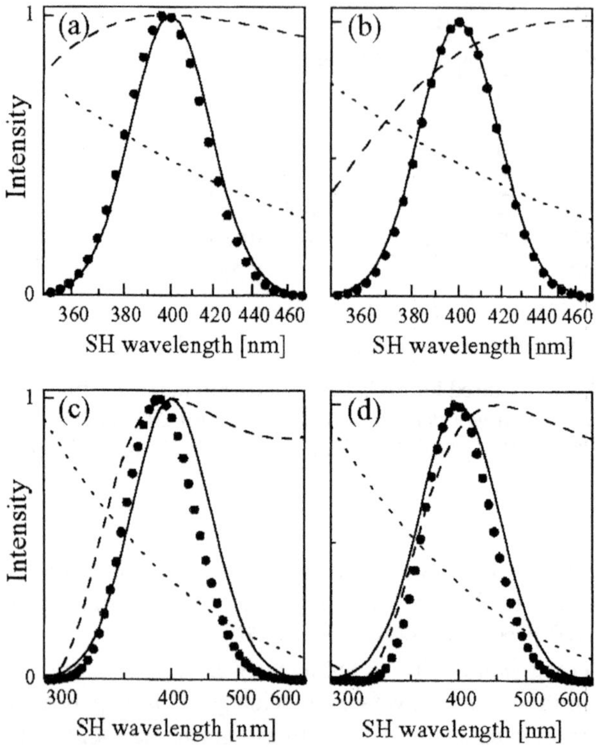

Fig. 5. Correction of frequency-conversion efficiency by crystal orientation for 8 fs (**a**,**b**) and 3 fs (**c**,**d**) bandwidth-limited Gaussian pulses. A type I 10 μm BBO crystal is oriented for the phase-matched wavelength of 800 nm (**a**,**c**) and 970 nm (**b**,**d**). The phase-matching curve and the Ω^3-dependence are shown as *dashed* and *dotted lines*, respectively. The *solid curves* depict the autoconvolution of the fundamental spectra, and the spectral marginals of FROG traces are given by *filled circles*. In (**b**), no spectral correction of the FROG trace is required for an 8 fs pulse because of the redshifted, phase-matched wavelength. In contrast, the use of the 970 nm phase-matched crystal irreparably corrupts the second-harmonic spectrum in a shorter 3 fs pulse (**d**). Note different horizontal scales in (**a**), (**b**) and (**c**), (**d**)

so low in the blue wing (Fig. 5d). This is in contrast to the 800 nm-cut case when correction is still possible (Fig. 6). Thus, to extend the phase-matching bandwidth in the blue, one should consider using a crystal with the phase-matching wavelength blueshifted with respect to the central frequency of the pulse. For example, a $L = 10$ μm BBO crystal oriented for peak conversion efficiency at 700 nm is more suitable for the measurement of sub-5 fs pulses centered at 800 nm than the same crystal tuned to 970 nm. Although the 700 nm-cut crystal has poorer conversion efficiency in the infrared, nonetheless, it allows the extension to phase matching below 600 nm. Consequently, this crystal has an appreciable SHG efficiency over the entire spectrum of

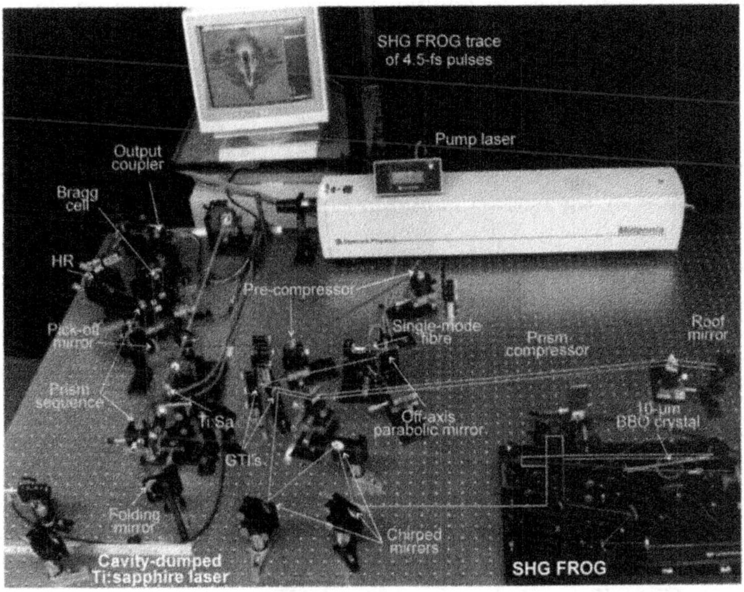

Fig. 6. The measurement of 4.5 fs pulses by SHG FROG (photo: Foppe de Haan)

a 5 fs pulse and, therefore, spectral correction can yield an adequate FROG trace. In contrast, information about the blue spectral wing is lost if a crystal cut for 970 nm is used.

These considerations show that measurement of few-cycle pulses is readily accomplished using thin standard crystals, but it should be pointed out that FROG measurements of such pulses can also be performed using more recently developed methods that involve thicker crystals and hence are more sensitive. These methods either remove the phase-matching bandwidth consideration from the problem or take advantage of it. These methods include angle-dithering the crystal (which can also be used in a single-shot beam geometry as long as the measurement is made over many pulses), focusing tightly into the crystal, and using the crystal phase-matching dispersion to act as the spectrometer [18]. These methods are sufficiently new, however, that they have not yet been used for the measurement of few-cycle pulses.

2.2.2 SHG FROG Apparatus

In our experiments, we used pulses from a self-mode-locked, cavity-dumped Ti:sapphire oscillator, which were then compressed after undergoing self-phase modulation in a single-mode, fused, silica fiber. We measured the amplitude and phase of the white-light continuum (WLC) directly at the fiber output and, again, upon their compression [13, 14].

Figure 6 shows a photograph of the all-solid-state sub-5 fs laser and the SHG FROG device. Unlike a conventional Ti:sapphire oscillator, this cavity-dumped version incorporates an additional mirror fold around an acousto-optic modulator. In this way, the intracavity pulse energy is stored in a relatively high-Q cavity, which can be switched out of the resonator at any desired repetition rate. The maximal pulse energy of a cavity-dumped Ti:sapphire laser is typically a factor of 10 higher than that of its non-cavity-dumped counterpart. Careful cavity design ensures that the Kerr-lens self-mode locking is not disturbed by the extra fold and by the added dispersion due to the Bragg cell [77].

With $\sim$ 4 W of pump power (Spectra Physics Millennia), the cavity-dumped laser routinely produced $\sim$13 fs, 40 nJ pulses at a 1 MHz repetition rate. Precompressed, these pulses had $\sim$ 75 nm spectral bandwidth around 790 nm. They were launched into an $\sim$ 2 mm long piece of a single-mode polarization-preserving fiber (Newport, 2.75 μm core diameter) through an 18/0.35 microscope objective. The fiber output was collected by an off-axis paraboloidal mirror, which ensured achromatic beam recollimation. The optimal pulse energy for injection into the fiber was $\sim$35 nJ, as judged by the quality of the generated continuum. The continuum pulse energy measured was about 18 nJ after recollimation. The continuum was compressed by a prism-chirped-mirror Gires–Tournois-interferometer (GTI) pulse compressor. The overall size of the sub-5 fs laser system was 1 m $\times$ 1.5 m, which made it extremely robust and ensures excellent stability over time.

The SHG FROG apparatus (Fig. 7) was based on a phase and amplitude balanced multishot autocorrelator designed for sub-5 fs short pulses [13, 14]. The input beam was split and recombined so that each of the beams traveled once through an identical 50% beam splitter with both reflections occurring on the same coating–air interfaces.[2] To match the beam splitters, the initial horizontal polarization of the laser beam was rotated by a periscope. The moving arm of the autocorrelator was driven by a piezo transducer (Physik Instrumente), which was controlled by a computer via a digital–analog converter and a high-voltage amplifier. Because such short delay increments ($\sim$ 0.1 fs) were necessary, the precise time calibration was provided by an auxiliary Michelson interferometer. The photodiode monitored the interference fringes that served as time calibration marks.

Fundamental pulses were focused in the nonlinear crystal with a spherical mirror (radius of curvature is -25 fs at near-normal incidence to minimize astigmatism. Due to the low curvature of the mirrors, delay variations within each beam were less than 0.1 fs. To achieve upconversion of the entire fundamental bandwidth, we used a 10 μm thick BBO crystal cut for a central wavelength of 700 nm (EKSMA Inc.). The dispersive lengthening of a 5 fs pulse by such a crystal is less than 0.02 fs. The blueshifted central wavelength permitted us to extend the phase-matching bandwidth below 600 nm, as shown in Fig. 4c. The crystal cut angle was verified with a tunable 100 fs

[2] For shorter pulses, one should use lower reflectivity beam splitters that have a broader reflectivity range and a flatter spectral phase

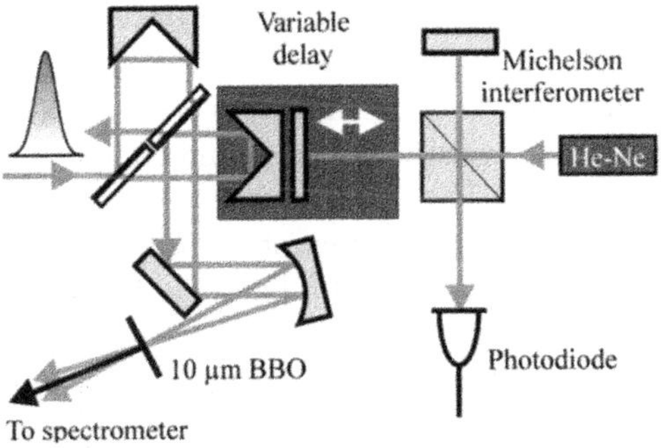

Fig. 7. Schematic of the SHG FROG apparatus. The spectrometer and its coupling optics are not shown

laser. Retro-reflection of the beams from the crystal surface provided an exact reference for the crystal orientation. This enabled us to calculate accurately $R(\Omega)$ required for data correction, according to (23). A visible-IR PC1000 (Ocean Optics) spectrometer was used to detect the fundamental spectra.

We used two different second-harmonic detection systems in these measurements of compressed and uncompressed pulses. For the compressed pulses, we used a well-characterized UV-vis PC1000 (Ocean Optics) spectrometer. Thus, the FROG traces could be readily corrected by $R(\Omega)$, as described above.

For strongly chirped pulses, a combination of a scanning monochromator and a photomultiplier tube provided the dynamic range necessary to measure the spectral wings (see the next section). The reason for this was that the dynamic range of the measurement in a CCD-based spectrometer is determined not only by the spectral sensitivity, which is adequately high, but also by the charge spreading all over the array due to overload of some channels. To further extend the dynamic range, a lock-in amplifier was used to detect the second-harmonic signal. Because of the unknown spectral sensitivity $Q(\Omega)$, spectral correction of the FROG traces in this case was performed according to the method of *Taft* et al. [11], i.e., by using the ratio of the autoconvoluted fundamental spectrum and the spectral marginal [18].

2.2.3 SHG FROG of a White-Light Continuum

Knowledge of the group delay, i.e., the spectral phase, of a chirped WLC is the key to its compression. The spectral-phase measurement of the pulses leaving the fiber allows us to assess the feasibility of pulse compression in general. The spectral phase must be sufficiently smooth to allow compensation by existing dispersion-control methods. A measurement of the spectral intensity, on the other hand, provides only a limited insight and reveals the minimum duration of the would-be compressed pulse. As an example of virtually uncompressible pulses, one might consider spectral broadening due to pure self-phase modulation. Furthermore, the task of building an appropriate pulse compressor is substantially simplified if the phase distortion of the pulse is measured beforehand. This becomes increasingly important with the growth of the pulse spectral bandwidth, which puts severe limitations on the tunability of the pulse compressor. Therefore, it is desirable to replace most of the "trial and error" work by measuring the phase distortion and computing the settings of the pulse compressor.

Somewhat counterintuitively, the FROG measurement of a strongly chirped pulse is considerably more complicated than that of a bandwidth-limited pulse with an identical spectrum. First, the upconversion signals are weaker due to lower peak power. This is evident, since the second-harmonic intensity of a pulse that is stretched to 10 times its initial duration drops by a factor of 100.

Second, a higher dynamic range is required due to the pulse structure in the spectral wings. This occurs due to high-order material dispersion. To explain this, we consider two spectral components with frequencies separated by $1000\,\mathrm{cm}^{-1}$. The group delay accumulated between them after passing 1 mm, if these components are situated around 1000 nm, exceeds 11 fs in the case of 600 nm. So, roughly speaking, the corresponding elements of the FROG trace scale approx. seven times in intensity. In our experiments, the bandwidth of the WLC that needs to be captured in the FROG trace is broader than $10\,000\,\mathrm{cm}^{-1}$, and, therefore, the signal intensity varies very strongly across the resultant FROG traces.

The third complication is purely numerical, since FROG inversion demands greater matrix sizes to provide adequate sampling in both the time and frequency domains. For the sake of speed, the FROG inversion algorithms employ fast Fourier transform (FFT) routines [72]. To avoid the loss of information in the change from the time to the frequency domain and vice versa, the FFT requires an equal number of points N in both domains. Therefore, if the FROG matrix covers the total delay of $N\delta\tau$ in the time domain, where $\delta\tau$ is the delay step, the spectral range of the trace is $1/\delta\tau$. Compared with bandwidth-limited pulses, the pulses stretched in time require a larger $\delta\tau$ to contain the whole time information of the FROG trace in the matrix used in the FROG inversion algorithm. This narrows the spectral window covered by the matrix, if N remains constant. Consequently, the number of points N, which should be a power of 2 in a FFT, must be increased to represent the FROG trace fully in the matrix used by the algorithm. This has an appreciable effect on calculation speed. The change of N from 2^n to 2^{n+1}, where n is an integral number, slows the FROG retrieval by a factor of $4(1+n^{-1})$. In other words, replacing a 128×128 matrix with a 256×256 matrix, increases the calculation time by a factor of ~ 4.5. However, FROG traces as large as 8192×8192 have been used (to measure a considerably broader continuum than discussed here) with personal computers with few complications.

The SHG FROG traces of the chirped WLC in our experiments were recorded in 2.5 fs delay steps and converted into 256×256 matrices for processing. To reveal the conditions best suited for compression of the WLC, we varied the parameters of the pulses entering the fiber by changing the settings of the prism precompressor. The intensity and the chirp of the input pulses, derived by SHG FROG, are shown in Fig. 8a. The measured and retrieved FROG traces of the WLC are depicted in Fig. 8b,c, and the retrieved WLC spectra and group delay are shown in Fig. 8d. The combined action of self-phase modulation and dispersion leads to a nearly linear group delay over most of the spectrum (Fig. 8b, solid curves).

The departure of the overall group delay from a linear asymptote can be partly explained by the bulk dispersion of the fiber, air, and beam splitters in the FROG apparatus. For instance, although the optimal fiber length was estimated at 1 mm [12], we used a longer piece for practical convenience and to improve the exiting mode structure.

The WLC spectrum changes dramatically with change in the input pulses (Fig. 8a, shaded contours). The widest and least modulated spectrum corresponds to an almost chirp-free input pulse (Fig. 8, the third from the top panel). Positive as well as negative chirping leads to substantial narrowing of the WLC spectrum. In contrast, the overall behavior of the group delays shown as solid lines in Fig. 8d, remains virtually unaffected. This ensures efficient pulse compression under different experimental conditions.

Group delay measurements of the generated continuum served as a target function for the design of the three-stage, high-throughput compressor (Fig. 6), consisting of a quartz 45° prism pair, broadband chirped mirrors, and thin-film Gires–Tournois dielectric interferometers [13, 14]. The spectral bandwidth of the compressor was 590 nm to 1100 nm, which was limited by the reflectivity of the chirped mirrors involved [78]. The phase characteristics of the compressor were analyzed using dispersive ray tracing and were mapped onto the measured group delay of the continuum. Figure 9 depicts the measured group delay for different pulses, entering the fiber (shown as broken curves), which were reproduced from Fig. 8d and the calculated group delay of the pulse compressor (solid line). As you can see, our design compensates for the group delay of the white light everywhere across the compressor bandwidth. Adjusting the length of material of the prism pair (by translating a prism into or out of the beam) allows final fine-tuning of the compressor dispersion, as judged from the FROG trace of the compressed pulses.

2.2.4 SHG FROG of Compressed Pulses

FROG traces compressed pulses were recorded by scanning the delay between the arms in steps of 0.5 fs. The two-dimensional arrays of points acquired were converted into a 128×128 FROG matrix. The experimental and retrieved FROG traces of compressed pulses are depicted in Fig. 10a,b. The FROG error amounted to 0.004 and is caused mainly by the noise in the spectral wings, which increased when the spectral correction of the FROG trace was performed.

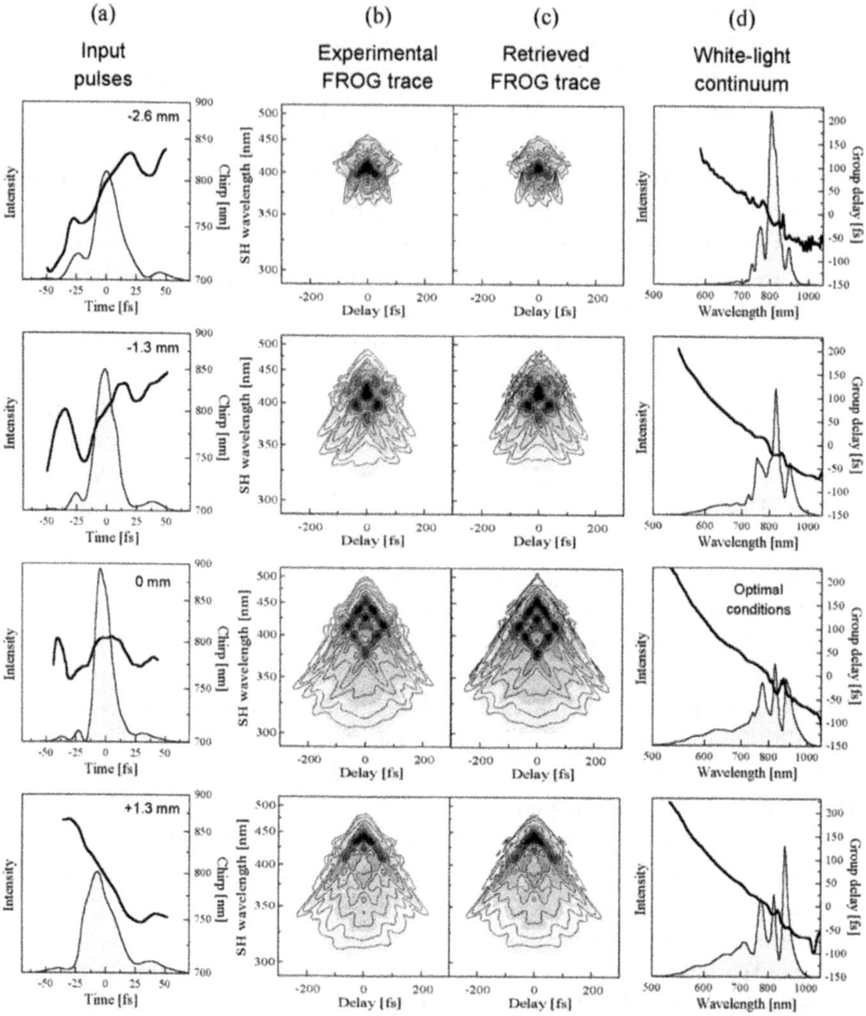

Fig. 8. Experimental results of FROG measurements of the strongly chirped white-light continuum (WLC). (**a**) Temporal intensity (*shaded contours*) and chirp (*solid curves*) of the pulses entering a single-mode fused-silica fiber. (**b**) Measured and (**c**) retrieved SHG FROG traces of the WLC. (**d**) Retrieved spectral intensity (*shaded contours*) and the group delay of the WLC (*solid curves*). The amount of bulk material (fused silica) used to prechirp the input pulses is indicated in the right top corners of (**a**). Note that the input pulse energy is kept constant, and the respective scaling of the WLC spectra in (**d**) is preserved

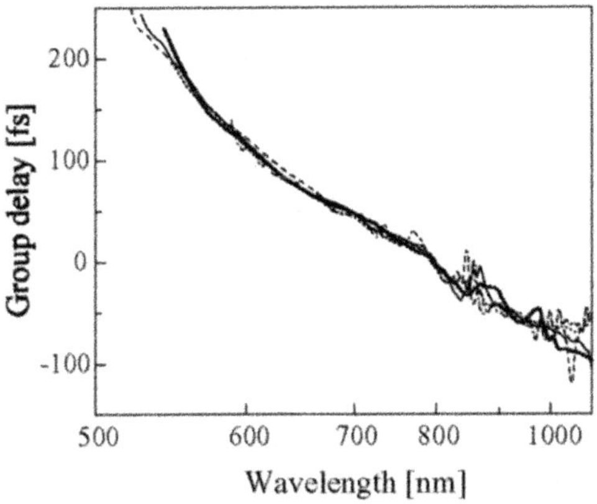

Fig. 9. Group delay of the designed pulse compressor. The *solid curve* was calculated by dispersive ray tracing and is depicted reversed in time. *Broken curves* are the measured group delay of the WLC reproduced from panels in Fig. 8d

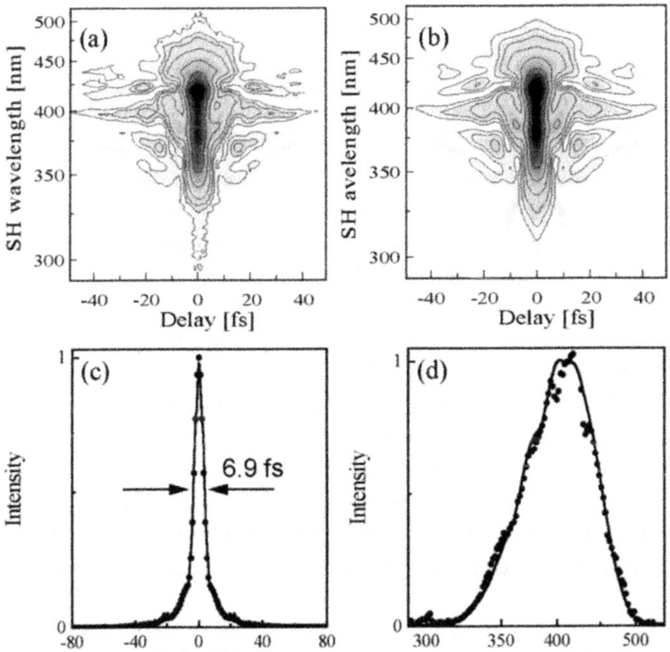

Fig. 10. The results of SHG FROG characterization of compressed pulses. (**a**) Experimental and (**b**) retrieved traces. (**c**) Temporal marginal (*filled circles*) and independently measured autocorrelation of 4.5 fs pulses (*solid curve*). (**d**) Frequency marginal (*filled circles*) and autoconvolution of the fundamental spectrum (*solid curve*)

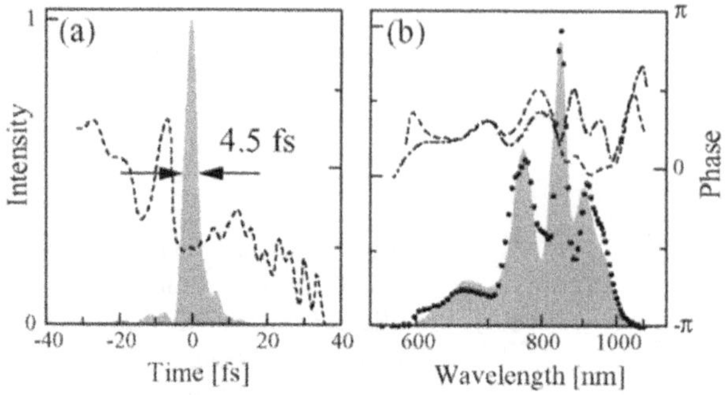

Fig. 11. Retrieved intensity and phase of 4.5 fs pulses in the time (**a**) and frequency (**b**) domains. The FROG-retrieved intensity and phase are shown as *shaded contours* and *dashed curves*, respectively. The independently measured spectrum (*filled circles*) and computed residual phase of the pulse compressor (*dash-dotted curve*) are given in (**b**) for comparison

The delay marginal of the measured FROG trace corresponds nicely to the independently measured intensity autocorrelation (Fig. 10c) obtained by detecting the unspectrally-resolved second-harmonic signal beam. This suggests that no spatial filtering of the second-harmonic beam has taken place. Comparison of the FROG frequency marginal and the autoconvolution of the fundamental spectrum (Fig. 10d) indicates that no loss of spectral information has occurred and that the spectral-filter correction was correct.

Figure 11 shows the retrieved intensity and phase in the time and frequency domains. To remove the time direction ambiguity in the measurement of the compressed pulses, we performed an additional FROG measurement introducing a known amount of dispersion (a thin fused silica plate) in front of the FROG apparatus. The resulting pulse duration was 4.5 fs. Variations of the spectral phase (dashed line in Fig. 11b) were less than $\pm\pi/4$ across the whole bandwidth. These results fully confirm our previous analysis using interferometric autocorrelation (which could not yield the pulseshape or phase) [13, 14].

To verify additionally both the self-consistency of our compressor calculations and the accuracy of the FROG retrieval, we compare the retrieved spectral phase of the 4.5 fs pulse (Fig. 11b, dashed curve) with the predicted residual phase of the pulse compressor (Fig. 11b, dash-dotted curve). The close similarity of the two reassures us of the correctness of all the procedures we used, including the measurement of the chirped WLC, knowledge of the dispersion of compressor constituent parts, the numerical routines employed for the ray tracing analysis, and, finally, the characterization of the compressed pulses.

SHG FROG traces are often considered unintuitive due to their symmetry along the delay axis [20, 22, 42]. We found that for nearly bandwidth-limited pulses, we can significantly increase the amount of information available from

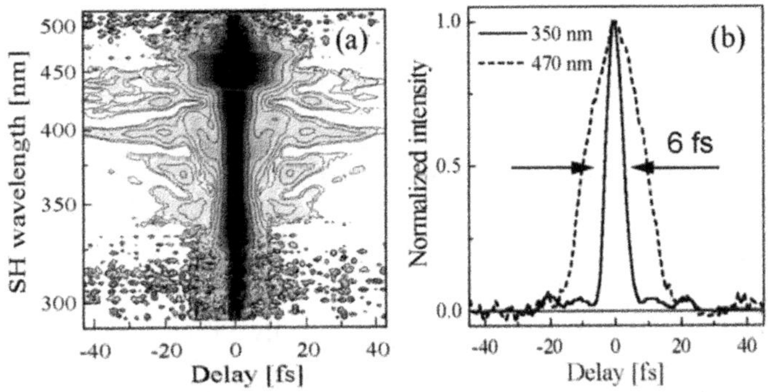

Fig. 12. Normalized FROG data of the 4.5 fs pulses. (**a**) SHG FROG trace of compressed pulses normalized along the delay axis. (**b**) Autocorrelation traces derived from the FROG trace at the second-harmonic wavelength of 350 nm (*solid*) and 470 nm (*dashed*). Note that because of spectral selection, the pulse length estimated from the autocorrelation width can be either lower or higher than actual and differ by as much as a factor of three

mere visual inspection of the trace. To do so, consider every constant-frequency slice of the trace (each of which is a function of delay only). Now normalize each slice to unity at its peak. Effectively, the SHG FROG trace has become a series of normalized autocorrelations. For a pulse with an arbitrary spectrum and flat spectral phase, such a representation of the SHG FROG trace would give a streak of uniform thickness around zero delay. The result of such an operation applied to the FROG trace of the 4.5 fs pulse is presented in Fig. 12a. The variation in thickness, that is, the width of autocorrelation at a given second-harmonic wavelength,[3] which can be seen in Fig. 12a, visually indicates any imperfect pulse compression without the need to run the FROG inversion algorithm.

Figure 12b shows two autocorrelation traces derived from the spectrogram in Fig. 12a at two separate wavelengths. The FWHM of the autocorrelation at 350 nm is merely 6 fs, indicative of a $\sim$ 4 fs pulse duration. However, the autocorrelation at 470 nm is three times broader. Such a difference clearly illustrates the effect of spectral filtering in the nonlinear crystal, as well as the detection of the autocorrelation width. This also underscores the importance of pulse characterization by frequency-resolved (e.g., FROG) rather than non-frequency-resolved (e.g., intensity autocorrelation) methods for such broadband pulses.

Finally, we note that the width of the autocorrelation traces, such as those shown in Fig. 12a, can be directly related to the instrument response in a spectroscopic experiment. For instance, the temporal resolution of a kinetic trace

[3] Here we use the term "autocorrelation" for a slice of a frequency-resolved autocorrelation purely for convenience. Such a slice is not necessarily an autocorrelation function of any real nonnegative distribution

in a frequency-resolved pump–probe experiment [79, 80] detected at 950 nm will be $\sim$ 12 fs, albeit the weighted average pulse duration is 4.5 fs [81, 82]. Therefore, the frequency-resolved measurement (as FROG) produces invaluable information, even if the correct estimation of the pulsewidth could be achieved by other, simpler means, such as autocorrelation measurement.

3 Conclusions

FROG is a powerful and accurate pulse diagnostic, ideally suited for the measurement of a vast variety of pulses. In particular, the convenience, accuracy, reliability, high sensitivity, and broadband response allow the measurement of the shortest pulses available to date. FROG is currently probably the only available means to measure accurately pulses shorter than 5 fs in a nonlinear-spectroscopic experiment, i.e., in a useful application. And because FROG can operate even with almost arbitrary noninstantaneous nonlinearities [18], it is an ideal choice for measuring subfemtosecond pulses.

In this chapter, we have developed the theory of FROG in the few-cycle regime and applied it to the SHG FROG measurement of 2.5-optical-cycle pulses with a central wavelength around 800 nm. These are some of the shortest pulses that have been completely characterized to date. We also described the measurement of uncompressed strongly non-transform-limited weak-intensity continuum pulses. These two key measurements, which were required to design, test, and optimize the pulse compressor, were both performed without a single change in the SHG FROG apparatus. No other pulse measurement technique known allows such simplicity and versatility.

This chapter has appeared in a more detailed form in R. Trebino, *Frequency-Resolved Optical Gating: The Measurement of Ultrashort Laser Pulses* (Kluwer, Boston 2002).

References

[1] J. Zhou, J. Peatross, M. M. Murnane, H. C. Kapteyn, I. P. Christov: Enhanced high-harmonic generation using 25 fs laser pulses, Phys. Rev. Lett. **76**, 752 (1996)
[2] C. J. Bardeen, Q. Wang, C. V. Shank: Selective excitation of vibrational wavepacket motion using chirped pulses, Phys. Rev. Lett. **75**, 3410 (1995)
[3] B. Kohler, V. V. Yakovlev, J. Che, J. L. Krause, M. Messina, K. R. Wilson, N. Schwentner, R. M. Whitnell, Y. Yan: Quantum control of wave packet evolution with tailored femtosecond pulses, Phys. Rev. Lett. **74**, 3360 (1995)
[4] V. V. Yakovlev, C. J. Bardeen, J. Che, J. Cao, K. R. Wilson: Chirped pulse enhancement of multiphoton absorption in molecular iodine, J. Comp. Phys. **108**, 2309 (1998)
[5] M. Fetterman, D. Goswami, D. Keusters, J. K. Rhee, X. J. Zhang, W. S. Warren: Generation of amplified shaped pulses for highly adiabatic excitation, in Ultrafast Phenomena XI (Springer, Berlin, Heidelberg 1998) paper MoA3

[6] C. J. Bardeen, J. Che, K. R. Wilson, V. V. Yakovlev, P. Cong, B. Kohler, J. L. Krause, M. Messina: Quantum control of NaI photodissociation reaction product states by ultrafast tailored light pulses, J. Phys. Chem. **101**, 3815 (1997)
[7] A. Assion, T. Baumert, M. Bergt, T. Brixner, B. Kiefer, V. Seyfried, M. Strehle, G. Gerber: Control of chemical reactions by feedback-optimized phase-shaped femtosecond laser pulses, Science **282**, 919 (1998)
[8] A. Assion, T. Baumert, M. Bergt, T. Brixner, B. Kiefer, V. Seyfried, M. Strehle, G. Gerber: Automated coherent control of chemical reactions and pulse compression by an evolutionary algorithm with feedback, in Ultrafast Phenomena XI (Springer, Berlin, Heidelberg 1998) postdeadline paper ThD9
[9] M. Bergt, T. Brixner, B. Kiefer, M. Strehle, G. Gerber: Controlling the femtochemistry of $Fe(CO)_5$, J. Comp. Phys. **103**, 10381 (1999)
[10] G. Taft, A. Rundquist, M. M. Murnane, H. C. Kapteyn: Ultrashort optical waveform measurement using frequency-resolved optical gating, Opt. Lett. **20**, 743 (1995)
[11] G. Taft, A. Rundquist, M. M. Murnane, I. P. Christov, H. C. Kapteyn, K. W. DeLong, D. N. Fittinghoff, M. A. Krumbügel, J. Sweetser, R. Trebino: Measurement of 10 fs pulses, IEEE J. Sel. Top. Quant. **2**, 575 (1996)
[12] A. Baltuška, Z. Wei, M. S. Pshenichnikov, D. A. Wiersma: Optical pulse compression to 5 fs at a 1 MHz repetition rate, Opt. Lett. **22**, 102 (1997)
[13] A. Baltuška, Z. Wei, M. S. Pshenichnikov, D. A.Wiersma, R. Szipöcs: All-solid-state cavity-dumped sub-5 fs laser, Appl. Phys. B-Lasers O. **65**, 175 (1997)
[14] M. Nisoli, S. D. Silvestri, R. Szipöcs, K. Ferencz, C. Spielmann, S. Sartania, F. Krausz: Compression of high-energy laser pulses below 5 fs, Opt. Lett. **22**, 522 (1997)
[15] M. Nisoli, S. Stagira, S. D. Silvestri, O. Svelto, S. Sartania, Z. Cheng, M. Lenzner, C. Spielmann, F. Krausz: A novel high-energy pulse compression system: Generation of multigigawatt sub-5 fs pulses, Appl. Phys. B-Lasers O. **65**, 185 (1997)
[16] D. J. Kane, R. Trebino: Characterization of arbitrary femtosecond pulses using frequency-resolved optical gating, IEEE J. Quantum Elect. **29**, 571 (1993)
[17] R. Trebino, D. J. Kane: Using phase-retrieval to measure the intensity and phase of ultrashort pulses: frequency-resolved optical gating, J. Opt. Soc. Am. A **10**, 1101 (1993)
[18] R. Trebino: *Frequency-Resolved Optical Gating: The Measurement of Ultrashort Laser Pulses* (Kluwer, Boston 2002)
[19] D. N. Fittinghoff, K. W. DeLong, R. Trebino, C. L. Ladera: Noise sensitivity in frequency-resolved optical-gating measurements of utrashort pulses, J. Opt. Soc. Am. B **12**, 1955 (1995)
[20] K. W. DeLong, R. Trebino, D. J. Kane: Comparison of ultrashort-pulse frequency-resolved-optical-gating traces for three common beam geometries, J. Opt. Soc. Am. B **11**, 1595 (1994)
[21] K. W. DeLong, D. N. Fittinghoff, R. Trebino: Practical issues in ultrashort-laser-pulse measurement using frequency-resolved optical gating, IEEE J. Quantum Elect. **32**, 1253 (1996)
[22] R. Trebino, K. W. DeLong, D. N. Fittinghoff, J. Sweetser, M. A. Krumbügel, B. Richman, D. J. Kane: Measuring ultrashort laser pulses in the time-frequency domain using frequency-resolved optical gating, Rev. Sci. Inst. **68**, 3277 (1997)

[23] J. K. Rhee, T. S. Sosnowski, A. C. Tien, T. B. Norris: Real-time dispersion analyzer of femtosecond laser pulses with use of a spectrally and temporally resolved up-conversion technique, J. Opt. Soc. Am. A **13**, 1780 (1996)
[24] J. P. Foing, J. P. Likforman, M. Joffre, A. Migus: Femtosecond pulse phase measurement by spectrally resolved up-conversion: application to continuum compression, IEEE J. Quantum Elect. **28**, 2285 (1992)
[25] C. Iaconis, I. A. Walmsley: Spectral phase interferometry for direct electric-field reconstruction of ultrashort optical pulses, Opt. Lett. **23**, 792 (1998)
[26] I. A. Walmsley, V. Wong: Characterization of the electric field of ultrashort optical pulses, J. Opt. Soc. Am. B **13**, 2453 (1996)
[27] V. Wong, I. A. Walmsley: Ultrashort-pulse characterization from dynamic spectrograms by iterative phase retrieval, J. Opt. Soc. Am. **14**, 994 (1997)
[28] L. Gallmann, D. H. Sutter, N. Matuschek, G. Steinmeyer, U. Keller: Characterization of sub-6 fs pulses with spectral phase interferometry for direct electric-field reconstruction, Opt. Lett. **24**, 1314 (1999)
[29] C. Dorrer: Influence of the calibration of the detector on spectral interferometry, J. Opt. Soc. Am. B **16**, 1160 (1999)
[30] C. Dorrer, B. d. Beauvoir, C. L. Blanc, S. Ranc, J. P. Rousseau, P. Rousseau, J. P. Chambaret, F. Salin: Single-shot real-time characterization of chirped-pulse amplification systems by spectral phase interferometry for direct electric-field reconstruction, Opt. Lett. **24**, 1644 (1999)
[31] C. Dorrer: Implementation of spectral phase interferometry for direct electric-field reconstruction with a simultaneously recorded reference interferogram, Opt. Lett. **24**, 1532 (1999)
[32] C. Iaconis, I. A. Walmsey: Self-referencing spectral interferometry for measuring ultrashort optical pulses, IEEE J. Quantum Elect. **35**, 501 (1999)
[33] M. S. Pshenichnikov, K. Duppen, D. A. Wiersma: Time-resolved femtosecond photon echo probes bimodal solvent dynamics, Phys. Rev. Lett. **74**, 674 (1995)
[34] P. Vöhringer, D. C. Arnett, T. S. Yang, N. F. Scherer: Time-gated photon echo spectroscopy in liquids, Chem. Phys. Lett. **237**, 387 (1995)
[35] T. Steffen, K. Duppen: Time resolved four- and six-wave mixing in liquids. II. Experiments, J. Comp. Phys. **106**, 3854 (1997)
[36] A. Tokmakoff, G. R. Fleming: Two-dimensional Raman spectroscopy of the intermolecular modes of liquid CS2, J. Comp. Phys. **106**, 2569 (1997)
[37] A. Tokmakoff, M. J. Lang, D. S. Larsen, G. R. Fleming, V. Chernyak, S. Mukamel: Two-dimensional Raman spectroscopy of vibrational interactions in liquids, Phys. Rev. Lett. **79**, 2702 (1997)
[38] K. Tominaga, K. Yoshihara: Fifth-order optical response of liquid CS_2 observed by ultrafast nonresonant six-wave mixing, Phys. Rev. Lett. **74**, 3061 (1995)
[39] W. P. d. Boeij, M. S. Pshenichnikov, D. A. Wiersma: Ultrafast solvation dynamics explored by femtosecond photon echo spectroscopies, Ann. Rev. Phys. Chem. p. 99 (1998)
[40] A. Shirakawa, I. Sakane, T. Kobayashi: Sub-5 fs pulse generation by pulse-front-matched optical parametric amplification, in Ultrafast Phenomena XI (Springer, Berlin, Heidelberg 1998) postdeadline paper ThD2
[41] K. W. DeLong, C. L. Ladera, R. Trebino, B. Kohler, K. R. Wilson: Ultrashort-pulse measurement using non-instantaneous nonlinearities: Raman effects in frequency-resolved optical gating, Opt. Lett. **20**, 486 (1995)

[42] K. W. DeLong, R. Trebino, J. Hunter, W. E. White: Frequency-resolved optical gating with the use of second-harmonic generation, J. Opt. Soc. Am. B **11**, 2206 (1994)

[43] P. O'Shea, M. Kimmel, X. Gu, R. Trebino: Highly simplified device for ultrashort laser pulse measurement, Opt. Lett. **26**, 932 (2001)

[44] P. O'Shea, et al.: Increased bandwidth in ultrashort-pulse measurement using an angle-dithered nonlinear-optical crystal, Opt. Express **7** (2001)

[45] T. Tsang, M. A. Krumbügel, K. W. DeLong, D. N. Fittinghoff, R. Trebino: Frequency-resolved optical gating measurements of ultrashort pulses using surface third-harmonic generation, Opt. Lett. **21**, 1381 (1996)

[46] K. Simoda: *Introduction to Laser Physics*, 2nd ed. (Springer, Berlin, Heidelberg 1996)

[47] L. Allen, J. H. Eberly: *Optical Resonance and Two-Level Atoms* (Dover, New York 1987)

[48] P. N. Butcher, D. Cotter: *The Elements of Nonlinear Optics* (Cambridge Univ. Press, Cambridge 1990)

[49] A. M. Weiner: Effect of group velocity mismatch on the measurement of ultrashort optical pulses via second harmonic generation, IEEE J. Quantum Elect. **19**, 1276 (1983)

[50] G. Taft, A. Rundquist, M. M. Murnane, I. P. Christov, H. C. Kapteyn, K. W. DeLong, D. F. Fittinghoff, M. A. Krumbuguel, J. N. Sweetser, R. Trebino: Measurement of 10 fs laser pulses, IEEE J. Sel. Top. Quant. **2**, 575 (1996)

[51] A. Baltuška, M. S. Pshenichnikov, D. A. Wiersma: Second-harmonic generation frequency-resolved optical gating in the single-cycle regime, IEEE J. Quantum Elect. **35**, 459 (1999)

[52] A. Baltuška, M. F. Emde, M. S. Pshenichnikov, D. A. Wiersma: Early-time dynamics of the photo-excited hydrated electron, J. Phys. Chem. A **103**, 10065 (1999)

[53] A. Baltuška, M. F. Emde, M. S. Pshenichnikov, D. A. Wiersma: Sub-5 fs pulses: generation, characterisation, and experiments, in Ultrafast Phenomena XI (Springer, Berlin, Heidelberg 1998)

[54] R. W. Boyd: *Nonlinear Optics* (Academic, San Diego 1992)

[55] Y. R. Shen: *The Principles of Nonlinear Optics* (Wiley, New York 1984)

[56] T. Brabec, F. Krausz: Nonlinear optical pulse propagation in the single-cycle regime, Phys. Rev. Lett. **78**, 3282 (1997)

[57] N. Bloembergen: *Nonlinear Optics* (Benjamin, New York 1997)

[58] Y. Wang, R. Dragila: Efficient conversion of picosecond laser pulses into second-harmonic frequency using group-velocity dispersion, Phys. Rev. A **41**, 5645 (1990)

[59] A. Umbrasas, J. C. Diels, J. Jacob, G. Valiulis, A. Piskarskas: Generation of femtosecond pulses through second-harmonic compression of the output of a Nd:YAG laser, Opt. Lett. **20**, 2228 (1995)

[60] A. M. Weiner, A. M. Kan'an, D. E. Leaird: High-efficiency blue generation by frequency doubling of femtosecond pulses in a thick nonlinear crystal, Opt. Lett. **23**, 1441 (1998)

[61] G. P. Agrawal: *Nonlinear Fiber Optics*, 2nd ed. (Academic, San Diego 1995)

[62] P. V. Mamyshev, S. V. Chernikov: Ultrashort-pulse propagation in optical fibers, Opt. Lett. **15**, 1076 (1990)

[63] J. T. Manassah, M. A. Mustafa, R. R. Alfano, P. P. Ho: Spectral extent and pulse shape of the supercontinuum for ultrashort laser pulse, IEEE J. Quantum Elect. **22**, 197 (1986)
[64] G. R. Fleming: *Chemical Applications of Ultrafast Spectroscopy* (Oxford Univ. Press, New York 1986)
[65] X. Gu, E. Zeek, M. Kimmel, R. Trebino: Suppression of geometrical smearing effects in noncollinear ultrashort-pulse measurement techniques, in press
[66] J. A. Squier, D. N. Fittinghoff, C. P. J. Barty, K. R. Wilson, M. Müller, G. J. Brakenhoff: Characterization of femtosecond pulses focused with high numerical aperture optics using interferometric- surface third-harmonic generation, Opt. Commun. **147**, 153 (1998)
[67] D. N. Fittinghoff, J. A. Squier, C. P. Barty, J. Sweetser, R. Trebino, M. Müller: Collinear type II second-harmonic-generation frequency-resolved optical gating for use with high-numerical aperture objectives, Opt. Lett. **23**, 1046 (1998)
[68] D. N. Fittinghoff, A. C. Millard, J. A. Squier, M. Müller: Frequency-resolved optical gating measurement of ultrashort pulses passing through a high numerical aperture objective, IEEE J. Quantum Elect. **35**, 479 (1999)
[69] L. Gallmann, G. Steinmeyer, D. H. Sutter, N. Matuschek, U. Keller: Collinear type II second-harmonic-generation frequency-resolved optical gating for the characterization of sub-10 fs optical pulses, Opt. Lett. **25**, 269 (2000)
[70] Z. Cheng, A. Fürbach, S. Sartania, M. Lenzner, C. Spielmann, F. Krausz: Amplitude and chirp characterization of high-power laser pulses in the 5 fs regime, Opt. Lett. **24**, 247 (1999)
[71] K. W. DeLong, D. N. Fittinghoff, R. Trebino, B. Kohler, K. Wilson: Pulse retrieval in frequency-resolved optical gating based on the method of generalized projections, Opt. Lett. **19**, 2152 (1994)
[72] W. H. Press, S. Teukolsky, W. T. Vetterling, B. P. Flannery: *Numerical Recipes in C* (Cambridge Univ. Press, New York 1996)
[73] A. Baltuška, M. S. Pshenichnikov, D. A. Wiersma: Amplitude and phase characterization of 4.5 fs pulses by frequency-resolved optical gating, Opt. Lett. **23**, 1474 (1998)
[74] F. Krausz: (1998), private communication
[75] L. Gallmann, D. H. Sutter, N. Matuschek, G. Steynmeyer, U. Keller: Techniques for the characterization of sub-10 fs pulses: A comparison, in (Conf. Ultrafast Optics 1999) paper Th12
[76] V. G. Dmitriev, G. G. Gurzadyan, D. N. Nikogosyan: *Handbook of Non-Linear Optical Crystals* (Springer, Berlin, Heidelberg 1991)
[77] M. S. Pshenichnikov, W. P. Boeij, D. A. Wiersma: Generation of 13 fs pulses from a cavity-dumped Ti:sapphire laser, Opt. Lett. **19**, 572 (1994)
[78] E. J. Mayer, J. Möbius, A. Euteneuer, W. W. Rühle, R. Szipöcs: Ultrabroadband chirped mirrors for femtosecond lasers, Opt. Lett. **22**, 528 (1997)
[79] C. Rullère (Ed.): *Femtosecond Laser Pulses* (Springer, Berlin, Heidelberg 1998) p. 310
[80] C. H. B. Cruz, R. L. Fork, W. H. Knox, C. V. Shank: Spectral hole burning in large molecules probed with 10 fs optical pulses, Chem. Phys. Lett. **132**, 341 (1986)
[81] A. Kummrow, M. F. Emde, A. Baltuška, D. A. Wiersma, M. S. Pshenichnikov: Hydrated electron dynamics at a five femtosecond time scale, Z. Phys. Chem. **212**, 153–160 (1999)

[82] M. F. Emde, A. Baltuška, A. Kummrow, M. S. Pshenichnikov, D. A. Wiersma: Hydrated electron dynamics on a 5 fs time scale, in Ultrafast Phenomena XI (Springer, Berlin, Heidelberg 1998) p. 586

Characterization of Ultrashort Optical Pulses in the Few-Cycle Regime Using Spectral Phase Interferometry for Direct Electric-Field Reconstruction

Ian A. Walmsley

Clarendon Laboratory, University of Oxford, Parks Rd., Oxford, OX1 3PU, UK
walmsley@physics.ox.ac.uk

Abstract. Interferometry provides a very sensitive and accurate means to measure the phase of an optical field. It has been applied with great effect in the past few years to measurements of the spectral phase of ultrashort optical pulses. Together with measurement of the pulse spectrum, which may be made simultaneously, this provides complete characterization of single pulses.

The conversion of phase to amplitude information that is the hallmark of interferometric measurement makes it very sensitive to small differences in phase and allows robust extraction of the spectral phase from measured data. The key features of spectral phase interferometry that make it useful for pulses in the few-cycle regime are the rapidity of data acquisition and the inversion and insensitivity of the measurement to wavelength-dependent apparatus response. Both of these properties are important for characterizing sources for which the pulseshape fluctuates and the pulses have extremely large bandwidths. In this chapter, we discuss the method of spectral phase interferometry for direct electric-field reconstruction and its application to the measurement of pulses in the sub-10 fs regime.

1 Introduction

Interferometry may be categorized as either test-plus-reference or as self-referencing. In the former category, one begins with a well-characterized reference pulse and gauges the unknown (or test) pulse against it [1]. The apparatus consists simply of a beam splitter, on which the two pulses are mixed, and a spectrometer, typically with an array detector, to record the spectral interferogram, as illustrated in Fig. 1. In contrast to spectrographic and tomographic methods, the data set is a function of only a single variable – the frequency – rather than of two. This means that the other dimension of the detector array, if a camera, may be used to record spatial variations in the spectral phase.

In Fourier transform spectral interferometry (FTSI), the spectral phase is extracted via a direct inversion that is both rapid and robust [2,3,4]. The test and reference pulse are delayed in time with respect to one another by τ. The detected signal (interferogram) is then $S(\omega) = |\widetilde{E}(\omega) + \widetilde{E}_{\mathrm{R}}(\omega)\mathrm{e}^{-\mathrm{i}\omega\tau}|^2$, where $\widetilde{E}_{\mathrm{R}}$ is the reference field (the Fourier transform of the reference pulse analytic

F. X. Kärtner (Ed.): Few-Cycle Laser Pulse Generation and Its Applications,
Topics Appl. Phys. **95**, 265–291 (2004)

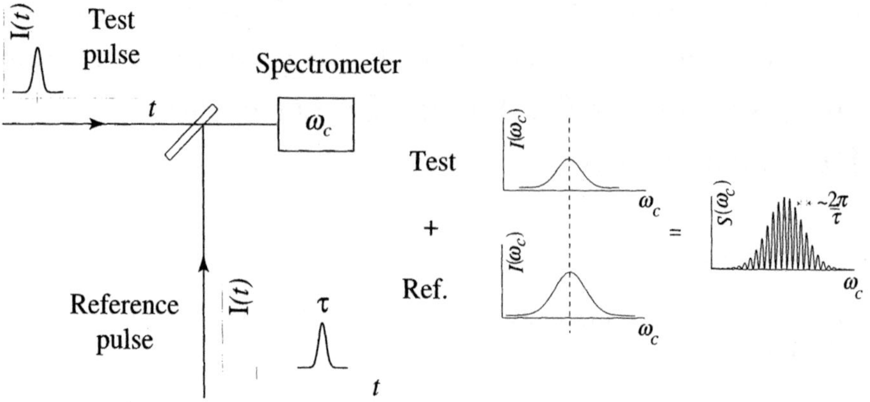

Fig. 1. The principle of operation of spectral interferometry. The test and reference pulse are mixed on a beam splitter, and the resulting joint spectrum is measured. The spectral phase difference between the two pulses is extracted from the interferogram using a direct algorithm

signal) and $\widetilde{E}(\omega)$ the test pulse field. The spectral phase difference between test and reference pulses is encoded in the relative positions of the spectral fringes with respect to the nominal spacing of $2\pi/\tau$. It can be extracted by using a three-step algorithm involving a Fourier transform to the pseudo-time domain, a filtering operation, and an inverse Fourier transform [5]. The interferogram may be written as

$$S(\omega) = D^{(\mathrm{dc})}(\omega) + \mathrm{e}^{-\mathrm{i}\tau\omega} D^{(-\mathrm{ac})}(\omega) + \mathrm{e}^{\mathrm{i}\tau\omega} D^{(+\mathrm{ac})}(\omega) , \tag{1}$$

where

$$D^{(\mathrm{dc})}(\omega) = \left|\widetilde{E}(\omega)\right|^2 + \left|\widetilde{E}_{\mathrm{R}}(\omega)\right|^2 , \tag{2}$$

and

$$D^{(\pm\mathrm{ac})}(\omega) = \left|\widetilde{E}(\omega)\widetilde{E}_{\mathrm{R}}(\omega)\right| \exp\left\{\mp\mathrm{i}\left[\phi_\omega(\omega) - \phi_\omega^{\mathrm{R}}(\omega)\right]\right\} . \tag{3}$$

The dc portion of the interferogram (2) is the sum of the individual spectra of the pulses and contains no phase information. The ac terms (3) are the result of the interference and contain all of the phase information.

There are three steps for reconstructing the spectral phase from the interferogram, as illustrated in Fig. 2. First, isolate one of the ac terms, and hence $\phi_\omega(\omega) - \phi_\omega^{\mathrm{R}}(\omega) - \tau\omega$, by means of a Fourier transform and filter technique. If τ is sufficiently large, the $t = 0$ and $\pm\tau$ components are well separated in time, and as a result the unwanted dc and negative ac components can be removed by filtering. For this purpose, we use a fourth-order super-Gaussian, $H(t)$, of full width τ centered at $t = \tau$. The filtered signal,

$$\widetilde{D}^{\mathrm{filter}}(t) = H(t-\tau)\widetilde{D}(t) = \mathrm{FT}\left\{D^{(+\mathrm{ac})}(\omega); \omega \to t - \tau\right\} , \tag{4}$$

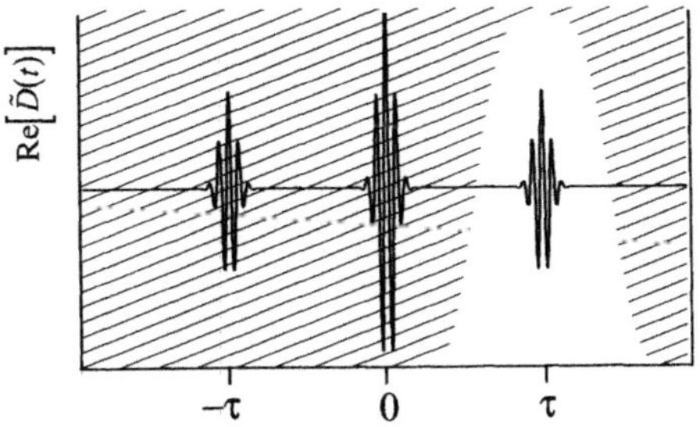

Fig. 2. Diagram of the inversion algorithm for Fourier transform spectral interferometry. After an initial Fourier transform to the pseudotime domain, an ac sideband is digitally filtered to isolate the interference term. An inverse Fourier transform is made, and the argument of the result is extracted

is simply the Fourier transform of the positive ac portion ($t = +\tau$) of the interferogram. The spectral phase difference is the argument of the inverse transform of $\widetilde{D}^{\text{filter}}(t)$,

$$\begin{aligned}\phi_\omega(\omega) - \phi_\omega^{\text{R}}(\omega) + \tau\omega &= \arg\left[D^{(+\text{ac})}(\omega)\right] \\ &= \arg\left(\text{IFT}\left\{\widetilde{D}^{\text{filter}}(t); t \to \omega\right\}\right). \end{aligned} \tag{5}$$

Second, remove $\tau\omega$ by subtracting a calibration phase. And third, reconstruct $\phi_\omega(\omega)$ by subtracting the reference phase.

Of course, FTSI begs the question, and therefore one must inquire whether there are suitable self-referencing interferometric methods that would not require a well-characterized reference pulse. Such methods are, possible, and they retain the advantageous features of FTSI [6, 7]. In the rest of the Chapter, we will confine the discussion to one of these methods – spectral phase interferometry for direct electric-field reconstruction (SPIDER) – that has been shown highly effective in the sub-two-cycle regime.

2 General Principles of SPIDER

Spectral phase interferometry for direct electric-field reconstruction is an implementation of shearing interferometry in the optical-frequency domain [8,9]. A schematic diagram of the principle of operation in shown in Fig. 3. Two delayed replicas of the (unknown) test pulse are generated in an interferometer, and one is frequency shifted with respect to the other. The combined spectrum of the pulse pair is measured using a spectrometer and a detector array [10]. The important feature is that the frequency shift, or spectral shear, allows two adjacent frequencies in the original pulse spectrum to interfere on an integrating detector. The resulting fringe pattern thus reflects the phase difference between pairs of spectral components of the pulse separated by the shear.

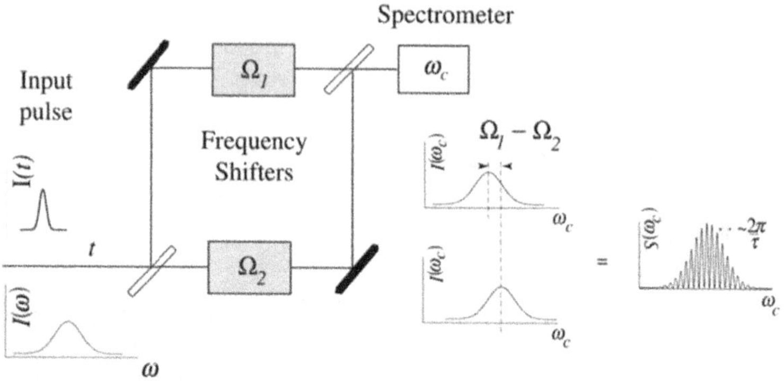

Fig. 3. The principle of operation of SPIDER. Two replicas of the test pulse are shifted in frequency by different amounts, and the resulting joint spectrum is measured. The phase difference between two points in the input spectrum separated in frequency by the difference of the shifts is extracted using the FTSI algorithm

SPIDER allows reconstruction of the electric field of an ultrashort light pulse from a phase gradient extracted directly from a one-dimensional interferogram. The spectral phase may be combined with a measured spectrum to yield complete characterization of the temporal or spectral field. The simplicity of the inversion means, that such characterization can be done at very rapid rates – up to a 1 kHz refresh rate has been reported to date, limited only by the detector readout time [11, 12, 13].

The SPIDER interferogram has the form of (1), except that the dc and ac terms contain different frequency arguments:

$$D^{(\mathrm{dc})}(\omega) = \left|\widetilde{E}(\omega - \Omega)\right|^2 + \left|\widetilde{E}(\omega)\right|^2 , \tag{6}$$

and

$$D^{(\pm\mathrm{ac})}(\omega) = \left|\widetilde{E}(\omega - \Omega)\widetilde{E}(\omega)\right| \exp\left\{\mp\mathrm{i}\left[\phi_\omega(\omega) - \phi_\omega(\omega - \Omega)\right]\right\} . \tag{7}$$

The spectral phase difference between two frequencies separated by the spectral shear Ω is extracted from the interferogram. Thus, a sampling of the spectral phase (to within a constant) across the pulse spectrum can be obtained by taking the difference of adjacent spectral phase differences at frequencies separated by integral multiples of the shear. The sampling interval is therefore just the spectral shear. According to the Shannon theorem, all pulses with compact support in the domain $[-T, T]$ may be completely characterized by sampling their spectrum on the interval $2\pi/T$. Thus, SPIDER is able to reconstruct all pulses that have support only in the temporal window $[-\pi/\Omega, \pi/\Omega]$.

The technique relies on the interference between two frequency-sheared replicas of the input (test) field. For short pulses, this shear is obtained by

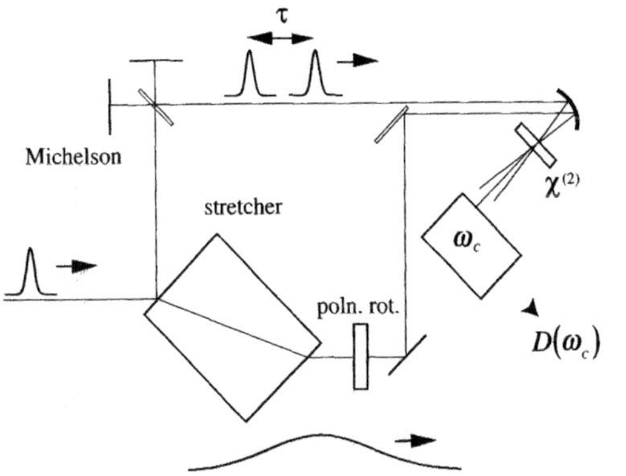

Fig. 4. The sub-10 fs SPIDER setup. A linearly polarized pulse enters the apparatus, and is divided by a reflection at the SF10 glass block which serves as a stretcher. Its polarization is then rotated by a periscope. The other beam is used to form two time delayed replicas of the input pulse in a low-dispersion, compensated Michelson interferometer. The beams are combined noncollinearly in a nonlinear $\chi^{(2)}$ crystal, arranged for Type-II upconversion. A spectrometer measures the resulting spectral interferogram on a CCD array

upconversion of two delayed replicas of the pulse by mixing them with an ancillary chirped pulse in a nonlinear crystal. This leads to a shift of each replica by a different frequency because of the change in the instantaneous frequency in the chirped pulse over the delay between the replicas. This, in turn, gives rise to relative shear between the two replicas.

An apparatus suitable for the measurement of pulses of less than 10 fs duration in the optical region of the spectrum is shown in Fig. 4 [14,15]. Two pulse replicas with a delay of $\tau = 300$ fs are generated in a Michelson-type interferometer. The input pulse for the interferometer is the reflection from the surface of a 6.5 cm long SF10 glass block. The signal transmitted through this block is used as the strongly chirped pulse for upconversion. The group delay dispersion of the block (10^4 fs^2) together with the delay of 300 fs results in a spectral shear Ω of 30 ps^{-1}. The chirp introduced by the glass block is adequate to ensure that each pulse replica is upconverted with a quasi-CW field.

The interferogram, as well as the relationship between the shear and delay in this apparatus, can be determined from the following analysis. The dispersion of the glass block is dominated by the quadratic spectral phase of the material. Its transfer function is thus approximately

$$\widetilde{S}_{\mathrm{q}}^{\mathrm{P}}(\omega) = \mathrm{e}^{-\mathrm{i}\frac{1}{2}\phi''\omega^2} . \tag{8}$$

The analytic signal of the stretched pulse after the block is thus,

$$E_{\mathrm{stretch}}(t) = E_{\mathrm{R}}(t) \otimes S_{\mathrm{q}}^{\mathrm{P}}(t) \propto \mathrm{e}^{-\mathrm{i}\frac{t^2}{2\phi''}} \int \mathrm{d}t' E_{\mathrm{R}}(t')\mathrm{e}^{-\mathrm{i}\frac{t'^2}{2\phi''}} \mathrm{e}^{\mathrm{i}\frac{tt'}{\phi''}} . \tag{9}$$

For large enough ϕ'', $t'^2/2\phi'' \approx 0$ for all t' over which the input pulse is nonzero. In this case, the analytic signal of the stretched pulse is simply

$$\begin{aligned} E_{\mathrm{stretch}}(t) &\approx \mathrm{e}^{-\mathrm{i}\frac{t^2}{2\phi''}} \int \mathrm{d}t' E_{\mathrm{R}}(t') \mathrm{e}^{\mathrm{i}\frac{tt'}{2\phi''}} \\ &= \mathrm{e}^{-\mathrm{i}\left(\omega_0 t + \frac{t^2}{2\phi''}\right)} \widetilde{E}_{\mathrm{R}} \left(\omega = \omega_0 + \frac{t}{\phi''}\right) . \end{aligned} \tag{10}$$

For large quadratic dispersion, the pulse stretcher effectively maps frequency into time.

As long as the large stretching approximation is satisfied, each pulse in the test pair is upconverted with a quasi-CW frequency slice of the stretched pulse. Thus, if the bandwidth of the nonlinear crystal is sufficiently broad, the analytic signals of the upconverted pulses are

$$E_1(t) \propto E(t)\widetilde{E}_{\mathrm{R}}(t)(\omega_0 + t/\phi'') \exp\left[-\mathrm{i}(\omega' + \Omega)t\right] \tag{11}$$

and

$$E_2(t) \propto E(t-\tau)\widetilde{E}_{\mathrm{R}}(\omega_0 + t/\phi'' + \Omega) \exp\left[-\mathrm{i}\omega'(t-\tau)\right] , \tag{12}$$

where $\omega' = \omega_0 + t_{\mathrm{s}}/\phi'' + \tau/\phi''$, t_{s} is the center of the time slice of the stretched pulse that the leading pulse in the test pair occupies and $\Omega = -\tau/\phi''$. Upconversion generates the spectral shear by simulating a linear temporal phase modulator.

The spectra of the sheared pulses are centered near twice the carrier frequency of the input pulse being characterized, i.e., $2\omega_0$. When the sheared pulses are interfered with, the frequency-resolved signal is related to the input pulse by

$$\begin{aligned} S(\omega) &= \left|\widetilde{E}_1(\omega) + \widetilde{E}_2(\omega)\right|^2 \\ &= \left|\widetilde{E}(\omega - \omega' - \Omega)\widetilde{E}_{\mathrm{R}}(\omega' + \Omega)\right|^2 + \left|\widetilde{E}(\omega - \omega')\widetilde{E}_{\mathrm{R}}(\omega')\right|^2 \\ &\quad + 2\left|\widetilde{E}(\omega - \omega' - \Omega)\right|\left|\widetilde{E}(\omega - \omega')\right|\left|\widetilde{E}_{\mathrm{R}}(\omega' + \Omega)\right|\left|\widetilde{E}_{\mathrm{R}}(\omega')\right| \\ &\quad \cdot \cos\left[\phi_\omega(\omega - \omega' - \Omega) - \phi_\omega^{\mathrm{R}}(\omega' + \Omega)\right. \\ &\qquad \left. -\phi_\omega(\omega - \omega') + \phi_\omega^{\mathrm{R}}(\omega') - \tau\omega\right] . \end{aligned} \tag{13}$$

In writing this equation, we have used notation consistent with a reference field $E_{\mathrm{R}}(\omega)$ to emphasize that the ancillary chirped pulse need not be the input pulse (although it is in the apparatus shown in the figure). It is important to understand that SPIDER may be implemented with a pulse for the nonlinear frequency conversion different from the input and that this pulse need not be known. This version of the technique is called X-SPIDER and has been used to characterize pulses in the blue [16] and visible regions of the spectrum [12, 13]. There are obvious technical advantages to using a high-energy pulse as this kind of ancilla.

3 Anatomy of a SPIDER

The two key parameters to be determined when designing a SPIDER apparatus are the spectral shear, Ω, and the relative delay between the pulses in the test pair τ. These are not independent since

$$\Omega = -\tau/\phi''. \tag{14}$$

Moreover, there are individual constraints on both parameters. For instance, since the fringes of the spectral interferogram are nominally spaced at $2\pi/\tau$, τ should not be so large that the fringes are not resolvable with the available spectrometer. But, τ should not be so small that the data inversion routine cannot separate the ac terms from the dc term. Likewise, Ω should not be so large that the phase sampling interval is larger than the Nyquist limit, but it should not be so small that the spectral phase difference is imperceptible. The stretcher dispersion, ϕ'', is also constrained. It must be sufficiently large to ensure that each pulse in the test pair is upconverted with a quasi-CW slice of the stretched pulse. Of course, the required dispersion depends upon the spectral width, $\Delta\omega$, and duration, Δt, of the input pulse [17].

Equation (14) is also used to determine the spectral shear in the experiment. The exact delay τ is obtained from the measurement of the calibration phase term $\omega\tau$ described in the previous section. And the GDD is either measured using spectral interferometry or calculated from the known dispersion properties of the pulse stretcher. As discussed earlier, the spectral shear determines the sampling interval in the frequency domain. This parameter therefore needs to be chosen in agreement with sampling theory.

Once a delay and a spectral shear have been determined, the required group delay dispersion ϕ'' follows from (14). Independently of this, ϕ'' has to be chosen large enough to ensure that the two short-pulse replicas are upconverted with quasi-CW slices of the strongly linearly chirped pulse. If all the conditions on shear, delay, and ϕ'' cannot be fulfilled simultaneously, the resolution of the spectrometer has to be increased. This allows for larger delays and thus for larger values of ϕ'' for a given shear. On the other hand, the SPIDER signal strength scales inversely proportional to ϕ'', which suggests the use of dispersions that are not orders of magnitude higher than necessary to satisfy the quasi-CW criterion.

An example will serve to illustrate the trade-offs. Consider an input pulse typical of an ultrafast Ti:sapphire laser system, with spectral width $\Delta\omega = 2\pi\times 10\,\text{THz}$ centered at $\omega_0 = 2\pi\times 375\,\text{THz}$ ($\Delta\lambda \approx 21\,\text{nm}$, $\lambda_0 = 800\,\text{nm}$). Such a pulse has a transform-limited duration $\Delta t \approx 50\,\text{fs}$. For large stretching, the duration of the stretched pulse is $T \approx \Delta\omega\phi''$. If the duration of the input pulse is Δt, then a stretched pulse of duration $100\times\Delta t$ ensures that each time slice of the stretched pulse has a range of frequencies no larger than $\Delta\omega/100$. In this example, since the input pulse might be longer than the transform limit, the stretched pulse should be, say, 20 ps long ($T = 400\times\Delta t$). The spectrometer must have adequate resolution to resolve the fringes at $2\pi/\tau$. If the spectrometer resolution is, say, $\gamma = \Delta\omega/100 = 2\pi \times 0.1\,\text{THz}$ (0.5 Å at $\lambda = 400\,\text{nm}$), the delay τ must be less than 5 picoseconds so that the spectrometer can resolve more than two points per fringe. The effects of uncorrelated noise diminish as the square root of the number of points per fringe, so one should choose optimally a delay less than 5 ps. On the other hand, the delay cannot be arbitrarily small because the inversion relies on separating the pulses sufficiently in time to distinguish the ac and dc terms of the interferogram. Since the ac and dc terms have roughly the same duration as the input pulse, we usually choose a delay at least 10 times the transform limit. Erring on the safe side, a delay of 1 to 2 picoseconds, $\tau \approx 1\,\text{ps}$ to $2\,\text{ps}$, resulting in 5 to 10 points per fringe, is adequate in this example. This gives a spectral shear, $\Omega = \Delta\omega\tau/T$, of 5% to 10% of the input pulse spectral width, which satisfies the sampling criterion. In general, the greater the number of sample points per fringe, the lower the signal-to-noise ratio needed for reconstruction.

For Gaussian-like spectra, good results are achieved with a ratio between shear and bandwidth of $\Omega/\Delta\omega = 1/4$. For more box-shaped spectra, this ratio has to be reduced, and values of $\Omega/\Delta\omega$ down to a few percent might be required. Smaller values of the shear allow measuring a larger range of pulse durations without changing the parameters of the apparatus. However, decreasing shear also reduces the phase differences $\theta(\omega) = \phi_\omega(\omega) - \phi_\omega^{\text{R}}(\omega - \Omega)$, which are ultimately obscured by the experimental noise.

All second-order nonlinear media that are phase-matchable over a sufficiently large spectral range may be used to implement SPIDER. It turns out, however, that Type-II phase-matched crystals are particularly well suited. A comparison of the different possible phase-matching geometries for a 60 µm thick beta barium borate (BBO) crystal is shown in Fig. 5. As can be seen, the broadest bandwidth is obtained for a given crystal thickness if Type-II phase matching is chosen with the strongly linearly chirped pulse oriented along the extraordinary axis and the short pulse replicas along the ordinary axis. The extraordinary axis does not require a large bandwidth because only a bandwidth of the stretched pulse corresponding to the spectral shear needs to be phase-matched for interaction with the two replicas of the input pulse. In Type-II phase matching, the solid curve is obtained for a CW-beam at 800 nm on the extraordinary axis and the dashed line with the same beam oriented along the ordinary axis. The situation for Type-I phase matching is shown by the dash-dotted curve.

3.1 Measurement Protocols and Calibration

The interferogram is measured ideally using a spectrometer with a step-function temporal response. The effect of a nonideal spectrometer is simply to reduce the signal-to-noise ratio of the interference fringes. The phase difference between the two electric fields is extracted using Fourier transform

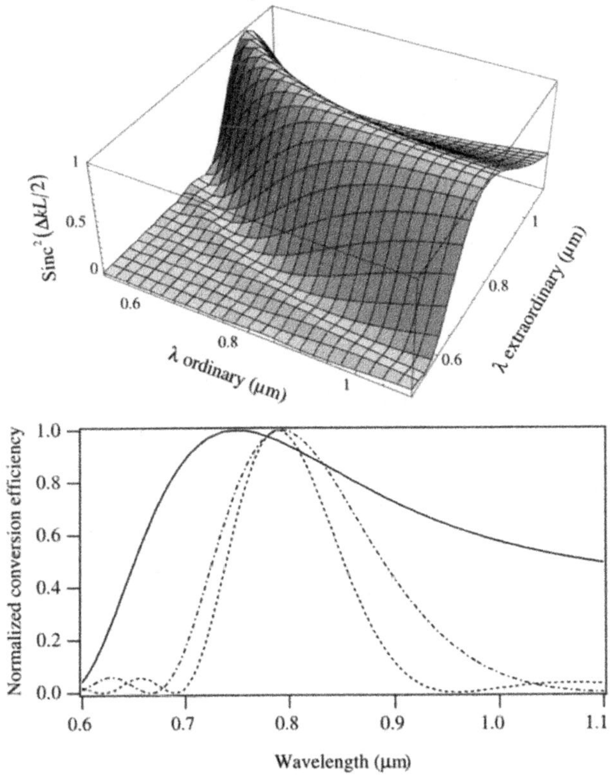

Fig. 5. Phase-matching curves (*upper diagram*) and sections (*lower diagram*) for sum-frequency generation in a 60 μm thick BBO crystal

spectral interferometry, as described earlier. This returns the spectral phase difference between the two upconverted replicas on an initial array of spectral sample points $\phi(\omega - \omega_0 + \Omega) - \phi(\omega - \omega_0) + \omega\tau$. The term linear in frequency $\omega\tau$ is removed using a reference phase either at the fundamental or harmonic wavelength [9, 10]. This reference phase exactly corrects any influence of the calibration of the spectrometer on the SPIDER interferogram. The device may be calibrated a priori (thus the signal can be integrated or averaged for a very long time) and is frequently a linear measurement. Furthermore, the phase can be measured at 800 nm where the detector sensitivity is high and the power is larger. The pulse spectrum can be measured simultaneously with the SPIDER interferogram and on the same experimental trace.

Inversion of the data is simple and robust. In the subsequent section, I discuss two numerical recipes for the reconstruction of the electric field from the experimental data. Both of them rely on the fact that the sampling interval of the spectrometer is much smaller than that of the spectral shear. This situation arises naturally in the experimental implementation of SPIDER.

3.2 Inversion Algorithms for Phase Reconstruction

The basic SPIDER algorithm is based on concatenation, that is, the phase at frequency $\omega + \Omega$ is found from the phase at frequency ω by simply subtracting the two SPIDER phases at these frequencies. This method provably reconstructs completely the spectral phase of any and all pulses in the temporal window $[-T, T]$. In practice, the phase at a particular frequency (usually chosen near the peak of the spectrum) is set equal to zero and the adjacent phases concatenated every Ω, following the formula:

$$\begin{aligned}\phi(0) &= 0\,,\\ \phi\left[(n+1)\Omega\right] &= \phi(n\Omega) + \psi(n\Omega)\,.\end{aligned} \tag{15}$$

Interpolation of the spectrum on the same grid completes the measurement in the spectral domain. This then gives the electric field in the spectral domain at frequencies $[0, \Omega, 2\Omega \ldots, N\Omega]$. The measured spectrum is therefore oversampled for reconstructing pulses on the interval $\pm\frac{\pi}{\Omega}$. Therefore, multiple sets of the concatenated phase can be used to estimate the precision of the reconstruction. Indeed, one can build a first set by starting at pixel 0, concatenating every Ω, a second set by starting at pixel 1, concatenating on Ω, and so on. The second set is sampled on a grid that is shifted from the previous one by $\delta\omega$:

$$\begin{aligned}\phi(\delta\omega) &= 0\,,\\ \phi\left[\delta\omega + (n+1)\Omega\right] &= \phi(\delta\omega + n\Omega) + \psi(\delta\omega + n\Omega)\,.\end{aligned} \tag{16}$$

Together with a determination of the spectrum on this sampling grid, this process yields a second measurement of the electric field in the spectral domain. The number of different independent determinations of the field M is of the order of $\Omega/\delta\omega$. Obviously, because of the initial hypothesis that Ω is sufficient sampling for the field, all these determinations are equivalent. These sets of data can be used to refine the measurement or reduce the sensitivity to noise [18]. Because the fields are on different sampling grids in the spectral domain, it is not possible to sum them directly to get an average retrieved field. Each of them is Fourier transformed to the temporal domain where they represent the same electric field on the same sampling grid using the discrete formula:

$$E_m(n\delta t) = \sum_j \widetilde{E}(m\delta\omega + j\Omega) \exp\left[\mathrm{i}(m\delta\omega + j\Omega)n\delta t\right], \tag{17}$$

which can be rewritten as

$$E_m(n\delta t) = \exp(\mathrm{i}m\delta\omega n\delta t) \sum_j \widetilde{E}(m\delta\omega + j\Omega) \exp(\mathrm{i}j\Omega n\delta t)\,. \tag{18}$$

Because we use a fast Fourier transform algorithm, a temporal linear phase $m\delta\omega n\delta t$ must be added to E_m.

The sum field can then be defined in the temporal domain using the formula

$$E(n\delta t) = \frac{1}{M}\sum_{m}^{M} E_m(n\delta t) \,. \tag{19}$$

Note that no shifting of temporal fields is needed. The reason is that all the fields retrieved from the same extracted phase $\psi(\omega)$ should have the same time fiducial, and thus be consistent. The constant phase of each of these fields is completely arbitrary; it depends only on the choice of the initial phase for each set before concatenation and is not determined by the phase $\psi(\omega)$. The constant phase is set at a point of high temporal intensity. A spectral representation can then be obtained by Fourier transforming the temporal representation, using the conventional discrete Fourier transform:

$$\widetilde{E}(n\Omega) = \sum_{j} E(j\delta t)\exp(-in\Omega j\delta t) \,. \tag{20}$$

This technique has several technical advantages. First, because it reconstructs several representations of the same electric field, it allows a test of the precision of the reconstruction. A precision diagnostic for SPIDER is developed in [18]. Second, because it uses the sampling rate Ω directly, it allows undersampling to be easily recognized. The reconstructed temporal electric field should be equal to zero at the edges of the time window to be consistent with the condition of finite temporal support compatible with the shear Ω.

3.3 SPIDER Parameters for the Two-Cycle Regime

SPIDER is particularly well suited for sub-10 fs pulse characterization for several reasons. In this regime, a piece of a highly dispersive glass is sufficient for the generation of the strong chirp of the ancillary pulse. Use of sum-frequency mixing results in a more economic use of detection bandwidth, compared to SHG based techniques. Furthermore, as pointed out in the previous section, the Type-II phase matching typically used in SPIDER offers larger bandwidth for a given crystal thickness than the Type-I phase matching conventionally used in SHG based characterization schemes. SPIDER has been used to characterize pulses with nanojoule energy, a transform limit of 5.3 fs, and a duration as short as 5.9 fs [14]. More recently, sub-6 fs pulses from an optical parametric amplifier and from hollow-core fiber compression systems have been measured with SPIDER [19, 20, 21, 22]. Several improvements that increase the sensitivity of the device have been shown [23] as well as real-time, closed-loop control of spectral phase using the rapid phase extraction of which it is capable [24, 25].

The apparatus shown in Fig. 4 is suitable for measuring pulses within a time window of about 200 fs, as long as the temporally overlapping slice of the stretched pulse can still be considered quasi-CW. A low-cost, uncoated glass block is used as the first beam splitter. This ensures a large bandwidth and an appropriate power splitting ratio, allowing for an optimal upconversion signal. The Michelson interferometer uses two 300 μm thick dielectric beam splitters in a symmetrical dispersion-balanced configuration. The symmetry allows for maximum fringe contrast. The two delayed pulses are mixed noncollinearly with the stretched pulse in a 30 μm thick Type-II BBO crystal. As indicated in the previous section, the phase-matching bandwidth on the ordinary axis of a Type-II crystal is usually much larger than that in a Type-I crystal of the same material and thickness, whereas the bandwidth on the extraordinary axis is smaller. Consequently, the narrow-band extraordinary axis of the crystal was oriented parallel to the polarization of the stretched pulse, and the short replicas were rotated into the broadband ordinary axis using a periscope. In this configuration, the upconversion efficiency in the crystal varied by less than 20% over the wavelength range from 660 nm to 1 μm. Note that using the other orientation would reduce this range to 750 nm to 880 nm. Noncollinear mixing geometry avoids the need for a second beam splitter. It is important to note that for SPIDER, the geometric smearing caused by this noncollinear geometry and integrating over the whole upconverted beam does not affect the accuracy of reconstruction, as long as the quasi-monochromatic approximation remains valid for the chirped pulse. This condition is easily fulfilled. The reconstruction is also unaffected by group velocity mismatch, under the same condition. The upconverted pulses were detected through a 0.3 m imaging spectrograph equipped with a 1200 groove per millimeter grating and a 1024×128 pixel UV-enhanced CCD array allowing for rapid acquisition. Only the horizontal axis of the CCD output is needed to measure the SPIDER interferogram. To calibrate the apparatus, the linear phase term arising from the delay between the replicas was measured using a 10 μm Type-I KDP in place of the BBO. An alternative method is simply to rotate the BBO crystal by 45°. The reduced bandwidth of this configuration causes significant shaping of the resulting interferogram but is still sufficient for accurate measurement of the linear phase.

A significant feature of SPIDER with regard to measurements of the extremely broad spectra of sub-10 fs pulses is that no correction is required for the spectral sensitivity of the apparatus. This holds as long as the detection efficiency varies slowly with frequency compared to the interference period $2\pi\tau$, which in most cases poses no restriction. Moreover, systematic errors in SPIDER measurements can be detected easily by the simple expedient of measuring a pulse before and after propagation through a material with known dispersion.

In principle, spectrographic and interferometric measurements require high-resolution spectrometers to ensure that the signal form constraint is satisfied. In practice, however, a higher resolution spectrometer is needed for SPIDER than with most spectrographic techniques, such as FROG because of the spectral fringes. For the measurements discussed here, 0.5 nm resolution was needed for the wavelength range from 350 nm to 450 nm.

No fundamental limitations of the SPIDER technique are known for the characterization of spectra covering more than a full octave. The noise that arises from background SHG at such extreme bandwidths can easily be eliminated. The upconverted beam is geometrically separated from both fundamental beams. Stray light of the fundamental pulse replicas can be reduced by use of a polarizer, and stray light from the upconverter pulse simply adds an unmodulated background that is removed in the inversion. If required, it is possible to use a larger spectral shift in the upconversion process. This could be done, for example, by frequency-doubling the strongly chirped upconverter pulse, which should push the measurement bandwidth to where it is ultimately limited by the transparency range of the nonlinear optical material used.

3.4 Space–Time Coupling

For pulses in the few-cycle regime, space–time coupling effects cannot be ignored. For example, ultrabroadband pulses may exhibit pronounced frequency-dependent mode sizes (FDMS). This leads to beams whose spectral content varies with position and, consequently, to spatial dependence of the temporal pulse shape. The implication of this for pulse measurement methods is that the assumption that the field is factorized into a space-dependent part and a time-dependent part fails. This means that methods that integrate signals over the beam spatial profile require careful handling.

Other examples of space–time coupling occur when ultrashort pulses are focused or shaped with a zero dispersion line, as well as in most nonlinear interactions. In fact, space–time coupling is widely used in ultrafast technology and in some coherent control and nonlinear optics experiments. Spatial distortions can be rather complex in broadband amplifier chains and in sub-10 fs Ti:sapphire lasers [26]. Generally, these effects become more severe with larger pulse bandwidth. Even free-space diffraction of broadband beams is a source of spatial chirp. For Gaussian beams, the beam size w as a function of the propagation distance z is given as

$$w(z) = \sqrt{\frac{\lambda z_c}{\pi}\left[1 + \left(\frac{z}{z_c}\right)\right]}\,, \tag{21}$$

with wavelength λ and confocal parameter z_c. If the laser cavity is linear, the confocal parameter z_c is constant. Therefore, the mode area is proportional to the wavelength at a fixed distance z from the beam waist. Sources generating spectra with nearly one octave of bandwidth can exhibit nearly 50% difference in mode size across the spectrum.

FDMS has been observed in Kerr-lens mode-locked (KLM) lasers [27]. In such lasers, the time-dependent Kerr-lens generated in the gain medium

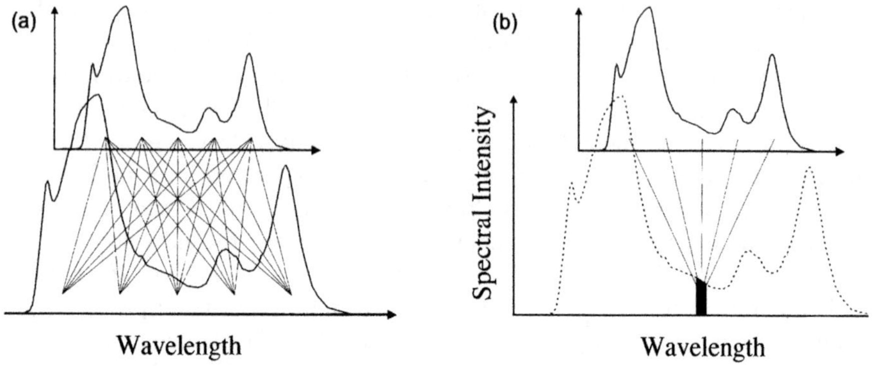

Fig. 6. Schematic picture of the mixing processes occurring in (**a**) broadband SHG and (**b**) SFG of a broadband pulse with a quasi-CW spectral slice. In SHG, many different input wavelengths with differing spatial patterns contribute to the signal at a given wavelength, whereas in SFG, each individual spectral component of the signal is generated by mixing with a single wavelength with fixed spatial mode (courtesy of Dr. L. Gallman)

maps the temporal intensity profile of the pulse into the spatial profile and vice versa. Thus the FDMS depends on the details of the pulse-temporal-shape. For such lasers, small adjustments of intracavity dispersion alter the temporal pulseshape in the Kerr medium, resulting in significant variations of the FDMS.

Aside from the phenomenon itself, spatial chirp affects the nonlinear processes that are required by all few-cycle pulse characterization techniques. In autocorrelation or cross-correlation-based methods using second-harmonic generation (SHG), each frequency component of the test pulse is mixed with every frequency component of a gating pulse, as shown in Fig. 6. Ideally, all these mixing processes occur with equal efficiency. However, even given perfect phase matching, the FDMS produces a frequency-dependent conversion efficiency. In the simple case of pure SHG of quasi-monochromatic slices of the spectrum, for example, a 41% increase in mode area (see Fig. 7) reduces the second-harmonic signal by a factor of 2.

Such efficiency variations can result in erroneous pulse characterization. In principle, the effect of the FDMS on an ideal Gaussian beam (21) may be accounted for in SHG-FROG measurements by suitable postprocessing in the reconstruction algorithm [28]. This requires, however, a model for the spatial chirp and may require ancillary measurements.

SPIDER uses sum-frequency generation (SFG). In this case, the two broadband test pulse replicas are mixed with two quasi-CW slices of a strongly chirped pulse. In the spectral domain, each of these SFG processes corresponds to the convolution of a broadband spectrum with a delta-function-like spectrum (Fig. 6b). As a result, the input beam is shifted by a constant frequency and multiplied by the spatial mode pattern of the particular quasi-CW

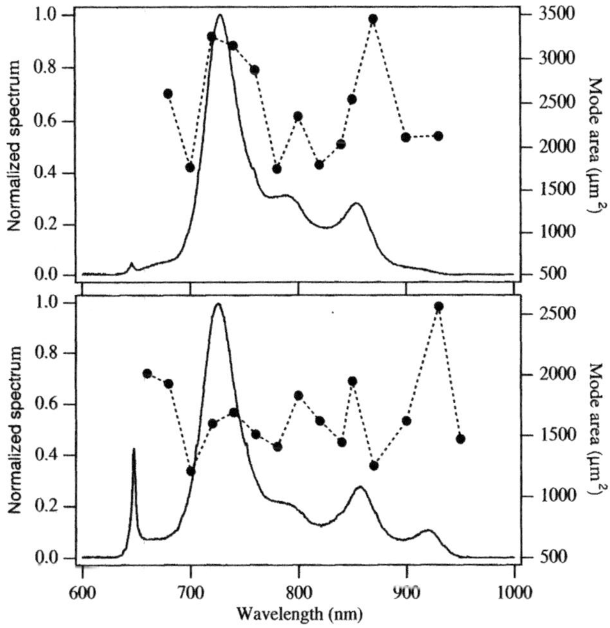

Fig. 7. Wavelength-dependent mode area of a sub-10 fs KLM Ti:sapphire laser. The spectral variation of the power density (*solid curve*) and the mode area (*dashed curve* and *filled circles*) reflect the dispersion of the double-chirped mirrors used inside the cavity. The *dash-dotted line* shows the qualitative behavior expected from (21) (up to a scaling factor) (courtesy of Dr. L. Gallman)

slice. Therefore, spatial chirp of the strongly temporally chirped pulse does not cause a frequency-dependent efficiency since the same CW slice is mixed with each frequency component of the test pulse. Thus, only the spatial intensity pattern, fringe contrast, and an undetermined phase constant of the SPIDER signal are affected. Importantly, the spectral fringe spacing is unaffected by spatial chirp. Because SPIDER uses only the latter for spectral phase reconstruction, SPIDER works correctly even in the presence of significant spatial chirp.

SPIDER is particularly well suited for measuring space–time coupling in ultrashort pulses because the temporal dependence of the field at a single point in the beam can be reconstructed from a one-dimensional measurement – a single spectral interferogram. Therefore, an imaging spectrometer and a two-dimensional detector array (a CCD camera, for example), enables measurements of the spatial dependence of the temporal field [29]. Moreover, the noniterative reconstruction algorithm enables rapid processing of the large amount of data resulting from the additional degree of freedom. This simple extension does not require any prior knowledge of the spatial chirp of the pulse before the apparatus, nor does it require the beam to be spatially filtered, as is often the case in autocorrelation-based measurements.

For the success of this method, it is important to preserve the spatial phase information during the nonlinear interaction and acquisition of the interferogram. This can be achieved by focusing the two time-delayed replicas of the input pulse into the nonlinear crystal, together with the unfocused chirped pulse. As the beam size of the focused replicas at the image plane is very small compared to the size of the unfocused chirped pulse, frequency conversion preserves the spatial information. A second recollimating lens then gives two frequency-shifted replicas $\widetilde{E}(x, \omega - \omega_0)$ and $\widetilde{E}(x, \omega - \omega_0 - \Omega)$. These fields may also be obtained by mixing the unfocused fundamental pulses with a spatially expanded chirped pulse. This eliminates any space–time coupling due to the focusing optics. The spectral interferogram is measured as a function of x and ω with a two-dimensional imaging spectrometer, whose entrance slit is oriented along x. The spatially resolved spectral phase gradient $\frac{\partial\phi}{\partial\omega}(x, \omega)$ is extracted by applying the Fourier transform spectral interferometry inversion algorithm to each spatial line of the interferogram. This gives the pulseshape at each location in the beam, but no wavefront information. The rapidity of the direct SPIDER inversion algorithm enables processing the two-dimensional interferogram in less than a second.

3.5 Space–Time SPIDER

It is possible to go beyond spatially resolved SPIDER measurements and relate the spectral phases at each point in the beam. This provides a unique capability, measurement of the complete spatiotemporal field, including spectrally dependent wavefront distortions [30]. The spectral phase gradient at each point in the beam $\frac{\partial\phi}{\partial\omega}(x, \omega)$ can be measured using a spatially resolved version of SPIDER, as described in the previous section. The spatial phase gradient is measured for each wavelength using a separate Michelson interferometer that generates two interfering spatially sheared replicas of the input field. This interferogram is spectrally and spatially resolved. From these two gradients, the complete spatiospectral phase is reconstructed. This procedure is diagrammed in Fig. 8.

The spectrally resolved spatial phase gradient $\frac{\partial\phi}{\partial x}(x, \omega)$ is measured by imaging the input beam at the fundamental frequency onto the slit of the two-dimensional spectrometer through the Michelson. This interferometer provides independent control of the shear, tilt, and delay between the two imaged pulses. In this configuration, the detected signal is proportional to $|\widetilde{E}(x, \omega) + \widetilde{E}(x + X, \omega) \exp(\mathrm{i}kx) \exp(\mathrm{i}\omega\tau)|^2$, where K is the tilt between the two wavefronts, X the spatial lateral shear, and τ the delay between the two pulses. The phase difference $\phi(x + X, \omega) - \phi(x, \omega) + Kx + \omega\tau$ can be extracted from this interferogram using standard Fourier transform algorithms, provided that the fringe periodicity due to the linear phase terms $Kx + \omega\tau$ is sufficiently small. A calibration trace to remove this phase can be recorded by setting the shear to zero while maintaining the tilt and delay between

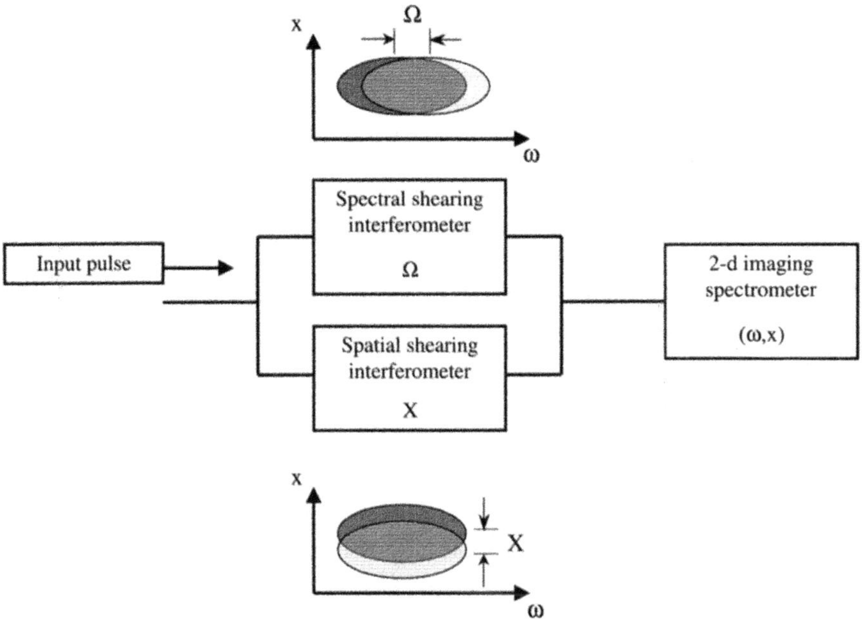

Fig. 8. The principle of space–time SPIDER. The input pulse is split into two. One-half is sent to a SPIDER apparatus in which the pulseshape at each location in the beam is measured. The other part is sent to a lateral shearing interferometer in which the wavefront is measured at every frequency in the pulse. The simultaneous acquisition of these two interferograms (which can be done using a single spectrometer and camera) gives the complete spatiospectral phase across the beam

the two replicas. This is most easily achieved by double-passing the interferometer to yield four replicas of the test pulse, among which two replicas are spatially overlapped (i.e., their relative spatial shear is equal to zero) but have a relative phase $2Kx + 2\omega\tau$. This phase difference can be extracted from the calibration interferogram and subtracted from all subsequent interferograms with these interferometer settings, and the required gradient can be reconstructed accurately.

The spatially resolved SPIDER interferogram, near the second harmonic wavelength of the input pulse, and the spectrally resolved lateral shearing interferogram, at the fundamental wavelength, can be recorded simultaneously on a single two-dimensional detector using the first and second diffraction order of the grating of the spectrometer. Single-shot operation of the device is therefore possible. Both phase gradients can be extracted simply from the single data set because they are encoded differently in the superimposed interferograms. The spectral phase gradient is extracted from spectral fringes that lie predominantly parallel to the spatial axis of the interferogram. The spatial phase gradient is obtained from spatial fringes that lie predominantly parallel to the spectral axis.

Methods capable of characterizing spatially chirped pulsed fields are likely to prove useful for optimization of pulses of extreme bandwidth, such as generated by nonlinear interactions in microstructured media.

4 Measurements of Few-Cycle Pulses Using SPIDER

4.1 Output of a Mode-locked Ti:Sapphire Laser with Double-Chirped Mirrors

SPIDER has been used to measure the amplitude and phase of sub-6 fs pulses from a state-of-the-art SESAM-assisted Kerr-lens mode-locked Ti:sapphire laser [26, 31]. Figure 9 shows the SPIDER interferogram of a pulse with a transform limit of 5.3 fs. The individual spectra of the upconverted pulses are also displayed. Note that they are identical but shifted, as expected. The spectral phase reconstructed from the SPIDER trace is plotted in Fig. 10, together with the independently measured pulse spectrum and the corresponding temporal intensity profile with a full-width at half-maximum (FWHM) of 5.9 fs. The oscillations in the central part of the phase are caused by the extracavity double-chirped mirrors [32], which are used to compensate for the dispersion of the output coupling mirror together with the higher order dispersion of the extracavity prism pair. Note the correspondence between the phase oscillations and the shape of the spectrum. This is expected since similar double-chirped mirrors are used in the laser cavity itself. The global 'S'-like shape of the phase stems from the residual uncompensated phase of the output coupler. This clearly demonstrates that SPIDER can be used to reconstruct directly the complicated phase distortions experienced by ultra-broadband pulses.

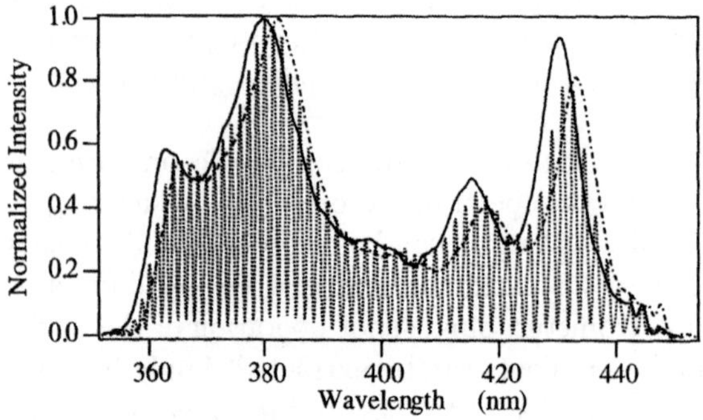

Fig. 9. SPIDER interferogram of a sub-6 fs pulse (*dotted curve*). Additionally, the spectra of the individual upconverted pulses are shown (*solid* and *dash-dotted curves*)

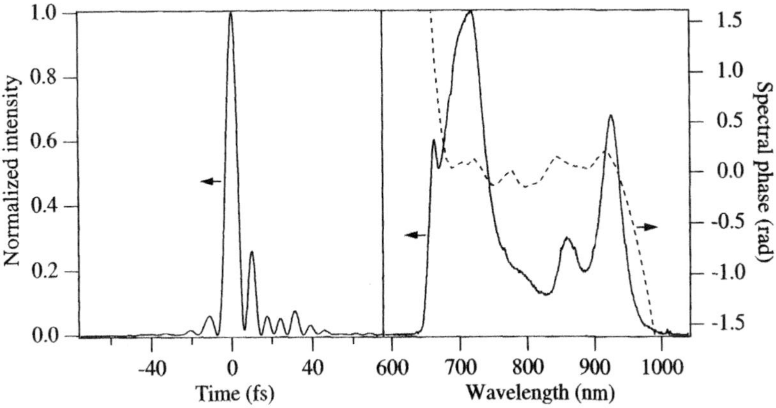

Fig. 10. Reconstructed temporal intensity profile (*left*) and spectral phase (*right*; *dashed curve*). The independently measured power spectrum of the pulse (*right*; *solid curve*) has a transform limit of 5.3 fs. The *solid curves* are referenced to the *left* and the *dashed curve* to the *right vertical axis*

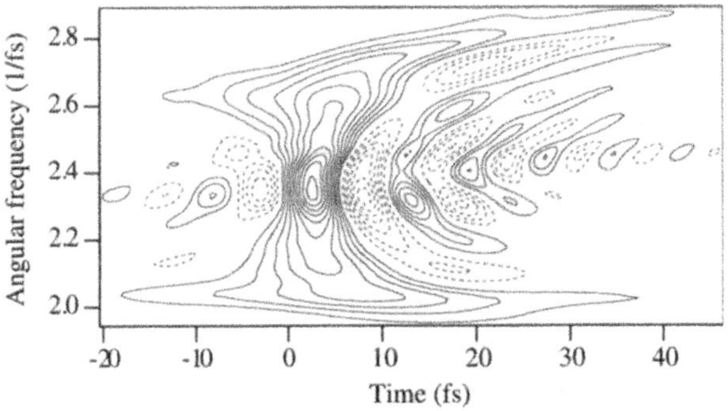

Fig. 11. Wigner trace of the reconstructed pulse. The contours are equidistant on a linear scale. The contours representing negative values are plotted as *dashed curves*. The positive values are represented by *solid curves* (courtesy of Dr. L. Gallman)

The Wigner trace calculated from the reconstructed pulse is shown in Fig. 11. It supports the conclusion that the phase structure in the spectral wings of the pulse prevents the pulse from being closer to its transform limit. As an independent check on the accuracy of the method, the interferometric autocorrelation (IAC) calculated from the SPIDER data was compared with a separately measured IAC (Fig. 12). The agreement is excellent even for the low-intensity structure in the wings of the IAC. The conventional but unjustified method of fitting a hyperbolic secant to the autocorrelation yields a deceivingly short pulse duration of 4.5 fs. This systematic underestimation of the pulse duration affirms the need for complete characterization methods.

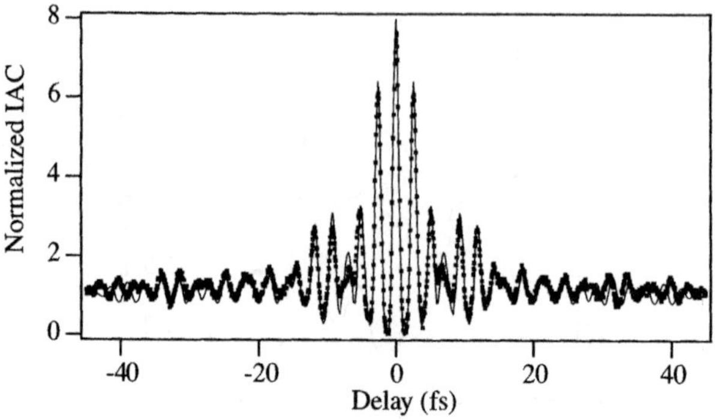

Fig. 12. Comparison of measured interferometric autocorrelation (IAC; *dots*) with the SPIDER-reconstructed IAC (*solid curve*)

4.2 Output of a Noncollinear Optical Parametric Amplifier

Very broad band radiation in the near infrared and visible may be generated using noncollinear optical parametric amplification (NOPA) [33, 34]. In several cases, bandwidths of over 150 THz have been observed. To compress such extreme spectra, detailed knowledge of the spectral phase is required. SPIDER is well suited to such applications. A typical source consists of a crystal of β-Barium Borate (BBO) arranged for Type-I phase matching for down-conversion of pump radiation at 400 nm or so. For pump pulse energies of one or two microjoules, when the device is seeded with a white-light continuum from a separate Ti:S crystal, the output spectrum can extend from 500 nm to 750 nm. Because of the chirp associated with the seed pulse and the inherent dispersion of the BBO, the output spectrum has a large group delay across its width. This can be compensated for by suitably designed dielectric mirrors [35]. The resulting pulse from such a system has been measured using upconversion SPIDER, though with a strong uncharacterized chirped pulse rather than one derived from the input test pulse itself. This cross-upconversion arrangement (or X-SPIDER) gives an improved signal-to-noise ratio.

The chirped pulse can be derived from the main laser that is used to generate the pump pulses for both the BBO and the Ti:S continuum source. In this case, the pulses are typically about 50 fs long, near 780 nm. In the experiment reported in [19], two replicas of the NOPA output pulse, delayed by 271 fs, were generated in a dispersion-compensated Michelson using a 100 μm thick broadband beam splitter, and mixed with a near-infrared chirped pulse in a 20 μm thick BBO crystal arranged for Type-II upconversion. The chirped pulse was stretched in a 5 cm long SF-10 glass block with dispersion 8270 fs^2 and polarized along the extraordinary axis of the BBO crystal. This resulted in a pair of upconverted pulses in a wavelength range of 300 nm to 380 nm. A 1200 lines/mm grating spectrometer with a 25 mm entrance slit was used to detect the interferogram. The calibration phase was measured using the

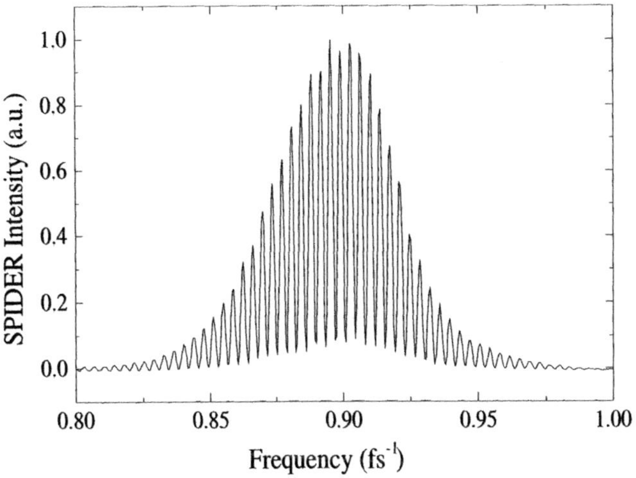

Fig. 13. SPIDER trace of the compressed pulse from a noncollinear NOPA. (Courtesy of Prof. S. Silvestri)

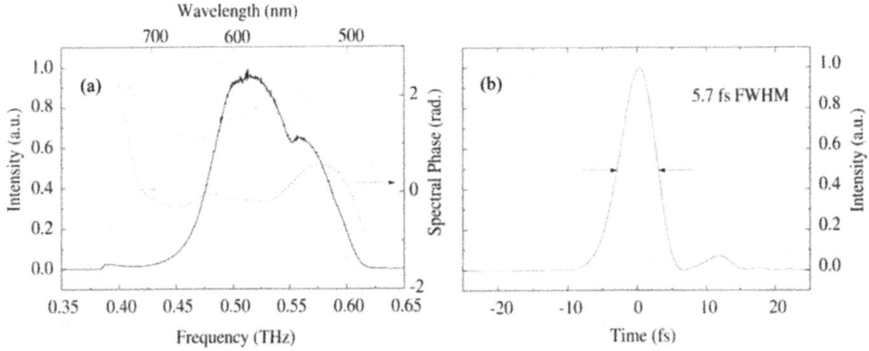

Fig. 14. (**a**) Reconstructed spectral phase and measured spectral amplitude and (**b**) reconstructed temporal electric-field envelope for a compressed NOPA pulse (courtesy of Prof. S. Silvestri)

second harmonic of the test pair replicas, which overlapped one end of the SPIDER interferogram. Because the overlap of the two spectra was not complete, it was necessary to calibrate the spectrometer using several lines from a mercury vapor lamp.

The spectral phase measurements reveal that the pulse had been almost optimally compressed, that is, the nearly flat spectral phase yields the shortest root-mean-square pulse duration possible for the given spectrum. The remaining oscillations in the spectral phase are due to residual dispersion in the mirror. The distortions for wavelengths longer than 720 nm are consis-

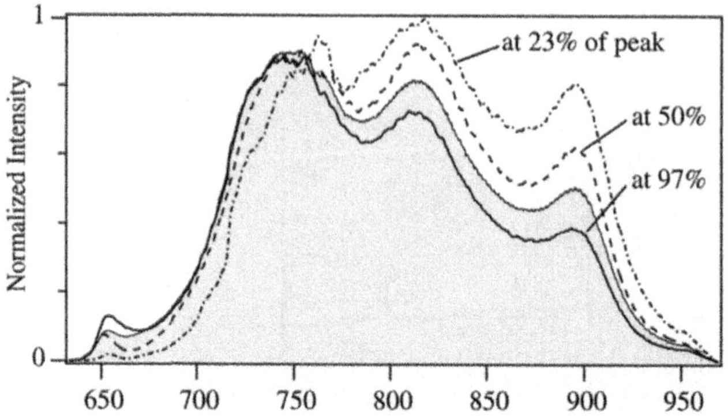

Fig. 15. Three spectra of a 10 fs pulse from a mode-locked Ti:sapphire laser measured at different lateral positions inside the beam from a mode-locked Ti:sapphire laser. For comparison, the spatially integrated spectrum is also shown (*shaded area*) (courtesy of Dr. L. Gallman)

tent with the mirror design parameters and allow compression of the pulse to a duration of 5.7 fs (compared with the transform-limited result of 5.2 fs).

4.3 Spatially Resolved Measurements from a Mode-locked Ti:Sapphire Laser

The spatial dependence of a few-cycle pulse beam has been measured by imaging the SPIDER signal beam and the fundamental beam from the SFG crystal on the entrance slit of a 0.3 m imaging spectrograph equipped with a 600 groove per millimeter grating and a two-dimensional 1024×128 pixel CCD camera [29]. In this experiment, the SPIDER signal and fundamental beam were measured independently. Simultaneous acquisition of both signals is also possible. The entrance slit provided spatial resolution along one axis, and the vertical CCD dimension resolves the beam in frequency. The magnification of the imaging optics gave an absolute spatial resolution of about 80 μm in each direction. With this setup, the spatially resolved spectral phase along one transverse slice of the beam was measured in a single acquisition. Other sections can be measured by either translation or rotation of the beam, or both. The beam can be rotated with a Dove prism or by out-of-plane reflection in an equivalent arrangement of mirrors. Figure 15 shows the fundamental spectrum of a sub-10 fs Ti:sapphire laser measured at three spatial positions relatively close to the beam center. With increasing distance from the center, the spectra shift to longer wavelengths, which agrees with the trend expected from (19). For different operating conditions of the laser, however, the mode-locking mechanism may even reverse the tendency expected from free space diffraction. Under typical operating conditions, the time-dependent Kerr lens was the dominant contributor to the FDMS. Similar measurements showed a highly irregular structure in the short-wavelength

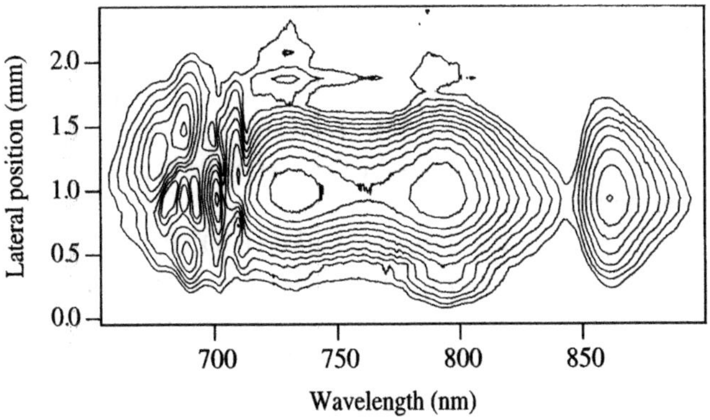

Fig. 16. Contour plot of the spatially resolved spectrum of a 10 fs pulse from a mode-locked Ti:sapphire laser. The contours are evenly spaced on a logarithmic scale and start at 0.7% of the maximum value. Note the complicated structure on the short-wavelength side (courtesy of Dr. L. Gallman)

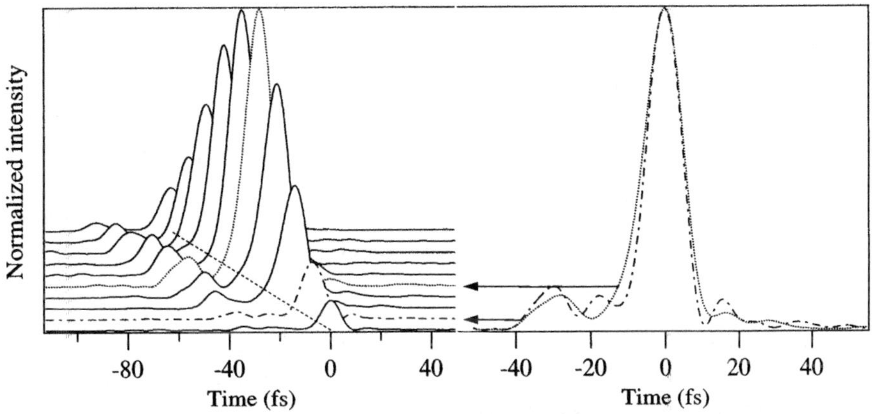

Fig. 17. The spatial dependence of the temporal pulse shape across the beam of a 10 fs pulse from a mode-locked Ti:sapphire laser (courtesy of Dr. L. Gallman)

part of the spectrum, as shown in the contour plot of the spatially resolved pulse spectrum (Fig. 16). This was attributed to diffraction effects due to clipping at one of the prisms used in the laser cavity and might be difficult to detect without spatially resolving the beam. The combination of the spatially resolved spectrum and the spectral phase measurement gives access to the full lateral dependence of the pulseshape (Fig. 17). On-center, the pulse has a duration of 12.3 fs, in contrast to an 11 fs duration in the wings. These variations result in systematic errors in spatially integrated measurements, the error depending on the specific technique and the power law of the nonlinear process employed.

5 Conclusions

Spectral interferometry provides a simple and robust way to characterize ultrashort optical pulses. In the form of SPIDER, it has been shown useful for rapid and accurate characterization of pulses in the few-cycle regime across a wide range of wavelengths [36].

There are a number of important features that make SPIDER particularly suitable for such applications. First, the rapidity of the data acquisition and inversion mean that the reconstruction is not compromised by the source stability. Moreover, the inversion algorithm returns the mean spectral phase when the signal is averaged over small random fluctuations in the pulseshape. The update rate for pulseshape reconstruction is usually limited by the time needed to write traces to the display apparatus: the algorithm itself runs at over 100 Hz, and interferograms can be acquired at 1 kHz.

Second, accurate measurement of the spectral phase does not require the recorded trace to be "corrected" for the phase-matching function of the nonlinear process or the detector sensitivity. The key to this remarkable robustness is that the phase information is contained in the fringe spacing rather than the visibility, and this is not compromised by wavelength-dependent responsivity in the apparatus, provided the sensitivity does not vary across one fringe. Moreover, systematic errors in SPIDER measurements can be easily identified by measuring a pulse before and after propagation through a material with known dispersion.

Third, the redundancy of data in a single SPIDER interferogram allows determining the precision of the measurement and also provides a built-in consistency check, analogous to the "FROG error" used in self-referencing spectrography.

Fourth, the inevitable presence of space–time coupling when using pulses of such extreme bandwidth means in many cases that measurements of the spatio-temporal field are warranted.

There are no fundamental limitations to the measurement of even shorter pulses using SPIDER. The major technical limitation lies in the nonlinear frequency conversion, which for all techniques, demands shorter crystals as the pulse bandwidth increases. This gives rise to noise in the form of non-phase-matched contributions to the signal. Nonetheless, in SPIDER, background signals from the fundamental of the test pulse and the upconverted pulse can be removed using filters, geometric multiplexing, or numerical filtering during inversion, and the apparatus measurement bandwidth is limited only by the transparency of the nonlinear materials used.

Acknowledgments

I am indebted to the many colleagues and students who contributed to the work described in this article. In particular, Christophe Dorrer, Chris Iaconis, and Lukas Gallman were instrumental in developing the SPIDER technique in its many forms. I enjoyed a very productive collaboration with the group of Prof. Dr. U. Keller in applications to the few-cycle regime, resulting in the excellent experimental results obtained by Dr. Gunter Steinmeyer and Dr. Gallman. I am indebted especially to Dr. Gallman for permission to use material from his Ph.D. thesis and for providing a number of the figures.

This work was supported by the National Science Foundation and the TRW Foundation.

References

[1] C. Froehly, A. Lacourt, J. C. Vienot: Notions de réponse impulsionelle et de fonction de transfert temporelles des pupilles optiques, justifications expérimentales et applications, Nouv. Rev. Optique **4**, 18 (1973)

[2] C. Dorrer, N. Belabas, J. P. Likforman, M. Joffre: Experimental implementation of Fourier-transform spectral interferometry and its application to the study of spectrometers, Appl. Phys. B-Lasers O. **70**, S9 (2000)

[3] D. N. Fittinghoff, J. L. Bowie, J. N. Sweetser, R. T. Jennings, M. A. Krumbügel, K. W. DeLong, R. Trebino, I. A. Walmsley: Measurement of the intensity and phase of ultraweak, ultrashort laser pulses, Opt. Lett. **21**, 88 (1996)

[4] L. Lepetit, G. Cheriaux, M. Joffre: Linear techniques of phase measurement by femtosecond spectral interferometry for applications in spectroscopy, J. Opt. Soc. Am. B **12**, 246 (1995)

[5] M. Takeda, H. Ina, S. Kobayashi: Fourier-transform method of fringe-pattern analysis for computer-based topography and interferometry, J. Opt. Soc. Am. A **72**, 15 (1982)

[6] V. A. Zubov, T. I. Kuznetsova: Solution of the phase problem for time-dependent optical signals by an interference system, Sov. J. Quantum Electron. **21**, 128 (1991)

[7] V. Wong, I. A. Walmsley: Analysis of ultrashort pulse-shape measurements using linear interferometers, Opt. Lett. **19**, 28 (1993)

[8] C. Iaconis, I. A. Walmsley: Spectral phase interferometry for direct electric-field reconstruction of ultrashort optical pulses, Opt. Lett. **23**, 79 (1998)

[9] C. Iaconis, I. A. Walmsley: Self-referencing spectral interferometry for measuring ultrashort optical pulses, IEEE J. Quantum Elect. **35**, 50 (1999)

[10] C. Dorrer: Implementation of spectral phase interferometry for direct electric-field reconstruction using a simultaneously recorded reference interferogram, Opt. Lett. **24**, 153 (1999)

[11] T. M. Shuman, M. E. Anderson, J. Bromage, C. Iaconis, L. Waxer, I. A. Walmsley: Real-time SPIDER: Ultrashort pulse characterization at 20 Hz, Opt. Express **5**, 13 (1999)

[12] W. Kornelis, J. Biegert, J. W. Tisch, U. Keller, M. Nisoli, G. Sansone, S. Stagira, C. Vozzi, S. D. Silvestri: Single-shot kHz SPIDER measurements of 3-cycle laser pulses, in CLEO (Opt. Soc. Am., Washington 2002) postdeadline paper # CPDC11

[13] W. Kornelius, J. Biegert, J. W. G. Tisch, M. Nisoli, G. Sansone, C. Vozzi, S. D. Silvestri: Single-shot kilohertz characterization of ultrashort pulses by spectral phase interferometry for direct electric-field reconstruction, Opt. Lett. **28**, 281 (2003)

[14] L. Gallmann, D. H. Sutter, N. Matuschek, G. Steinmeyer, U. Keller, C. Iaconis, I. A. Walmsley: Characterization of sub-6-fs optical pulses with spectral phase interferometry for direct electric-field reconstruction, Opt. Lett. **24**, 131 (1999)
[15] L. Gallmann: Ph.D. thesis, ETH Zürich (2001), unpublished
[16] P. Londero, M. E. Anderson, E. Kosik, I. A. Walmsley: Measuring ultrafast pulses in the near-ultraviolet using spectral phase interferometry for direct electric field reconstruction, J. Mod. Opt. **49** (2002)
[17] C. Dorrer, I. A. Walmsley: Accuracy criterion for ultrashort pulse characterization techniques: Application to spectral phase interferometry for direct electric field reconstruction, J. Opt. Soc. Am. B **19**, 101 (2002)
[18] C. Dorrer, I. A. Walmsley: Precision and consistency criteria in spectral phase interferometry for direct electric-field reconstruction, J. Opt. Soc. Am. B **19**, 103 (2002)
[19] M. Zavelani-Rossi, G. Cerullo, S. D. Silvestri, L. Gallmann, N. Matuschek, G. Steinmeyer, U. Keller, G. Angelow, V. Scheuer, T. Tschudi: Pulse compression over a 170-THz bandwidth in the visible by use of only chirped mirrors, Opt. Lett. **26**, 115 (2001)
[20] M. Zavelani-Rossi, D. Polli, G. Cerullo, S. D. Silvestri, L. Gallmann, G. Steinmeyer, U. Keller: Few-optical-cycle laser pulses by OPA: Broadband chirped mirror compression and SPIDER characterization, Appl. Phys. B-Lasers O. **74**, S24 (2002)
[21] P. Villoresi, P. Ceccherini, L. Poletto, G. Tondello, C. Altucci, R. Bruzzee, C. de Lisio, M. Nisoli, S. Stagira, G. Cerullo, S. D. Silvestri, O. Svelto: Spectral features and modeling of high-order harmonics generated by sub-10-fs pulses, Phys. Rev. Lett. **85**, 249 (2000)
[22] L. Li, S. Kusaka, N. Karasawa, R. Morita, H. Shigekawa, M. Yamashita: Amplitude and phase characterization of 5.0 fs optical pulses using spectral phase interferometry for direct electric-field reconstruction, Jpn. J. Appl. Phys. **40**, L68 (2001)
[23] M. Hirasawa, N. Nakagawa, K. Yamamoto, R. Morita, H. Shigekawa, M. Yamashita: Sensitivity improvement of spectral phase interferometry for direct electric-field reconstruction for the characterization of low-intensity femtosecond pulses, Appl. Phys. B-Lasers O. **74**, S22 (2002)
[24] M. Adachi, M. Hirasawa, A. Suguro, N. Karasawa, S. Kobayashi, R. Morita, M. Yamashita: Spectral-phase characterization and adapted compensation of strongly chirped pulses from a tapered fiber, Jpn. J. Appl. Phys. Part 2 - Lett. **42**, 2 (2003)
[25] P. Baum, S. Lochbrunner, L. Gallmann, G. Steinmeyer, U. Keller, E. Riedle: Real-time characterization and optimal phase control of tunable visible pulses with a flexible compressor, Appl. Phys. B-Lasers O. **74**, S21 (2002)
[26] D.H. Sutter, L. Gallmann, N. Matuschek, F. Morier-Genoud, V. Scheuer, G. Angelow, T. Tschudi, G. Steinmeyer, U. Keller: Sub-6-fs pulses from a SESAM-assisted Kerr-lens modelocked Ti:sapphire laser: At the frontiers of ultrashort pulse generation, Appl. Phys. B-Lasers O. **70**, S5 (2000)
[27] S. T. Cundiff, W. H. Knox, E. P. Ippen, H. A. Haus: Frequency-dependent mode size in broadband Kerr-lens mode locking, Opt. Lett. **21**, 662 (1996)
[28] A. Baltuska, M. S. Pshenichnikov, D. A. Wiersma: Second-harmonic generation frequency-resolved optical gating in the single-cycle regime, IEEE J. Quantum Elect. **35**, 459 (1999)

[29] L. Gallmann, G. Steinmeyer, D. H. Sutter, T. Rupp, C. Iaconis, I. A. Walmsley, U. Keller: Spatially resolved amplitude and phase characterization of femtosecond optical pulses, Opt. Lett. **26**, 98 (2001)

[30] C. Dorrer, E. M. Kosik, I. A. Walmsley: Direct space-time characterization of the electric-field of ultrashort optical pulses, Opt. Lett. **27**, 548 (2002)

[31] D. H. Sutter, G. Steinmeyer, L. Gallmann, N. Matuschek, F. Morier-Genoud, U. Keller, V. Scheuer, G. Angelow, T. Tschudi: Semiconductor saturable-absorber mirror-assisted Kerr-lens mode-locked Ti:sapphire laser producing pulses in the two-cycle regime, Opt. Lett. **24**, 631 (1999)

[32] F. X. Kärtner, N. Matuschek, T. Schibli, U. Keller, H. A. Haus, C. Heine, R. Morf, V. Scheuer, M. Tilsch, T. Tschudi: Design and fabrication of double-chirped mirrors, Opt. Lett. **22**, 831 (1997)

[33] M. Nisoli, S. D. Silvestri, O. Svelto, R. Szipöcs, K. Ferenz, C. Spielmann, S. Sartania, F. Krausz: Compression of high energy laser pulses below 5 fs, Opt. Lett. **22**, 522 (1997)

[34] A. Shirakawa, I. Sakane, M. Takasaka, T. Kobayashi: Sub-5-fs visible pulse generation by pulse-front-matched noncollinear optical parametric amplification, Appl. Phys. Lett. **74**, 2268 (1999)

[35] G. Cerullo, M. Nisoli, S. Stagira, S. D. Silvestri: Sub-8-fs pulses from an ultrabroadband optical parametric amplifier in the visible, Opt. Lett. **23**, 1283 (1998)

[36] L. Gallmann, D. H. Sutter, N. Matuschek, G. Steinmeyer, U. Keller: Techniques for the characterization of sub-10-fs optical pulses: A comparison, Appl. Phys. B-Lasers O. **70**, S67 (2000)

Part III

Applications

Optical Frequency-Comb Generation and High-Resolution Laser Spectroscopy

Thomas Udem, Ronald Holzwarth, Marcus Zimmermann, Christoph Gohle, and Theodor Hänsch

Max-Planck-Institut für Quantenoptik, 85748 Garching, Germany
thomas.udem@mpq.mpg.de

Abstract. The new alliance between the time and frequency domains in laser spectroscopy has made it possible to develop ultrafast counting schemes that can keep track of single optical oscillations. With this, counting optical oscillations of more than 10^{15} cycles in one second has become a simple task. High-resolution spectroscopy for basic research as well as metrology greatly benefits from this technology as it has allowed the highest possible precision. This development has also led to the construction of a all-optical atomic clocks that are expected to outperform current state-of-the-art cesium clocks eventually.

1 Introduction

In the frequency domain, a train of short pulses from a mode-locked laser is the result of the phase coherent superposition of many continuous wave (CW) longitudinal cavity modes. These modes form a series of frequency spikes that is called a frequency comb. In 1998, our lab demonstrated [1] that

- individual modes can be tracked by phase locking other CW lasers to them;
- the mode spacing is uniform across the comb within 3 parts in 10^{17};
- the mode spacing equals the repetition rate within 6 parts in 10^{16};
- the frequency comb can be self-referenced, thus fixing its position in frequency space.

With this, it became clear that the modes of a mode-locked femtosecond laser can be used as a precise ruler in frequency space to measure large frequency differences and absolute frequencies in the optical region. However, this ruler is free to drift as a whole, so that the frequency ω_n of the nth cavity mode may be expressed by

$$\omega_n = n\omega_\mathrm{r} + \omega_\mathrm{CE} \,. \tag{1}$$

Here, ω_r and ω_CE are the pulse repetition rate and some frequency offset ω_CE where n is chosen such that $\omega_\mathrm{CE} \leq \omega_r$. The frequency offset ω_CE can be determined by self-referencing, i.e., by measuring the frequency gap between *different* harmonics derived from the *same* CW-laser. Due to the limited bandwidth of our frequency comb, in a first crude experiment [1] we used the

F. X. Kärtner (Ed.): Few-Cycle Laser Pulse Generation and Its Applications,
Topics Appl. Phys. **95**, 295–316 (2004)

4th and the 3.5th harmonic of a 3.39 μm laser to determine ω_{CE}, according to $4\omega_n - 3.5\omega_{n'} = (4n - 3.5n')\omega_{\mathrm{r}} + 0.5\omega_{\mathrm{CE}} = 0.5\omega_{\mathrm{CE}}$, by choosing the mode numbers n and n' such that $(4n - 3.5n') = 0$. For a Ti:sapphire mode-locked laser operating around 900 nm, this requires a comb bandwidth of roughly 50 THz. As discussed in more detail in the Chapter by *Jones* et al., this has, at the same time, fixed the pulse carrier wave relative to the pulse envelope in the time domain [1, 2, 3]. The same idea has subsequently been discussed in [4], but the simplest and thus real practical approach to fixing the absolute position of the frequency comb is to measure the gap between ω_n and ω_{2n} of frequency components taken from the frequency comb itself [5,6,7,8,9,10]. In this case, the carrier–envelope (CE) offset frequency ω_{CE} is directly produced by beating the frequency-doubled red wing of the comb $2\omega_n$ with the blue side of the comb at ω_{2n}: $2\omega_n - \omega_{n'} = (2n - n')\omega_{\mathrm{r}} + \omega_{\mathrm{CE}} = \omega_{\mathrm{CE}}$ where again the mode numbers n and n' are chosen such that $(2n - n') = 0$. This approach requires a comb bandwidth of 375 THz if centered at 800 nm.

Once both radio frequencies ω_{r} and ω_{CE} have been phase locked to a precise reference, ideally an atomic clock, the frequency of every individual optical mode in (1) is phase locked as well to the radio-frequency reference. Thus, by virtue of the large number n, some 100 000 precisely known laser lines, nicely ordered, are available at once. This device is truly an optical synthesizer. Due to the large multiplication factor, it has initially been predicted that the modes would be buried in excess phase noise fuzz [11], but it turned out that the mode-locked laser itself provides a stable flywheel in the optical region and thereby greatly reduces this problem.

After the comb properties expressed by (1) were precisely verified [12], frequency combs were used to measure various optical frequencies of fundamental and practical interest (see [13] for an almost complete list up to 2001). Among them are the $1S - 2S$ transition frequency in atomic hydrogen [14], the cesium D_1 line [15], and the $^2S_{1/2} \rightarrow {}^2D_{5/2}$ quadrupole transition in a single trapped Hg^+ ion [16]. The $1S - 2S$ transition frequency is used to derive precise values for the $1S$-Lamb shift to test QED and to determine the Rydberg constant, and the cesium D_1 transition frequency is used to derive the fine structure constant from a measurement of the photon recoil of that line [17].

The optical synthesizer can be operated in reverse to produce a countable radio-frequency output (ω_{r}) that is phase coherently linked to a sharp optical transition. The whole setup, the optical "pendulum," and the frequency comb optical clockwork that counts its oscillations, make up an optical timepiece that is predicted to outperform even the current state-of-the-art cesium fountain clocks. The first optical clock of this kind has been constructed at the *National Institute of Standards and Technology* (NIST) [18]. It uses the above mentioned clock transition in a single trapped Hg^+. Among other well-suited ion standards, the "clock" transition in a single trapped $^{115}\mathrm{In}$ ion is particularly interesting because of the absence of most systematic shifts that affect other ions [19].

2 Frequency Combs and Pulse Trains

The transition from the time-domain description of a pulse train to the frequency-domain description of (1) can be understood by considering the properties of the pulse-creating laser. Let us assume that the laser cavity stores a soliton-like pulse of which a copy becomes available at the output coupler after each round-trip time T. Now, the laser cavity contains dispersive elements that make the round-trip group delay T different from the round-trip phase delay. The carrier wave will therefore continuously shift with respect to the pulse envelope by, say, $\Delta\phi$ per round-trip, as shown in Fig. 1. When the pulses leave the cavity, the relative CE-phase no longer changes. The Fourier transform of such a pulse train is easy to calculate if the electric field $E(t)$, at, say, the output coupling mirror can be written as the product of a carrier wave $\exp(-\mathrm{i}\omega_c t)$ and a periodic envelope function $A(t)$

$$E(t) = A(t)\mathrm{e}^{-\mathrm{i}\omega_c t} . \tag{2}$$

The envelope function defines the pulse repetition time $T = 2\pi/\omega_r$ by demanding $A(t) = A(t+T)$. Even though $E(t)$ is not necessarily periodic in time, we can still Fourier transform the envelope function $A(t)$ to obtain the frequencies of the CW waves that make up $E(t)$:

$$E(t) = A(t)\mathrm{e}^{-\mathrm{i}\omega_c t} = \sum_{m=-\infty}^{+\infty} A_m \mathrm{e}^{-\mathrm{i}(m\omega_r + \omega_c)t} . \tag{3}$$

Like (1), this simple picture suggests that all modes are separated from their neighboring modes by ω_r. We can even allow a periodic frequency chirp imposed on the pulses by choosing a complex envelope function $A(t)$.

Following a derivation in [23], it is instructive to consider the spectrum as it builds up from a series of N pulses with a given energy:[1]

$$E(t) = \frac{E_0}{\sqrt{N}} \sum_{m=0}^{N-1} \mathrm{e}^{\mathrm{i}m\Delta\phi} \mathcal{E}(t - mT) . \tag{4}$$

The Fourier transform of a single pulse $\widetilde{\mathcal{E}}(\omega)$ and the shift theorem

$$\mathcal{FT}\left[\mathcal{E}(t-\tau)\right] = \mathrm{e}^{-\mathrm{i}\omega\tau} \mathcal{FT}\left[\mathcal{E}(t)\right] \tag{5}$$

may be used to Fourier transform the right-hand side of (4):

$$\widetilde{E}(\omega) = \frac{E_0 \widetilde{\mathcal{E}}(\omega)}{\sqrt{N}} \sum_{m=0}^{N-1} \mathrm{e}^{-\mathrm{i}m(\omega T - \Delta\phi)} = \frac{E_0 \widetilde{\mathcal{E}}(\omega)}{\sqrt{N}} \frac{1 - \mathrm{e}^{-\mathrm{i}N(\omega T - \Delta\phi)}}{1 - \mathrm{e}^{-\mathrm{i}(\omega T - \Delta\phi)}} . \tag{6}$$

[1] The pulse to pulse carrier–envelope phase shift has been left out in [23]

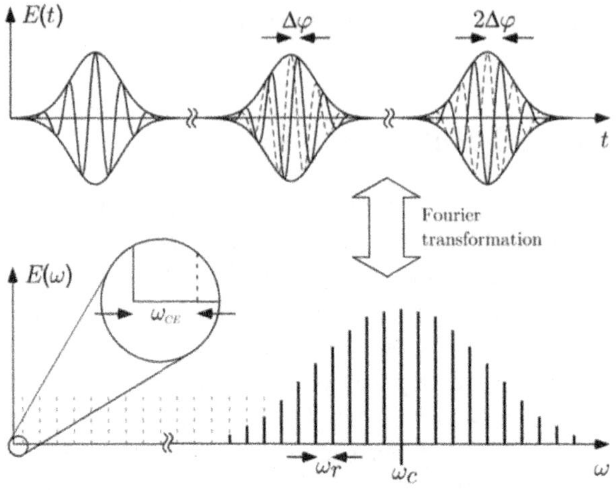

Fig. 1. *Top:* Consecutive pulses of the pulse train emitted by a mode-locked laser and the corresponding spectrum. As the carrier wave at ω_c moves with the phase velocity while the envelope moves with a different group velocity, the carrier wave shifts by $\Delta\phi$ after each round-trip with respect to the pulse envelope. *Bottom:* This continuous shift results in a frequency offset $\omega_{CE} = \Delta\phi/T$ of the comb consisting of exact harmonics of the pulse repetition frequency ω_r [1, 2, 3, 4, 20, 21, 22]

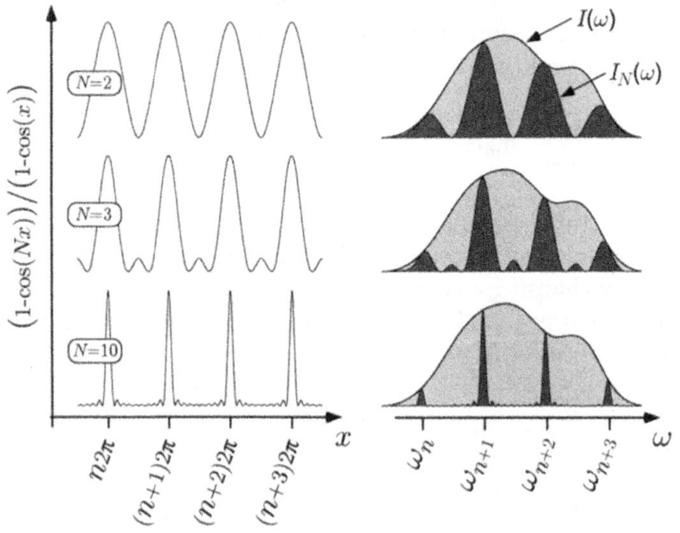

Fig. 2. *Left*: The function $[1 - \cos(Nx)] / [1 - \cos(x)]$ (peak normalized) for two, three, and ten pulses. The spectrum of the pulse train at the *right* is the product of the continuous spectrum of a solitary pulse $I(\omega)$ with that function. As the number of pulses increases, the modes build up at $x = \omega_n T - \Delta\phi = n2\pi$, where n again is an integer. From this, we derive $\omega_n = n\omega_r + \Delta\phi/T$

Thus the intensity spectrum $I_N(\omega)$ for N pulses can now be calculated from the intensity spectrum of a single pulse $I(\omega) \propto |\widetilde{\mathcal{E}}(\omega)|^2$:

$$I_N(\omega) \propto |\widetilde{E}(\omega)|^2 \propto \frac{1-\cos[N(\omega T - \Delta\phi)]}{N[1-\cos(\omega T - \Delta\phi)]} I(\omega)\,. \tag{7}$$

Figure 2 sketches this situation. Though the spectrum of a single pulse is truly a continuum, the modes appear as soon as the second pulse arrives and become sharper as the number of pulses increases. The spectral width of each mode for a large number of pulses is roughly given by $\Delta\omega \approx \sqrt{24/T^2(N^2-1)}$. It decreases as N^{-1}, such that after one second (i.e., $N = 1\,\mathrm{sec}/T$) it is down by $\sqrt{24}/2\pi\,\mathrm{Hz} \approx 0.78\,\mathrm{Hz}$ and continues to go down with the inverse observation time. As $N \to \infty$, the spectrum becomes a series of delta functions,

$$\frac{1}{2\pi} \lim_{N\to\infty} \frac{1-\cos(Nx)}{N\,[1-\cos(x)]} = \sum_n \delta(x - 2\pi n)\,, \tag{8}$$

with its peaks located at $x = \omega T - \Delta\phi = 2\pi n$. Thus, neglecting other broadening processes, an array of δ-function spectra laser modes, i.e., CW modes, arise. In addition, by comparison with (1), we conclude that $\omega_{\mathrm{r}} = 2\pi/T$ and $\omega_{\mathrm{CE}} = \Delta\phi/T$.

3 Testing the Mode Spacing Constancy

To be able to use the property of (1) for high precision measurements, one has to make sure that the simple intuitive picture given above is correct with very high accuracy. To test the mode spacing constancy, we used two laser diodes as an input for an optical frequency interval divider (OFID) [24,25], as sketched in Fig. 3. These two laser diodes are locked to distant modes of the comb by forcing the beat notes to stay in phase with a precisely known radio-frequency reference. Well-established servo techniques from rf phase-locked loop systems can be adopted to that [26]. The OFID produces an output beam at the precise midpoint of the frequency interval given by the two input laser diodes which is used to produce a third beat note with the comb. If the modes are distributed equally spaced on the frequency axis and if there is an odd number of modes between the two OFID inputs, then one would expect this beat note to appear at zero hertz or at some offset determined by the known radio-frequency references used for phase locking. We have averaged this beat signal for about 72 000 s worth of data and obtained a mean deviation of 3 parts in 10^{17} from the beat note expected under the assumption of a perfectly regular mode grid [12]. To rule out some possible deviations from the grid symmetrical to the center of the frequency comb, we modified the experiment such that the two OFID input laser frequencies would lie inside the blue wing of the comb. The average deviation was of the same order and, as in the previous test, consistent with zero.

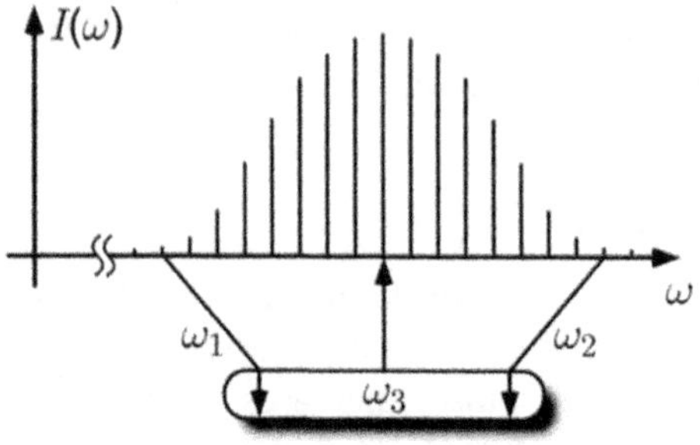

Fig. 3. To test the constancy of the mode spacing, we used an optical frequency interval divider (*oval symbol*) that locks the laser with the frequency ω_3 at the precise midpoint $(\omega_1 + \omega_2)/2$ between the two laser frequencies ω_1 and ω_2 by phase locking the second harmonic of ω_3 to the sum frequency $\omega_1 + \omega_2$ thus fixing $2\omega_3 = \omega_1 + \omega_2$

By comparing with another frequency comb generated by an electro-optic modulator, which has been tested previously [27], we could also verify that the mode spacing equals the pulse repetition rate within 6 parts in 10^{16} [12].

4 Self-Referencing

As already pointed out in the introduction, self-referencing a femtosecond frequency comb, as first suggested in [3], fixes its pulse to pulse CE-phase shift and thereby the frequency offset ω_{CE}. So far, self-referencing has been done in two different ways. Lacking an octave-spanning frequency comb, our first version was rather complicated (see Fig. 4).

The frequency comb fixes the frequency difference between $3.5\omega_0$ and $4\omega_0$ via mode spacing. The rest of the setup is used to create the frequencies $3.5\omega_0$ and $4\omega_0$ from a 3.39 µm HeNe laser at ω_0. The nonlinear steps necessary include second-harmonic generation (SHG) and one OFID. The latter is used here to lock the HeNe laser using the laser beams at $4\omega_0$ and $7\omega_0$ as input. Around 580 000 modes have been used to stabilize $0.5\omega_0$. Along with that frequency $4\omega_0$ is also an integral multiple of ω_r, and at the same time, it represents one of the modes. According to (1), this means that there is no CE-phase slippage[2] in the pulses when locked up as shown in Fig. 4. However, it should be pointed out that this ensures identical pulses only, without knowledge of the CE-phase.[3]

After the introduction of photonic crystal fibers [28,29] (also called microstructured fibers), self-referencing a frequency comb derived from a low peak

[2] Of course, this is true only if the reference frequencies for phase locking are neglected. However, their sum can always be set to zero

[3] Sometimes called the "absolute phase" even though a phase is, by definition, always measured relative to something else

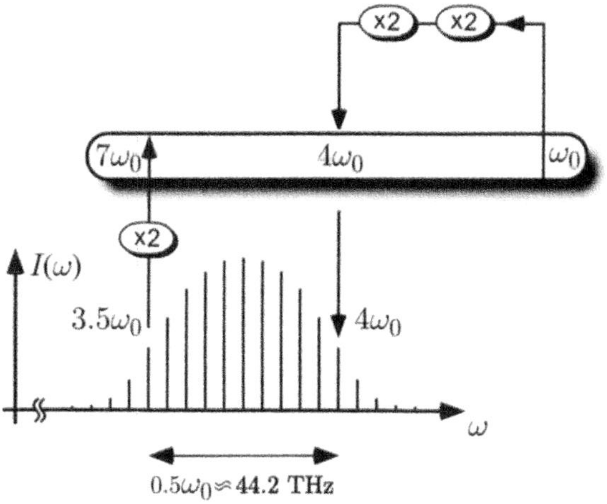

Fig. 4. The first direct radio-frequency to optical-frequency conversion using a femtosecond laser. As explained in the text, the optical-frequency interval divider (*oval symbol*) fixes the frequency ratios at precisely $7\omega_0 : 4\omega_0 : \omega_0$. With the frequency-quadrupling stage, that was already used to measure other optical transition frequencies relative to a HeNe reference, the frequency comb fixes the interval $4\omega_0 - 3.5\omega_0 = 0.5\omega_0$ and therefore ω_0 and any other frequency in the setup

power, high repetition rate fs laser has become a simple task [5, 6, 7, 8, 9, 10]. The fiber's pure silica core with a diameter of approximately 1.5 μm is surrounded by an array of air holes that run along the length of the fiber. The remarkable dispersion characteristics attainable with the large effective index step (including zero group velocity dispersion well below 800 nm), the high peak intensities associated with the short pulses, and the small core size enable effective self-phase modulation and therefore spectral broadening to more than an optical octave. The nonlinear action of the fiber, i.e., self-phase modulation, can be expressed by imposing a nonlinear phase $\Phi_{\mathrm{NL}}(t)$ on the pulse envelope, which is the same for all pulses. In this case, the pulse envelope is modified according to $A(t) \longrightarrow A(t)\mathrm{e}^{\mathrm{i}\Phi_{\mathrm{NL}}(t)}$, and the above derivation of the frequency comb remains valid. Even though a fiber can raise the noise level far above the shot noise limit [30], it generally maintains the comb structure.

A much simpler way of self-referencing is possible when an octave-spanning comb is produced with a photonic crystal fiber, as depicted in Fig. 5. By frequency doubling the modes directly, no additional laser is necessary [5]. In addition, this SHG can be efficient because the peak power after the fiber can still be significant. Another advantage of using the frequency comb itself arises from the fact that the CE-offset beat frequency $\omega_{\mathrm{CE}} = 2\omega_n - \omega_{2n}$ is produced by a large number of modes depending on the bandwidth of the SHG crystal. To bring all these beat notes in phase, both pulses, the original "blue" pulse (mostly around 540 nm) and the frequency-doubled "red"

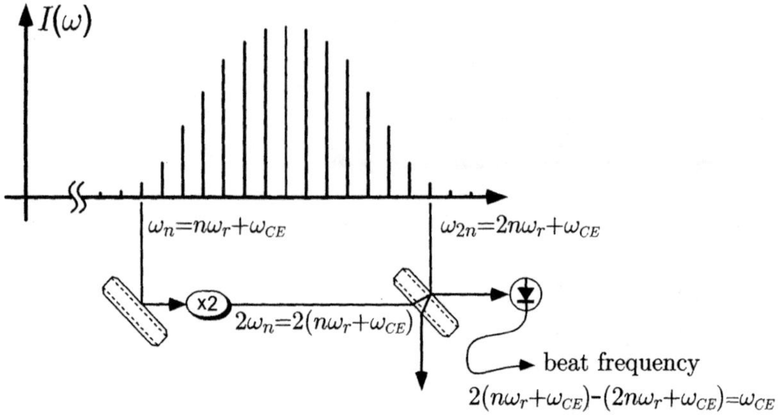

Fig. 5. The principle of the single laser optical synthesizer. A mode with the mode number n at the red wing of the comb and whose frequency is given according to (1) by $\omega_n = n\omega_\mathrm{r} + \omega_\mathrm{CE}$ is frequency doubled in a nonlinear crystal. If the frequency comb covers a full optical octave, a mode with the number $2n$ should oscillate simultaneously at $\omega_{2n} = 2n\omega_\mathrm{r} + \omega_\mathrm{CE}$. The beat note between the frequency-doubled mode and the mode at $2n$ yields the CE-offset frequency $2\omega_n - \omega_{2n} = 2(n\omega_\mathrm{r} + \omega_\mathrm{CE}) - (2n\omega_\mathrm{r} + \omega_\mathrm{CE}) = \omega_\mathrm{CE}$

pulse (mostly around 1080 nm) have to arrive at the ω_CE detector at the same time. This is accomplished by an optical delay line, as shown in Fig. 6. With this setup, the two radio frequencies that enter (1) are determined. To obtain a stable comb for absolute optical-frequency measurements, it is advantageous to phase lock both ω_r and ω_CE to a precise radio-frequency reference such as a cesium atomic clock or GPS-controlled quartz oscillator. The absolute frequency ω_l of an unknown CW laser is then determined by creating a beat note ω_b with the nearest mode:

$$\omega_\mathrm{l} = n\omega_\mathrm{r} + \omega_\mathrm{CE} \pm \omega_\mathrm{b} \,. \tag{9}$$

The correct sign of ω_b may be determined by a small change in either of the frequencies, and the mode number n may be determined by a coarse measurement of ω_l, for example, with a wavemeter. Figure 6 shows the current setup in our lab in more detail.

To control the two comb parameters ω_r and ω_CE necessary for phase locking, different approaches have been used. One obvious "knob" is to change the cavity length with a piezodriven folding mirror. This changes the pulse round-trip time T but leaves the CE-phase slippage for one round-trip as is, as long as the dispersion of the extra air can be neglected. According to $\omega_\mathrm{CE} = \Delta\phi/T$, it, nevertheless, has an impact on ω_CE, so that a cavity length change shifts the frequency of every mode proportional to their absolute frequencies. As the second "knob," we used the tilt of the mirror at the dispersive cavity end [15] for a prism-controlled laser. Suppose that the

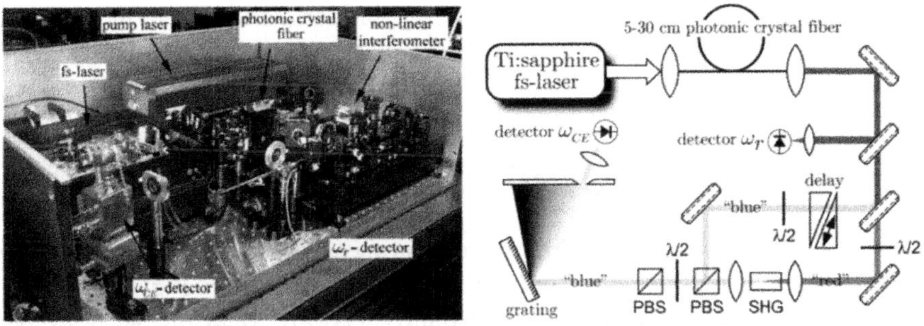

Fig. 6. The setup that is sketched in Fig. 5 in more detail. It consists of a high repetition rate ring laser $\omega_r = 2\pi \times 625$ MHz [32] (GigaOptics, model GigaJet) powered by a green pump laser (Coherent, model Verdi), a photonic crystal fiber from the University of Bath, England [28], and a nonlinear interferometer as sketched on the right-hand side. The half-wave plates are used to match the polarizations and to optimize the signal-to-noise ratio [20]

vertical tilt axis has its pivot at the reflection point of mode n'. Then modes with higher frequencies see a longer cavity, whereas modes with a lower frequency will have a shorter cavity, or vice versa, depending on the direction of the tilt [31]. Therefore, the mode at the tilt pivot keeps its frequency while the whole comb is increasing its mode separation. Again, this action causes both comb parameters to change, but in a different way. We used the slow tilt to control ω_r in order not to act too fast on this parameter (bandwidth $\approx$ 100 Hz). The faster (bandwidth $\approx$ 10 kHz) folding mirror was then used to position the comb as a whole, for example, to steer a particular ω_n or to control ω_{CE}.

For a laser based on chirped mirrors without intracavity prisms, this approach does not work. In this case, we used the power of the pump laser as the second "knob" [10]. It turned out that this effectively changes $\Delta\phi$ and therefore ω_{CE} [33]. Thus we used the power of the pump laser to shift the CE-offset frequency and a piezodriven folding mirror to control the repetition rate. To keep the noise level down when locking the repetition rate, it proved useful not to lock ω_r itself but a high harmonic of it (in the 10 GHz range) to a synthesizer [12]. Again the two controls are not independent, but they affect the round-trip group delay T and the round-trip phase delay differently. Keeping the servo bandwidths for ω_r as low as possible and for ω_{CE} as high as possible usually decouples the two servo systems.

In the meantime, there are several alternatives to spectral broadening in photonic crystal fibers. The group of *Russell* at the University of Bath, England, has shown that conventional telecom fibers tapered in a flame can have the same effect [34]. Also, as detailed in the contribution of *Kärtner* et al. in this volume, it is now possible to obtain octave-spanning spectra directly from a laser and high efficient spectral broadening in ZnO_2 has been reported by the group of *Wegener* at the University of Karlsruhe, Germany (see contribution by *Mücke* et al. in this volume). It remains to be shown which approach is suited best for frequency metrology applications.

5 Testing Optical Synthesizers

To check the integrity of a broad frequency comb and evaluate the overall performance of the $\omega : 2\omega$–optical synthesizer shown in Fig. 5, we compared one of its optical outputs at around 353.5 THz (848 nm) with a mode of the comb stabilized according to Fig. 4 [10]. For this, we phase locked a laser diode to $n\omega_r/2\pi + \omega_{CE} - \omega_{LO} = 565\,725 \times 624.87\,\text{MHz} + 64\,\text{MHz} - 20\,\text{MHz} = 353\,504\,624\,750\,000\,\text{Hz}$ relative to the quartz oscillator where ω_{LO} is a radio-frequency offset used as the local oscillator in the phase-locked loop. The $3.5\omega : 4\omega$-optical synthesizer was used to phase lock a second diode laser to 353 504 494 400 000 Hz; in a similar way, we expect a beat note of 130.35 MHz between them (see Fig. 7). In both cases, the ω_r and ω_{CE} were locked to the same radio-frequency source, a 10 MHz quartz oscillator, which is also used to reference a radio-frequency counter that is used to measure the beat frequency. After averaging our data worth a total of 4.5 h, we obtained a mean deviation from the expected beat frequency of 71 ± 179 mHz. This corresponds to a relative uncertainty of 5.1×10^{-16} [10]. Since we do not see a systematic shift at this level of accuracy, the measurement precision is dominated by the stability of the rf to optical conversion process. Using only two optical synthesizers, it is in general not possible to determine which one is more stable, but there is quite some evidence that this would be the more complicated $3.5\omega : 4\omega$ setup of Fig. 4.

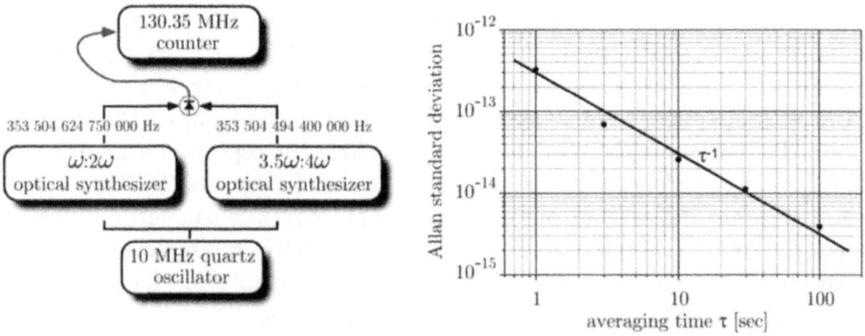

Fig. 7. *Left*: Checking the accuracy and stability by comparison of the two optical synthesizers in Fig. 4 and Fig. 5. A 10 MHz quartz oscillator is used by both synthesizers and multiplied to 354 THz. *Right*: The measured stability in terms of the standard Allan variance (SAV) [35]. Given the accuracy of 5.1×10^{-16}, the SAV gives the averaging time necessary to yield a certain accuracy

Even more accurate testing was recently performed by *Diddams* et al. [36]. In this work, individual modes of two independent frequency combs where locked to a stabilized 456 THz (657 nm) diode laser while their CE-offset frequency was locked to a hydrogen maser. Then the beating between the combs at 275 THz, 350 THz, and 550 THz was observed. The average uncertainty in the frequency of the comb lines was 4×10^{-17}. Quite remarkable in this experiment, the authors observed stability more than two orders of magnitude better than that shown in Fig. 7.

Other tests so far include an OFID experiment that is sketched in Fig. 3 with a fiber-broadened frequency comb [37]. In this case, the input to the OFID was taken from modes that were not emitted by the laser itself, but were created in the fiber. The relative deviation from the regular grid of modes was smaller than 3×10^{-18}.

So far none of these experiments did show a significant systematic shift. So it remains to be shown what limits the accuracy of these optical synthesizers and at which level this limit sets in.

6 Self-Differencing the Comb

Another possibility to produce stable frequency combs might be the creation of frequency combs without the CE-offset frequency ω_{CE} through difference-frequency generation (DFG). Subtracting lower order modes (n_2) from higher order modes (n_1),

$$\omega_{n2} - \omega_{n1} = n'\omega_{\mathrm{r}} \qquad \text{with} \quad n' = n_1 - n_2 > 0\,, \tag{10}$$

produces a frequency comb where all the modes are integral multiples ($n' = n_1 - n_2$) of the pulse repetition rate. This also produces a pulse with vanishing CE-phase slippage, i.e., a train of identical pulses. Such passive self-stabilization of the carrier–envelope phase has recently been demonstrated by *Baltuška* et al. [38] (also in this volume). Motivated by the observation that CE-offset frequency can be quite unstable compared to ω_{r}, it is hoped that self-differencing would yield more stable combs for frequency metrology applications.

A self differenced frequency comb can be created with a small modification of the setup in Fig. 6. The crystal that is phase matched to frequency double the "red" wing of the comb is, at the same time, phase matched to substract the "red" from the "blue" wing, if there is an optical octave between the two. This means that the self-differenced frequency comb is created by placing the SHG crystal of Fig. 6 outside the interferometer instead of the second polarizing beam splitter (PBS). As a first test of this method, we checked the position of the self-differenced frequency comb with respect to the original comb. The two should deliver a beat note at ω_{CE}, as claimed by (1) and (10). We compared this beat frequency with the self-referenced beat note created the usual way and found, after averaging almost 9 hours worth of data, a mean deviation of $0.21 \pm 2.1\,\mu$Hz or 6.6×10^{-21} relative to the infrared part of the frequency comb [39]. Together with the results from the previous section, this means that a self-differenced frequency comb operates with the same accuracy. Whether or not this comb is more stable remains to be shown.

7 Spectroscopy

7.1 Cesium

The first absolute optical-frequency measurement using a femtosecond laser was even done without self-referencing [15]. In that experiment, we determined the frequency of the ^{133}Cs D_1 line of 335 THz (895 nm) relative to the fourth harmonic of a precisely calibrated methane-stabilized HeNe laser at 3.39 µm.[4] The two laser frequencies, the well-known fourth harmonic of the HeNe/CH_4 laser and the "unknown" D_1, were 18 THz apart. To measure this difference, we locked one of the modes of a frequency comb, produced by a 73 fs laser, to the fourth harmonic of the CH_4 laser standard and observed a beat note with a diode laser that was employed in a cesium saturation spectrometer. By measuring the repetition rate at the same time, we could determine the D_1 frequency modulo the repetition rate. Unfortunately, we could neither resolve the modes with our wavemeter, nor did we want to rely solely on previous measurements to exclude the $\omega_r \approx 2\pi \times 75\,$MHz ambiguity. The observation of a frequency shift $n \times \delta\omega_r$ of one of the modes after changing ω_r by $\delta\omega_r$ may be difficult because it demands a resolution of $\delta\omega_r$ to distinguish between the mode numbers n and $n \pm 1$. The shift $n \times \delta\omega_r$ could not be chosen too large if one has to track it with a phase-locked diode laser. Therefore, we coupled the pulses into an additional cavity that was locked to have a free spectral range of exactly 20 times the repetition rate [15, 40], effectively multiplying the mode spacing by 20. The resulting repetition rate of $\omega_r \approx 2\pi \times 1.5\,$GHz was well above the resolution limit of our wavemeter. In this way, we recorded a total of 112 lines of the four hyperfine components to determine improved values for the hyperfine centroid and

$$f_{D1} = 335\,116\,048\,807\,(41)\,\text{kHz} \tag{11}$$

and the hyperfine splitting of the $6P_{1/2}$ state:

$$\Delta f_{6P_{1/2}}^{\text{HFS}} = 1\,167\,688\,(81)\,\text{kHz}\,. \tag{12}$$

7.2 Determination of the Fine Structure Constant

For a long time, the most precise value for the fine structure constant α was derived from a measurement of the anomalous magnetic moment $g-2$ of the

[4] This laser was designed and constructed in the group of *Bagayev* at the Institute of Laser Physics, Novosibirsk, Russia

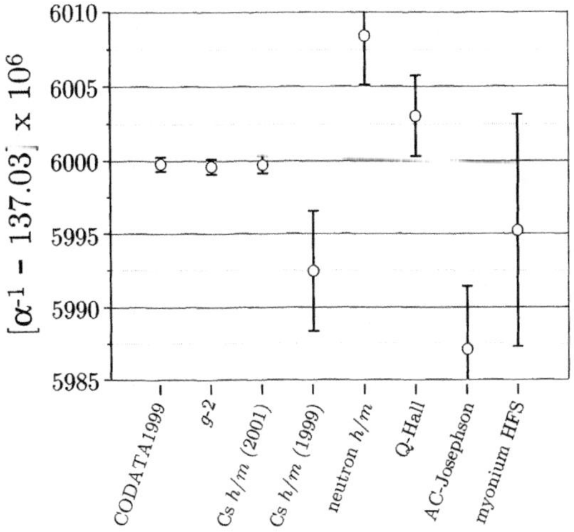

Fig. 8. The fine structure as determined by different methods. The CODATA99 recommended follows closely the $g-2$ value. From cesium measurements (recoil, D_1 line, and mass), one derives $\alpha^{-1} = 137.03599971\ (60)$. This may be considered the first independent confirmation of the $g-2$–value with comparable accuracy [42]. Other measurements are based on m_n/h for neutrons, the quantum Hall effect, the ac Josephson effect, and the myonium hyperfine structure [43]

electron and comparison of this value with its theoretical value. The extensive QED calculations [41] are largely unverified and other less precise values were partially in disagreement. Therefore, an independent determination with an accuracy at the ppb level was desperately needed. An effort started by *Chu* and co-workers (Stanford University) [17] aimed to solve that problem by using the recoil shift of the Cs D_1 line $f_{\text{rec}} = f_{\text{D1}}^2 h/2m_{\text{Cs}}c^2$ and the very well known Rydberg constant $R_\infty = m_e c\alpha^2/2h$:

$$\alpha^2 = \frac{2R_\infty}{c}\frac{h}{m_e} = 2R_\infty \times \frac{2cf_{\text{rec}}}{f_{\text{D1}}^2} \times \frac{m_p}{m_e} \times \frac{m_{\text{Cs}}}{m_p}\,. \tag{13}$$

The mass ratios on the right-hand side can be measured very precisely in penning traps by their cyclotron frequencies. It is noteworthy that every term in (13) is determined by a frequency measurement, including R_∞ (see Sect. 7.4) taking advantage of this intrinsically most precisely measurable quantity. A comparison of the most recent result [42] and the electron $g-2$ value along with values obtained by various other methods is given in Fig. 8.

7.3 Indium

In quite a similar way, the Cs D_1 line was measured. In collaboration with the group of *Walther* at our institute in Garching, we have also determined a sharp transition in a single captured $^{133}In^+$. This transition may be used in an all-optical clock as the "pendulum" in combination with an optical synthesizer that would count its oscillations. The $5s^2\ ^1S_0 \rightarrow 5s5p\ ^3P_0$ "clock" transition is excited by the fourth harmonic of a narrow line Nd:YAG laser at 946 nm and detected using the shelving method [44]. This time, a 36.65 THz frequency difference had to be measured to the fourth harmonic of our HeNe/CH_4 standard at 848 nm. At that time, this was out of the range of our fs laser system, so we used self-phase modulation in a single-mode fiber to achieve a usable[5] bandwidth of about 50 THz. Unlike the Cs D_1 line, the accuracy of the In^+ clock transition was now limited by the accuracy of the HeNe/CH_4 reference [45]:

$$f_{\mathrm{In}} = 1\,267\,402\,452\,899.92\,(23)\,\mathrm{kHz}\,. \tag{14}$$

7.4 Hydrogen

The initial purpose of constructing the optical synthesizer shown in Fig. 4 was to measure the hydrogen $1S-2S$ transition frequency at 2466 THz (121 nm). To observe this narrow transition in a beam of atomic hydrogen, a frequency-doubled dye laser was used. A large suppression of the Doppler effect is made possible by a two-photon excitation scheme. Therefore, the dye laser operates at one-fourth of the $1S-2S$ transition frequency. This frequency is close to $7\omega_0$ of Fig. 4, so that the dye laser could be used as part of the OFID stage. As explained in Sect. 4, this links the dye laser frequency directly to the Cs clock that controls the mode spacing of the mode-locked laser. Here, the 10 MHz output from the Cs clock is multiplied in one step to 44.2 THz and with only a few more nonlinear steps to vacuum UV.

For the first set of measurements [7], we used a commercial Cs atomic clock (Hewlett Packard, model 5071A) that was continuously monitored by comparing it with a time signal broadcast by the Global Positioning System (GPS). To improve on the accuracy, we subsequently invited the group of *Clairon* from the *Laboratoire Primaire du Temps et des Fréquences* (BNM-LPTF), Paris, to bring their Cs fountain clock [46] to our laboratory. This clock belongs to the current most precise clocks. Because of its superior stability, it also allows us to reduce the statistical uncertainty below the systematic uncertainty of about 10^{-14} in a much shorter measurement time. With this clock, the uncertainty of the $1S-2S$ transition frequency could be reduced by another order of magnitude, making it the best known optical frequency at the time [14]. For the hyperfine centroid, we obtained

$$f_{1S2S} = 2\,466\,061\,413\,187\,103\,(46)\,\mathrm{Hz}\,. \tag{15}$$

[5] The range where other lasers can be locked to the comb is usually much broader than the FWHM

8 Rydberg Constant and $1S$-Lamb Shift

With precisely measured transition frequencies in hydrogen, one can determine the Rydberg constant and the Lamb shifts of various levels [47]. If a transition involving the ground state is involved, the largest, and thus most interesting $1S$-Lamb shift may be determined. The Rydberg constant represents an important cornerstone for the determination of the most probable set of the natural constants [43], but the latter provides one of the most stringent testing grounds for the theory of quantum electrodynamics (QED).

Hydrogen is the most simple of all atoms, and its properties have been calculated very precisely [48]. For highest precision, its energy levels are conventionally described as the sum of three contributions: the hyperfine interaction, which is very well known for the states discussed here, and the Dirac energy $R_\infty e(nl)$ which includes all recoil corrections[6] up to the order α^4. All that is left, e.g., QED contributions, remaining recoils contributions, nuclear size effects etc., is by definition called the Lamb shift L_{nl}. The Lamb shift scales roughly as n^{-3} and is much smaller for P and D states than for S states. For the $1S-2S$ hyperfine centroid we write

$$f_{1S-2S} = R_\infty \left[e(2S) - e(1S)\right] + L_{2S} - L_{1S} \,. \tag{16}$$

In this equation, the left-hand side is determined experimentally to extract some values of the quantities on the right. To obtain the interesting numbers, R_∞, L_{2S}, and L_{1S}, other transition frequencies and combinations of them are used. For the most precise determination, a general adjustment method that uses all available data is used. For a more detailed overview and reference to the measurements mentioned, check [47] and references therein. The result of such an adjustment for the $1S$-Lamb shift is

$$L_{1S} = 8172.840\,(22)\,\mathrm{MHz}\,. \tag{17}$$

Unfortunately, the theory uses the proton charge radius as a parameter that is not known sufficiently accurate. Depending on which of the partially contradicting values from the literature we use to determine the theoretical value, we find a discrepancy reaching from 2.4 to 5.7 combined standard deviations [47]. For this reason there are ongoing preparations at the *Paul-Scherrer-Institut* Villingen, Switzerland, to determine this quantity much more precisely from the Lamb shift of muonic hydrogen.

In contrast to the $1S$-Lamb shift, the Rydberg constant has seen dramatic improvements over the years as detailed in Fig. 9. Today, the Rydberg constant is the most precisely known natural constant with the current value of

$$R_\infty = 10\,973\,731.568\,550\,(84)\,\mathrm{m}^{-1}\,. \tag{18}$$

[6] Unlike the Schrödinger theory, the Dirac theory does not allow us to account for a finite mass nucleus simply by replacing m_e by the reduced mass. Instead, an expansion in α is used

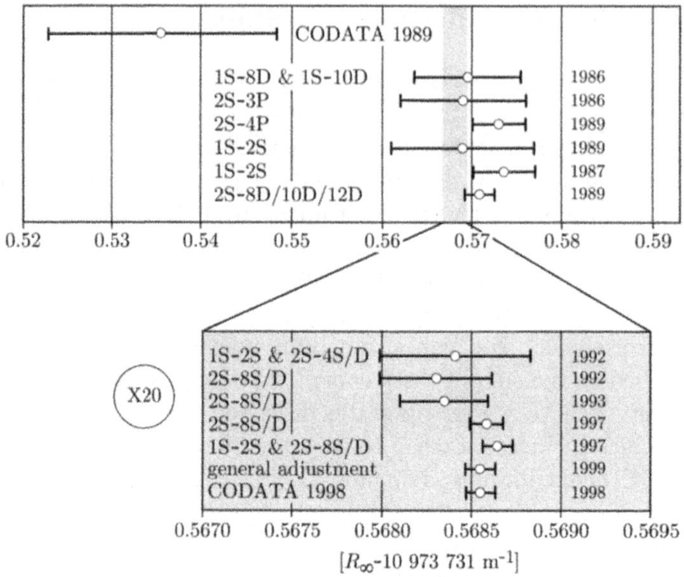

Fig. 9. History of the values of the Rydberg constant determined from the spectroscopy of the hydrogen atom along with the transition frequencies used. The general adjustment uses all hydrogen and deuterium data available

Because it is made up of other constants, according to $R_\infty = m_e c\alpha^2/2h$, its value is very useful for adjusting the complete set of natural constants to their most likely values compatible with a large variety of measurements [43].

9 Optical Clocks

An optical clock consists, like any other clock, of an oscillator (the "pendulum") that defines the ticks in time and a counter that keeps track of these cycles (the "clockwork mechanism"). In a cesium clock, for example, the oscillations are those of the nuclear spin precessing in the field of the electrons. An electronic counter advances the second hand every time the counter has completed 9 192 631 770 oscillations. This number was chosen when the SI second was redefined for the last time in 1967 to match the previous definition. If we look at the history of timekeeping and compare clocks as different as sun dials, pendulum clocks, and quartz clocks, it is obvious that they get more accurate as the oscillation frequency increases simply because a higher oscillation frequency slices time into smaller pieces. In general terms, the fractional frequency instability is usually measured in terms of the standard Allan deviation $\sigma(\tau)$. Roughly speaking, this quantity gives the mean deviation of the results of two consecutive measurements of duration τ [35]. For a clock that uses a "pendulum" at a frequency ω_0 the quantum noise limited instability is calculated by [49]

$$\sigma(\tau) = \frac{\Delta\omega_0}{\pi\omega_0}\sqrt{\frac{T_c}{2N\tau}}, \tag{19}$$

where $\Delta\omega_0$ gives the linewidth of the transition, and T_c the cycle time required to interrogate N atoms. The statistical uncertainty expressed by (19) decreases as the inverse square root of the measurement time and eventually reaches below the systematic uncertainties not included in (19). Even if these systematics can be reduced to a relative level of 10^{-18}, a conventional cesium clock that uses a microwave transition would require unrealistically long averaging times. For this reason, using an optical transition as the clock "pendulum" increasing ω_0 by about five orders of magnitude has been discussed for a long time.

After tremendous advances in laser spectroscopy in the 1970s that ultimately resulted in trapped atom standards in the 1980s (for trapped ions, see [50], and for trapped neutrals, see [51]), systematic uncertainties may be reduced to 10^{-18} for some of these standards [19]. With the femtosecond frequency synthesizer at hand, counting optical oscillations has become a simple task. Still a rather new technology, these optical "clockworks" can run continuously for hours with much technical improvement yet to come. Currently, the setup that probably comes closest to what we believe deserves the term "optical clock" uses a transition at 1064 THz in a trapped single mercury ion and is operated at NIST [18]. This clock already challenges today's best cesium fountain clocks in accuracy. The measurement of the stability of this clock was essentially limited by the comparison with the NIST hydrogen maser ensemble which is one of the most stable radio frequencies available. Besides the mercury clock, there are other optical clocks in preparation based on neutral calcium [49,52] and molecular iodine [53]. Other national standard institutions are following close in the same direction, such as the *Physikalisch-Technische Bundesanstalt* (PTB) in Braunschweig, Germany, on Yb^+ [54], the *National Research Council of Canada* (NRC) in Ottawa, Canada [55], and the *National Physics Laboratory* (NPL) in Teddington, England, on Sr^+ [56]; and the *Advanced Industrial Science and Technology* (AIST) in Tsukuba, Japan, on I_2 and Rb [57]. In our institute, in the group of *Walther*, a clock based on trapped indium is being prepared [19].

10 Further Applications

Hopefully these new optical clocks will open a new window to nature where we can expect new discoveries and phenomena. One example is the quest for natural constants that spatially vary or that would drift slowly in time as the universe evolves, as discussed by some theoreticians [58, 59]. So far, experimental tests have not been able to detect such behavior [16,60,61] and recent astronomical observations suggest evidence for cosmological evolution of the fine structure constant [62]. These clocks may also help to refine general relativity (GR) which still poses one of the major problems in physics as it refuses proper quantization. This theory still remains untested beyond a level of a few parts in 10^5 [63,64], which might be the reason why small corrections due to a quantized theory have not been discovered so far.

Among the industrial applications is the realization of the unit of length. Since the speed of light has a defined value, the wavelength of a laser is as precisely known as its frequency. For practical realization, iodine-stabilized lasers have been used together with interferometers for the conversion to a macroscopic length. The frequencies of those lasers have been determined with harmonic frequency chains and international comparisons. The precision of this approach can easily be reached with a laser directly stabilized to a GPS receiver making laser transports dispensable.

Finally applications such as satellite navigation, very long baseline interferometry, communication, and network synchronization could also benefit from improved clock performance.

Acknowledgements

We would like to thank our collaborators P. Lemonde, G. Santarelli, M. Abgrall, Ph. Laurent, and A. Clairon from BNM-LPTF, Paris, and J. Knight, W. Wadsworth, and P. St. J. Russell from the University of Bath, England, for their support.

References

[1] T. Udem, J. Reichert, R. Holzwarth, T. W. Hänsch, M. Kourogi: The measurement of large optical frequency differences and the design of a new type of frequency chain, in Proc. Joint Meeting Eur. Freq. Time Forum and the IEEE International Frequency Control Symposium (IEEE-press 1999) pp. 620–625

[2] L. Xu, C. Spielmann, A. Poppe, T. Brabec, F. Krausz, T. W. Hänsch: Route to phase control of ultrashort light pulses, Opt. Lett. **21**, 2008–2010 (1996)

[3] T. Udem: *Phasenkohärente optische Frequenzmessungen am Wasserstoffatom. Bestimmung der Rydberg-Konstanten und der 1S Lamb-Verschiebung*, Ph.D. thesis, Ludwig-Maximilians-Universität, Munich (1997)

[4] H. R. Telle, G. Steinmeyer, A. E. Dunlop, J. Stenger, D. H. Sutter, U. Keller: Carrier-envelope offset phase control: A novel concept for absolute optical frequency measurement and ultrashort pulse generation **69**, 327–332 (1999)

[5] S. A. Diddams, D. J. Jones, S. T. Cundiff, J. L. Hall, J. K. Ranka, R. S. Windeler, A. J. Stentz: A direct rf to optical frequency measurement with a femtosecond laser comb spanning 300 THz, in OSA Tech. Dig. Ser. (Opt. Soc. Am. 2000) pp. 109–110

[6] R. Holzwarth, J. Reichert, T. Udem, T. W. Hänsch, J. C. Knight, W. Wadsworth, P. S. J. Russell: Broadening of femtsecond frequency combs and compact optical to radio frequency conversion, in OSA Tech. Dig. Ser. (Opt. Soc. Am. 2000) p. 197

[7] J. Reichert, M. Niering, R. Holzwarth, M. Weitz, T. Udem, T. W. Hänsch: Phase coherent vacuum-ultraviolet to radio frequency comparison with a mode-locked laser, Phys. Rev. Lett. **84**, 3232–3235 (2000)
[8] S. A. Diddams, D. J. Jones, J. Ye, S. T. Cundiff, J. L. Hall, J. K. Ranka, R. S. Windeler, R. Holzwarth, T. Udem, T. W. Hänsch: Direct link between microwave and optical frequencies with a 300 THz femtosecond laser comb, Phys. Rev. Lett. **84**, 5102–5105 (2000)
[9] D. J. Jones, S. A. Diddams, J. K. Ranka, A. Stentz, R. S. Windeler, J. L. Hall, S. T. Cundiff: Carrier-envelope phase control of femtosecond mode-locked lasers and direct optical frequency synthesis, Science **288**, 635–639 (2000)
[10] R. Holzwarth, T. Udem, T. W. Hänsch, J. C. Knight, W. J. Wadsworth, P. S. J. Russell: Optical frequency synthesizer for precision spectroscopy, Phys. Rev. Lett. **85**, 2264–2267 (2000)
[11] H. R. Telle: Absolute measurement of optical frequencies, in M. Ohtsu (Ed.): *Frequency Control of Semiconductor Lasers* (Wiley, New York 1996) pp. 137–167
[12] T. Udem, J. Reichert, R. Holzwarth, T. W. Hänsch: Accurate measurement of large optical frequency differences with a mode-locked laser, Opt. Lett. **24**, 881–883 (1999)
[13] P. Gill (Ed.): *Frequency Standards and Metrology*, Proc. 6th Symposium, St. Andrews (World Scientific, Singapore 2002)
[14] M. Niering, R. Holzwarth, J. Reichert, P. Pokasov, T. Udem, M. Weitz, T. W. Hänsch, P. Lemonde, G. Santarelli, M. Abgrall, P. Laurent, C. Salomon, A. Clairon: Measurement of the hydrogen $1S$–$2S$ transition frequency by phase coherent comparison with a microwave cesium fountain clock, Phys. Rev. Lett. **84**, 5496–5499 (2000)
[15] T. Udem, J. Reichert, R. Holzwarth, T. W. Hänsch: Absolute optical frequency measurement of the cesium D_1 line with a mode-locked laser, Phys. Rev. Lett. **82**, 3568–3571 (1999)
[16] T. Udem, S. A. Diddams, K. R. Vogel, C. W. Oates, E. A. Curtis, W. D. Lee, W. M. Itano, R. E. Drullinger, J. C. Bergquist, L. Hollberg: Absolute frequency measurements of the Hg^+ and Ca optical clock transitions with a femtosecond laser, Phys. Rev. Lett. **86**, 4996–4999 (2001)
[17] A. Peters, K. Y. Chung, B. Young, J. Hensley, S. Chu: Precision atom interferometry, Phil. Trans. R. Soc. Lond. A **355**, 2223–2233 (1997)
[18] S. A. Diddams, T. Udem, J. C. Bergquist, E. A. Curtis, R. E. Drullinger, L. Hollberg, W. M. Itano, W. D. Lee, C. W. Oates, K. R. Vogel, D. J. Wineland: An optical clock based on a single trapped ^{199}Hg ion, Science **293**, 825–828 (2001)
[19] T. Becker, J. v. Zanthier, A. Y. Nevsky, C. Schwedes, M. N. Skvortsov, H. Walther, E. Peik: High-resolution spectroscopy of a single In^+ ion: Progress towards an optical frequency standard, Phys. Rev. A **63**, 051802(R) (2001)
[20] J. Reichert, R. Holzwarth, T. Udem, T. W. Hänsch: Measuring the frequency of light with mode-locked lasers, Opt. Commun. **172**, 59–68 (1999)
[21] A. I. Ferguson, J. N. Eckstein, T. W. Hänsch: Polarization spectroscopy with ultrashort light pulses, Ann. Phys. (NY) **18**, 257–260 (1979)
[22] D. J. Wineland, J. C. Bergquist, W. M. Itano, F. Diedrich, C. S. Weimer: Frequency standards in the optical spectrum, in T. W. H. G. F. Bassani, M. Inguscio (Ed.): Proc. of The Hydrogen Atom (Springer, Berlin, Heidelberg 1989) pp. 123–133

[23] A. Siegman: *Lasers* (Univ. Science Books, Mill Valley) p. 1044

[24] T. W. Hänsch: High resolution spectroscopy of hydrogen, in G. F. Bassani, M. Inguscio, T. W. Hänsch (Eds.): Proc. of The Hydrogen Atom (Springer, Berlin, Heidelberg 1989) pp. 93–102

[25] H. R. Telle, D. Meschede, T. W. Hänsch: Realization of a new concept for visible frequency division: phase-locking of harmonic and sum frequencies, Opt. Lett. **15**, 532–534 (1990)

[26] M. Prevedelli, T. Freegarde, T. W. Hänsch: Phase locking of grating-tuned diode lasers, Appl. Phys. B-Lasers O. **60**, S241–S248 (1995)

[27] T. Udem, J. Reichert, T. W. Hänsch: Accuracy of optical frequency comb generators and optical frequency interval divider chains, Opt. Lett. **23**, 1387–1389 (1998)

[28] S. Coen, A. H. L. Chan, R. Leonhardt, J. D. Harvey, J. C. Knight, W. J. Wadsworth, P. S. J. Russell: White-light supercontinuum generation with 60-ps pump pulses in a photonic crystal fiber, Opt. Lett. **26**, 1356–1358 (2001)

[29] J. K. Ranka, R. S. Windeler, A. J. Stentz: Visible continuum generation in air–silica microstructure optical fibers with anomalous dispersion at 800 nm, Opt. Lett. **25**, 25–27 (2000)

[30] M. Nakazawa, K. Tamura, E. Y. H. Kubota: Coherence degradation in the process of supercontinuum generation in an optical fiber, Opt. Fiber. Technol. **4**, 215–223 (1998)

[31] K. F. Kwong, D. Yankelevich, K. C. Chu, J. P. Heritage, A. Dienes: 400-Hz mechanical scanning optical delay line, Opt. Lett. **18**, 558–560 (1993)

[32] A. Bartels, T. Dekorsy, H. Kurz: Femtosecond Ti:sapphire ring laser with a 2-GHz repetition rate and its application in time-resolved spectroscopy, Opt. Lett. **24**, 996 (1999)

[33] H. A. Haus, E. P. Ippen: Group velocity of solitons, Opt. Lett. **26**, 1654 (2001)

[34] T. A. Birks, W. J. Wadsworth, P. S. J. Russell: Supercontinuum generation in tapered fibers, Opt. Lett. **25**, 1415–1417 (2000)

[35] J. A. Barnes, A. R. Chi, L. S. Cutler, D. J. Healey, D. B. Leeson, T. E. McGunigal, J. Mullen, J. A., W. L. Smith, R. L. Sydnor, R. F. C. Vessot, G. M. R. Winkler: Characterization of frequency stability, IEEE Trans. Instrum. Meas. **20**, 105–120 (1971)

[36] S. A. Diddams, L. Hollberg, L. S. Ma, L. Robertsson: Femtosecond-laser-based optical clockwork with instability $\leq 6.3 \times 10^{16}$ in 1 s, Opt. Lett. **27**, 58 (2002)

[37] R. Holzwarth: *Measuring the frequency of light using femtosecond laser pulses*, Ph.d. thesis, Ludwig-Maximilians-Universität, Munich (2001)

[38] A. Baltuška, T. Fuji, T. Kobayashi: Controlling the carrier-envelope phase of ultrashort light pulses with optical parametric amplifiers, Phys. Rev. Lett. **88**, 133901 (2002)

[39] M. Zimmermann, C. Gohle, R. Holzwarth, T. Udem, T. W. Hänsch: Optical clockwork with an offset-free difference-frequency comb: accuracy of sum- and difference-frequency generation, Opt. Lett. **29** (2004)

[40] A. I. Ferguson, R. A. Taylor: Active mode stabilization of a synchronously pumped mode locked dye laser, Opt. Commun. **41**, 271–276 (1982)

[41] T. Kinoshita: The fine structure constant, Rep. Prog. Phys. **59**, 1459–1492 (1996)

[42] J. M. Hensley: *A precision measurement of the fine structure constant*, Ph.D. thesis, Stanford University, Standford (2001)

[43] P. J. Mohr, B. N. Taylor: Codata recommended values of the fundamental constants: 1998, Rev. Mod. Phys. **72**, 351–496 (2000)

[44] H. G. Dehmelt: Mono-ion oscillator as potential ultimate laser frequency standard, IEEE Trans. Instrum. Meas. **31**, 83–87 (1982)

[45] J. von Zanthier, T. Becker, M. Eichenseer, A. Nevsky, C. Schwedes, E. Peik, H. Walther, R. Holzwarth, J. Reichert, T. Udem, T. W. Hänsch, P. V. Pokasov, M. N. Skvortsov, S. N. Bagayev: Absolute frequency measurement of the In^+ clock transition with a mode-locked laser, Opt. Lett. **25**, 1729–1731 (2000)

[46] G. Santarelli, P. Laurent, P. Lemonde, A. Clairon, A. G. Mann, S. Chang, A. N. Luiten, C. Salomon: Quantum projection noise in an atomic fountain: A high stability cesium frequency standard, Phys. Rev. Lett. **82**, 4619–4622 (1999)

[47] B. de Beauvoir, C. Schwob, O. Acef, L. Jozefowski, L. Hilico, F. Nez, L. Julien, A. Clairon, F. B. M. of the hydrogen, deuterium atoms: Determination of the Rydberg constant and Lamb shifts, Eur. Phys. J. D **12**, 61–93 (2000)

[48] S. Mallampalli, J. Sapirstein: Perturbed orbital contribution to the two-loop lamb shift in hydrogen, Phys. Rev. Lett. **80**, 5297–5300 (1998)

[49] L. Hollberg, C. W. Oates, E. A. Curtis, E. N. Ivanov, S. A. Diddams, T. Udem, H. G. Robinson, J. C. Bergquist, R. J. Rafac, W. M. Itano, R. E. Drullinger, D. J. Wineland: Optical frequency standards and measurements, IEEE J. Quantum Elect. **37**, 1502–1513 (2001)

[50] A. A. Madej, J. E. Bernard: Single-ion optical frequency standards and measurement of their absolute optical frequency, in A. N. Luiten (Ed.): *Frequency measurement and control: advanced techniques and future trends*, Top. Appl. Phys. **79** (Springer, Berlin, Heidelberg 2001) pp. 153–186

[51] F. Riehle, J. Helmcke: Optical frequency standards based on neutral atoms and molecules, in A. N. Luiten (Ed.): *Frequency measurement and control: advanced techniques and future trends*, Top. Appl. Phys. **79** (Springer, Berlin, Heidelberg 2001) pp. 95–123

[52] J. Stenger, T. Binnewies, G. Wilpers, F. Riehle, H. R. Telle, J. K. Ranka, R. S. Windeler, A. J. Stentz: Phase-coherent frequency measurement of the Ca intercombination line at 657 nm with a Kerr-lens mode-locked femtosecond laser, Phys. Rev. A **63**, 21802(R) (2001)

[53] J. Ye, L. S. Ma, J. L. Hall: Molecular iodine clock, Phys. Rev. Lett. **87**, 270801 (2001)

[54] J. Stenger, C. Tamm, N. Haverkamp, S. Weyers, H. R. Telle: Absolute frequency measurement of the 435.5 nm $^{171}Yb^+$ clock transition with a Kerr-lens mode-locked femtosecond laser, Opt. Lett. **26**, 1589–1591 (2001)

[55] P. Dub, L. Marmet, J. E. Bernard, K. J. Siemsen, A. A. Madej: Progress towards and improved $^{8}8Sr^+$ single ion optical frequency standard, in P. Gill (Ed.): *Frequency Standards and Metrology*, Proc. 6th Symposium, St. Andrews (World Scientific, Singapore 2002)

[56] S. N. Lea, H. S. Margolis, G. Huang, W. R. C. Rowley, D. Henderson, G. P. Barwood, H. A. Klein, S. A. Webster, P. Blythe, P. Gill, R. S. Windeler: Femtosecond optical frequency comb measurements of lasers stabilised to transitions in $^{8}8Sr^+$, $^{1}71Yb^+$ and I_2 at NPL, in P. Gill (Ed.): *Frequency Standards and Metrology*, Proc. 6th Symposium, St. Andrews (World Scientific, Singapore 2002) pp. 144–151

[57] K. Sugiyama, A. Onae, F. Hong, H. Inaba, S. N. Slyusarev, T. Ikegami, J. Ishikawa, K. Minoshima, H. Matsumoto: in P. Gill (Ed.): *Frequency Standards and Metrology*, Proc. 6th Symposium, St. Andrews (World Scientific, Singapore 2002)

[58] P. A. Dirac: The cosmological constants, Nature **139**, 323 (1937)

[59] S. G. Karshenboim: Some possibilities for laboratory searches for variations of fundamental constants, Can. J. Phys. **78**, 639–678 (2000)

[60] C. Salomon, Y. Sortais, S. Bize, M. Abgrall, S. Zhang, C. Nicolas, C. Mandache, P. Lemonde, P. Laurent, G. Santarelli, A. Clairon, N. Dimarcq, P. Petit, A. Mann, A. Luiten, S. Chang: Cold atom clocks, in E. Arimondo, P. deNatale, M. Inguscio (Eds.): *Atomic Physics*, vol. 551, AIP Conf. Proc. (17th Int. Conf. at Florence 2000) pp. 23–40

[61] P. Sisterna, H. Vucetich: Time variation of fundamental constants: Bounds from geophysical and astronomical data, Phys. Rev. D **41**, 1034–1046 (1990)

[62] J. K. Webb, M. T. Murphy, V. V. Flambaum, V. A. Dzuba, J. D. Barrow, C. W. Churchill, J. X. Prochaska, A. M. Wolfe: Further evidence for cosmological evolution of the fine structure constant, Phys. Rev. Lett. **87**, 091301 (2001)

[63] R. F. C. Vessot, M. W. Levine, E. M. Mattison, E. L. Blomberg, T. Hoffmann, G. U. Nystrom, B. F. Farrell, R. Decher, P. B. Eby, C. R. Baugher, J. W. Watts, D. L. Teuber, F. D. Wills: Test of relativistic gravitation with a space-borne hydrogen maser, Phys. Rev. Lett. **45**, 2081–2084 (1980)

[64] B. Bertotti, L. Iess, P. Tortora: A test of general relativity using radio links with the Cassini spacecraft, Nature **425**, 374–376 (2003)

Carrier–Envelope Phase Stabilization of Single and Multiple Femtosecond Lasers

David J. Jones, Steve T. Cundiff, Tara M. Fortier, John L. Hall, and Jun Ye

JILA, University of Colorado and National Institute of Standards and Technology, Boulder, CO 80309-0440, USA
djjones@jila.colorado.edu

Abstract. The basic concepts, technical implementation, and known limitations of actively stabilizing the carrier–envelope phase of a few-cycle pulse train are discussed. The route toward determining the "absolute" carrier–envelope phase, thereby enabling electronic waveform synthesis at optical frequencies, is reviewed. Lastly, techniques and applications of stabilizing the relative carrier–envelope phase between two (or more) femtosecond lasers are also covered.

1 Introduction

The advent of few-cycle laser pulse generation has heightened interest in measuring and controlling the phase between the optical carrier wave and the pulse intensity envelope. There are a number of physical processes that are dependent on the electric field, rather than just the intensity envelope of a pulse [1]. Accordingly, such processes, including coherent (quantum) control of atomic and molecular systems [2], optimization of high-harmonic (soft X-ray) generation [3], and investigation of atomic systems on femtosecond and attosecond timescales, will benefit from control over the carrier–envelope phase. Combined with well-established methods of conventional amplitude and chirp pulse shaping [4], control over the carrier–envelope phase will enable us to synthesize electronic waveforms at optical frequencies. This type of waveform synthesis could be used with the above mentioned physical investigations, but it can also be employed in more of an application-oriented manner. Phase-coherent operations, such as analog signal processing, that have historically been operating at microwave frequencies can now be performed at optical frequencies. Some of these possibilities are discussed in more detail in other chapters of this book as well as in other books on high-field physics [5].

1.1 Definition of the Carrier–Envelope Phase for a Few-Cycle Pulse

Figure 1 displays a few-cycle pulse with the carrier–envelope phase (CEP) defined as ϕ_{CE}. Mathematically, the electric field of a pulse can be expressed

F. X. Kärtner (Ed.): Few-Cycle Laser Pulse Generation and Its Applications,
Topics Appl. Phys. **95**, 317–342 (2004)

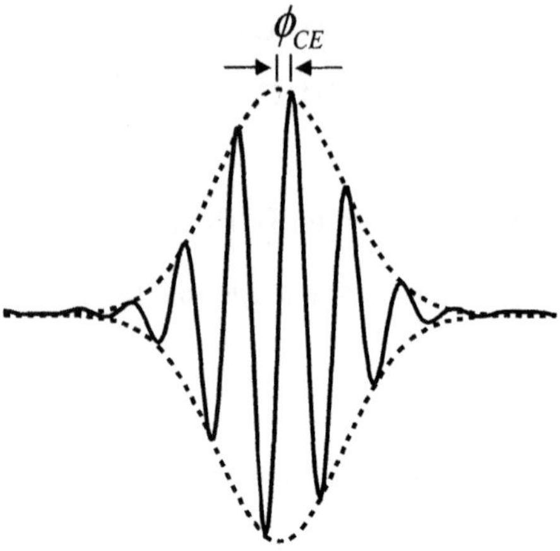

Fig. 1. Definition of carrier–envelope phase. Ultrashort pulse where the period of the carrier frequency approaches the pulse width. The pulse intensity envelope is shown as a *dotted line*, and the *solid line* is the oscillating electric field. ϕ_{CE} is the phase between the peak of the pulse intensity envelope and the peak of the carrier wave

as

$$E(t) = A(t)\cos(\omega_{\mathrm{c}}t + \phi_{\mathrm{CE}})\,, \tag{1}$$

where $A(t)$ is the pulse envelope, ω_{c} is the carrier frequency, and ϕ_{CE} is the carrier–envelope phase. The reliable periodicity of a train of optical pulses generated by a mode-locked laser allows identification of a phase referenced to the pulse envelope. Relative to this frame, the phase (ϕ_{CE}) of the oscillating electric field can vary, depending on conditions both within and outside the laser cavity. For a clear understanding of the dynamics of ϕ_{CE}, the CEP can be broken into two components,

$$\phi_{\mathrm{CE}} = \phi_{\mathrm{o}} + \Delta\phi_{\mathrm{CE}}\,, \tag{2}$$

where ϕ_{o} is the "static" offset CEP and $\Delta\phi_{\mathrm{CE}}$ represents the pulse-to-pulse change in CEP due to conditions inside the cavity of the laser oscillator.[1] As the pulse propagates through any medium outside the laser cavity (except vacuum), a difference between the phase and group velocities (caused by dispersion) will cause ϕ_{o} to vary; so in reality, ϕ_{o} is not truly static. In a similar vein, the physical origin of $\Delta\phi_{\mathrm{CE}}$ results from dispersion of the optical elements inside a laser cavity. In the case of $\Delta\phi_{\mathrm{CE}}$, the pulse is sampled once per round-trip when it hits the output coupler, and it is only the phase change modulo 2π that matters. Specifically,

$$\Delta\phi_{\mathrm{CE}} = \left(\frac{1}{v_{\mathrm{g}}} - \frac{1}{v_{\mathrm{p}}}\right) l_{\mathrm{c}}\omega_{\mathrm{c}} \bmod [2\pi]\,, \tag{3}$$

[1] Often the term absolute phase has been used to label ϕ_{o}, which can be misleading as there is nothing absolute about the peak of the pulse envelope that is used as the reference. This terminology has probably arisen to help distinguish between ϕ_{o} and $\Delta\phi_{\mathrm{CE}}$ and to emphasize the fact that ϕ_{o} is not relative to a second optical reference beam

where $v_g(v_p)$ is the group (phase) velocity and l_c is the round-trip length of the laser cavity.

As discussed in detail in Sect. 2, it is now possible to detect and control $\Delta\phi_{CE}$. However, as (2) indicates, to define ϕ_{CE} completely, ϕ_o must also be measured (and stabilized). Although not yet completely realized, steps towards this latter requirement are covered in Sect. 3. Recent work will be described in locking together the carrier–envelope phases of two independent fs lasers in Sect. 4. Finally, an outlook on future developments and applications is presented in Sect. 5.

2 Pulse-to-Pulse Carrier–Envelope Phase

The techniques used to stabilize the pulse-to-pulse evolution of the CEP are best understood in the frequency domain. Figure 2a displays three pulses that are part of an infinite train that has a constant $\Delta\phi_{CE}$. The frequency-domain representation of this pulse train, given in Fig. 2b, is a frequency comb with tooth spacing equal to the pulse repetition rate (f_r). The entire comb is offset from exact harmonics of f_r by an offset frequency (f_o). From a careful derivation [6], the relation between f_o and $\Delta\phi_{CE}$ can be expressed as

$$\Delta\phi_{CE} = 2\pi\frac{f_o}{f_r}. \tag{4}$$

Thus, the task of stabilizing $\Delta\phi_{CE}$ is reduced to stabilization of f_o.

2.1 Detection of the Offset Frequency

As each comb element is shifted by the (same) offset frequency by optically heterodyning different harmonics of the frequency comb together, it is not possible to extract the value of the offset frequency. Instead, scaling of the comb spectrum must be implemented before the heterodyne comparison. A straightforward method is to frequency double the red end of the comb spectrum and compare it with the existing spectrum at the blue end where these two beams spectrally overlap. Thus, the simplest heterodyne procedure requires an octave of optical bandwidth and is known as ν-to-2ν self-referencing. The initial demonstrations of this technique [7, 8] used the continuum generated from microstructure (photonic crystal) fiber [9] and normal optical fiber, respectively. More recently, an octave of bandwidth has been generated directly from a mode-locked laser [10, 11, 12]; the latter result has enough energy in the spectral extremes to provide a suitable signal-to-noise ratio for a tight phase lock of f_o.

Consider a single comb component on the red side of the spectrum. The electrical field of this (the nth) comb line will have a phase,

$$\phi_n = 2\pi(f_n + f_o)t + \phi_{on} = 2\pi(nf_r + f_o)t + \phi_{on}\,, \tag{5}$$

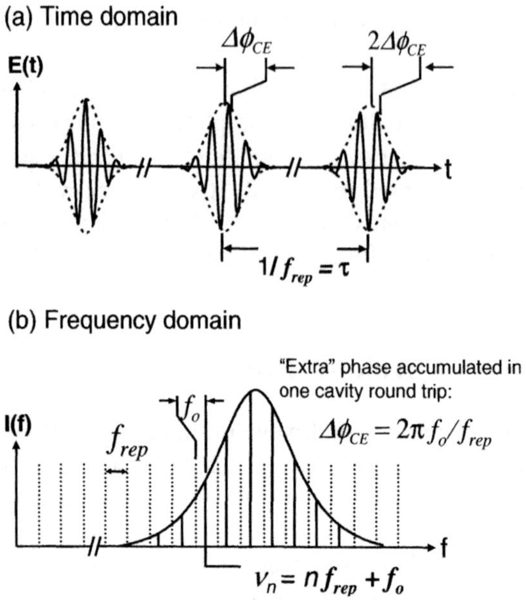

Fig. 2. Time-frequency correspondence and relationship between $\Delta\phi_{\mathrm{CE}}$ and f_{o}. (**a**) In the time domain, the relative phase between the carrier and the envelope evolves from pulse to pulse by the amount $\Delta\phi_{\mathrm{CE}}$ due to an inequality of intracavity group and phase velocities. (**b**) In the frequency domain, the elements of the frequency comb of a mode-locked pulse train are spaced by f_{r}. The entire comb is offset from integral multiples of f_{r} by an offset frequency f_{o}. Without active stabilization, f_{o} is a dynamic quantity, which is sensitive to perturbations of the laser. Hence $\Delta\phi_{\mathrm{CE}}$ changes in a nondeterministic manner from pulse to pulse in an unstabilized laser

where $\phi_{\mathrm{o}n}$ is the residual optical spectral phase constant of the nth comb line. Similarly, an octave away (2 times the frequency) on the blue side of the spectrum, the $2n$th comb line will have the phase,

$$\phi_{2n} = 2\pi(f_{2n} + f_{\mathrm{o}})t + \phi_{\mathrm{o}2n} = 2\pi(2nf_{\mathrm{r}} + f_{\mathrm{o}})t + \phi_{\mathrm{o}2n}\,. \tag{6}$$

If the electric field of the nth line is doubled (with any standard second-harmonic nonlinear crystal) and the optical heterodyne beat between the doubled signal and the original field at $2n$ is detected on a photodiode, the signal will have an interference term with the phase,

$$\phi_{\mathrm{detect}} = 2\pi f_{\mathrm{o}}t + 2\phi_{\mathrm{o}n} - \phi_{\mathrm{o}2n}\,. \tag{7}$$

From the photodiode signal, simple filtering in the radio-frequency domain yields f_{o}. The meaning and use of the phase term $(2\phi_{\mathrm{o}n} - \phi_{\mathrm{o}2n})$ in (7) is discussed in Sect. 3, and a complete derivation of (5–7) can be found elsewhere [6].

The actual experimental setup implementing the ν-to-2ν interferometer is shown in Fig. 3. In reality, a group of comb lines contributes phase coher-

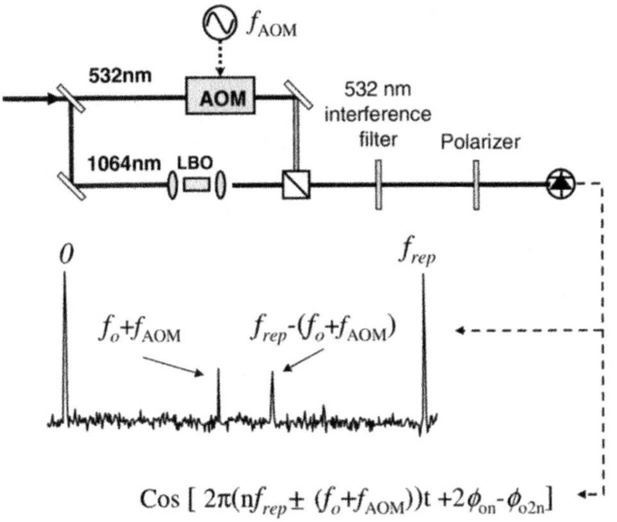

Fig. 3. ν-to-2ν interferometer used to measure f_o. The incoming octave-spanning comb is spectrally separated using a dichroic mirror. Typically, we used wavelengths centered at 1064 nm for ν and 532 nm for 2ν. In reality, the S/N of f_o is optimized by experimentally adjusting ν and 2ν. The infrared portion (ν) is frequency doubled with a lithium triborate (LBO) crystal and then polarization multiplexed with the existing 2ν signal. The combined beam passes through an interference filter (centered at 2ν) to reject any nonspectrally overlapped comb components. An acousto-optic modulator (AOM) is placed in the visible arm to enable measuring the heterodyne beat unambiguously. An example of an observed rf spectrum is also shown

ently to the photodiode signal, so even with at most 10 nW per comb line (a typical amount for a 100 MHz spaced comb after broadening in microstructure fiber) there is enough signal-to-noise (S/N) to lock f_o properly. In our experience, the S/N should be $>$ 25 dB to 30 dB (in a 100 kHz bandwidth) to achieve a tight phase lock. Normally, a collection of comb lines spanning approximately 10 nm centered at ν_{2n} is used for the heterodyne beat. There are two other important characteristics regarding the ν-to-2ν interferometer shown in Fig. 3. First, to observe the heterodyne beat, the path lengths of each arm must be matched so that the field at $2n$ and the (doubled) field at $2 \times n$ overlap in time. Second, by including an acousto-optical modulator (AOM) in one arm of the interferometer, the comb lines at $2n$ are shifted in frequency by f_{AOM}. Thus, the observed beats (7) are shifted by f_{AOM} as well. This condition allows us to avoid processing f_o around the troublesome dc or f_r frequency range when f_o is locked to a value of zero. With the AOM, f_o is easily locked to zero by mixing the photodiode signal with f_{AOM} to generate an error signal, although one does have to be careful to avoid electronic pickup noise at f_{AOM}. Locking f_o to zero forces every pulse to have the same

CEP and allows one to let f_r float while still generating pulses with a stable carrier–envelope phase. Recently, an alternate implementation of an ν-to-2ν interferometer has also been used [12] which has less differential noise.

There are other methods capable of measuring f_o, such as detecting the beat from 2ν and 3ν [11]. This latter method requires only two-thirds of an octave. However, a cascaded nonlinear process is required and producing the offset frequency with suitable signal-to-noise is rather difficult. Recently, using a high repetition rate mode-locked laser (1 GHz), an S/N of 25 dB (in a 100 kHz bandwidth) was observed on f_o derived from a 2ν-to-3ν scheme [13]. A more thorough examination of these and other (similar) techniques to obtain f_o is presented in [14]. Heterodyne beating of two harmonics from a CW laser stabilized against the comb can also be used to detect f_o [15].

2.2 Stabilization of the Offset Frequency

To complete the stabilization loop, an accessible adjustment (or "knob") on the laser is required that can be used to adjust f_o. In this case, such a knob must be capable of changing the difference between the intracavity group and phase velocities. One such technique is to swivel the end mirror in the arm of the laser cavity that contains the prism sequence [16], as shown in Fig. 4. Since the spectrum is spatially dispersed on this mirror, a small tilt produces a linear phase delay with frequency, which is equivalent to a group delay [17]. An alternate method of controlling f_o is via modulation of the pump power [8,18]. Empirically, this, it was found, causes a change in f_o [19], although at the present time the exact physical mechanism is not clearly understood. Several possibilities may cause the sensitivity of f_o to change with respect to the laser pump power, including a nonlinear phase shift in the Ti:sapphire crystal [8, 19], spectral shifts combined with wavelength-dependent group velocity dispersion, [8] or a differential change in the phase and group velocities [20, 21]. The pump power is usually modulated with an AOM or electro-optic modulator (EOM) which typically has greater bandwidths than a mirror mounted on a piezoelectric transducer (PZT). However, the modulation depth required to lock f_o successfully can induce a significant amount of amplitude noise on the laser output; the consequences of this added noise are discussed in Sect. 3.3. By carefully designing the mechanical structure of the PZT, a 3 dB bandwidth of approximately 50 kHz can be realized, which can be as effective as an AOM for stabilization. Indeed, when we have directly compared both PZT- and AOM-based stabilization schemes, little difference in locking performance was observed. These latter results are discussed in more detail in Sect. 2.4.

An alternate method of generating CEP stable pulses, recently reported by *Baltuška* and co-workers [22], is discussed in the Chapter by *Baltuška* et al. in this book. Using the phase relations of optical parametric amplification, they demonstrated an elegant technique capable of passively producing pulses with zero pulse-to-pulse CEP change ($\Delta\phi_{CE} = 0$). As this is an amplified

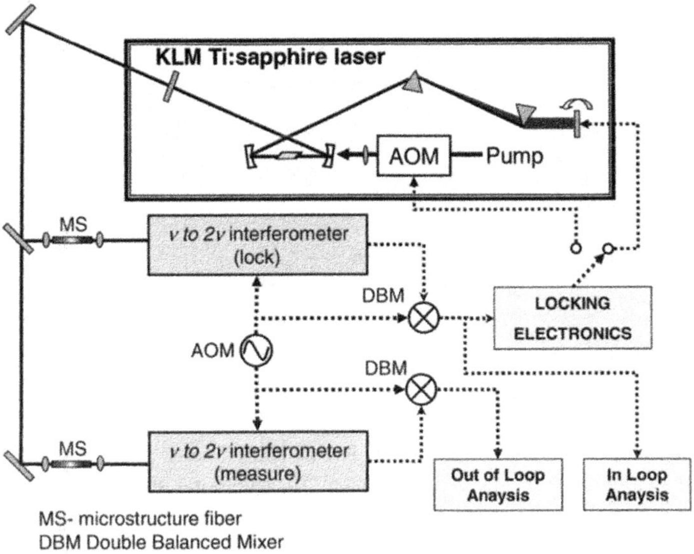

Fig. 4. Experimental setup showing stabilization of $\Delta\phi_{\mathrm{CE}}$. As the pulse spectrum is spatially resolved after the second prism, feedback via a small tilt on the end mirror produces a linear phase delay with frequency (a group delay) which changes $\Delta\phi_{\mathrm{CE}}$. A second ν-to-2ν interferometer provides an out-of-loop measurement of f_{o} which is critical for measuring the true coherence time of ϕ_{CE} (see Sect. 2.4)

system with kilohertz repetition rates, determination of ϕ_{o} is difficult in the frequency domain (as discussed in Sect. 3). Rather, a time-domain, high-field process [3, 23] or a measurement using spectral interferometry [24] is necessary.

2.3 Time-Domain Measurement of Phase-Stable Pulses

The relation between the offset frequency f_{o} and the pulse-to-pulse evolution of the CEP given in (4) was confirmed by examining the second-order cross-correlation between pulse i and $i+2$. By locking f_{o} to various rational fractions of f_{r}, the subsequent change in $\Delta\phi_{\mathrm{CE}}$, it was found, accurately follows the prediction of (4) [7], as shown in Fig. 5. However, such a time-domain measurement with a single time delay does not yield a timescale (beyond two successive pulses or 20 ns) over which the CEP remains coherent. Methods and results for determining the coherence time of the CEP are addressed in Sect. 2.4.

2.4 Coherence of the Carrier–Envelope Phase

Although control of a pulse train's $\Delta\phi_{\mathrm{CE}}$ is a critical first step, the utility of CEP-stable pulses in many actual nonlinear experiments will be significantly enhanced by long-term phase coherence of the CEP. As seen from (4),

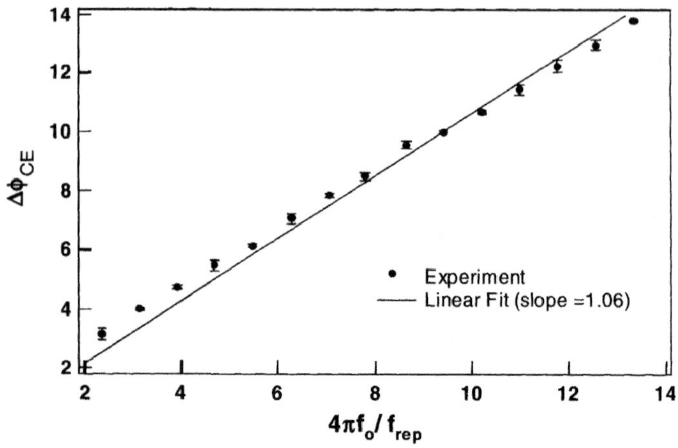

Fig. 5. Experimental measurement of $\Delta\phi_{\mathrm{CE}}$ between the ith and ith + 2 versus various rational fractions of $4\pi f_{\mathrm{o}}/f_{\mathrm{r}}$. According to (2), the slope should = 1 (an extra factor of 2 comes from correlating every second pulse). A linear fit of the data yields a slope of 1.06

slight excursions from $f_{\mathrm{o}} = 0$ (or from a rational fraction with respect to f_{r}) cause an accumulated phase error in $\Delta\phi_{\mathrm{CE}}$. Thus, phase noise of $\Delta\phi_{\mathrm{CE}}$ is manifested as frequency noise of f_{o} and leads to broadening of its linewidth. The rms fluctuations in the carrier–envelope phase, $\Delta\phi_{\mathrm{CErms}}$, can be found by integrating the frequency noise power spectral density (PSD), $S_{f_{\mathrm{o}}}(\nu)$, of f_{o} [25],

$$\Delta\phi_{\mathrm{CErms}}\,|_{\tau_{\mathrm{obs}}} = 2\sqrt{\int_{-\infty}^{1/2\pi\tau_{\mathrm{obs}}} \frac{S_{f_{\mathrm{o}}}(\nu)\,\mathrm{d}\nu}{\nu^2}} = 2\sqrt{\int_{-\infty}^{1/2\pi\tau_{\mathrm{obs}}} S_{\phi}(\nu)\,\mathrm{d}\nu}\,, \qquad (8)$$

where τ_{obs} is the observation time, $S_{\phi}(\nu)$ is the phase noise spectral density of f_{o}, and ν is an offset frequency relative to the optical carrier. Physically $S_{f_{\mathrm{o}}}(\nu)$ is the frequency-domain representation of the fluctuations in f_{o} at frequencies about its carrier. Thus, $S_{f_{\mathrm{o}}}(\nu)$ gives the power spectrum of the frequency-noise sidebands present on the offset frequency's linewidth, which may be converted to the phase noise spectrum using the relation $S_{\phi}(\nu) = S_{f_{\mathrm{o}}}(\nu)/\nu^2$. Following a convention from the frequency metrology field, the coherence time is defined as

$$\tau_{\mathrm{coh}} \equiv \tau_{\mathrm{obs}}\,|_{\Delta\phi_{\mathrm{CErms}} \approx 1\ \mathrm{rad}}\,. \qquad (9)$$

The experimental setup used to measure $S_{\phi}(\nu)$ is shown in Fig. 4. Two ν-to-2ν interferometers are used simultaneously, each with its own piece of microstructure fiber. One interferometer is employed to lock $\Delta\phi_{\mathrm{CE}}$; the second interferometer is used for an authentic evaluation of $S_{\phi}(\nu)$ independent of the feedback loop (out-of-loop). An in-loop measurement of $S_{\phi}(\nu)$ (from the locking interferometer) quantifies only the stabilization capability of the

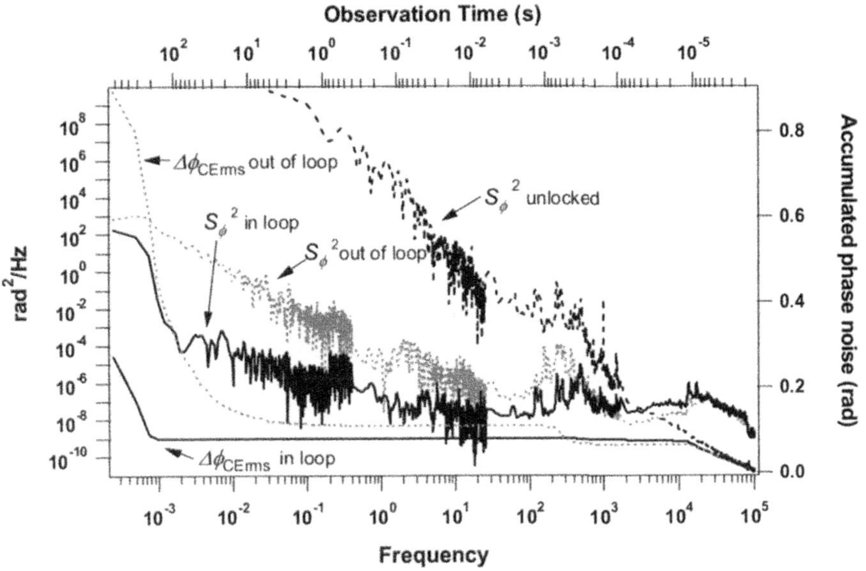

Fig. 6. Phase power spectral density (S_ϕ) vs. offset frequency (ν) from the carrier (f_o), *left axis*. Accumulated phase noise as a function of observation time is obtained via integration of $S_\phi(\nu)$, *right axis*. The in-loop (*black*) and out-of-loop (*gray*) spectra (0.488 mHz to 102 kHz) were compiled from five different spectra of decreasing span and increasing resolution to obtain greater resolution close to the carrier (displayed here as zero frequency). The stabilization process adds noise past roughly 5 kHz, and roll-off in the out-of-loop spectrum at approximately 30 kHz is consistent with the stabilization servo bandwidth. An unlocked spectrum (spanning 31.25 mHz to 102 kHz) is included to indicate the effectiveness of the stabilization loop. The first data points for the two spectra are artifacts as they include the dc offset given by the carrier

feedback loop. Furthermore, the loop (in particular, the locking interferometer itself) could possibly write noise onto $\Delta\phi_{\mathrm{CE}}$ that can be diagnosed only with an out-of-loop measurement. Both the in-loop and out-of-loop $S_{f_o}(\nu)$ can be measured with a dynamic signal analyzer (such as the Stanford Research Systems SR785, Tektronics RSA 3303A or the Agilent 35670A).[2]

Figure 6 displays $S_\phi(\nu)$ for the unlocked, in-loop locked, and out-of-loop locked cases (left axis). The difference between the locked and unlocked cases shows an excellent improvement in the frequency deviations of f_o. As expected, the in-loop measurement has a lower phase noise spectral density compared to the out-of-loop case. We attribute most of this difference to differential noise between the in-loop and out-of-loop ν-to-2ν interferometers.

[2] Mentioning of product names is for technical communication only and doesn't constitute an endorsement by the authors or NIST

Any interferometer noise in the in-loop setup will be written on the laser by the feedback loop, creating phase noise in the carrier–envelope phase.

Also shown in Fig. 6 is $\Delta\phi_{\mathrm{CErms}}$ (right axis) for in-loop and out-of-loop cases, calculated as prescribed by (8) and displayed as a function of τ_{obs}. At an observation time of 320 s, the out-of-loop $\Delta\phi_{\mathrm{CErms}}$ has accumulated only 0.8 rad of phase corresponding to a coherence time of at least 320 s. At present, this is a lower limit on the coherence time that is limited by measurement details. It should be specifically noted that the coherence time quoted above is not a coherence time of the optical carrier wave with successive pulses; rather it is the coherence time of the carrier–envelope phase. The repetition rate of the laser was not stabilized, so the carrier frequency is free to shift, when f_{r} changes, to maintain coherence of the CEP. However, if f_{r} is stabilized, then optical coherence can be realized. Considering that there is a rather large multiplication factor of 10^6 from radio to optical frequencies, simply locking f_{r} to an rf synthesizer will not be adequate. The phase noise of the synthesizer will be written on the pulse envelope and effectively multiplied and written onto the optical carrier as well (because the CEP remains locked). This will severely limit any chance of realizing optical coherence. Rather, locking f_{r} to an optical transition or optical cavity will most likely be necessary to achieve this goal.

The contribution of fluctuations in the ν-to-2ν interferometers to the $\Delta\phi_{\mathrm{CErms}}$ was investigated by threading a single-frequency HeNe laser through one interferometer. From the resulting transmission signal, a measurement of the phase spectral density can quantify the interferometers' effect on the CEP noise. And in a manner analogous to (8), the phase noise added to the $\Delta\phi_{\mathrm{CErms}}$ by the ν-to-2ν interferometer can be calculated. Figure 7 displays these results. If fluctuations in the in-loop and out-of-loop interferometers are assumed to be uncorrelated, then they add roughly $\sqrt{2} \times 0.12 = 0.17$ rad of phase noise to $\Delta\phi_{\mathrm{CErms}}$ at a frequency range of 0.01 Hz to 104 Hz, an amount consistent with the results in Fig. 6.

The results shown in Fig. 6 were obtained by controlling (swiveling) the laser cavity mirror following the prism sequence that was mounted on a specially designed, high-speed PZT that had a resonance frequency of 50 kHz. The design consisted of two 1/8 in PZT discs that were mounted with opposite polarity to a large copper mass (to reduce recoil and damp mechanical resonances of the mirror mount). On the front side of the PZTs, a rectangular mirror (with as little mass as possible) was mounted with a stiff wax. We observed slightly worse performance using an AOM to modulate the pump power. Between the two techniques (AOM vs. PZT), we prefer the PZT-based method as the AOM tends to place amplitude noise on the output of the laser. However, if a prismless laser is stabilized, then, at the present time, an AOM is the only option to control f_{o} with a reasonably large bandwidth. However, when a laser (with intracavity prisms) that is locked with an AOM is included

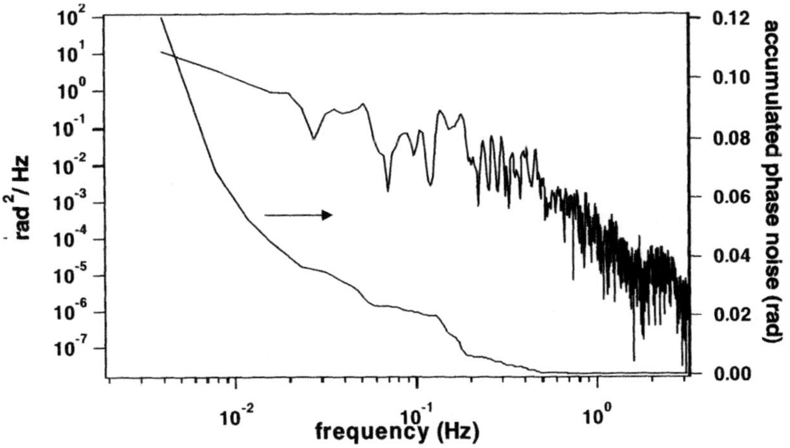

Fig. 7. Phase spectral density and accumulated phase noise resulting from an unstabilized ν-to-2ν interferometer. This data is measured by examining the fluctuations of an independent laser (a single-frequency HeNe) through the interferometer

in a system with microstructure fiber, there can be severe consequences for the coherence of ϕ_{CE} as discussed in Sect. 3.3.

To obtain a tight lock on f_o and realize long-term coherence of ϕ_{CE}, it is necessary to pay close attention to the mechanical construction of the laser cavity. Anything that can passively reduce the environmental perturbations to the laser relieves the burden placed on servoloops and leads to more successful stabilization. Cavity designs with dispersion-compensating mirrors (DCMs) used in place of prisms are reportedly more stable [26]. This is understandable as DCM-based cavities are free of beam-pointing fluctuations through the prism sequence (which lead to CEP fluctuations). However, DCM-based cavities are by no means a requirement to realize long-term coherence of ϕ_{CE}. The long-term coherence results reported above were obtained by carefully constructing a laser using intra-cavity prisms. Some of the more obvious measures include using the lowest practical beam height (we use 2.5 in); employing high quality, solid mounts with stiff springs; enclosing the cavity in a sealed Plexiglass^{TM}($>$ 0.5 in thick) box; and using a single-frequency, diode-pumped, solid-state, 532 nm laser as the pump source. Building the cavity on a solid *cast* aluminum breadboard (as opposed to a honeycomb or rolled aluminum breadboard) helps enormously. Plate modes of the cast breadboard can be reduced by attaching lead plates with suitable impedance matching. Low-frequency vibrations (to 10 Hz) are attenuated by any standard optical table with an air flotation system. Frequencies up to 200 Hz to 300 Hz can be attenuated by placing the breadboard on rubber stoppers or other suitable low-frequency springs. Frequencies above 500 Hz are typically airborne; thus, it is helpful to place the Plexiglass box around (rather than on top of) the breadboard. Elimination (or attenuation) of any noise sources in the audio-frequency range (such as chillers or air compressors) is also a crucial step.

2.5 Application to Optical-Frequency Metrology

In addition to providing control over the CEP, the successful stabilization of f_o has led to a revolution in optical-frequency metrology by using the frequency combs from femtosecond (fs) lasers as optical rulers. Building on proposals and preliminary experimental demonstrations with mode-locked picosecond pulses first made in the 1970s [27, 28, 29], in 1999, workers from *Hänsch*'s group at MPQ used a frequency comb from a femtosecond laser to span 20 THz in a frequency chain [30]. Following their lead, at JILA, a femtosecond comb-spanning 104 THz was demonstrated [31]. Our JILA work culminated in an octave-spanning frequency comb [32]. This latter development enabled the demonstration of an optical-frequency synthesizer [7] that provided the first self-referenced, phase-coherent, direct link between optical and radio frequencies. This work was independently confirmed a short time later by the MPQ group [33]. A complete discussion of stabilized fs frequency combs and their application to optical frequency metrology can be found in [34] and in the Chapter by *Udem* et al. in this book.

3 Absolute Carrier–Envelope Phase

Consider a train of pulses with high carrier–envelope phase coherence, as described in the previous section. The logical next step is to measure and control ϕ_o. A careful derivation [6] shows that, in principle, the phase of the signal detected from an ν-to-2ν interferometer is ϕ_o when $\Delta\phi_{CE} = 0$. However, an interferometer, such as that shown in Fig. 3, that has distinct arms for ν and 2ν light introduces arbitrary phase shifts that make the detected phase no longer equal to ϕ_o. In the following, we will present three alternatives that preserve the phase relation between the detected signal and ϕ_o. At this point, none of them has been demonstrated as a viable means for measuring ϕ_o.

3.1 Chirp Compensation

The two-arm ν-to-2ν interferometer shown in Fig. 3 is used to achieve temporal overlap between the ν and 2ν spectral components. Since the required broad spectrum is often obtained by nonlinear broadening in microstructure fiber, there is often substantial third-order chirp on the pulse. This means that there can be significant temporal separation between the ν and 2ν components. If an octave-spanning laser is used [10, 12], the cubic chirp can be compensated for by standard techniques [12]. If close to a flat phase can be obtained, an interferometer is not needed at all, but just passing the beam through a second harmonic crystal will produce ν-to-2ν beats [11].

Although chirp compensation removes the need for an explicit interferometer, phase shifts in the second-harmonic (SHG) crystal can still prevent the phase of the ν-to-2ν signal from being an accurate measure of the pulse carrier–envelope phase because dispersion in the SHG crystal causes the CEP to evolve in a known manner, but with a large degree of uncertainty. Although second-harmonic generation requires phase matching, which in this case is obtained via birefringence to cancel dispersion, this is true only within a very limited range of wavelengths. Since the pulse inherently has a broad bandwidth, sum-frequency generation is truly the relevant process when the ν-to-2ν signal is generated. Calculations show that even a small discrepancy between the phase matching and detection wavelengths can cause large errors in the phase [35]. Of course, a very thin second-harmonic crystal can overcome this uncertainty, although the resulting signal may be unusably weak.

3.2 Quantum Interference

An alternative approach to optical interference [14, 36] from the detection of f_o is to use quantum interference. If a final state can be reached by both one- and two-photon transitions, quantum mechanical interference between the pathways will produce a phase-dependent population. However, if these are discrete states, say, in an atom, parity results in selection rules that prevent both one- and two-photon transitions from a given initial state to any final state. However, if the final state does not have parity as a good quantum number, this is a possible scenario. Continuum states are an example of such states. Thus, this scheme can be implemented if the final state is in the ionization continuum of an atom [37] or is in band states in a solid.

Such quantum interference has been demonstrated in semiconductors at the University of Toronto [38, 39]. In this work, two laser pulses with a factor of 2 difference in frequency and adjustable relative phase were generated. They were then combined and illuminated a low-temperature-grown GaAs sample. The resulting interference between one- and two-photon absorption produces a detectable current with a relative phase-dependent direction. If a single pulse with an octave-spanning spectrum is used, the current generated will depend on the carrier–envelope phase. Simulations show that this should generate a weak, but detectable signal. Since the absorption occurs in a very thin semiconductor layer, approximately 1 μm, unmeasured phase slippage of ϕ_o due to propagation effects is minimized. Recently, using this technique, we detected the CEP using a GaAs sample [40]. An alternate quantum interference approach in the multiphoton regime uses photoemission from a gold cathode [41] and has also been recently demonstrated.

3.3 Effects of External Broadening

At the present time, two Ti:sapphire (Ti:s) oscillators have been demonstrated which generate enough energy in the spectral wings to provide sufficient S/N for a good ν-to-2ν lock [12, 42]. However, it is useful to consider the possible deleterious effects on the stabilization of $\Delta\phi_{\mathrm{CE}}$ and ϕ_{o} arising from external broadening.

One area of obvious concern is amplitude-to-phase conversion in the microstructure fiber. When the feedback loop for stabilizing the CEP is closed, phase noise generated in the fiber will be written back onto the laser by the action of the loop as it tries to correct for this extracavity phase error. As the origin of this noise is outside the cavity, this represents a noise term added to the laser. In Fig. 8, the spectral density of relative power fluctuations $[S_{\mathrm{P}}(\nu)]$ is shown on the left axis for the unlocked laser, PZT-locked laser, and AOM-locked laser. It is clear that though there is little difference between unlocked and PZT-locked lasers (indicating that the PZT lock does not increase the amplitude noise of the laser), there is a significant increase in amplitude noise with an AOM-locked laser. Using a setup similar to Fig. 4 with two, parallel ν-to-2ν interferometers, the amplitude-to-phase coefficient of the microstructure fiber was measured [43] at[3] 6.0 rad/mW. With this coefficient, the accumulated fiber phase noise on the CEP can be easily calculated [using an expression similar to (8)] as a function of frequency. The result of this integration is shown in Fig. 8 for both the PZT-lock and AOM-lock. Though the amplitude fluctuations in the PZT-locked case contribute a negligible amount of phase noise, it is entirely a different story for the AOM-locked laser. After integration from $+\infty$ to 100 Hz, the accumulated phase noise has already reached 2π. Thus after an observation time of only $(2\pi 100\,\mathrm{Hz}^{-1}) =$ 1.6 ms, coherence of the CEP is lost when an AOM is used to stabilize f_{o} in a laser with intracavity prisms in combination with microstructure fiber. In contrast, it should be noted that with DCM cavities, the amplitude noise of the laser *decreases* when f_{o} is locked via feedback to an AOM located in the pump beam [8]. This difference in behavior between prism and DCM fs lasers clearly indicates variations in noise processes and stabilization dynamics of f_{o} between the two cavity designs and is a topic of other work [45].

Another matter of possible concern is the degree of coherence within the continuum that is generated via a microstructure fiber. Without a high degree of coherence throughout the broadened spectra, the true values of both ϕ_{o} and $\Delta\phi_{\mathrm{CE}}$ could not be determined from the detected heterodyne beat at f_{o}. However, the experimental time-domain results presented in Fig. 5 display the expected linear relationship between f_{o} and $\Delta\phi_{\mathrm{CE}}$ within 6%. Furthermore, a lack of coherence in the comb would be manifested as uncorrelated broadening of individual comb lines. If coherence in the broadened frequency

[3] As a side note, this value of an amplitude-to-phase coefficient is significantly larger than that of standard optical fiber. Consequently, the microstructure fiber may be an ideal medium to generate amplitude-squeezed light (e.g. [44])

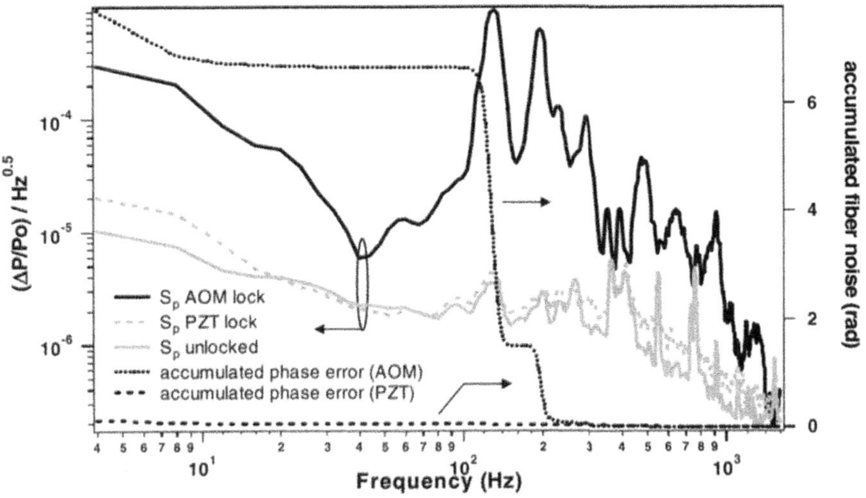

Fig. 8. Spectral density of normalized power fluctuations (*left axis*) for unlocked (*solid gray*), PZT-locked (*dashed gray*), and AOM-locked (*solid black*) f_o. The accumulated phase noise due to amplitude-to-phase conversion in a microstructure fiber is shown on the *right axis.* Note that using an AOM to lock f_o leads to decoherence of the CEP after only an observation time of $(2\pi 100\,\mathrm{Hz}^{-1}) = 1.6\,\mathrm{ms}$, for a laser using intracavity prisms

comb were a limiting factor, our measurement of the linewidth of f_o would have produced a finite linewidth that could not be corrected (narrowed) by feedback to the laser. As the linewidth of f_o is presently measurement-limited at 0.488 mHz, coherence of the comb does not appear to be a limiting factor for establishing stable CEP pulses or determining ϕ_o from the phase of the ν-to-2ν signal.

4 Synchronizing the Carrier–Envelope Phase of Two Independent Femtosecond Lasers

A natural extension beyond establishing the long-term CEP coherence of a pulse train generated by a single mode-locked laser is to lock the relative CEP of two or more fs lasers coherently. In the frequency domain, this is equivalent to coherently stitching together each independent frequency comb into a single comb. One application of this technology is immediately obvious: the resulting comb represents a single coherent pulse stream that has a broader bandwidth than the individual combs by themselves. Thus, it is possible to synthesize a shorter pulse than can be generated by a single laser (with the appropriate spectral coverage). This concept is discussed in more detail in the Chapter on few-cycle pulse generation by *Kärtner* et al. Other applications include significant (and possibly enabling) flexibility to control

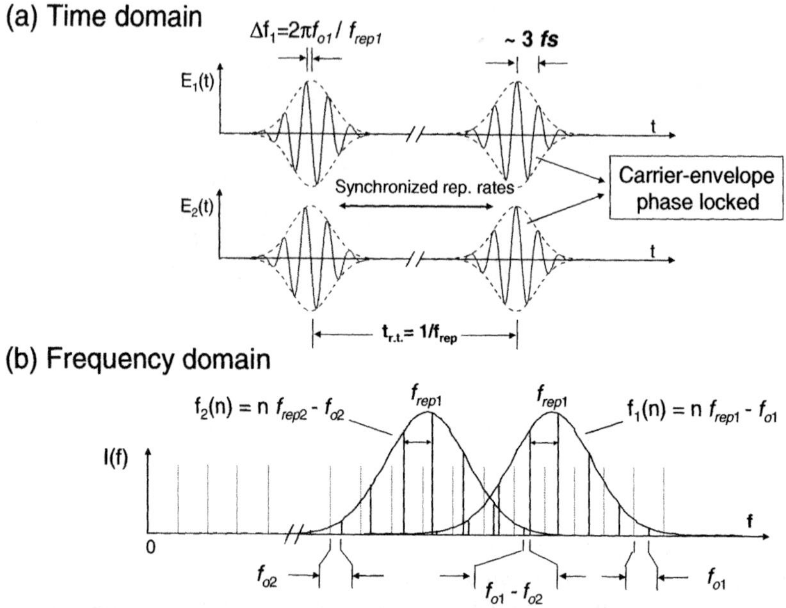

Fig. 9. (**a**) Time- and (**b**) frequency-domain representations of the required conditions on pulse synchronization and carrier–envelope phase lock to establish phase coherence between two independent femtosecond lasers

molecular systems [2] and other two-photon processes coherently that are best implemented using two independent, time-synchronized lasers such as coherent anti-Stokes Raman spectroscopy [46].

4.1 Repetition Rate Synchronization

To establish phase coherence between two separate ultrafast lasers, it is necessary first to achieve a level of pulse repetition rate synchronization between the two lasers such that the remaining timing jitter is less than the oscillation period of the optical carrier wave, namely, 2.7 fs for Ti:sapphire lasers centered around 800 nm. This requirement is illustrated in Fig. 9a. Though other techniques are available for synchronization, such as using cross-phase modulation to synchronize passively two mode-locked lasers that share the same intracavity gain medium [47, 48, 49], we employed a flexible all-electronic approach for active stabilization of repetition rates to achieve an unprecedented level of synchronization for fs lasers [50, 51]. An even tighter lock can be realized by first using this electronic approach and then switching to an error signal generated by an optical cross-correlation [52].

Two Kerr-lens, mode-locked Ti:sapphire lasers are located in a mechanically and thermally stable environment for the synchronization experiment. To synchronize the two lasers, two phase-locked loops (PLL) are employed that work at different timing resolutions, as shown in Fig. 10. One PLL

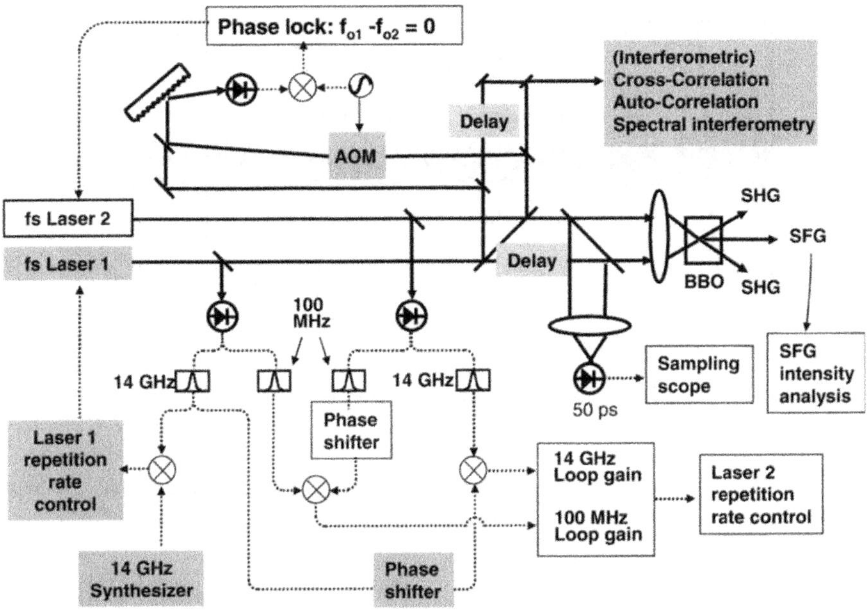

Fig. 10. Experimental schematic for pulse synchronization and carrier–envelope phase locking

compares and locks the fundamental repetition frequencies (100 MHz) of the lasers. An rf phase shifter between the two 100 MHz signals can be used to control the (coarse) timing offset between the two pulse trains with a full dynamic range of 10 ns. The second, high-resolution PLL compares the phase of high-order harmonics of the two repetition frequencies, for example, the 140th harmonic at 14 GHz. This second loop provides enhanced phase stability of the repetition frequency when it supplements and then replaces the first PLL. A transition of control from the first PLL to the second PLL can cause a jump in the timing offset by at most 35.7 ps (1/2 of one 14 GHz cycle), whereas the adjustable range of the 14 GHz phase shifter is 167 ps. The servo action on the slave laser is carried out by a combination of transducers, including a small mirror mounted on a fast PZT, a regular mirror mounted on a slow piezo with a large dynamic range, and an acousto-optic modulator placed in the pump beam to help with fast noise. The unity gain frequency of the servoloop is about 200 kHz and the loop employs three integrator stages in the low-frequency region [51].

To characterize the timing jitter, we focus the two pulse trains so that they cross in a thin β-barium borate (BBO) crystal cut for Type-I sum-frequency generation (SFG). The crossed-beam geometry produces an intensity SFG cross-correlation signal. A Gaussian fit to the cross-correlation (obtained when the two lasers are free-running) yielded a pulsewidth of approximately 160 fs full-width half-maximum. (No extracavity dispersion compensation is

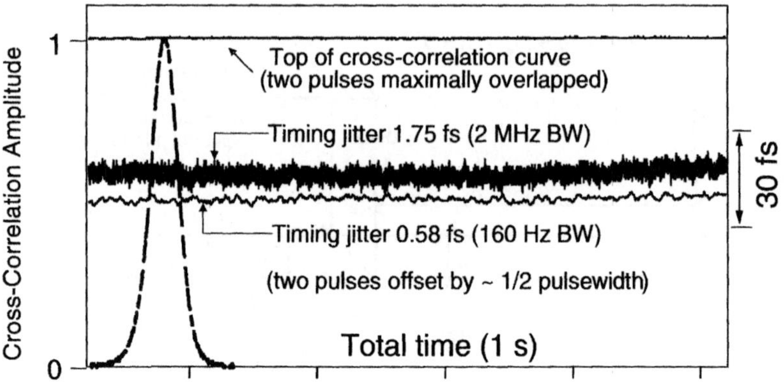

Fig. 11. Timing jitter between two synchronized fs lasers. The *dotted curve* is the cross-correlation signal of the two lasers when the relative pulse timing is scanned across the overlap region. Timing jitter determined from the intensity fluctuations of the SFG intensity is shown over a period of 1 s, using two different low-pass bandwidths

used, and the pulses would be 20 fs in the transform limit.) The top trace in Fig. 11 shows that the SFG signal, recorded in a 2 MHz bandwidth, has no detectable intensity fluctuations when the two laser pulses are maximally overlapped (at the top of the cross-correlation peak). The two middle traces are recorded with 2 MHz and 160 Hz bandwidths, when the timing offset between the two pulse trains is adjusted to yield the half-maximum intensity level of the SFG signal. The slope of the cross-correlation signal near half-maximum can be used to determine the relative timing jitter between the two lasers from the corresponding intensity fluctuations. Timing jitter is calculated from the intensity noise using the slope of the correlation peak, with the scale of the jitter indicated on the vertical axis of Fig. 11. The rms timing noise thus determined is 1.75 fs at a 2 MHz bandwidth and 0.58 fs at a 160 Hz bandwidth. For detection bandwidths above 2 MHz, the observed jitter does not increase. We have recorded such stable performance over several seconds. The synchronization lock can be maintained for several hours. However, the intensity stability of the SFG signal strongly correlates with the temperature variations in the microwave cables located in the high-speed PLL. A careful study of the servo error signal inside the feedback loop reveals that a major limitation on the present performance is actually due to the intrinsic noise of the 14 GHz phase detector, a double balanced mixer and the rf amplifiers. Integration of the intrinsic noise level in the mixer produces the lowest possible rms timing jitter limit for the synchronization loop. From 1 Hz to 160 Hz frequency range it was calculated as

$$\sqrt{2.6 \times 10^{-3}\,\mathrm{fs}^2/\mathrm{Hz} \times 160\,\mathrm{Hz}} \approx 0.64\,\mathrm{fs},$$

which is the approximate jitter performance observed. To achieve even better performance, one must leave the rf domain and use either a single highly stable CW laser [53] or a stable optical cavity to control a high-order harmonic of the repetition frequency, well into the terahertz or tens and hundreds of terahertz frequency range. Timing noise below 0.1 fs should be achievable [53].

4.2 Coherent Phase Locking of Mode-Locked Lasers

Coherent phase locking of the CEP of two separate fs lasers requires a step beyond tight synchronization of the two pulse trains. One also needs to detect and stabilize the phase difference between the two optical carrier waves underlying the envelope of the pulses [54]. As illustrated in Fig. 10b, after the synchronization procedure discussed in the previous section matches the repetition rates ($f_{r1} = f_{r2}$), phase locking requires maintaining the spectral combs of the individual lasers exactly coincident in the region of spectral overlap so that the two sets of optical-frequency combs form a continuous and phase-coherent entity. In other words, the offset frequencies of the lasers must be set such that $f_{o12} = f_{o1} - f_{o2} = 0$. f_{o12} is easily detected by a coherent heterodyne beat signal between overlapping comb components of the two lasers. By phase locking f_{o12} to a frequency of zero mean value, the two pulse trains evolve with identical CEP, i.e., $\Delta\phi_{CE1} - \Delta\phi_{CE2} = 0$.

To demonstrate this coherent comb "stitching" experimentally, two independent mode-locked Ti:sapphire lasers are operated at a 100 MHz repetition rate, one centered at 760 nm and the other at 810 nm. The bandwidth of each laser corresponds to a sub-20 fs transform-limited pulse. When synchronized, the heterodyne beat between the two combs can be recovered with a S/N of 60 dB in a 100 kHz bandwidth, as shown in Fig. 12. Hundreds of comb pairs contribute to the heterodyne beat signal. Figure 12 also indicates that before the phase lock loop is activated, fluctuations in the relative difference between the offset frequencies (f_{o12}) of the two combs easily exceed megahertz levels on timescales as short as several tens of seconds. By stabilizing f_{o12} at a mean value of zero hertz, the carrier–envelope phase slip per pulse of slave laser will accurately match the master laser.

f_{o12} is locked to zero Hz using an AOM. One of the laser beams is passed through the AOM, thereby shifting the entire comb by the drive frequency of the AOM. This procedure avoids the need to process the beat signal in the troublesome frequency range around dc or f_r. The detected beat signal is then phase locked to the drive frequency of the AOM, effectively removing the AOM frequency. When unlocked, the intercomb beat frequency has a standard deviation of a few megahertz with 1 s averaging time. Figure 12 shows the recorded beat frequency signal under a phase-locked condition. With an averaging time of 1 s, the standard deviation of the beat signal is 0.15 Hz. The established phase coherence between the two femtosecond lasers is also revealed via a direct time-domain analysis, as depicted in Fig. 13. For this example, we have employed spectral interferometric analysis of the joint spectra of the two pulses to produce interference fringes that correspond to phase

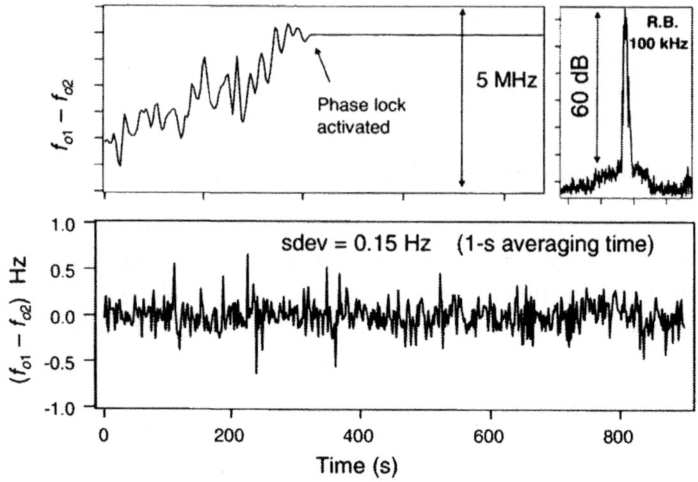

Fig. 12. Heterodyne beat signals can be detected between the two fs lasers after they are tightly synchronized. However, with phase locking, the fluctuations of the beat signal can exceed several megahertz in a short time period. The beat frequency under the locked condition shows a standard deviation of 0.15 Hz at 1 s averaging time

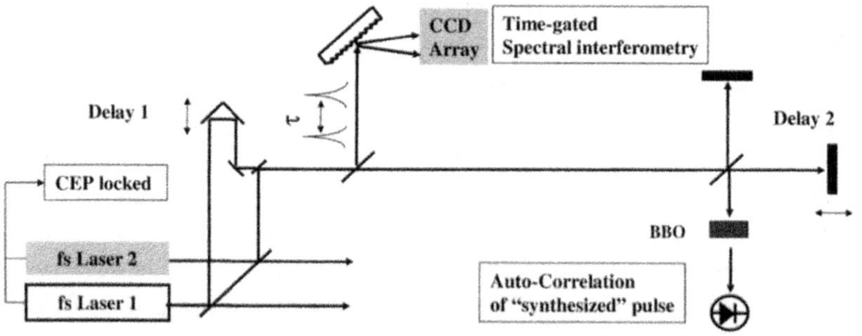

Fig. 13. Time-domain analysis of the established mutual coherence between the two fs lasers. Shown are spectral interferometer and second-order autocorrelator

coherence between the two pulse trains persisting over the measurement time period.

The result is displayed in Fig. 14. We note that the fringe visibility is reduced when the measurement time is increased, due to the increased phase noise between the two lasers. A cross-correlation measurement between the two pulse trains also manifests the phase coherence in the display of persistent fringe patterns.

A powerful demonstration of the "coherently synthesized" aspect of a combined pulse is through a second-order autocorrelation measurement of the combined pulse. For this measurement, the two pulse trains were maximally

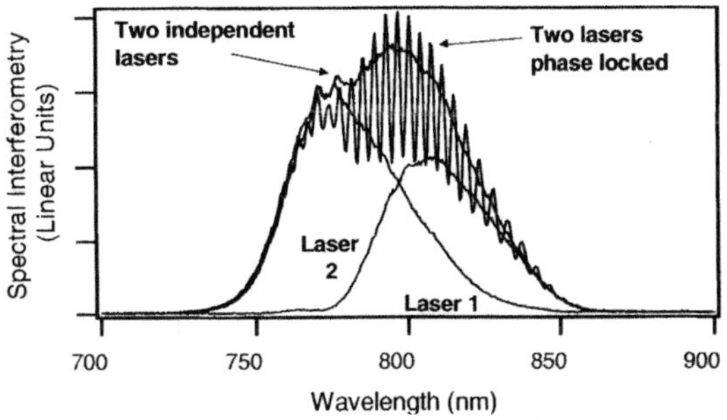

Fig. 14. Spectral interferometric data of two individual lasers, along with the cases when both lasers are present under locked and unlocked conditions. The interference fringes in the spectrally overlapping region between the two lasers clearly indicate phase coherence between the two pulse trains when they are locked together

overlapped in the time domain before the autocorrelator. The autocorrelation curves of each individual laser are shown (Fig. 15a,b, respectively). The spectra of the lasers are centered around 760 nm and 810 nm. An interesting autocorrelation measurement is obtained when the two lasers are not even synchronized (Fig. 15c). Basically, we obtain an autocorrelation that is an average of the two individual lasers, with a sharp spike in the data at a random position. The spike appears because, at that particular instant, the pulses from the two lasers overlapped in time and the two electric fields came into phase and coherently added together. The timescale of this random interference is related to the offset frequency difference between the two repetition rates and is usually less than a few nanoseconds. When the two lasers are synchronized but not phase locked, the resulting autocorrelation measurement indicates increased signal amplitude compared to the unsynchronized case, typically by a factor of 2.7. However, as expected, this signal displays considerable random phase noise within the autocorrelation interference fringes. When the two femtosecond lasers are phase locked, the autocorrelation reveals a clean pulse that is often shorter in apparent duration and larger in amplitude (Fig. 15d). Note that we did not attempt to recompress the light pulses outside the laser cavities to an (optimal) short duration and the pulses are dispersively broadened to 50 fs to 70 fs. The width of the central fringe pattern in an interferometric autocorrelation is more characteristic of the overall bandwidth of the pulse than of the pulse duration and can result in a trace that appears deceptively short. However, from Fig. 15, it is clear that both amplitude enhancement and pulsewidth reduction are present as a result of the combined synchronization and carrier phase locking. We have therefore demonstrated successful implementation of coherent light synthesis: the coherent combination of output from more than one laser so that the combined output can be viewed as a coherent, femtosecond pulse emitted from a single source [54].

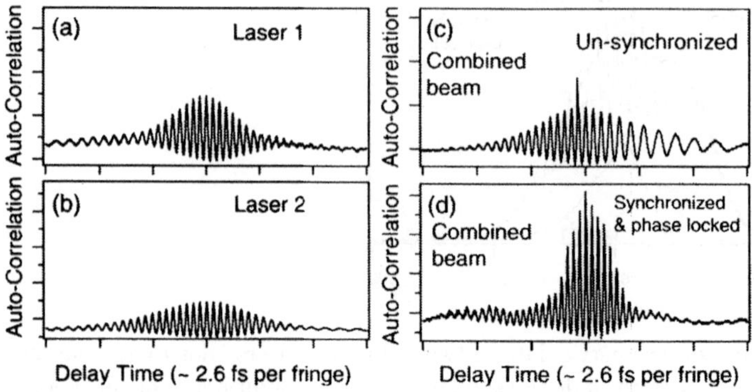

Fig. 15. Demonstration of coherent pulse synthesis. The second-order autocorrelation data show that the combined pulse has a narrower width and a higher amplitude compared with the two original laser pulses

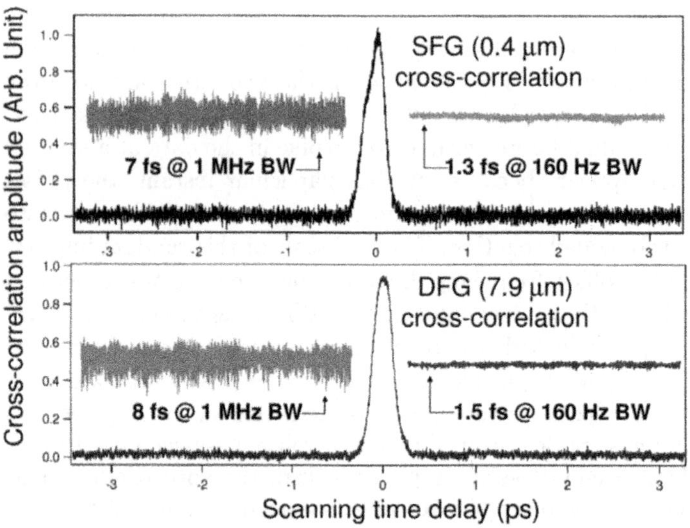

Fig. 16. Simultaneous sum- and difference-frequency generation from two stabilized femtosecond lasers

The capability of stabilizing the pulse repetition rate and the CEP evolution of two mode-locked lasers to such a high degree enables many possible applications. It may be particularly important in the generation of tunable femtosecond sources in otherwise previously unreachable spectral regions. Previous work in electronic synchronization of two mode-locked Ti:sapphire lasers demonstrated timing jitter of a few hundred fs at best. Therefore, the present level of synchronization would make it possible to take full advantage of this time resolution for applications such as novel pulse generation and shaping [54], high-power sum- and difference-frequency mixing [55], new generations of laser/accelerator based light sources [56], or experiments requiring synchronized laser light and X-rays or electron beams from synchrotrons [57]. Figure 16 shows the cross-correlation measurement of the two stabilized mode-locked Ti:sapphire lasers using both (SFG) sum- and difference-frequency generation (DFG). The DFG signal produced by a GaSe crystal can be tuned from 6 μm to beyond 12 μm with a high repetition rate (the same as the original lasers) and a reasonable average power (tens of microwatts). Arbitrary amplitude waveform generation and rapid wavelength switching in these nonlinear signals are simple to implement [58]. Another important application is in the field of nonlinear-optics-based spectroscopy and nanoscale imaging. For example, using two picosecond lasers with tightly synchronized repetition rates ($\approx$ 20 fs of jitter), we are able to achieve significant improvements in experimental sensitivity and spatial resolutions for coherent anti-Stokes Raman scattering (CARS) microscopy [59].

5 Outlook

Over the last three years, the timely convergence of ultrafast lasers and high-precision spectroscopy (with accompanying highly stable CW lasers) has generated a number of advances in both fields. Though the most immediate (and in itself, a truly revolutionary) effect has been realized in optical-frequency metrology, other applications are beginning to emerge based on the unprecedented degree of control now possible over few-cycle optical pulses. This control capability is now having an impact on time-domain experiments and promises to bring about dramatic advances in this area just as it has in optical-frequency metrology and optical clocks. The ability to synthesize arbitrary electronic waveforms at optical frequencies is expected to impact both fundamental physics and more application-specific technologies such as high-speed analog signal processing. As we have described, the ability to generate a pulse train with very high carrier–envelope phase coherence forms the technological basis for waveform synthesis. The next step is to develop a method of measuring the "absolute" or static carrier–envelope phase. Work on several methods for doing so is well underway. Once this is achieved, the final step is to adapt current pulseshaping technology to include the carrier–envelope phase. Several applications of this technology have been mentioned in this chapter, and others are expected to emerge as the technology matures.

Acknowledgments

The authors acknowledge support from NIST, DARPA, NASA, NSF, and ONR. The authors acknowledge contributions from S. A. Diddams, S. Foreman, L.-S. Ma, R. Shelton, H. Kapteyn and M. Murnane.

References

[1] T. Brabec, F. Krausz: Rev. Mod. Phys. **72**, 545–591 (2000)

[2] R. S. Judson, H. Rabitz: Phys. Rev. Lett. **68**, 1500–1503 (1992)

[3] A. Baltuška, T. Udem, M. Uiberacker, M. Hentschel, E. Goulielmak, C. Gohle, R. Holzwarth, V. S.Yakovlev, A. Scrinzi, T. W. Hänsch, F. Krausz: Nature **421**, 611–615 (2003)

[4] A. M. Weiner: Rev. Sci. Inst. **71**, 1929–1960 (2000)

[5] T. Brabec, H. C. Kapteyn (Eds.): *Strong Field Physics* (Springer, Berlin, Heidelberg 2002)

[6] S. T. Cundiff: J. Phys. D. **35**, R43–R59 (2002)

[7] D. J. Jones, S. A. Diddams, J. K. Ranka, R. S. Windeler, J. L. Hall, S. T. Cundiff: Science **288**, 635–639 (2000)

[8] A. Poppe, R. Holzwarth, A. Apolonski, G. Tempea, C. Spielmann, T. W. Hänsch, F. Krausz: Appl. Phys. B-Lasers O. **72**, 373–376 (2001)

[9] J. Ranka, R. Windeler, A. Stentz: Opt. Lett. **25**, 25–27 (2000)

[10] R. Ell, U. Morgner, F. X. Kärtner, J. G. Fujimoto, E. P. Ippen, V. Scheuer, G. Angelow, T. Tschudi, M. J. Lederer, A. Boiko, B. Luther-Davies: Opt. Lett. **26**, 373–375 (2001)

[11] U. Morgner, R. Ell, G. Metzler, T. R. Schibli, F. X. Kärtner, J. G. Fujimoto, H. A. Haus, E. P. Ippen: Phys. Rev. Lett. **86**, 5462–5465 (2001)

[12] T. M. Fortier, D. J. Jones, S. T. Cundiff: Opt. Lett. **28**, 2198–2200 (2003)

[13] T. M. Ramond, S. A. Diddams, L. Hollberg, A. Bartels: Opt. Lett. **27**, 1842–1844 (2002)

[14] H. R. Telle, G. Steinmeyer, A. E. Dunlop, J. Stenger, D. H. Sutter, U. Keller: Appl. Phys. B-Lasers O. **69**, 327–333 (1999)

[15] J. Ye, L. S. Ma, J. L. Hall: Phys. Rev. Lett. **87**, 270801 (2001)

[16] J. Reichert, R. Holzwarth, T. Udem, T. W. Hänsch: Opt. Commun. **172**, 59–68 (1999)

[17] K. F. Kwong, D. Yankelevich, K. C. Chu, J. P. Heritage, A. Dienes: Opt. Lett. **18**, 558–560 (1993)

[18] A. Apolonski, A. Poppe, G. Tempea, C. Spielmann, T. Udem, T. W. Hänsch, F. Krausz: Phys. Rev. Lett. **85**, 740–743 (2000)

[19] L. Xu, C. Spielmann, A. Poppe, T. Brabec, F. Krausz, T. W. Hänsch: Opt. Lett. **21**, 2008–2010 (1996)

[20] H. A. Haus, E. P. Ippen: Opt. Lett. **26**, 1654–1656 (2001)

[21] P. M. Goorjian, S. T. Cundiff: Opt. Lett. **29**, 1363–1365 (2004)

[22] A. Baltuška, T. Fuji, T. Kobayashi: Phys. Rev. Lett. **88**, 133901 (2002)

[23] G. G. Paulus, F. Grasbon, H. Walther, P. Villoresi, M. Nisoli, S. Stagira, E. Priori, S. D. Silvestri: Nature **414**, 182–184 (2001)

[24] M. Kakehata, H. Takada, Y. Kobayashi, K. Torizuka, Y. Fujihira, T. Homma, H. Takahashi: Opt. Lett. **26**, 1436–1438 (2001)

[25] J. Ye, S. T. Cundiff, S. Foreman, T. M. Fortier, J. L. Hall, K. W. Holman, D. J. Jones, J. D. Jost, H. C. Kapteyn, K. A. V. Leeuwen, L. S. Ma, M. M. Murnane, J. L. Peng, R. K. Shelton: Appl. Phys. B-Lasers O. **74**, 27–34 (2002)

[26] F. W. Helbing, G. Steinmeyer, U. Keller, R. S. Windeler, J. Stenger, H. R. Telle: Opt. Lett. **27**, 194–196 (2001)

[27] E. V. Baklanov, V. P. Chebotaev: Sov. J. Quant. Electron. **7**, 1252–1255 (1977)

[28] R. Teets, J. Eckstein, T. W. Hänsch: Phys. Rev. Lett. **38**, 760–764 (1977)

[29] J. N. Eckstein, A. I. Ferguson, T. W. Hänsch: Phys. Rev. Lett. **40**, 847–850 (1978)
[30] T. Udem, J. Reichert, R. Holzwarth, T. W. Hänsch: Phys. Rev. Lett. **82**, 3568–3571 (1999)
[31] S. A. Diddams, D. J. Jones, L. S. Ma, S. T. Cundiff, J. L. Hall: Opt. Lett. **25**, 186–188 (2000)
[32] S. A. Diddams, D. J. Jones, J. Ye, S. T. Cundiff, J. L. Hall, J. K. Ranka, R. S. Windeler, R. Holzwarth, T. Udem, T. W. Hänsch: Phys. Rev. Lett. **84**, 5102–5105 (2000)
[33] R. Holzwarth, T. Udem, T. W. Hänsch, J. C. Knight, W. J. Wadsworth, P. S. J. Russell: Phys. Rev. Lett. **85**, 2264–2267 (2000)
[34] T. Udem, J. Reichert, R. Holzwarth, S. Diddams, D. Jones, J. Ye, S. Cundiff, T. Hänsch, J. Hall: A new type of frequency chain and its application to fundamental frequency metrology, in S. Karshenboim (Ed.): *Hydrogen II: Physics of Simple Atomic Systems* (2001) pp. 125–145
[35] D. J. Jones, T. M. Fortier, S. T. Cundiff: J. Opt. Soc. Am. B **21**, 1098–1101 (2004)
[36] O. Mücke, T. Tritschler, M. Wegener, U. Morgner, F. Kärtner: Phys. Rev. Lett. **89**, 127401 (2002)
[37] E. Cormier, P. Lambropoulos: Eur. Phys. J. D **2**, 15–20 (1998)
[38] R. Atanasov, A. Hache, J. L. Hughes, H. M. van Driel, J. E. Sipe: Phys. Rev. Lett. **76**, 1703–1706 (1996)
[39] A. Hache, Y. Kostoulas, R. Atanasov, J. L. Hughes, J. E. Sipe, H. M. van Driel: Phys. Rev. Lett. **78**, 306–309 (1997)
[40] T. Fortier, P. Roos, D. Jones, S. Cundiff: Carrier-envelope phase controlled quantum interference of injected photocurrents in semiconductors, Phys. Rev. Lett. **92**, 147403 (2004)
[41] A. Apolonski, P. Dombi, G. G. Paulus, M. Kakehata, R. Holzwarth, T. Udem, C. Lemell, K. Torizuka, J. Burgdorfer, T. W. Hänsch, F. Krausz: Phys. Rev. Lett. **92**, 073902 (2004)
[42] F. Kärtner: (2003), personal communication
[43] T. M. Fortier, J. Ye, S. T. Cundiff, R. S. Windeler: Opt. Lett. **27**, 445–447 (2002)
[44] M. Fiorentino, J. E. Sharping, P. Kumar, A. Porzio, R. S. Windeler: Opt. Lett. **27**, 649–651 (2002)
[45] K. Holman, R. J. Jones, A. Marian, S. Cundiff, J. Ye: Opt. Lett. **28**, 851–853 (2002)
[46] J. X. Cheng, A. Volkmer, L. D. Book, X. S. Xie: J. Phys. Chem. B **105**, 1277–1280 (2001)
[47] A. Leitenstorfer, C. Furst, A. Laubereau: Opt. Lett. **20**, 916–918 (1995)
[48] Z. Wei, Y. Kobayashi, Z. Zhang, K. Torizuka: Opt. Lett. **26**, 1806–1808 (2001)
[49] M. Betz, F. Sotier, F. Tauser, S. Trumm, A. Laubereau: Opt. Lett. **29**, 629–631 (2004)
[50] L. S. Ma, R. K. Shelton, H. C. Kapteyn, M. M. Murnane, J. Ye: Phys. Rev. A **64**, 021802(R) (2001)
[51] R. K. Shelton, S. M. Foreman, L. S. Ma, J. L. Hall, H. C. Kapteyn, M. M. Murnane, M. Notcutt, J. Ye: Opt. Lett. **27**, 312–314 (2002)
[52] T. R. Schibli, J. Kim, O. Kuzucu, J. T. Gopinath, S. N. Tandon, G. S. Petrich, L. A. Kolodziejski, J. G. Fujimoto, E. P. Ippen, F. X. Kaertner: Opt. Lett. **28**, 947–949 (2003)

[53] A. Bartels, S. A. Diddams, T. M. Ramond, L. Hollberg: Opt. Lett. **28**, 663–666 (2003)
[54] R.K. Shelton, L.S. Ma, H.C. Kapteyn, M.M. Murnane, J.L. Hall, J. Ye: Science **293**, 1286–1289 (2001)
[55] R.A. Kaindl, M. Wurm, K. Reimann, P. Hamm, A.M. Weiner, M. Woerner: J. Opt. Soc. Am. B **17**, 2086–2094 (2000)
[56] Lawrence Berkeley National Laboratory: LUX at LBNL. Synchronization URL `http://lux.lbl.gov/sync.html`
[57] R.W. Schoenlein, W.P. Leemans, A.H. Chin, P. Volfbeyn, T.E. Glover, P. Balling, M. Zolotorev, K.J. Kim, S. Chattopadhyay, C.V. Shank: Science **274**, 236–238 (1996)
[58] S. Foreman, D.J. Jones, J. Ye: Opt. Lett. **28**, 370–372 (2003)
[59] D.J. Jones, E.O. Potma, J.X. Cheng, B. Burfeindt, Y. Pang, J. Ye, X.S. Xie: Rev. Sci. Inst. **73**, 2843–2848 (2002)

Subfemtosecond XUV Pulses: Attosecond Metrology and Spectroscopy

Reinhard Kienberger and Ferenc Krausz

Institut für Photonik, Technische Universität Wien, Gusshausstr. 27/387, 1040 Wien, Austria
reinhard.kienberger@tuwien.ac.at

Abstract. The generation of ever shorter pulses is a key to exploring the dynamic behavior of matter on ever shorter timescales. Recent developments have pushed the duration of laser pulses close to its natural limit, to the wave cycle, which lasts somewhat longer than one femtosecond ($1\,\text{fs} = 10^{-15}\,\text{s}$) in the visible spectral range. Time-resolved measurements with these pulses are able to trace dynamics of molecular structure but fail to capture electronic processes occurring in atoms on an attosecond ($1\,\text{as} = 10^{-18}\,\text{s}$) timescale. The generation of high-order harmonic radiation in the extreme ultraviolet and soft X-ray regime from atoms exposed to intense few-femtosecond laser pulses comprising just a few wave cycles opened the way to the generation of isolated XUV/X-ray pulses shorter than 1 fs. This chapter will focus on the generation and measurement of these pulses (metrology) and discuss possible ways of using these atomic-timescale bursts of electromagnetic radiation for tracking ultrafast electronic processes with attosecond resolution (spectroscopy).

1 Generation of Atomic-Timescale Pulses

Electrons inside atoms and molecules can be excited and subsequently relax or interact with each other within attoseconds, requiring subfemtosecond pulses for capturing these ultrafast motions. As nature limits the shortness of light pulses to one wave cycle, which lasts somewhat longer than one femtosecond in the visible spectral range, one has to draw on XUV or X-ray radiation for producing attosecond pulses. Excellent coherence and relatively low pump-energy requirements, which can be readily met by table-top-femtosecond laser systems, make high-order harmonic generation (HHG) an attractive approach to developing a laboratory short-wavelength source [1, 2, 3]. The XUV radiation emerges as an electron, is detached from an atom by a high laser field, and then driven back to its parent ion where it can recombine into its original bound state. This process is repeated quasi-periodically, each half cycle of the driving laser wave, resulting in a discrete frequency spectrum made up of high-order odd harmonics of the driving laser. The highest order harmonics are produced near the intensity peak of the laser pulse, as the recollision energy of the electrons is highest when the laser field is largest. Using a few-cycle driver [4, 5, 6] (see also Chapter by *De Silvestri* in this volume) and passing the harmonics through a filter that transmits only the highest energy photons, we select radiation produced within only a fraction of the laser

F. X. Kärtner (Ed.): Few-Cycle Laser Pulse Generation and Its Applications, Topics Appl. Phys. **95**, 343–378 (2004)

cycle near the peak of the pulse. This method, it has been predicted, is capable of generating single XUV/X-ray pulses of ~ 100 as duration emitted in a collimated, laser-like beam [7, 8, 9].

1.1 High-Order Harmonic Generation

High-order harmonic generation is a process resulting from the interaction of an electric field provided by laser pulses with atoms [1,2,10,11,12,13,14,15,16, 17,18,19], atom clusters [20,21], and molecules [22,23]. High intensity, linear polarization, and ultrashort pulse duration are indispensable requirements for implementing such an extreme frequency transformation from the visible spectral range (~ 400 nm to 800 nm) into the XUV- or X-ray regime. If the pulses reach a duration of a few light field oscillation cycles, the highest order harmonics merge into a continuum, as will be discussed in Sect. 1.3.

1.1.1 Interaction of a Laser with a Single Atom

Although an accurate description of HHG requires a full quantum-mechanical treatment [24], the most important properties of HH radiation can be understood in terms of a semiclassical model [3, 25]. This is based on two approximations first introduced by Keldysh for the calculation of strong-field ionization processes [26, 27].

1. After its liberation, the electron is treated as a free particle by neglecting the effect of the Coulomb potential in positive-energy (continuum) states.
2. The contribution of all bound states, other than the ground state, to the evolution of the system is neglected. This is fullfilled in the so-called nonperturbative regime, characterized by $\gamma < 1$, where γ is the Keldysh parameter

$$\gamma = \frac{\omega_{\mathrm{L}}\sqrt{2W_{\mathrm{b}}}}{E_{\mathrm{a}}}, \tag{1}$$

W_{b} is the binding potential, ω_{L} is the laser frequency, and E_{a} is the amplitude of the laser electric field strength. Ultrashort, high-power laser pulses focused on a small spot provide access to this nonperturbative regime.

Under the two assumptions above, HHG can be understood and described in terms of three elementary processes depicted in Fig. 1:

1. A free electron is "born" by tunnel ionization at instant τ_{b}. This happens when the Coulomb potential of the atom is bent down so strongly by the laser electric field that the most weakly bound electron can tunnel through or escape above the barrier and is freed.
2. The electron is accelerated by the laser electric field and removed from the ionized atom. When the laser field reverses its sign, the electron will start to slow down, come to rest, and accelerate toward the ion. As Fig. 2

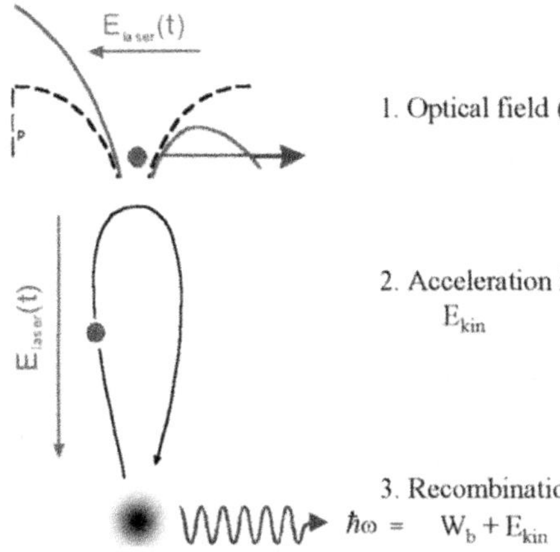

Fig. 1. Principle of high-order harmonic generation in the framework of the semi-classical model described in the text

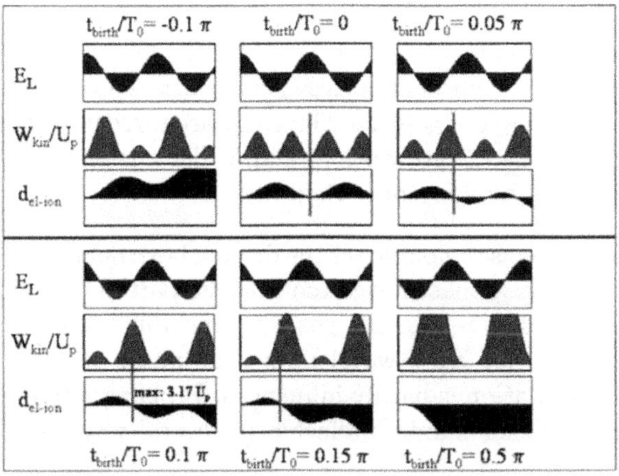

Fig. 2. Evolution of the freed electron's kinetic energy ($W_{\rm kin}$) and distance ($d_{\rm el\text{-}ion}$) as a function of time for different instants of release $\tau_{\rm b}$ (coinciding with the *left* margin of the chart) in the laser field ($E_{\rm L}$)

shows, the trajectories sensitively depend on the time of "birth" of the electron with respect to the oscillation part of the laser electric field. If the electron is born *before* the field oscillation peak, it never returns to the atomic core ($d_{\rm el} > 0$ for any time $t > t_{\rm birth}$). For $t_{\rm birth} = 0$ (release at the instant of maximum electric field), the electron returns with zero kinetic energy to its parent ion. Only if $t_{\rm birth} > 0$, i.e., the electron is released *after* the field oscillation maximum, will it return to the atomic core with nonzero kinetic energy. This energy is maximum for $t_{\rm birth}/T_0 = 17/360$

and reaches the value

$$[W_{\mathrm{kin}}(\tau)]_{\max} = 3.17\,U_{\mathrm{p}}(\tau)\,, \tag{2}$$

where U_{p} is the cycle-averaged kinetic energy of a wiggling electron in the laser field

$$U_{\mathrm{p}}(t) = \frac{e^2 E_{\mathrm{a}}^2(t)}{4m\omega_{\mathrm{L}}^2}\,. \tag{3}$$

3. Upon the electron's recombination into its ground state, a photon is emitted with an energy of $\hbar\omega_{\mathrm{x}} = W_{\mathrm{b}} + W_{\mathrm{kin}}(t_{\mathrm{birth}})$, where W_{b} is the binding energy. The highest photon energy emerging from the process (at the cut-off of the radition spectrum) is ultimately limited by the peak intensity of the laser pulse, i.e., $(\hbar\omega_{\mathrm{x}})_{\max} = W_{\mathrm{b}} + 3.17(U_{\mathrm{p}})_{\max}$. A shift of the cut-off toward higher energies evidently requires increasing the peak intensity, which is ultimately limited by propagation and ensemble effects.

1.1.2 Propagation and Ensemble Effects

Single-atom theory makes many important features of HHG understandable, but a detailed comparison with experiments calls for a macroscopic theory, including propagation and ensemble effects [28,29]. There are three important propagation effects: 1. dephasing, 2. absorption, and 3. defocusing, setting a limit to the maximum achievable harmonic yield rate.

1. The difference between the phase velocities of the driving laser field and the high harmonic wave results in a phase mismatch between partial harmonic waves emitted at $z = 0$ and $z > 0$ in the gas medium. Since the index of refraction is almost unity for XUV wavelengths, the phase velocity of the generated harmonic radiation is hardly influenced. Propagation of the driving laser field, on the other hand, is affected in several ways by the interaction.
   (a) The so-called Guoy phase shift is a simple geometric effect occuring at the waist of a laser beam where the wavefront undergoes a transition from a negative radius of curvature over a plane front to a positive radius of curvature. Due to this geometric effect, the coherent copropagation of the fundamental (λ_0) and the Nth harmonic field is limited to the coherence length

$$L_{\mathrm{g}} = \frac{\pi z_0}{N} = \frac{\pi^2 w_0^2}{N\lambda_0}\,, \tag{4}$$

   where z_0 is the Rayleigh range and w_0 the radius of the laser beam waist [28]. The coherence length is defined as the propagation length, at which the phase mismatch between harmonic and fundamental radiation reaches π.

(b) Ionization is inextricably linked to HHG. As the probabilities for recombination are many orders of magnitude less than one, many more atoms are ionized and hence result in free electrons than contribute to HHG. Free electrons introduce a negative contribution to the refractive index resulting in a dephasing length

$$L_{\rm fe} = \frac{2\pi c\omega_0}{N\omega_{\rm p}^2}, \tag{5}$$

where

$$\omega_{\rm p} = \sqrt{\frac{e^2 n_{\rm e}}{m\epsilon_0}} \tag{6}$$

is the plasma frequency and $n_{\rm e}$ is the instantaneous electron density [9]. If the positive refractive index of the neutral gas can be balanced with the contributions 1.(a) and 1.(b), e.g., by adjusting the pressure of the gas target properly, identical phase velocities can be ensured for the driving field and selected harmonic waves.

(c) The single-atom dipole moment for HHG depends on the instantaneous laser intensity. As the intensity changes during propagation due to dispersion and diffraction, the phase of the dipole moment decreases, providing another contribution to dephasing.

2. Due to its high photon energy, harmonic radiation can excite core electrons during propagation in the gas and be reabsorbed [27]. The conversion will saturate after the absorption length $L_{\rm abs} = 1/\sigma\rho$, where σ is the absorption cross section and ρ is the gas density [30].
3. The radial variation of the free electron density crates a negative lens, resulting in self-defocusing of the driving laser beam. Compared to others, this effect usually plays a minor role.

The experimental results presented in Sect. 4.1, reveal that dispersion tends to broaden the laser pulse, and thus impair conversion efficiency in the cutoff region. The dynamic change in the refractive index of the ionized gas leads to an instantaneous frequency upshift in the driving pulse. This effect appears to benefit subfemtosecond XUV-pulse characterization (see Sect. 4.2).

1.1.3 Finding Optimum Parameters for HHG

Limitations arising from geometric effects, such as dephasing by the Gouy shift and defocusing, depend on the confocal parameter and therefore can always be suppressed by increasing the pulse energy and the beam radius such that the peak intensity remains constant. In neon, conclusive evidence for the limiting role of absorption in few-cycle-driven HHG at the border of the XUV/soft-X-ray regime has been obtained in previous studies. Owing to the steep front edge of the few-cycle driver, harmonics up to the 100 eV photon energy range can be produced at fractional ionization levels as low as 0.5% using few-cycle (sub-10) femtosecond driver pulses [31]. This results in a coherence length L_{fe} (5) that exceeds the XUV absorption length. At these low ionization levels and for a sufficiently large laser beam waist, the phase advance of the driving laser induced by free electrons and the Guoy phase shift, which limit the coherence length, can be compensated for by the phase delay caused by the positive contribution of neutral atoms to the refractive index. This compensation, which can be optimized by fine adjustment of the density n_a (i.e., pressure) of the atomic gas, results in substantial enhancement of the coherence length over a limited fraction of the interaction time.

Whereas few-cycle drivers allow reaching the absorption limit in the entire XUV regime, dephasing plays a major role at energies of > 100 eV even for the shortest driver pulses available [13]. The free-electron density in the generation medium (gas) leading to a phase mismatch between the fundamental and the harmonic beams, limits the propagation length over which harmonic radiation can grow. As shorter wavelengths are generated in the presence of a higher free-electron density, the phase mismatch increases rapidly with the harmonic order, setting a limit to the highest achievable harmonics. In the subkiloelectronvolt energy regime, dephasing becomes the dominant limitation.

A promising route to "switching off" dephasing may be quasi-phase-matching (QPM) [32, 33]. At the simplest realization, a periodic gas–vacuum structure can be investigated by using, e.g., an array of gas jets. According to the theoretical analysis, QPM with few-femtosecond pulses works properly over 35 gas–vacuum periods and enhances the harmonic yield by approximately three orders of magnitude. Another implementation of QPM is generating harmonics in a hollow core fiber using a longitudinal modulation of the core diameter [34, 35].

1.2 Attosecond Pulse Trains

HHG by femtosecond laser pulses comprising many wave cycles has been predicted [36, 37] and recently resulted in a train of sub-femtosecond or attosecond bursts of XUV radiation. This time structure emerges by selecting a number of harmonics over a substantial band near the highest photon energies (cutoff) from the high-harmonic spectrum.

The first possible indication of harmonic radiation containing an attosecond substructure was observed in a high-order autocorrelation of the driving laser pulse [38].

A more recent experiment [39] shows that a group of neighbouring harmonics can beat together to form a sequence of attosecond VUV bursts. Due to the constant frequency spacing of the harmonics, this occurs twice per optical cycle of the driving laser. With N denoting the number of phase-locked

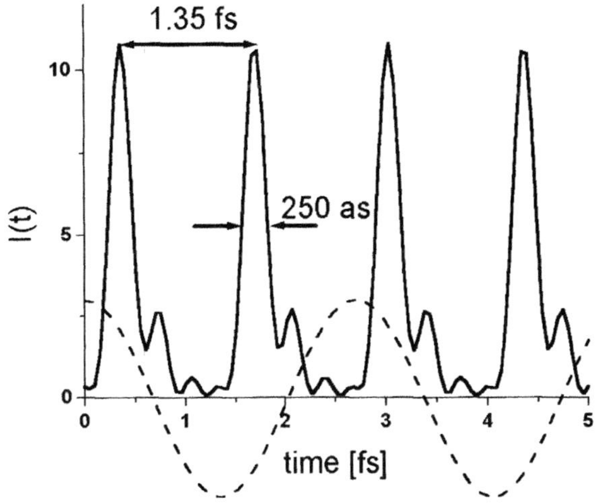

Fig. 3. Temporal intensity profile of the resultant field of five high-order (VUV) harmonics, as reconstructed from measured phases and amplitudes. The FWHM of each peak is ≈ 250 as. Reprinted from [39] by permission of the authors

harmonics and T_0 the laser field oscillation cycle, the burst can be as short as $T_0/2N$. Therefore, a number of $N = 5$ harmonics generated by an 800 nm laser (here in Ar) can form a train of pulses in the attosecond regime. The main condition for the appearance of an attosecond time structure is – similar to the well known mode-locking process – phase locking of the contributing harmonics, which are selected by a band-pass filter. In the experiment described in [39], five VUV harmonics of 800 nm laser pulses were investigated in the frequency domain. The relative phases of the harmonics found were fairly constant, yielding the half oscillation period of the laser field in a train of 250 as pulses spaced by 1.35 fs (Fig. 3).

However, the high repetition rate (= twice the laser frequency) may pose problems in spectroscopic applications because the interpretation of pump–probe experiments tends to be ambiguous. Such a train for either triggering or probing a dynamic process that does not completely come to a standstill within 1.3 fs gives rise to multiple pumping or probing, making data evaluation difficult.

1.3 Single Attosecond Pulses

1.3.1 Spectral Analysis

Driven with sub-100 fs laser pulses, high-order harmonic radiation typically lasts a few tens of femtoseconds [40, 41] and exhibits a quasi-periodic substructure of attosecond spikes at the highest (cutoff) harmonic photon energies [38, 39]. The spectrum of this radiation consists of discrete lines at

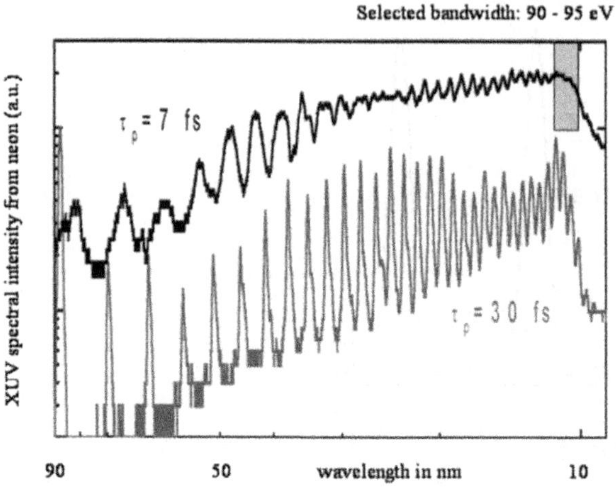

Fig. 4. High-order harmonic spectra produced by multicycle ($\tau_\mathrm{p} = 30\,\mathrm{fs}$) and few-cycle ($\tau_\mathrm{p} = 7\,\mathrm{fs}$) 780 nm laser pulses. The highly modulated region near the cutoff for $\tau_\mathrm{p} = 30\,\mathrm{fs}$ merges to a quasi-continuum for few-cycle driver pulses

odd multiples of the laser frequency, as a spectral manifestation of temporal periodicity with a period of $T_0/2$. In the absence of significant frequency dependence of the spectral phase, the overall emission time is related to the linewidth of the individual harmonics, whereas the duration of individual spikes is determined by the overall spectral width of lines filtered out of the harmonic spectrum. The cutoff harmonics of a few-cycle light pulse, it was predicted, would merge into a continuum, forming an isolated subfemtosecond pulse [7, 9, 42, 43].

Figure 4 shows two different spectra produced by multicycle (30 fs) and a few-cycle (7 fs) 780 nm laser pulses in Ne. The highly modulated region near the cutoff ($> 90\,\mathrm{eV}$) turns into a quasi-continuum in few-cycle driver pulses. The reasons and the consequences will be discussed in the next section.

The harmonic beam can be detected by a thinned, back-illuminated X-ray CCD camera for optimizing the radial intensity profile of the XUV beam (Fig. 5). This can be done by fine adjustment of the position and pressure of the neon gas target and of the intensity of the laser pulses before HHG. For spectral analysis, an XUV transmission grating can be inserted into the XUV beam.

The grating used here has 10 000 lines per millimeter or, in other words, a distance of $a = 100\,\mathrm{nm}$ between the lines. This leads to an angle of $\theta_\mathrm{out} = 7.5°$ between zeroth and first order, according to the grating equation

$$a(\sin\theta_\mathrm{in} - \sin\theta_\mathrm{out}) = m\lambda_\mathrm{x}\,, \tag{7}$$

where $\theta_\mathrm{in} = 0$ and θ_out are the angles of the in- and outcoming beams, respectively; m is the diffraction order; and λ_x is the XUV wavelength, in the present case about 13 nm.

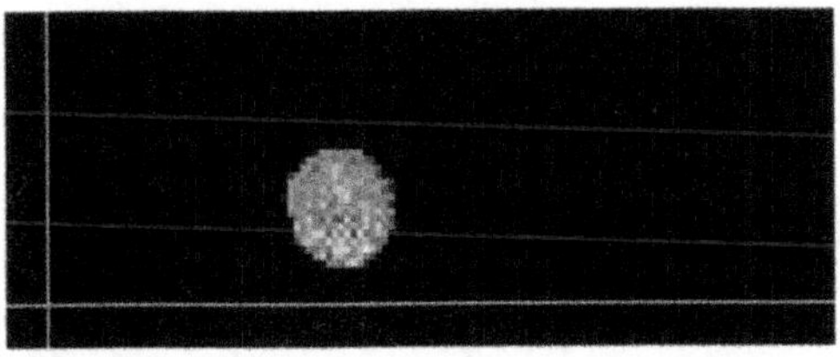

Fig. 5. On-line measured profile of the XUV beam used for adjustment optimization

Fig. 6. On-line measured spectrum of XUV radiation

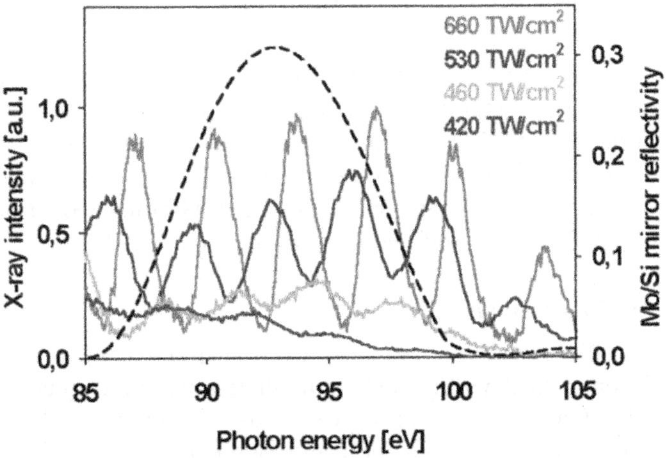

Fig. 7. High-order harmonic spectra emitted from a 2 mm long, 110 mbar neon gas source exposed to linearly-polarized 7 fs, 780 nm laser pulses with peak intensities given in the figure. The degree of modulation at a certrain wavelength range can be tuned by the intensity of the driving laser pulses

The CCD camera is flanged to the XUV beamline through a flexible tube so that it can be adjusted to monitor either the zeroth or first order. In the latter case, the XUV harmonic spectrum can be monitored on-line (Fig. 6). The spectra shown in Fig. 7 have also been recorded with this apparatus.

As the cutoff of the HHG process sensitively depends on the fundamental laser intensity, adjusting the laser intensity, e.g., by an iris, provides a simple means of controlling the photon energy of cutoff harmonics. In this way, the

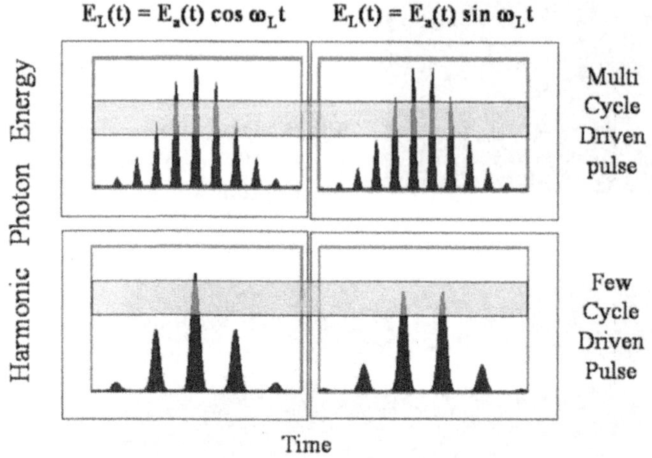

Fig. 8. Comparison of the temporal structure of XUV pulses emerging from an HHG driver by multicycle or few-cycle pulses for different values of the carrier–envelope phase. The peaks sketch the temporal structure (*horizontal axis*) of emitted XUV radiation as a function of XUV photon energy (*vertical axis*)

cutoff region and – in the case of a few-cycle pulse – the quasi-continuum can be shifted relatively to a band-pass filter. Figure 7 shows the high-order harmonic spectra produced by 7 fs, 780 nm laser pulses. The highest peak laser intensities resulted in well-resolved "plateau" harmonics within the high-reflectivity range of the multilayer mirror (dashed line). Gradual reduction of the laser intensity shifts the "cutoff" continuum into the filter band.

1.3.2 Time-Domain Analysis

Harmonic radiation within a limited frequency band near the cutoff is confined to a small fraction of $T_0/2$ within one laser oscillation period. In general, this results in a train of subfemtosecond or even attosecond bursts of short-wavelength radiation [36, 37]. Although the conversion efficiency of HHG is lower than that of other schemes for attosecond pulse generation [44], the advantage is that several possibilities exist for cutting a single subfemtosecond pulse out of the pulse train. One straightforward method is using few-cycle fundamental pulses. Under these conditions, emission of the high-energy end of the harmonic spectrum (near the cutoff region) can be confined to one half-oscillation period near the pulse peak. By selecting a well-defined bandwidth near the spectral cutoff, it is possible to filter out one single subfemtosecond burst. Figure 8 shows the importance of few-cycle pulses used for this method in comparison with multicycle pulses, as well as the sensitivity of isolated pulse generation to the absolute phase ϕ (carrier–envelope phase) of the fundamental laser pulse

$$E_{\mathrm{L}}(t) = E_{\mathrm{a}}(t) \cos(\omega_{\mathrm{L}} t + \phi). \quad (8)$$

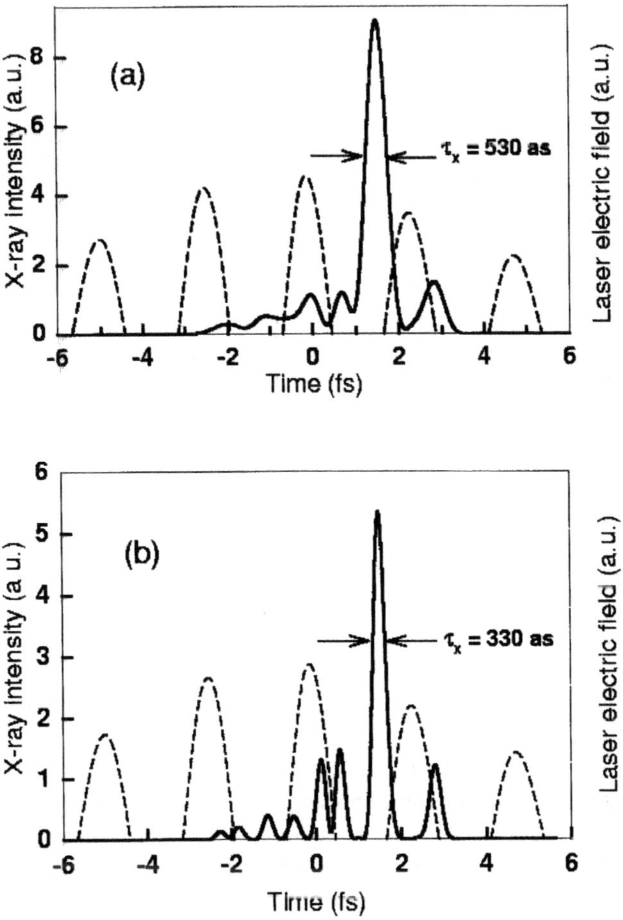

Fig. 9. Calculated far-field, near-axis temporal intensity profile of XUV harmonic radiation (*full line*) selected within the spectral range of (**a**) 90 eV ± 2.5 eV and (**b**) 90 eV ± 5 eV, produced in a 3 mm long 200 mbar neon gas volume by a 7 fs, 780 nm Gaussian laser pulse $E(t) \propto \exp(-t^2/\tau_L^2)\cos(\omega_0 t + \phi)$ with $\phi = 0$ (cosine pulse) and an on-axis peak intensity of (8×10^{14}) W/cm^2. The *dashed line* shows the on-axis electric field of the laser pulse exiting the interaction region

Only "cosine" pulses ($\phi = 0$) have the potential for producing a single burst. Within a preselected frequency band, the fundamental pulse intensity can be adjusted to produce an isolated burst for $\phi = 0$. If the carrier–envelope phase is not stabilized, the selected frequency band must be narrow enough so that only driver pulses with $\phi \approx 0$ are intense enough to generate photons in the filter band (Fig. 8).

The attosecond experiments performed so far and described in the following sections have been carried out under these conditions. Meanwhile, phase-stable, few-cycle laser pulses are available using frequency-domain techniques (see also the Chapters by *Udem* et al. and *Jones* et al.) and allow subfemtosecond XUV pulse generation at each laser shot [45].

The feasibility of a single subfemtosecond pulse produced by few-cycle-driven high-order harmonic generation was corroborated by numerical studies based on a computer code [24] solving Maxwell's wave equations in three dimensions and calculating the radiation of strongly-driven atomic dipoles using the quantum theory of *Lewenstein* et al. [3]. These numerical calculations predict a near bandwidth-limited, isolated, 530 as XUV pulse emitted from neon within the spectral range of 90 eV $\pm$ 2.5 eV (Fig. 9a) under the conditions summarized in the caption of Fig. 9. They also indicate the potential for even shorter pulses. In fact, increasing the XUV mirror bandwidth by a factor of 2 in our simulations results in an 330 as XUV pulse, accompanied by somewhat more pronounced satellites (Fig. 9b).

Section 4.1 and Sect. 4.2 will reveal good agreement between experimental data and these theoretical predictions.

2 Basic Concepts for XUV Pulse Measurement

The most direct information about the duration of short flashes of electromagnetic radiation can be gained from time-domain measurements. To this end, the burst to be characterized must be probed by a sampling pulse that is comparable to or shorter than the duration of the burst. In the optical regime, this concept has been implemented by nonlinear autocorrelation techniques. Because of the unfavorable scaling of the nonlinear polarizability with wavelength and the low energy of the XUV pulse to be measured, these techniques can not be applied currently to radiation at wavelengths much shorter than 100 nm. The only alternative for characterizing subfemtosecond XUV/X-ray pulses is to correlate the short-wavelength burst with the electric field evolution of a strong visible or near-infrared laser pulse in a cross-correlation scheme, where the quarter-wave cycle provides an attosecond probe for sampling the XUV/X-ray pulse.

2.1 Autocorrelation at Short Wavelengths

Autocorrelation techniques for measuring the pulse duration of VUV harmonics are based on two-photon absorption used as a nonlinear effect. In the first experiments [40], harmonics generated by a 10 mJ laser beam were used to ionize He atoms. The number of He^+ ions was measured as a function of the delay between two identical ninth-harmonic pulses (14.1 eV) generated by two replica of the fundamental pulse. A pulsewidth of 27 fs was obtained by assuming a $sech^2$ pulseshape (Fig. 10).

In a more recent experiment, a two-photon absorption-based autocorrelation was implemented for a frequency-resolved optical gating (FROG) measurement [47]. Here the two replica of the fifth harmonic of a 7 mJ Ti:sapphire

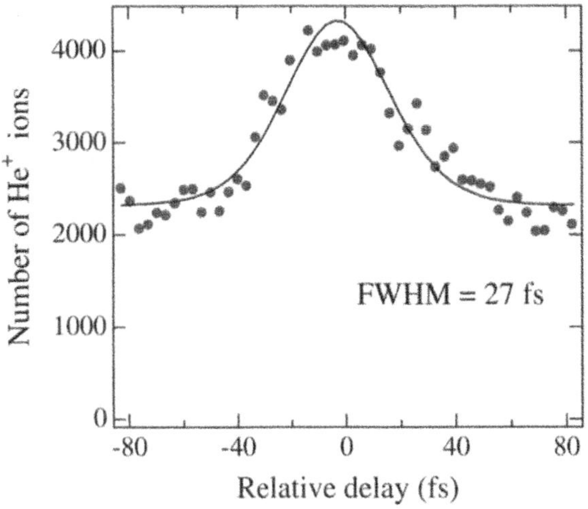

Fig. 10. Autocorrelation trace of the ninth harmonic by two-photon ionization of He [46]. Reprinted by permission of the authors

pulse were focused on a Xe gas jet. From the recorded FROG trace, the intensity profile and chirp of a high harmonic pulse could be retrieved for the first time.

However, the atomic cross section of the absorption process σ scales with λ^6, implying a dramatic decrease of the two-photon transition probability with decreasing wavelength. As a consequence, autocorrelation schemes require photon fluences orders of magnitude higher than those available from existing HHG sources in the XUV regime. The latest experiments demonstrated two-photon absorption in Kr and Xe [48], but the extension of autocorrelation techniques into the XUV/soft-X-ray regime remains a formidable challenge.

Recently, nonlinear XUV time-domain metrology was carried out measuring a second-order autocorrelation trace of a train of attosecond pulses [49].

2.2 X-Ray-Intensity/Laser-Field Cross-Correlation

Several cross-correlation measurements have been proposed and carried out experimentally, e.g., using femtosecond dynamics of field ionization [50]. Here, we review a novel cross-correlation technique based on XUV photoionization in the presence of an intense few-cycle laser field [51, 52]. The method exploits the dependence of the final kinetic energy of an XUV-pulse-generated photoelectron on the phase (and strength) of a laser field at the instant of photoionization. The width and position of the resultant photoelectron spectra are measured as functions of the relative delay between XUV and laser pulses and yield a convolution of the laser light field with the XUV pulse envelope. Careful deconvolution allows us to determine the duration of the

XUV pulse and its timing jitter with respect to the light field with attosecond resolution.

The experimental realization of this technique falls within a class of experiments that began with the work of *Schins* and co-workers [41, 53, 54, 55, 56]. Here, the XUV pulse is "cross-correlated" against a short visible pulse. An XUV photon excites a bound atomic electron into a positive-energy state in the presence of a laser field. Laser-induced transitions from this state or a laser-induced shift in its energy (Stark shift) provide the nonlinearity linking the XUV to the laser pulse, enabling us to compare the XUV pulse duration with that of the visible pulse. In the experiments reported so far, the number of laser photons increasing rapidly with the laser intensity is absorbed and emitted by the freed electron (forced to oscillate at ω_L). To identify the Stark shift amidst the resulting sidebands in the photoelectron energy spectrum or measure the amplitude of the sidebands, the XUV spectral width must be kept significantly below the laser photon energy $\hbar\omega_L \approx 1.6\,\text{eV}$ to avoid overlap of the sidebands with the main peak. Without modification, Schins' approach cannot be extended to sub-laser-cycle XUV pulses. Because we are to measure XUV pulses with durations of $\tau_x < T_0$, the photoelectron energy distribution must become broader than the laser photon energy, so that it is no longer possible to identify individual sidebands and measure their amplitudes or determine the Stark shift.

Recently, it was shown that choosing the geometry of observation carefully and confining the (dressing) laser field to a few oscillation cycles opens the door to broadband ($> \hbar\omega_L$) XUV pulse measurement with subfemtosecond resolution [51, 52]. To this end, photoelectrons from atoms exposed simultaneously to a strong light field and an XUV pulse are detected with a time-of-flight (TOF) spectrometer. The basic idea is to overlap the two pulses with variable delay with respect to each other in a gas medium and measure, as a function of delay, the kinetic energy distribution of the photoelectrons stripped off the atoms by the XUV pulse.

3 Laser-Field-Assisted XUV Photoionization

The effect of a light field on the XUV photoelectron energy spectrum is now well understood within a quantum mechanical framework [57, 58]. In the strong field limit, the most conspicuous features can also be accounted for by a quasi-classical model treating the interaction as a two-step process [51]. According to this model, the photoelectron is first ejected by a short XUV pulse with a distribution of initial momenta known from conventional photoionization studies [27]. Subsequently, it is accelerated (or decelerated) by the light field. For XUV pulse durations τ_x, very short compared to $T_0/2$, the model predicts that, depending on the oscillation phase of the light field at the instant of "birth" of the electron, a momentum component is added to the initial momentum of the electron.

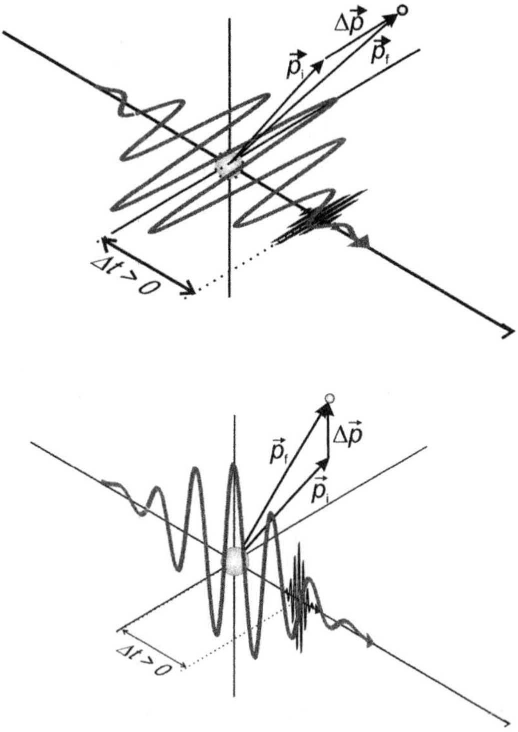

Fig. 11. Two different geometries: The electric fields of the fundamental and XUV pulses can be polarized parallel or orthogonally with the direction of detection of the photoelectrons. This leads to different effects in the photoelectron's momentum shift

The concept of laser-field-assisted XUV photoionization can be implemented in various detection geometries. The two most important are shown in Fig. 11. The photoelectrons are detected within a cone aligned parallel and orthogonally to the electric field vector of the linearly polarized laser field (henceforth referred to as parallel and orthogonal detection geometry, respectively). Here we present a detailed analysis of the othogonal geometry, because the first subfemtosecond pulse measurement used this configuration [52]. In the next section, we shall demonstrate that parallel geometry can also be employed for efficient attosecond diagnostics.

Figure 12 depicts the final momentum distribution of the photoelectron for different instants of release in the strong field. In general, a laser-induced momentum component is added to the electric field vector, resulting in a shift of the photoelectron angular distribution up or down in momentum space. The momentum transferred from the light field is largest if the electric field of the light pulse is zero, i.e., its slope is maximum, at the instant of birth of the photoelectron. The width of the XUV photoelectron energy spectrum

ΔW, which is equal to the bandwidth of the XUV pulse spectrum in the absence of the light field, increases with increasing momentum shift within a finite detection cone aligned orthogonally to the laser polarization.

As the analysis in the next section will show, scanning the instant of birth of the photoelectron through the light field oscillations by changing the relative delay t_d between the light pulse and XUV pulse results in modulation of the center of gravity as well as the width $\Delta W(t_d)$ of the XUV photoelectron spectrum with a period equal to one-half of the light oscillation period T_0. For subfemtosecond XUV pulse measurement, we shall exploit the field-induced spectral broadening [52] because this effect is enhanced by increasing the detection aperture, as revealed by Fig. 12, resulting in a significantly enhanced signal yield and modulation amplitude of $\Delta W(t_d)$. By contrast, modulation in the kinetic energy shift is increasingly washed out with increasing detection angle, deteriorating the temporal resolution.

Taking advantage of the semiclassical approach outlined above, the motion of an XUV-induced photoelectron "born" at a delay time t_d with respect to the peak of the laser pulse is governed by the classical equation of motion. In the direction of laser polarization (x) this equation reads

$$m\dot{v}_{\parallel} = qE_L = -eE_a(t)\cos\omega_L t\,, \tag{9}$$

where m and $-e$ are the electron mass and charge and $E_a(t)$ represents the amplitude envelope of the light pulse. In the adiabatic limit $\mathrm{d}E_a(t)/\mathrm{d}t \ll E_a(t)\omega_L$ (which is a remarkably good approximation even for 7 fs pulses), the femtosecond laser pulse changes the velocity of the electron by

$$\Delta v_{\parallel} = \frac{e}{m}\frac{E_a(t_d)\sin\omega_L t_d}{\omega_L} = \sqrt{\frac{4U_p(t_d)}{m}}\sin\omega_L t_d\,, \tag{10}$$

$$\Delta v_{\perp} = 0 \tag{11}$$

in directions parallel and perpendicular to the laser polarization, respectively. Here,

$$U_p = \frac{e^2E_a^2(t_d)}{4m\omega_L^2} \tag{12}$$

is the electron wiggling energy averaged over an optical cycle. This change of the initial velocity of the electron results in modification of the final kinetic energy W_f. For $U_p \ll W_0$, the final energy is given by

$$W_f \approx W_0 + 2U_p(t_d)\sin^2\omega_L t_d\cos 2\theta + \sqrt{8W_0U_p(t_d)}\sin\omega_L t_d\cos\theta\,, \tag{13}$$

where $W_0 = \hbar\omega_x - W_b$ is the initial kinetic energy of the photoelectron, ω_x denotes the XUV photon energy, W_b stands for the atomic binding energy of the electron liberated ($\hbar\omega_x \gg W_b$), and θ is the angle between the final momentum of the electron and the laser electric-field vector.

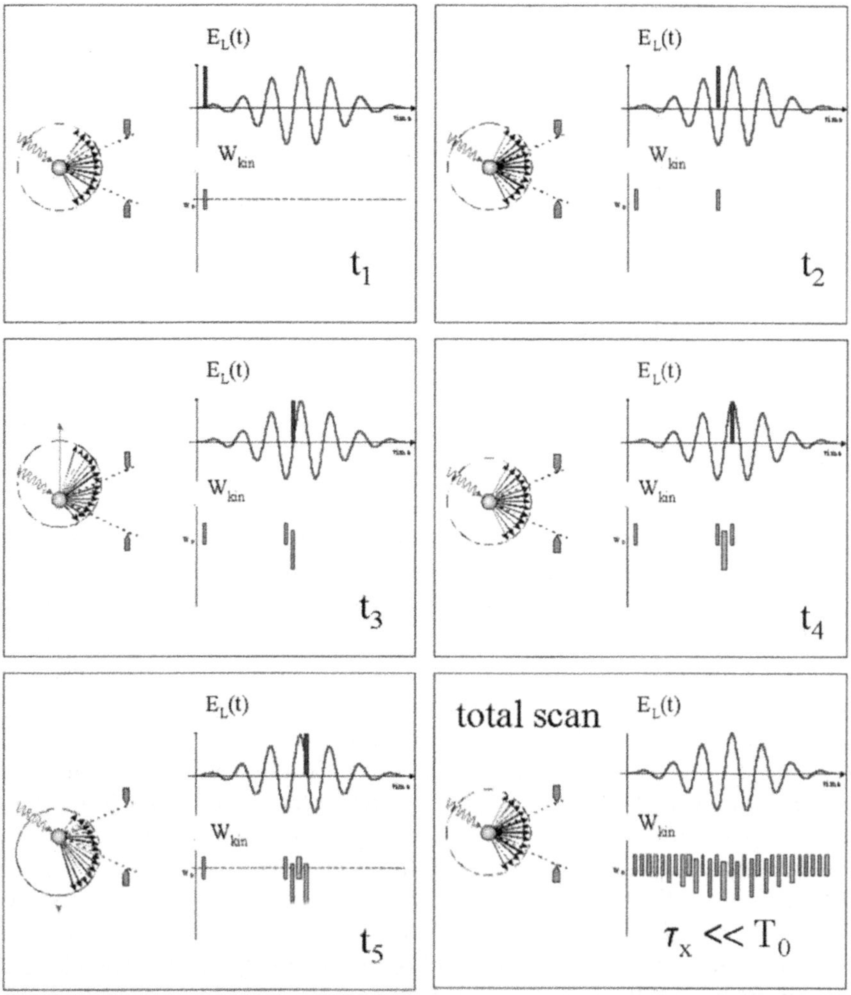

Fig. 12. Principle of measuring light-field/XUV-intensity cross-correlation with attosecond resolution. The photoelectrons created initially with isotropic momentum distribution by the absorption of XUV photons pick up momentum from the strong laser light field. The light-induced momentum change Δp_{x} deforms the final photoelectron momentum distribution at instants t_1, t_2, $t_3 = t_2 + T_0/4$, $t_4 = t_2 + T_0/2$, and $t_5 = t_2 + 3T_0/4$ as shown, where $E_{\mathrm{L}}(t_3) = E_{\mathrm{L}}(t_5) = 0$. Photoelectrons detected within a cone aligned orthogonally to the direction of the light field vector (x direction) display a kinetic energy spread at t_3 and t_5 but remain unaffected by the light field at t_1, t_2 and t_4. The total scan shows the expected modulation of the spectra if $\tau_{\mathrm{x}} < T_0/2$

To compare the predicted variation of photoelectron kinetic energy as a function of t_{d} with measurements, (13) needs to be generalized for a finite

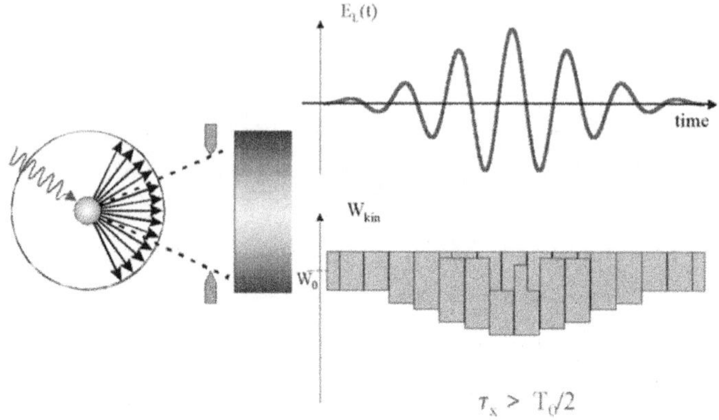

Fig. 13. XUV pulse having a duration or a timing jitter exceeding $T_0/2$ ($\sim$ 1.25 fs in our case) used for the cross-correlation to smear out the modulation completely

emission time slot of the photoelectron (corresponding to the finite XUV pulse duration) and for distribution of initial kinetic energies over a finite energy range (reflecting the finite bandwidth of the XUV pulse). In addition, the distribution of final kinetic energies as a function of θ has to be integrated over the finite detection cone to obtain a realistic prediction for the measured XUV photoelectron spectrum modified by the laser field.

For an XUV pulse that is short compared to half the laser period, i.e., $\tau_x \ll T_0/2$, and a fast (attosecond) response of the electronic bound-free transitions induced by the XUV photons, the predicted modulations in the final kinetic energy distribution versus t_d survive these integrations and provide a sub-femtosecond probe for the measurement of the XUV pulse duration. A finite XUV pulse duration or a finite timing jitter of the XUV pulse (relative to the phase of the light field) of any origin results in a reduced depth of the resultant modulation of the width and center of gravity of the photoelectron energy spectrum (illustrated in Fig. 12).

An XUV pulse duration or a timing jitter exceeding $T_0/2$ ($\sim$ 1.25 fs in our case) would smear out the modulation completely, as shown in Fig. 13 [51]. Hence, the modulation depth of $\Delta W(t_d)$ provides a reliable upper limit, on an attosecond timescale, for XUV pulse duration and timing jitter.

4 Experimental Implementation of Attosecond Metrology

Figure 14 displays the experimental setup for the generation of subfemtosecond XUV pulses (by HHG) and attosecond metrology (by laser-assisted XUV photoionization). The XUV pulses exiting the neon harmonic source copropagate with the laser pulses down the beam delivery tube. Some 150 cm after the XUV pulse source, they hit a 200 nm thick zirconium foil with an

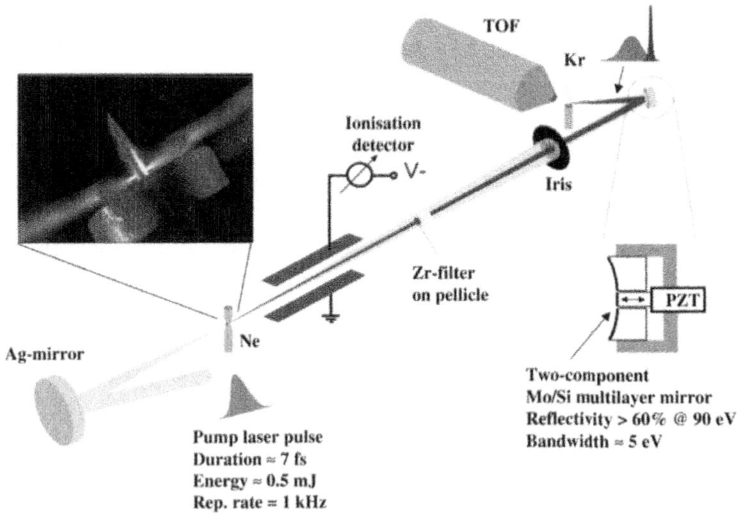

Fig. 14. The schematic of the experiment. The focused 7 fs laser beam interacts with neon atoms to produce high-harmonic radiation. The laser and the highly collimated XUV beam copropagate collinearly through a 2 m beamline toward the measurement. In the beamline, the laser and harmonic beams pass through a 200 nm thick, 3 mm diameter zirconium foil placed on a 5 μm thick nitrocellulose pellicle to cover a hole of 2 mm diameter. The energy transported by the resulting annular beam can be adjusted with a motorized iris between a fraction of a microjoule and few tens of microjoules. The Mo/Si multilayer consists of an annular part having an outer diameter of 10 mm with a concentric hole of 3 mm diameter hosting a miniature mirror of slightly smaller diameter. Both parts originate from the same substrate, ensuring identical radii of curvature ($R = 70$ mm). The miniature central mirror is mounted on a wide range, nanometer precision piezo stage, allowing alignment and translation with respect to the external part

aperture of 2 mm. The aperture is matched to the beam diameter of the harmonic beam, which has a low divergence of ~ 0.7 mrad. This filter is placed in the beam to block the laser and low-order harmonics, transmitting only photons with an energy of more than 70 eV. It is mounted on a nitrocellulose membrane 5 μm thick that has a hole in the middle, which must have equal thickness over the whole area to avoid phase errors in the laser beam, which would wash out the temporal structure. This device is virtually dispersion-free at the laser wavelength and produces an annular laser beam with the XUV beam in the center. The energy in the laser beam is adjusted by a motorized iris and measured by a photodiode. Both the laser and the colinearly propagating XUV beam are focused with a special multipurpose mirror with a focal length of 35 mm. The mirror is mounted on a motorized stage so that it can be removed from the beamline for on-line diagnostics of the beam profile and spectrum by a CCD camera (see Sect. 1.3).

The XUV mirror is coated with a Mo/Si multilayer stack designed to reflect photons with 90 eV energy within a 5 eV bandwidth. This bandwidth is large enough to support XUV pulses as short as 0.4 fs. The mirror substrate consists of two concentric parts, matching the impinging laser and XUV beam. The outer part acts as a metallic mirror, reflecting the entire spectrum of the sub-10 fs laser pulses. The central piece is sitting on a piezo stage adjustable in the transverse and longitudinal directions. In this manner, the two pulses can be overlapped spatially and temporally in the focal plane, where a nozzle supplying the target atoms is situated. The wide-range piezo stage is able to introduce a delay of maximum 100 μm which corresponds to a time delay of several tens of femtoseconds. Further, it offers very good precision in positioning with reproducibility of better than 5 nm, implying a delay setting with sub-50 as accuracy.

The profile of the laser focus can be imaged by a lens on a CMOS camera for monitoring and course preadjustment of spatial overlap between the beams reflected by the two components of the Mo/Si mirror. The folding angle is kept below 5° to minimize astigmatism. Spherical aberration gives rise to time smearing of < 0.05 fs for the harmonic and of $\ll 1$ fs for the laser beam. The focal spots of the XUV and the laser beam have beam diameters of approximately 2 μm and more than 20 μm, respectively. In other words, the XUV beam spot size in the target was more than an order of magnitude smaller than that of the laser beam, ensuring that all the XUV photoelectrons are exposed to the same light field, which is important for comparison with theory.

Krypton atoms effusing from the orifice of the nozzle (a metal-coated glass tube having an inner diameter of 50 μm) at a rate of 0.02 mbar l/s are photoionized by XUV pulses in the presence of the laser field. To avoid a constant photoelectron background due to ionization by the unfocused XUV beam, a thin shield is placed next to the krypton target. All parts in the vicinity of the spectrometer are conductively coated and properly grounded to avoid electrostatic charging.

Before the XUV/laser-field cross-correlation can be recorded by measuring the XUV photoelectron spectrum as a function of t_d, the temporal overlap between the two pulses has to be found. To this end, we generate ATI (above threshold ionization) electrons upon removing the Zr filter that stops the central fraction of the laser beam (Fig. 14) from the beamline and by focusing both the inner and the annular part of the laser beam with the two-component Mo/Si-multilayer mirror onto the krypton target. Scanning the delay between the outer and inner parts of the mirror with a feedback-controlled piezo translator, an "interferometric" ATI autocorrelation of the 7 fs laser pulse can be recorded. Figure 15 shows the detected electron count rate as a function of the delay between the two replicas of the laser pulse.

Owing to the strongly nonlinear dependence of the ionization rate on the electric-field strength and the few-femtosecond pulse duration, the trace is

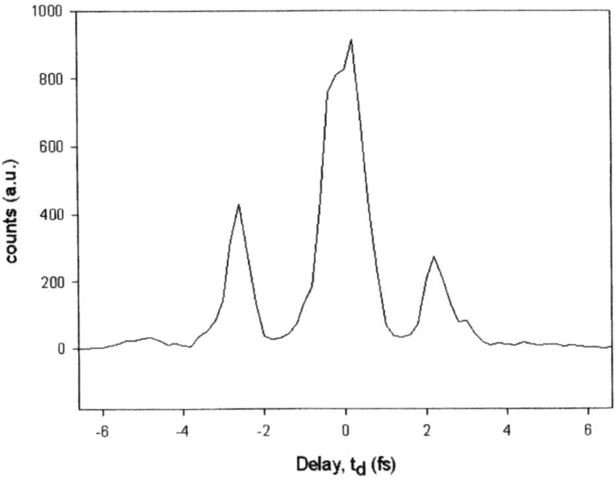

Fig. 15. ATI-electron autocorrelation

confined virtually to a single laser cycle. The central peak indicates the delay of optimum temporal overlap between the two beams. From this reference, the zero delay between the XUV and laser pulse can be inferred by taking into account the finite ($\sim 5\,\mu$m) thickness of the pellicle holding the Zr foil, which is reintroduced for the XUV/laser-field cross-correlation measurements.

To record the cross-correlation depicted in Sect. 2.2, the delay t_d is varied in steps of $\Delta t_d = 150\,$as in the central range of temporal overlap ($-5\,$fs $\leq t_d \leq 5\,$fs, with $t_d = 0$ approximately at the peak of the light pulse) and of increased step size outside this range.

4.1 Measurement of Subfemtosecond XUV Pulse Duration

Figure 16 shows two representative Kr-4p spectra corrected for (a constant) ATI background and recorded at $t_d = -450\,$as and $t_d = 0\,$as, along with respective asymmetrical Gaussian fits. This model function yields excellent fits to our data, characterized by a fidelity parameter $R^2 > 0.95$. It has to be emphasized that the significant broadening and shift of the center of gravity between these two spectra takes place within only 450 as. The enhanced noise at lower energies can be attributed to ATI electrons.

The contour plot in Fig. 17 comprises some 120 spectra of the 4p feature (processed as described above) as a function of delay t_d. Each spectrum was normalized to the same number of counts, resulting in a constant area under the spectral distribution functions. Owing to the relatively large solid angle of detection, providing a high number of electron counts, the overall measurement time could be limited to $\sim 5\,$h. The data clearly bring to light a quasi-periodic evolution of the photoelectron energy spectrum with a period of $\sim T_0/2$.

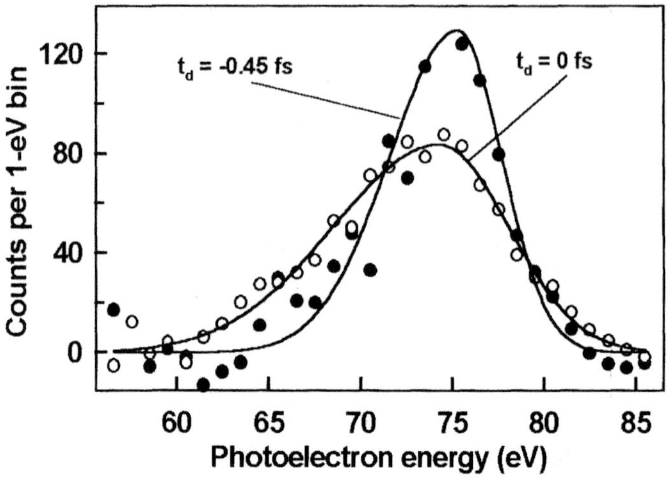

Fig. 16. Kr-4p photoelectron spectra produced by $\sim$ 90 eV XUV pulses in the presence of a strong visible light field at two different delays of the XUV pulse with respect to the light wave. The *dots* represent spectra corrected for some ATI background from the measurements performed by a time-of-flight electron spectrometer. The *lines* show asymmetrical Gaussian fits to the data

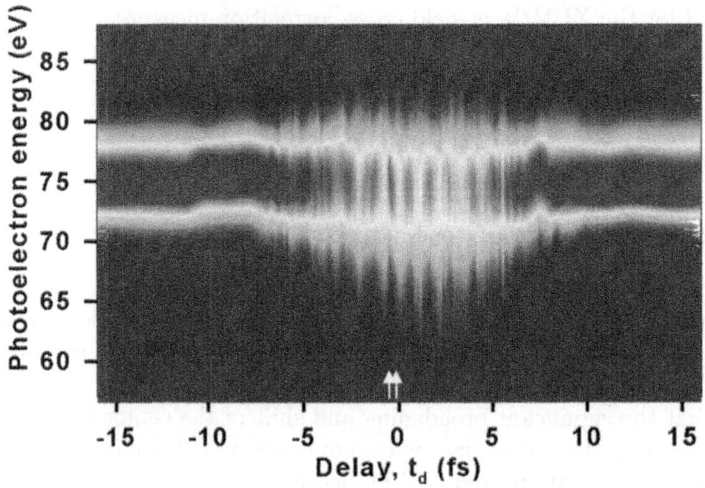

Fig. 17. Energy distibutions of the Kr-4p photoelectrons as a function of the delay t_d between the light wave and the XUV pulse over some tens of femtosecond (evaluated from asymmetrical Gaussian fits). The *two white arrows* point at the positions of the two spectra shown in Fig. 16

To evaluate the XUV pulse duration from our XUV-intensity/visible-light-field cross-correlation data, we introduce a measure for the depth of

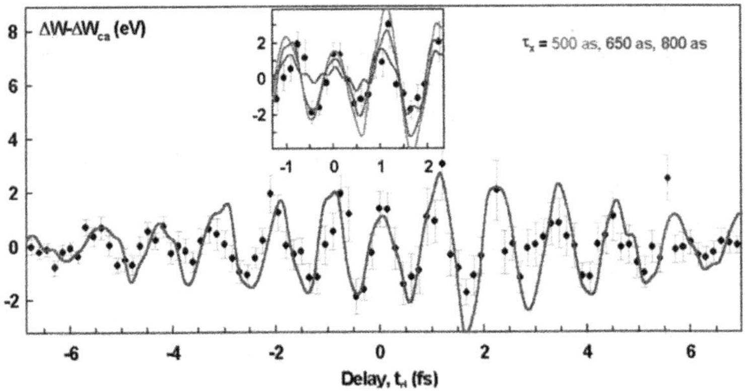

Fig. 18. Oscillating component of the spectral width, $\Delta W - \Delta W_{\mathrm{ca}}(t_{\mathrm{d}})$ of the Kr-4p photoelectron spectra as a function of the delay t_{d} between the light wave and the XUV pulse, where $\Delta W_{\mathrm{ca}}(t_{\mathrm{d}})$ stands for the cycle-averaged value of $\Delta W(t_{\mathrm{d}})$. The *dots* are the measured results. The *full line* is the result of simulations based on the quasi-classical theory of two-color XUV photoionization by assuming a frequency-swept (see Sect. 1.1.2 and Sect. 4.2), 7 fs, 780 nm linearly polarized light field of $(5 \times 10^{13})\,\mathrm{W/cm^2}$ peak intensity and a 650 as, 90 eV Gaussian XUV pulse. The *insert* shows these simulations for different XUV pulse durations τ_{x}

modulation of $\Delta W(t_{\mathrm{d}})$, the fringe visibility

$$\mathrm{FV} = (\Delta W_{\max} - \Delta W_{\min})/(\Delta W_{\max} + \Delta W_{\min} - 2\Delta W_{\infty}), \tag{14}$$

where $\Delta W_{\max}$ and $\Delta W_{\min}$ stand for adjacent local extremal values of $\Delta W(t_{\mathrm{d}})$ as a function of the delay time and $\Delta W_{\infty} = \Delta W(t_{\mathrm{d}} \to \infty)$, and measured the local fringe spacing (FS). The central portion of the correlation function $\Delta W(t_{\mathrm{d}})$, with maximum FV and minimum FS, provides optimum resolution for determining the XUV pulse duration. A sinusoidal fit to the two central modulation periods (see red line in Fig. 18) yields FS $= 960\,\mathrm{as} \pm 30\,\mathrm{as}$ (corresponding to half an oscillation cycle of the laser light see below) and $\mathrm{FV} = 0.30 \pm 0.03$.

For comparison with the measured data (dots in Fig. 18), we simulated the light-field-induced variation of the Kr-4p photoelectron spectrum versus t_{d} based on the quasi-classical model described above. The duration τ_{x} of our Gaussian model XUV pulse was used as the only fit parameter. The visible light pulse was also modeled with a Gaussian intensity envelope peaking at $10^{14}\,\mathrm{W/cm^2}$, carrying a quadratic frequency sweep to account for the frequency upshift present at the center of the measured correlation function $\Delta W(t_{\mathrm{d}})$. Best agreement between measured (dots) and calculated data (red line in Fig. 18) was achieved for XUV pulse durations in the range of $600\,\mathrm{as} \leq \tau_{\mathrm{x}} \leq 700\,\mathrm{as}$, yielding a fringe visibility of $\mathrm{FV} = 0.30 \pm 0.05$ in the correlation function near $t_{\mathrm{d}} = 0\,\mathrm{as}$. Assuming pulse durations of $\tau_{\mathrm{x}} = 500\,\mathrm{as}$ and $\tau_{\mathrm{x}} = 800\,\mathrm{as}$, respectively, the FV is enhanced and suppressed with respect to the measured value well beyond the experimental error of 10%. Therefore, $\tau_x = 650\,\text{as} \pm 150\,\text{as}$ represents a safe estimate for the duration of our 90 eV XUV pulse. The sensitivity of the method is underlined by the computed FV decreasing by a factor as large as 3 (from 0.30 to 0.10) as the XUV pulse duration is increased from 650 as to 900 as, with the modulation in $\Delta W(t_d)$ disappearing for $\tau_x \geq 1$ fs. Strong evidence that the subfemtosecond XUV pulse is substantially isolated (i.e., for satellite pulses ccarrying negligable fractional energy) is provided by the observation that the substantial frequency sweep of the fundamental electric field near the pulse peak (see next subsection) does not affect the modulation depth of $\Delta W(t_d)$. Should the XUV emission be made up of two (or more) bursts of comparable energy, the observed 30% change in the instantaneous laser frequency (see next subsection) would imply suppression of the depth of modulation of $\Delta W(t_d)$ in some range within $-2\,\text{fs} < t_d < 2\,\text{fs}$. The nearly constant modulation depth of $\Delta W(t_d)$ over this delay range can be reconciled only with an isolated XUV burst.

4.2 Probing Dynamic Frequency Shifts of Light

With its duration extracted from the central part of $\Delta W(t_d)$ in Fig. 18, the XUV pulse may now also be used to probe the evolution of the electric field in the few-cycle light pulse. The sweep of instantaneous frequency ν_{inst} (or wavelength λ_{inst}) in the visible light pulse can be evaluated from the modulation in $\Delta W(t_d)$ by fitting a sinusoidal half-oscillation of adjustable period to the data in Fig. 18. The sampling interval was scanned through the range of $-8\,\text{fs} \leq t_d \leq 8\,\text{fs}$. The dots in Fig. 19 show the carrier frequency sweep evaluated in this manner, revealing a dynamic blueshift from a carrier wavelength of ~ 780 nm to ~ 550 nm. To understand this finding, we have to remember that the light pulse used here is derived from the one generating high harmonics.

The line in Fig. 19 is obtained from propagating a bandwidth-limited 7 fs laser pulse through the volume of neon gas emitting high harmonics in a numerical experiment [24] and exhibits a dynamic frequency shift at the pulse center originating from self-phase modulation due to ionization of neon gas in the harmonic source. The measured $\sim 30\%$ dynamic frequency blueshift at the center of the pulse is larger than predicted by our numerical study but reflects the predicted behavior qualitatively.

This qualitative agreement suggests that the observed blueshift can be attributed to ionization-induced self-phase modulation in the high harmonic generation process. The direct probing of the field oscillations achieved in a light wave, as implemented here, is another demonstration of attosecond metrology. It will permit complete measurement of the electric field of carrier–envelope-phase-stabilized few-cycle light [3, 59].

The observed subfemtosecond rise time of the ionization-induced dynamic blueshift provides other conclusive evidence for the isolated nature of the

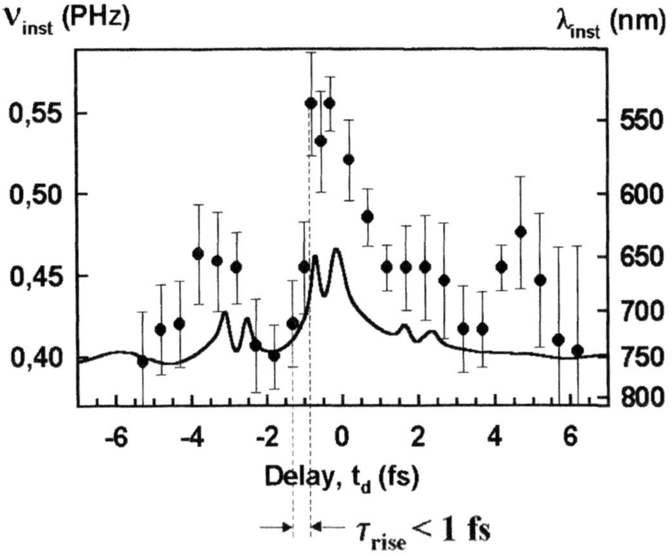

Fig. 19. Calculated (*line*) and measured blueshift of the fundamental pulse, probed by the XUV pulse. The steep rising edge indicates the presence of single XUV pulses without significant satellites

subfemtosecond XUV burst. Any satellite of substantial energy spaced by some 1.3 fs from the main burst would broaden this rise time to more than 1 fs.

The agreement of the measured XUV pulse duration and that obtained from our numerical calculations [58] suggests within experimental error that the timing jitter of the XUV pulse with respect to the phase of the visible light wave must be small compared to 1 fs. As a consequence, our experiment bears evidence that the subfemtosecond XUV pulse is locked to the carrier wave of its generating few-cycle light pulse with attosecond precision. This indicates a surprising robustness of the high-order harmonic generation process against random shifts of the absolute phase [42], i.e., of the position of the carrier with respect to the amplitude envelope. The attosecond timing stability of our subfemtosecond XUV pulse to a few-cycle visible light wave makes these pulses a unique tool for investigating the dynamic behavior of matter on an attosecond timescale. Our light-field-controlled photoemission experiment already demonstrates this capability. In the measured energy distribution $\Delta W(t_d)$, it is implicit that in the spectral range investigated near 90 eV, bound-free electronic transitions from the 4p state in krypton respond to XUV excitation within less than 500 as. This has been, we believe, the first truly attosecond measurement.

4.3 Simplified Attosecond Diagnostics

Another straightforward method of measuring the duration of subfemtosecond pulses is shown in Fig. 20 [60]. Here, we draw on the parallel detection

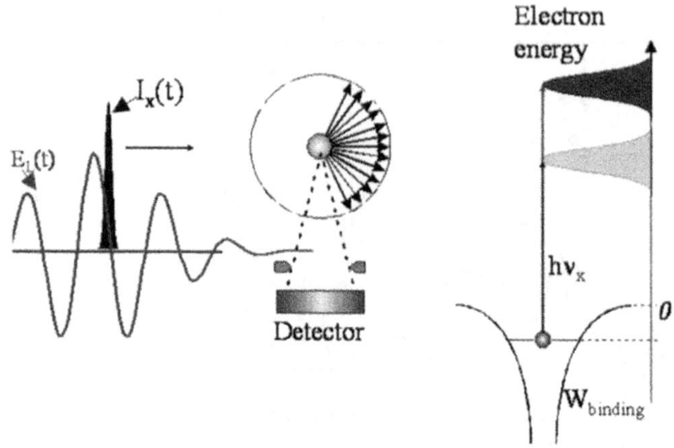

Fig. 20. XUV photoelectrons generated by a few-cycle pulse moving along the laser polarization

geometry introduced in Sect. 3 (Fig. 11). If the photoelectron is ejected near the zero crossing of the laser electric field, its energy spectrum may be up- or downshifted by many electron volts. This is in strong contrast with the orthogonal geometry used before, where the primary effect of the laser field is a broadening, rather than a shift of the photoelectron energy spectrum.

The essential point now is that this shift without substantial broadening happens only if the XUV burst generated (and therefore the generated photoelectron wave packet) is very short compared to the laser field half-oscillation period $T_0/2$ (Fig. 20). If the XUV pulse and therefore the electron wave packet approaches or exceeds $T_0/2$, different portions of the wave packet experience different momentum transfer, resulting in a large spread of energies (Fig. 21).

The false-color contour plot of the photoelectron spectra in Fig. 22 shows the calculated effect on the electron spectra depending on the ratio of XUV pulse duration τ_x to the laser oscillation cycle T_0. For $\tau_x \geq T_0/5$, the multiple shift by $\hbar\omega_L$ can be seen beautifully. For $\tau_x < T_0/5$ and for the electron wave packet launched near the zero transition of the laser electric field, a considerable shift is achieved, resulting in a final electron energy of

$$W_f = W_0 + 2U_p(t_d) \pm \Delta W \,, \qquad (15)$$

with

$$\Delta W = [8W_0U_p(t_d)]^{1/2} \,, \qquad (16)$$

where $W_0 = \hbar\omega_x - W_b$ is the initial kinetic energy of the photoelectron.

Figure 23 and Fig. 24 display electron energy spectra as a function of the measured relative change of E_L in a false-color representation. In the absence of a light field, the photoelectron energy spectrum (lowest plot on the right

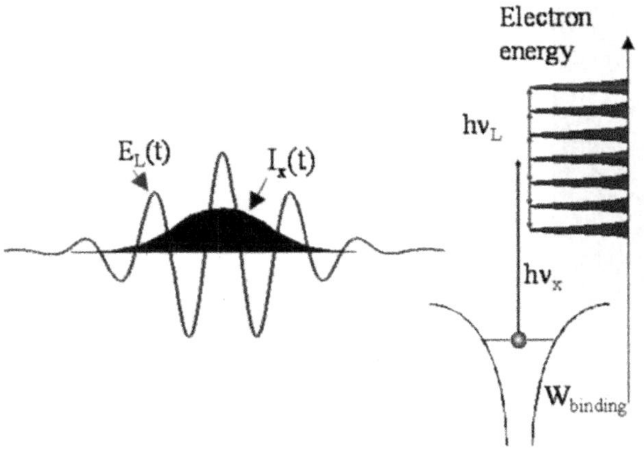

Fig. 21. XUV photoelectrons generated by a multicycle pulse moving along the laser polarization

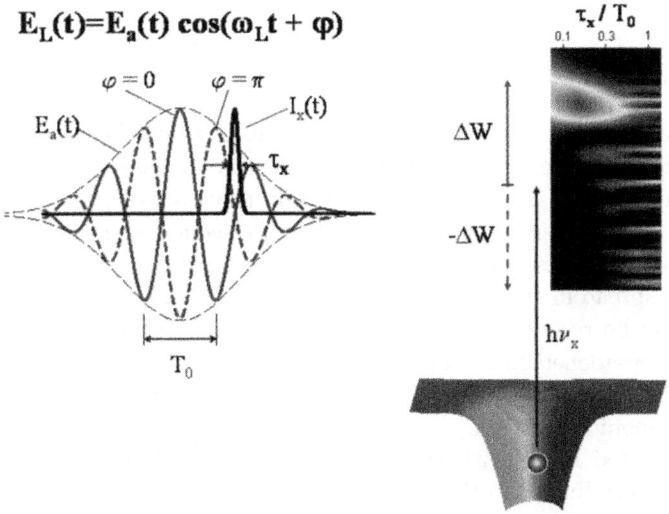

Fig. 22. Effect on electron spectra depending on the relation between XUV pulse duration τ_x and the fundamental laser oscillation cycle T_0

of Fig. 24) mimics that of the XUV pulse downshifted by the ionization potential of neon ($W_b = 21.5\,\text{eV}$). Figure 23 summarizes data obtained when the peak intensity of the harmonic driver was adjusted to $650\,\text{TW/cm}^2$ to $700\,\text{TW/cm}^2$ to produce discrete harmonic spectra within the range of 85 eV to 100 eV (Fig. 7). The quasi-periodic time structure of the emission with a period length evaluated as $\sim 1.25\,\text{fs}$ from the violet spectrum in Fig. 7 suggests that the timing of each spike in the emerging pulse sequence might

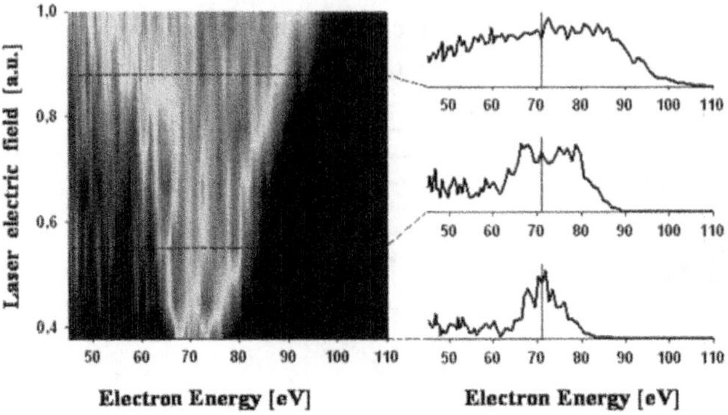

Fig. 23. Photoelectron spectra generated by soft XUV pulses having a discrete harmonic spectrum in the 85 eV to 100 eV spectral range. For increasing field strength, the relative spectral intensity for $W < W_0 \approx 71\,\mathrm{eV}$ decreases and spreads due to an increasing number of electrons deflected out of the detection cone due to decelerating momentum transfer

fulfill

$$\omega_{\mathrm{L}} t_{\mathrm{d}} + \phi = \pi/2 + m\pi\,. \tag{17}$$

Hence only an up- and downshifted spectral feature would be expected, if the spikes were not longer than $T_0/5$. A blueshift near the laser pulse peak (induced by ionization in the harmonic generation process, see Sect. 1.1.2) tends to impair synchronism between the XUV pulse sequence and the zero transitions of the probe laser field. However, it cannot fully account for the dramatic energy spread in the highest applied fields in Fig. 23. The observed energy spread can be reconciled only with the spikes carrying a substantial chirp and hence broadened significantly beyond $T_0/5$. Intuition and simulations support the presence of enhanced phase distortions in the "plateau" region of the harmonic spectrum.

The data depicted in Fig. 24 were collected when the intensity of the harmonic driver was adjusted in the range of $400\,\mathrm{TW/cm^2}$ to $450\,\mathrm{TW/cm^2}$ to generate a continuous harmonic spectrum near 100 eV. In a striking contrast with the behavior shown in Fig. 23, with increasing field strength, the electron energy spectrum is upshifted as a whole without a dramatic distortion of its shape. The simultaneous appearance of an up- and downshifted spectral feature is the consequence of pump pulses with $\phi \approx 0$ and $\phi \approx \pi$ contributing to the electron spectra accumulated over many laser shots, i.e., of the *lack of phase stabilization* of our sub-10 fs laser pulses. The downshifted feature is increasingly washed out for increasing E_{L} due to modifications in the angular momentum distribution by the decelerating momentum transfer within the finite detection cone.

This series of upshifted photoelectron spectra indicates that the electron wave packet must be temporally confined to a small fraction of $T_0/2$ and

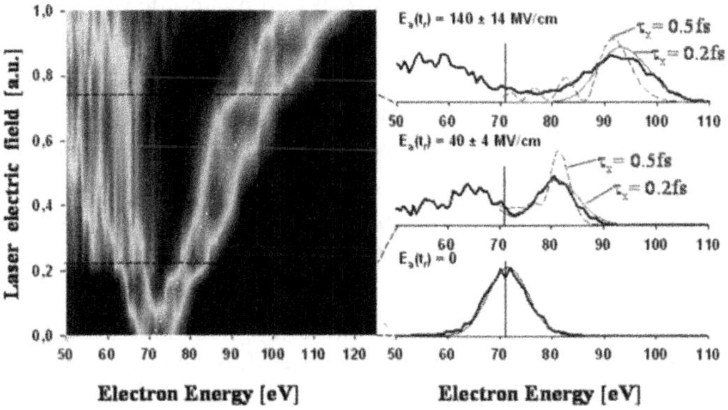

Fig. 24. Photoelectron spectra generated by XUV pulses having a continuous harmonic spectrum in the 85 eV to 100 eV spectral range. $E_L(t_d)$ has been evaluated from the observed energy shift. The *bottom red line* is the model XUV spectrum (downshifted by the electron binding energy of 21.5 eV) used in the simulations resulting in the central and *upper red lines* for bandwidth-limited 200 as (*full lines*) and chirped 500 as soft XUV pulses (*dashed lines*)

released when $E_L(t_d) \approx 0$, in accordance with the results of numerical simulations [52]. The dotted regression line in Fig. 24 displays a linear dependence of the energy shift on the light field strength and corroborates the classical prediction (15,16). The measured spectra can be well reproduced by simulating the electron wave packet as a classical ensemble of electrons having a distribution of initial energies and release times, whose motions are governed by the laws of classical mechanics and electrodynamics. Approximating the initial energy distribution with a Gaussian function (bottom red line in Fig. 24) and assuming a bandwidth-limited XUV pulse ($\tau_x \approx 200$ as) that peaks at $E_L(t_d) = 0$, the classical (just as the quantum) analysis delivers results (central and upper full red lines in Fig. 24) in good agreement with the spectra measured for different strengths of $E_a(t_d)$. Attosecond confinement appears to mask quantum effects in the interaction of photoelectrons with strong light.

The generation of isolated subfemtosecond XUV pulses synchronized with a strong few-cycle light pulse [52] opened up a route to time-resolved (pump–probe) inner shell atomic spectroscopy by employing the XUV pulse as a pump and the light pulse as a probe [61]. Indispensable to the routine application of this new tool is reliable, simple, and fast diagnostics of its key characteristics. These include the strength and phase stability of the few-cycle probe field, the timing jitter between the XUV pulse and light wave, and the duration of the XUV pulse. The sensitivity of the energy spectrum of the XUV-induced, light-accelerated photoelectrons to these parameters makes it an ideal diagnostic tool for attosecond spectroscopy.

Owing to the favorable timing of the subfemtosecond harmonic pulse to its few-cycle driver, $E_L(t_d) \approx 0$, the photoelectron energy shift in Fig. 24 directly probes the instantaneous electric-field amplitude $E_a(t_d)$ with an accuracy of better than 10%, providing direct experimental access to the peak electric-field strength of an intense light pulse. Moreover, the light-steered electron proves a sensitive probe of the absolute phase stability of intense few-cycle light. For ϕ firmly locked to zero, the photoelectrons would be exclusively accelerated (with the convention of $E_L > 0$ implying a field vector that points towards the electron detector). This would imply disappearance of the downshifted spectral feature, offering a reliable diagnostic of the phase stability of intense few-cycle pulses for future phase-stabilized systems.

Timing jitter between the XUV pulse and the probe light phase affects the temporal resolution to the same extent as τ_x and T_0 [61]; hence, its measurement is imperative for attosecond spectroscopy. The periodic harmonic structure can be perceived in several downshifted spectra, e.g., the central plot in Fig. 24 (the structure tends to get blurred in the upshifted features because time-of-flight resolution worsens for increasing electron energy). An analysis of these spectra results in the conclusion that smearing of the electron spectrum accumulated over many laser shots due to fluctuations in $E_L(t_d)$ (by either laser intensity variations or XUV pulse timing jitter) is less than $\pm 10\%$ of the energy shift ΔW. From simulations of the electron energy spectrum for slightly different values of t_d, we inferred a shot-to-shot timing stability of the XUV pulse to the phase of its few-cycle driver of better than ± 200 as.

Figure 23 and Fig. 24 confirm that the energy spectrum of the light-accelerated XUV photoelectron responds sensitively to temporal broadening of the ionizing XUV pulse. Although it is predicted that a conspicuous energy spread will come into play only as τ_x becomes longer than $T_0/3$, a considerable narrowing of the energy-shifted spectral feature already sets in for $\tau_x > T_0/10$, accompanied by the appearance of a prominent downshifted satellite for $\tau_x > T_0/5$. Figure 24 shows the energy distributions calculated under the assumption that the measured XUV spectrum carries a quadratic spectral phase adjusted to yield $\tau_x = T_0/5 = 500$ as (dashed lines). By the same token, simulations with higher order phase distortions adjusted to broaden the XUV pulse to $\sim T_0/5$ gave rise to significant discrepancies in the entire series of spectra recorded at different field strengths. The deviations cannot be accounted for by spectral smearing effects that are limited to $\leq \Delta W/10$, as concluded above. Contrasting these discrepancies with the good agreement of the series of upshifted spectra measured and computed under the assumption of a bandwidth-limited XUV pulse (two of which are depicted with full red lines in Fig. 24), $\tau_x < T_0/5$, implies a safe upper limit of $\tau_x < T_0/5 = 500$ as on the XUV pulse duration.

Most recent experiments with a further improvement of the measurement technique and an increased bandwidth of the XUV reflecting mirror showed

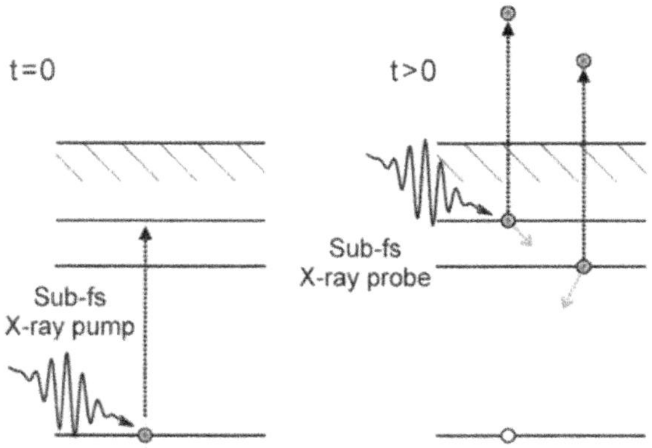

Fig. 25. Tracing inner shell relaxation processes by X-ray-pump/X-ray-probe spectroscopy. The kinetic energy of photoelectrons detached by the probe pulse is analyzed as a function of delay between the pump and the probe pulse

nearly bandwidth-limited XUV pulses whose duration is as long as 250 as [62, 63].

5 Attosecond Time-Resolved Spectroscopy of Atomic Processes

In this section, we show how pump–probe spectroscopy of atomic excitation and relaxation processes – such as, e.g., optical-field ionization and inner shell relaxation processes – can be traced directly in the time domain by drawing on low-energy isolated X-ray pulses in combination with synchronized, strong, few-cycle laser pulses, tools that are available now.

pump–probe experiments turned out to be the most direct approach to tracing dynamics in the time domain. The extension of time-resolved (pump–probe) spectroscopy to ultrafast electronic processes taking place deep inside atoms has so far been frustrated by the simultaneous requirements of short wavelengths (i.e., high photon energy) and subfemtosecond pulse duration, constituting challenging demands. The generation of subfemtosecond XUV pulses was not the only hurdle. Unfortunately, attosecond pulses from HHG often emerge in trains at a very high repetition rate (= twice the laser frequency). The very short ($\sim$ 1 fs to 2 fs) spacing between adjacent bursts within the train prevents an unambiguous interpretation of pump–probe studies. Straightforward interpretation of spectroscopic data requires *isolated, single* pulses [7, 64]. Using few-cycle pulses as a driver for HHG, the demand of single, isolated pulses could be satisfied; so it is now, in principle, possible to extend time-resolved spectroscopy into the attosecond domain. Figure 25 shows the principle of a XUV-pump/XUV-probe experiment, where these pulses are used for both triggering and probing bound–bound or bound-free transitions in atoms or molecules.

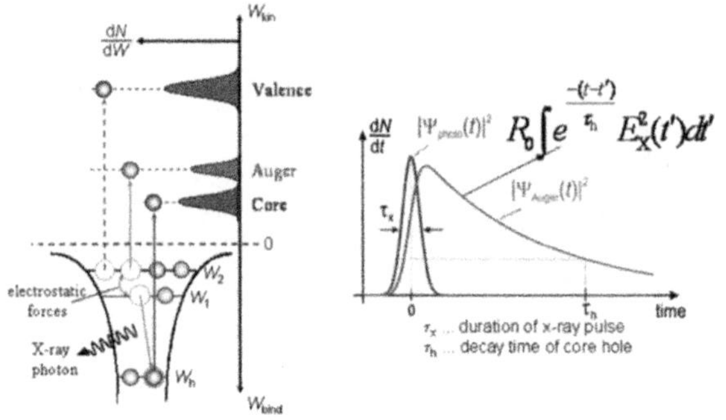

Fig. 26. Schematic and temporal evolution of the Auger process

However, the subfemtosecond XUV pulses currently available do not yet have sufficient flux for XUV-pump/XUV-probe spectroscopy because in this kind of experiment, the physical quantity measured as a function of delay between pump and probe pulse relies on a two-photon transition, the probability of which scales with

$$\sigma \cdot I_{\text{pump}} \cdot I_{\text{probe}} , \tag{18}$$

where σ is the atomic cross section of the absorption process and I_{pump} and I_{probe} are the intensities of the XUV bursts. Both I_{pump} and I_{probe} are, by orders of magnitude, less intense than in the optical regime. Moreover, σ is also reduced by many orders of magnitude because it scales with λ^6.

Extension of the concept of laser-field-assisted XUV photoelectron emission to sampling the emission of secondary (Auger) electrons offers an alternative solution to time-resolving atomic inner shell processes with attosecond accuracy. The principle of the measurement is illustrated in Fig. 26.

The subfemtosecond XUV pulse excites a core electron and produces thereby a short-lived inner shell vacancy. This is rapidly filled by an electron from a higher energy level (outer shell). The energy lost by the electron undergoing this transition is carried away either by an energetic (XUV/X-ray) photon or by a secondary (Auger) electron. The emission time of this Auger electron corresponds exactly to the lifetime of the inner shell vacancy. Hence, sampling the Auger electron emission in the same way as we sampled the photoelectron emission with the oscillating laser field enables researchers to gain direct time-domain access to inner shell atomic processes with attosecond resolution. The first proof-of-principle experiment was recently demonstrated [65]. Although the decay time measured in this experiment can also be inferred from energy-domain measurements, it served as a benchmark process for testing the feasibility of the methods described above for attosecond time-resolved spectroscopy.

Insight into transient states of inner shell processes as well as into the dynamic behavior of atoms exposed to short-lived optical perturbation, for instance, can be gained only by time-domain techniques. The evolution of time-resolved atomic physics, what might be dubbed attophysics, and its impact on other fields will – beyond experimental advances – depend critically on progress in the theoretical understanding of the dynamics of excited multi-electron systems and their interaction with strong fields.

References

[1] A. L'Huillier, P. Balcou: High-order harmonic generation in rare gases with a 1-ps 1053 nm laser, Phys. Rev. Lett. **70**, 774–777 (1993)

[2] J. J. Macklin, J. D. Kmetec, I. C. L. Gordon: High-order harmonic generation using intense femtosecond pulses, Phys. Rev. Lett. **70**, 766–769 (1993)

[3] M. Lewenstein, P. Balcou, M. Y. Ivanov, A. L'Huillier, P. B. Corkum: Theory of high-harmonic generation by low frequency laser fields, Phys. Rev. A **49**, 2117 (1994)

[4] C. Spielmann, N. H. Burnett, S. Sartania, R. Koppitsch, M. Schnürer, C. Kan, M. Lenzner, P. Wobrauschek, F. Krausz: Generation of coherent X-rays in the water window using 5-femtosecond laser pulses, Science **278**, 661 (1997)

[5] C. Spielmann, C. Kan, N. H. Burnett, T. Brabec, M. Geissler, A. Scrinzi, M. Schnürer, F. Krausz: Near-keV coherent X-ray generation with sub-10 fs lasers, IEEE J. Sel. Top. Quant. **4**, 249 (1998)

[6] M. Schnürer, Z. Cheng, M. Hentschel, F. Krausz, T. Wilhein, D. Hambach, G. Schmahl, M. Drescher, Y. Lim, U. Heinzmann: Few-cycle-driven XUV laser hamonics: Generation and focusing, Appl. Phys. B-Lasers O. **70**, S227 (2000)

[7] I. P. Christov, M. M. Murnane, H. C. Kapteyn: High-harmonic generation of attosecond pulses in the "single-cycle" regime, Phys. Rev. Lett. **78**, 1251–1254 (1997)

[8] K. J. Schafer, K. C. Kulander: Theory of resonant multiphoton population transfer in xenon, Phys. Rev. Lett. **78**, 638 (1997)

[9] C. Kan, N. H. Burnett, C. Capjack, R. Rankin: Coherent XUV generation from gases by several cycle optical pulses, Phys. Rev. Lett. **79**, 2971 (1997)

[10] A. McPherson, T. S. Luk, M. H. Hutchinson, H. Jara, U. Johann, I. A. McIntyre, A. P. Schwarzenbach, K. Boyer, C. K. Rhodes: VUV fluorescence and harmonic generation with intense picosecond 248 nm KrF* radiation, in (SPIE Proceedings 1987) p. 103

[11] X. Li, A. L'Huillier, M. Ferray, L. A. Lompre, G. Mainfray: Multiple-harmonic generation in rare gases at high laser intensity, Phys. Rev. A **39**, 5751 (1989)

[12] N. Sarukura, K. Hata, T. Adachi, R. Nodomi, M. Watanabe, S. Watanabe: Coherent soft X-ray generation by the harmonics of an ultrahigh-power KrF laser, Phys. Rev. A **43**, 1669 (1991)

[13] J. K. Crane, M. D. Perry, S. Hermann, R. W. Falcone: High-field harmonic generation in helium, Opt. Lett. **917**, 1256 (1992)

[14] M. E. Faldon, M. H. R. Hutchinson, J. P. Marangos, J. E. Muffett, R. A. Smith, J. W. G. Tisch, C. G. Wahlström: Studies of time-resolved harmonic generation in intense laser fields in xenon, J. Opt. Soc. Am. B **9**, 2094 (1992)

[15] K. Kondo, N. Sarukura, K. Sajiki, S. Watanabe: Terawatt class Ti:sapphire laser, Phys. Rev. A **47**, R2480 (1993)

[16] M. D. Ferry, J. K. Crane: High-order harmonic emission from mixed fields, Phys. Rev. A **48**, R4051 (1993)
[17] J. W. Tisch, R. A. Smith, J. E. Muffet, M. Ciarocca, J. P. Marangos, M. H. Hutchinson: Angularly-resolved high-order harmonic generation in helium, Phys. Rev. A **49**, R28 (1994)
[18] K. Myazaki, H. Takada: High-order harmonic generation in the tunneling regime, Phys. Rev. A **52**, 3007 (1995)
[19] C. G. Wahlström, J. Larsson, A. Persson, T. Starczewski, S. Svanberg, P. Salieres, P. Balcou, A. L'Huillier: High-order harmonic generation in rare gases with an intense short-pulse laser, Phys. Rev. A **48**, 4709 (1993)
[20] T. D. Donnelly, T. Ditmire, K. Neumann, M. D. Perry, R. W. Falcone: High-order harmonic generation in atom clusters, Phys. Rev. Lett. **76**, 2472 (1996)
[21] S. X. Hu, Z. Z. Xu: Enhanced harmonic emission from ionized clusters in intense laser pulses, Appl. Phys. Lett. **71**, 2605 (1997)
[22] M. Y. Ivanov, P. B. Corkum: Generation of high-order harmonics from inertially confined molecular ions, Phys. Rev. A **48**, 580 (1993)
[23] Y. Liang, S. August, S. L. Chin, Y. Beaudoin, M. Chaker: High harmonic generation in atomic and diatomic molecular gases using intense picosecond laser pulses-a comparison, Phys. Rev. B **27**, 5119 (1994)
[24] N. Milosevic, A. Scrinzi, T. Brabec: Ab initio numerical calculation of attosecond pulse generation, Phys. Rev. Lett. **88**, 093905 (1996)
[25] P. B. Corkum: Plasma perspective on strong-field multiphoton ionization, Phys. Rev. Lett. **71**, 1994 (1993)
[26] L. V. Keldysh: Ionization in the field of a strong electromagnetic wave, Sov. Phys. JETP-USSR **20**, 1307 (1965)
[27] U. Becker, D. A. Shirley: *VUV and Soft X-Ray Photoionization* (Plenum Press, New York, London 1996)
[28] A. L'Huillier, K. J. Schafer, K. C. Kulander: High-order harmonic generation in xenon at 1064 nm: The role of phase matching, Phys. Rev. Lett. **66**, 2200 (1991)
[29] G. Tempea, M. Geissler, M. Schnürer, T. Brabec: Self-phase-matched high harmonic generation, Phys. Rev. Lett. **84**, 4329 (1999)
[30] E. Constant, D. Garzella, P. Breger, E. M'evel, C. Dorrer, C. L. Blanc, F. Salin, P. Agostini: Optimizing high harmonic generation in absorbing gases: Model and experiment, Phys. Rev. Lett. **82**, 1668 (1999)
[31] M. Schnürer, Z. Cheng, M. Hentschel, G. Tempea, P. Kálmán, T. Brabec, F. Krausz: Absorption-limited generation of coherent ultrashort soft-X-ray pulses, Phys. Rev. Lett. **83**, 722 (1999)
[32] M. Geissler, G. Tempea, T. Brabec: Phys. Rev. A **62**, 0333817 (2000)
[33] K. Z. Hatsagortsyan, C. H. Keitel: Phase-matched high-harmonic generation from laser-driven crystals **35**, L175–L180 (2002)
[34] I. P. Christov: Control of high harmonic and attosecond pulse generation in aperiodic modulated waveguides, J. Opt. Soc. Am. B **18**, 1877–1881 (2001)
[35] E. A. Gibson, A. Paul, N. Wagner, R. Tobey, D. Gaudiosi, S. Backus, I. P. Christov, A. Aquila, E. M. Gullikson, D. A. Attwood, M. M. Murnane, H. C. Kapteyn: Coherent soft X-ray generation in the water window with quasi-phase matching, Science **302**, 95–98 (2003)
[36] G. Farkas, C. Toth: Proposal for attosecond light pulse generation using laser induced multiple-harmonic conversion processes in rare gases, Phys. Lett. A **168**, 447 (1992)

[37] P. Antoine, A. L'Huillier, M. Lewenstein: Attosecond pulse trains using high-order harmonics, Phys. Rev. Lett. **77**, 1234 (1996)
[38] N. A. Papadogiannis, B. Witzel, C. Kalpouzos, D. Charalambidis: Observation of attosecond light localization in higher order harmonic generation, Phys. Rev. Lett. **83**, 4289 (1999)
[39] P. M. Paul, E. S. Toma, P. Breger, G. Mullot, F. Augé, P. Balcou, H. G. Muller, P. Agostini: Observation of a train of attosecond pulses from high harmonic generation, Science **292**, 1689–1692 (2001)
[40] Y. Kobayashi, T. Sekikawa, Y. Nabekawa, S. Watanabe: 27 fs extreme ultraviolet pulse generation by high-order harmonics, Opt. Lett. **23**, 64 (1998)
[41] E. S. Toma, H. G. Muller, P. M. Paul, P. Breger, M. Cheret, P. Agostini, C. L. Blanc, G. Mullot, G. Cheriaux: Ponderomotive streaking of the ionization potential as a method for measuring pulse durations in the XUV domain with fs resolution, Phys. Rev. A **62**, 061801(R) (2000)
[42] T. Brabec, F. Krausz: Intense few-cycle laser fields: Frontiers of nonlinear optics, Rev. Mod. Phys. **72**, 545 (2000)
[43] A. de Bohan, P. Antoine, D. B. Milosevic, B. Piroux: Phase-dependent harmonic emission with ultrashort laser pulses, Phys. Rev. Lett. **81**, 1837 (1998)
[44] S. E. Harris, J. J. Macklin, T. W. Hänsch: Atomic scale temporal structure inherent to high-order harmonic generation, Opt. Commun. **100**, 487 (1993)
[45] A. Baltuska, T. Udem, M. Uiberacker, M. Hentschel, E. Goulielmakis, C. Gohle, R. Holzwarth, V. S. Yakovlev, A. Scrinzi, T. W. Hnsch, F. Krausz: Attosecond control of electronic processes by intense light fields, Nature **421**, 611–615 (2003)
[46] Y. Kobayashi, T. Ohno, T. Sekikawa, Y. Nabekawa, S. Watanabe: Pulse width measurement of high-order harmonics by autocorrelation, Appl. Phys. B-Lasers O. **70**, 389–394 (2000)
[47] T. Sekikawa, T. Katsura, S. Miura, S. Watanbe: Measurement of the intensity-dependent atomic dipole phase of a high harmonic by frequency-resolved optical gating, Phys. Rev. Lett. **88**, 193902 (2002)
[48] D. Descamps, L. Roos, C. Delfin, A. L'Huillier, C. G. Wahlström: Two- and three-photon ionization of rare gases using femtosecond harmonic pulses generated in a gas medium, Phys. Rev. A **64**, 031404 (2001)
[49] P. Tzallas, D. Charalambidis, N. A. Papadogiannis, K. Witte, G. D. Tsakiris: Direct observation of attosecond light bunching, Nature **426**, 267 (2003)
[50] Y. Kobayashi, O. Yoshihara, Y. Nabekawa, K. Kondo, S. Watanabe: Femtosecond measurement of high-order harmonic pulse width and electron recombination time by field ioization, Opt. Lett. **21**, 417–419 (1996)
[51] M. Drescher, M. Hentschel, R. Kienberger, G. Tempea, C. Spielmann, G. Reider, P. B. Corkum, F. Krausz: X-ray pulses approaching the attosecond frontier, Science **291**, 1923 (2001)
[52] M. Hentschel, R. Kienberger, C. Spielmann, G. A. Reider, N. Milosevic, T. Brabec, P. B. Corkum, U. Heinzmann, M. Drescher, F. Krausz: Attosecond metrology, Nature **414**, 509–513 (2001)
[53] J. M. Schins, P. Breger, P. Agostini, R. C. Constantinescu, H. G. Muller, G. Grillon, A. Antonetti, A. Mysyrowicz: Observation of laser-assisted Auger decay in argon, Phys. Rev. Lett. **73**, 2180 (1994)

[54] J.M. Schins, P. Breger, P. Agostini, R.C. Constantinescu, H.G. Muller, A. Bouhal, G. Grillon, A. Antonetti, A. Mysyrowicz: Cross-correlation measurements of femtosecond extreme-ultraviolet high-order harmonics, J. Opt. Soc. Am. B **13**, 197 (1996)
[55] T.E. Glover, R.W. Schoenlein, A.H. Chin, C.V. Shank: Observation of laser assisted photoelectric effect and femtosecond high order harmonic radiation, Phys. Rev. Lett. **76**, 2468 (1996)
[56] A. Bouhal, R. Evans, G. Grillon, A. Mysyrowicz, P. Breger, P. Agostini, R.C. Constantinescu, H.G. Muller, D. von-der Linde: Cross-correlation measurement of femtosecond noncollinear high-order harmonics, J. Opt. Soc. Am. B **14**, 950 (1997)
[57] M. Kitzler, N. Milosevic, A. Scrinci, T. Brabec: Theory of attosecond streak camera, Phys. Rev. Lett. **88**, 173904 (2002)
[58] J. Itatani, F. Quéré, G.L. Yudin, M.Y. Ivanov, F. Krausz., P.B. Corkum: Attosecond streak camera, Phys. Rev. Lett. **88**, 173903 (2002)
[59] A. Poppe, R. Holzwarth, A. Apolonski, G. Tempea, C. Spielmann, T.W. Hänsch, F. Krausz: Few-cycle optical waveform synthesis, Appl. Phys. B-Lasers O. **72**, 373–376 (2001) doi:10.1007/s003400000526 published online
[60] R. Kienberger, M. Hentschel, M. Uiberacker, C. Spielmann, M. Kitzler, A. Scrinzi, M. Wieland, T. Westerwalbesloh, U. Kleineberg, U. Heinzmann, M. Drescher, F. Krausz: Steering attosecond electron wave packets with light, Science **297**, 1144–1148 (2002)
[61] F. Krausz: Attosecond spectroscopy comes of age, Opt. Photonics News **13**, 62 (2002)
[62] R. Kienberger, E. Goulielmakis, A. Baltuška, M. Uiberacker, T. Westerwalbesloh, U. Kleineberg, U. Heinzmann, M. Drescher, F. Krausz: Reproducible generation and measurement of isolated sub-fs XUV pulses with phase-controlled few-cycle light, in F. Krausz, G. Korn, P. Corkum, I. Walmsley (Eds.): Ultrafast Optics IV (Springer, New York 2004)
[63] R. Kienberger, E. Goulielmakis, M. Uiberacker, A. Baltuška, V. Yakovlev, F. Bammer, A. Scrinzi, T. Westerwalbesloh, U. Kleineberg, U. Heinzmann, M. Drescher, F. Krausz: Atomic transient recorder, Nature **427**, 817 (2004)
[64] M. Ivanov, P.B. Corkum, T. Zuo, A. Bandrauk: Routes to control of intense-field atomic polarizability, Phys. Rev. Lett. **74**, 2933–2936 (1995)
[65] M. Drescher, M. Hentschel, R. Kienberger, M. Uiberacker, V. Yakovlev, A. Scrinzi, T. Westerwalbesloh, U. Kleineberg, U. Heinzmann, F. Krausz: Time-resolved atomic inner-shell spectroscopy, Nature **419**, 803–807 (2002)

Resonant Nonperturbative Extreme Nonlinear Optics with Two-Cycle Pulses: Carrier-Wave Rabi Flopping and Role of the Carrier–Envelope Offset Phase

Oliver D. Mücke, Thorsten Tritschler, and Martin Wegener

Institut für Angewandte Physik, Universität Karlsruhe (TH),
Wolfgang-Gaede-Straße 1, 76131 Karlsruhe, Germany
martin.wegener@physik.uni-karlsruhe.de

Abstract. Carrier-wave Rabi flopping occurs when the Rabi frequency becomes comparable to the light frequency, while maintaining electronic coherence. Exciting the model semiconductor GaAs, which has a band-gap period of 2.9 fs, with optical pulses which are both extremely short (5 fs) and extremely intense (estimated Rabi periods < 3 fs), we can meet this highly unusual condition. After reviewing corresponding experimental data, we compare the experiments with model calculations, especially focusing on the dependence of this example of resonant nonperturbative extreme nonlinear optics on the carrier–envelope offset phase.

1 Introduction

Nonlinear optics with optical pulses as short as a few cycles of light has many interesting and novel aspects [1]. Much of the previous work has been devoted to off-resonant effects, where the center frequency of the laser pulses and the relevant resonance frequency of a system are well separated. In contrast to this, in this Chapter, we discuss resonant effects in solids excited with intense pulses. What do we mean by intense?

If a two-level system is excited by a resonant light field, electrons absorb photons which pump them from the ground state into the excited state (Fig. 1). It is sometimes stated that one cannot reach inversion by optical pumping in a two-level system. This statement is, however, true only in the incoherent case, where one can only reach transparency, i.e., 50% of the electrons are in the ground state, 50% are in the excited state. In contrast to this, if the system remains fully coherent in the quantum mechanical sense,[1] complete inversion can be reached. If the light field remains switched on, stimulated emission brings the electrons back into the ground state (Fig. 1).

[1] As we will discuss in more detail later, generally electrons are not *either* in the ground state *or* in the excited state but they are both, i.e., their wave function is a linear combination (superposition) of these two states. In this sense, the dots in Fig. 1 are somewhat misleading. Figure 1 does, however, correctly suggest that the inversion oscillates with time

F. X. Kärtner (Ed.): Few-Cycle Laser Pulse Generation and Its Applications,
Topics Appl. Phys. **95**, 379–408 (2004)

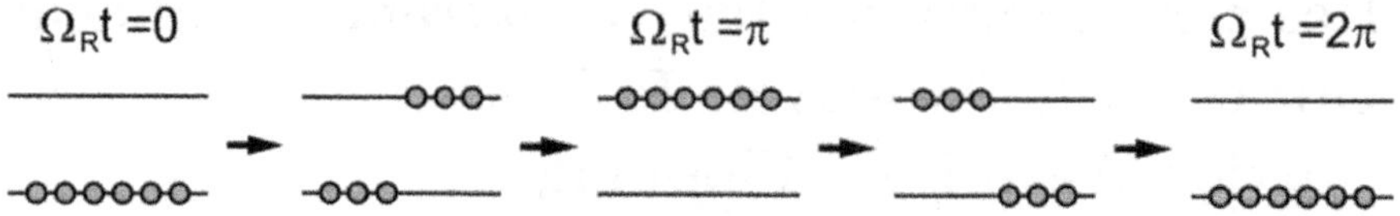

Fig. 1. A constant light intensity can lead to Rabi oscillations of the inversion in a two-level system. The *lower horizontal line* is the ground state; the *upper horizontal line* is the excited state. Several snapshots of the electron distribution (*dots*) during one Rabi oscillation period are shown for illustration

This oscillation of the inversion is known as Rabi oscillation or Rabi flopping [2, 3]. In a (direct gap) semiconductor, the ground state is replaced by the valence band, the excited state by the conduction band, and Rabi oscillations can also be observed [4, 5, 6, 7, 8, 9]. The frequency of this oscillation, the envelope Rabi frequency Ω_{R}, with

$$\hbar\Omega_{\mathrm{R}} = d\tilde{E}\,, \tag{1}$$

is proportional to the envelope of the light field $\tilde{E}$ and to the dipole matrix element d of the optical transition. Back to our question: What do we mean by intense? Intensity can quite naturally be measured in terms of the Rabi frequency which has to be compared with the light frequency. A cycle of light, $2\pi/\omega_0$, corresponding to the room-temperature band-gap energy, $E_{\mathrm{g}} = 1.42\,\mathrm{eV}$, of the model semiconductor GaAs has a period of

$$h/E_{\mathrm{g}} = 2\pi/\omega_0 = 2.9\,\mathrm{fs}\,, \tag{2}$$

with Planck's quantum[2]

$$h = 2\pi \times 0.658\,\mathrm{eV}\cdot\mathrm{fs}\,. \tag{3}$$

What happens if the light intensity ($\propto |\tilde{E}|^2$) is so large that the period of one Rabi oscillation becomes as short as a cycle of light?

This question brings us into a highly unusual regime of light–matter interaction in solids as well as into an unexplored regime of nonlinear optics. The situation, Rabi frequency equal to light frequency, for parameters corresponding to the model semiconductor GaAs, actually means that the semiconductor turns into a metal after half an optical cycle ($1.45\,\mathrm{fs} = 2.9/2\,\mathrm{fs}$), returning to a semiconductor after another 1.45 fs. Moreover, this carrier-wave regime [10, 11] connects two effects which are usually thought of as unrelated, namely, (envelope) Rabi flopping, on the one hand, and Zener tunneling [12] on the other hand. For envelope Rabi flopping, one assumes that the light (center) frequency ω_0 is much larger than the Rabi frequency Ω_{R}, i.e.,

$$\Omega_{\mathrm{R}} \ll \omega_0\,, \tag{4}$$

[2] In units that are helpful throughout this chapter

which allows treating the problem within a frame rotating with the frequency of light (rotating wave approximation) [2, 3]. For a semiconductor, this can lead to periodic oscillation of the occupation of electron states within the conduction (valence) band. Zener tunneling of electrons from the valence to the conduction band through the energy gap, on the other hand, occurs for static electric fields, i.e., for $\omega_0 = 0$ or, equivalently, for

$$\Omega_{\mathrm{R}} \gg \omega_0 \,. \tag{5}$$

In the regime of carrier-wave Rabi flopping, i.e., for

$$\Omega_{\mathrm{R}} \approx \omega_0 \,, \tag{6}$$

these two known pictures merge. For GaAs parameters, the condition,

$$\Omega_{\mathrm{R}} = \omega_0 \Longleftrightarrow_{\mathrm{GaAs}} \tilde{E}_0 = 4 \times 10^9\,\mathrm{V/m} = 2\,\mathrm{V}/a \,, \tag{7}$$

corresponds to a peak of the electric-field envelope of the order of $\tilde{E}_0 = 4 \times 10^9\,\mathrm{V/m}$, equivalent to about two volt per lattice constant a, with $a \approx 0.5\,\mathrm{nm}$.

Beside these transitions between valence and conduction band, i.e., interband transitions, transitions within the bands, i.e. intraband effects, also occur. In classical terms, they describe the acceleration and motion of electrons as a result of the electric field of a light pulse. In atomic physics, this aspect is usually expressed by the cycle-averaged kinetic energy of the electrons with effective mass m_{e}, the ponderomotive energy, which is given by

$$\langle E_{\mathrm{kin}} \rangle = \frac{e^2 \tilde{E}_0^2}{4 m_{\mathrm{e}} \omega_0^2} \,. \tag{8}$$

For $\tilde{E}_0 = 4 \times 10^9\,\mathrm{V/m}$ and GaAs parameters ($m_{\mathrm{e}} = 0.07 \times m_0$, with the free electron mass $m_0 = 9.1 \times 10^{-31}\,\mathrm{kg}$ and $\hbar\omega_0 = E_{\mathrm{g}}$), one obtains $\langle E_{\mathrm{kin}} \rangle = 2.4\,\mathrm{eV}$, which is even larger than the photon energy corresponding to the GaAs band-gap energy of $E_{\mathrm{g}} = 1.42\,\mathrm{eV}$. Unfortunately, for semiconductors, the concept of ponderomotive energy is meaningful only within the range of validity of the effective mass approximation, which obviously fails for large values of $\langle E_{\mathrm{kin}} \rangle$, typically already above several $0.1\,\mathrm{eV}$.

Thus, for solids, it is more appropriate to express the strength of the intraband effects in terms of the Bloch frequency, which is directly related to the potential drop over one lattice constant a. When electrons are accelerated so much that their wave number k reaches the end of the first Brillouin zone, i.e., $k = \pm\pi/a$, they are Bragg-reflected to $k = \mp\pi/a$, i.e., their momentum changes sign. This leads to a real space oscillation of the electron position, known as Bloch oscillation. As in our case, the electric field of the laser pulse oscillates in time, the Bloch frequency also oscillates. What is the appropriate quantum mechanical picture? Without an electric field, the electron wave functions of the atoms forming a solid overlap, which lifts their degeneracy, leading to the bands describing delocalized electron wave functions. In the presence of a strong electric field, the potential drop over one lattice constant lifts the degeneracy, and the wave functions become localized again. For an electronic wave packet, one obtains a quantum beating between these Wannier–Stark states, evenly separated in energy by $ae\tilde{E}_0$ (Wannier–Stark ladder). This quantum beating is the quantum mechanical analogue of Bloch oscillations. Thus, the beat frequency, the Bloch frequency Ω_{Bloch}, is given by

$$\hbar\Omega_{\mathrm{Bloch}} = ae\tilde{E}_0\,, \tag{9}$$

with the elementary charge $e = 1.6 \times 10^{-19}\,\mathrm{A \cdot s}$. For $\tilde{E}_0 = 4 \times 10^9\,\mathrm{V/m}$, this leads to a Bloch period of $2\pi/\Omega_{\mathrm{Bloch}} = 2\,\mathrm{fs}$. As this is already shorter than one cycle of light, one approaches the point at which the carriers experience Bragg reflections within an optical cycle, i.e., the electron kinetic energy reaches the top of the conduction band or the end of the first Brillouin zone.

Thus, in this unusual regime of short times and high field amplitudes, four different energies become comparable: the photon energy, the Rabi energy, $\hbar$ times the Bloch frequency, and the ponderomotive energy. We will see later that, for resonant excitation conditions, the interband effects, i.e., effects arising from carrier-wave Rabi flopping, dominate over intraband effects. The latter become more prominent for off-resonant excitation and lead to phenomena such as the dynamic Franz–Keldysh effect [13], which has been observed experimentally [14], or Bloch oscillations [15, 16], which have not been observed experimentally in "real" crystals (they have, however, been observed in artificial superlattices [3]).

2 Carrier-Wave Rabi Flopping

The notion of carrier-wave Rabi flopping was introduced by *Hughes*, who discussed an ensemble of identical and uncoupled two-level systems in vacuum [10]. As one is interested in the system's dynamics on a timescale of one period of light or less, both the rotating wave approximation and the slowly varying envelope approximation [2] must obviously *not* be used. His theoretical work as well as that of others [17] is based on the theoretical framework of [18].

What are the anticipated signatures of carrier-wave Rabi flopping? The condition of the Rabi period equal to the light period corresponds to a huge intensity (for a solid). It is clear that the pulse energy increases when increasing the pulse duration for a fixed value of the peak electric field (equivalent to a fixed value of the Rabi frequency). This will lead to destruction of a sample for very large pulse energies. Thus, it seems favorable to study excitation with very short pulses, ideally with only one or two cycles of light in duration. Remember that, for GaAs parameters, the period of light corresponding to the room-temperature band-gap energy is 2.9 fs. To highlight

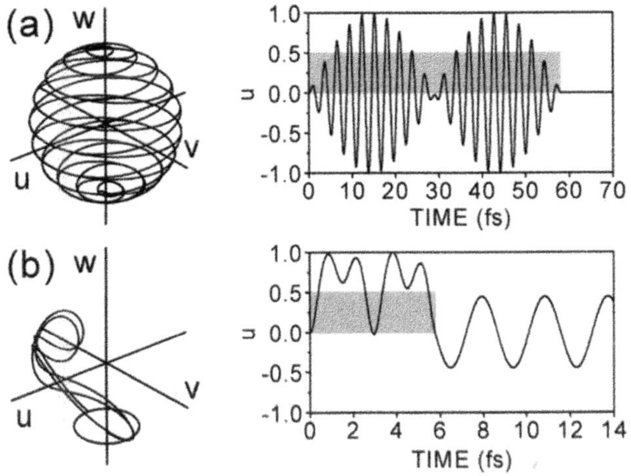

Fig. 2. (**a**) Scheme of the trace of the Bloch vector for conventional Rabi flopping. Pulse duration is 20 optical cycles, envelope pulse area is $\Theta = 2\pi$. (**b**) Same for carrier-wave Rabi flopping. Pulse duration is 2 optical cycles, $\Theta = 4\pi$. The optical pulse envelopes are indicated by the *gray areas*. Taken from [11]

the general aspects of carrier-wave Rabi flopping, let us first review the behavior for an ensemble of uncoupled and identical two-level systems, which is the level of sophistication of [10, 17, 18]. For reference, Fig. 2a schematically depicts conventional Rabi flopping plotted on the Bloch sphere, i.e., the Rabi period is much larger than the light period. For clarity, we neglect any damping at this point. The components u and v of the Bloch vector (u, v, w) correspond to twice the real and imaginary parts of the optical transition amplitude, respectively; w is the inversion of the two-level system [2]. In this representation, the optical oscillation corresponds to orbiting of the Bloch vector parallel to the equatorial plane (uv-plane) with the optical transition frequency Ω (here, $\Omega = \omega_0 = 2\pi/2.9\,\text{fs}$), the oscillation of the inversion to a motion in the vw-plane. For a square-shaped pulse with envelope pulse area $\Theta = 2\pi$ starting from the south pole, i.e., all electrons are in the ground state (valence band), the Bloch vector spirals up to the north pole, i.e., all electrons are in the excited state (conduction band) and back to the south pole. This leads to modulation of the optical transition amplitude (the real part, u, is shown in Fig. 2a), which leads to sidebands in the spectrum centered around the transition frequency. Figure 2b shows results for $\Theta = 4\pi$ and for a much shorter pulse, *such that the Rabi period equals the light period.* Two related aspects are obvious. First, though $\Theta = 4\pi$, the Bloch vector does not come back to the south pole. In this sense, the usual definition of the envelope pulse area Θ fails. Hence, the area theorem of nonlinear optics, which is based on this definition, also fails. Despite this failure, we quote Θ for reference in this article. Second, it is obvious that the optical polarization becomes strongly

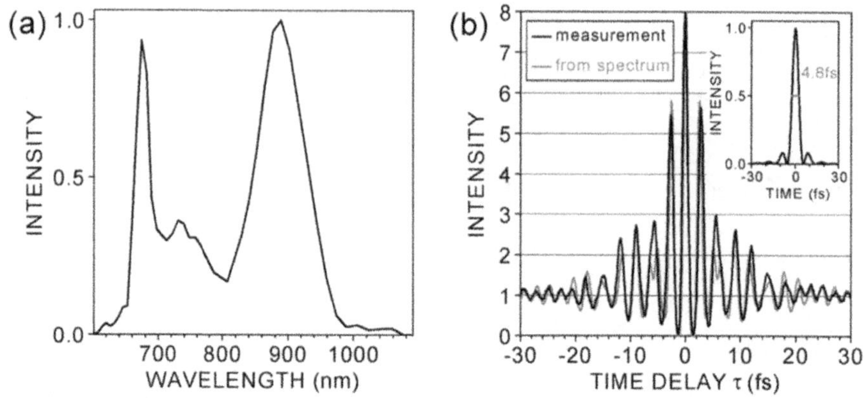

Fig. 3. Experiment: (**a**) Measured laser spectrum, (**b**) measured interferometric autocorrelation. The *gray curve* in (**b**) is the autocorrelation computed from the spectrum, (**a**), under the assumption of a constant spectral phase (no chirp). The *inset* in (**b**) depicts a 4.8 fs full-width at half-maximum real-time intensity profile computed under the same assumption

distorted during the two cycles of the optical pulse (see u versus time in Fig. 2b). Thus, harmonics are being generated, the most prominent of which, for an inversion symmetrical medium, is the third harmonic. For low intensities, this is nothing but the resonantly enhanced third-harmonic generation. *For very high intensities, i.e., for carrier-wave Rabi flopping, one also expects sidebands around the third harmonic of the transition frequency.* Note that no harmonics are generated *after* the two cycles of the optical pulse (Fig. 2b). Here, one merely has free oscillation of the optical polarization at the optical transition frequency of the two-level system Ω.

2.1 Experiments

From the above, it has become clear that carrier-wave Rabi flopping needs short, i.e., one or two optical cycles long, and very intense optical pulses. From the anticipated signal levels and the anticipated damage thresholds of thin films of semiconductors, high repetition rate mode-locked lasers are strongly favored. Thus, we perform our experiments with 5 fs linearly polarized (p-polarization) optical pulses at an 81 MHz (= 1/12 ns) repetition rate, which have recently become available [19]. Our home-built copy of this laser system very nearly reproduces the pulse properties described in [19]. The typical average output power of the laser is 120 mW. Figure 3a shows a typical laser spectrum, which has been obtained via Fourier transform of an interferogram taken with a pyroelectric detector, which is spectrally extremely flat. The Michelson interferometer used at this point and for all results throughout this article is carefully balanced and employs home-made beam splitters fabricated by evaporating a thin film of silver on a 100 μm thin glass substrate. The Michelson interferometer is actively stabilized by means of a Pancharatnam screw [20], which allows continuous scanning of the time delay while maintaining active stabilization. The remaining fluctuations in the time delay between the two arms of the interferometer are around $\pm 0.05\,\text{fs}$. The spectral wings, which can be seen in Fig. 3a, result from the spectral characteristics of the output coupler. The measured interferometric autocorrelation depicted in Fig. 3b is very nearly identical to that computed from the spectrum (Fig. 3a) under the assumption of a constant spectral phase. This shows that the pulses are nearly transform-limited. The temporal intensity profile computed under the same assumption is shown as an inset in Fig. 3b and reveals a duration of about 5 fs. As a result of the strongly structured spectrum (a square function to zeroth order), the intensity versus time shows satellites (a sinc^2 function to zeroth order). Using a high numerical aperture reflective microscope objective [21], we can focus these pulses tightly to a profile which is very roughly Gaussian with a 1 μm radius. This value has been carefully measured by a knife-edge technique at the sample position. This sample position is equivalent to that of the second-harmonic (SHG) crystal used for autocorrelation in terms of group delay dispersion. In front of the sample, each arm of the interferometer typically has an average power of about 8 mW. The resulting peak intensity of one arm can be estimated as

$$I_0 = \frac{8\,\text{mW}}{\pi(10^{-4}\,\text{cm})^2}\,\frac{12\,\text{ns}}{5\,\text{fs}} = 0.6 \times 10^{12}\,\text{W/cm}^2\,. \tag{10}$$

This corresponds to a peak of the field envelope (in vacuum)

$$\tilde{E}_0 = \sqrt{2\sqrt{\frac{\mu_0}{\varepsilon_0}}\,I_0} = 2.1 \times 10^9\,\text{V/m}\,, \tag{11}$$

or about 1 volt per lattice constant a, as mentioned in the introduction ($a \approx 0.5\,\text{nm}$). To estimate the envelope pulse area Θ, one furthermore needs the dipole matrix element d of the optical dipole transition. From the literature for GaAs, we find $d = 0.3e\,\text{nm}$ [22] and $d = 0.6e\,\text{nm}$ [7]. *kp*-perturbation theory delivers $d = 0.65e\,\text{nm}$ via the following rule of thumb:

$$|d|^2 = \frac{\hbar^2 e^2}{2E_\text{g}}\left(\frac{1}{m_\text{e}} - \frac{1}{m_0}\right). \tag{12}$$

Choosing $d = 0.5e\,\text{nm}$ in this Chapter, this translates into an envelope pulse area,

$$\Theta = \hbar^{-1} d\tilde{E}_0 \times 5\,\text{fs} = 8.1 > 2\pi\,, \tag{13}$$

for one arm ($I = 0.601 \times I_0$ corresponds to 2π pulse area) and $> 4\pi$ (two Rabi periods) for two constructively interfering arms of the interferometer. For a 5 fs pulse and a 2.9 fs band-gap period, this corresponds to a Rabi frequency which even slightly exceeds the light frequency. It is also interesting to give a very rough estimate for the excited carrier density under these conditions. The GaAs band-to-band absorption coefficient is $\alpha = 10^4\,\mathrm{cm}^{-1}$. If all the light was absorbed according to this number – certainly an upper limit – one arrives at a carrier density of

$$n_{\mathrm{eh}} = \alpha I_0 \times 5\,\mathrm{fs}/1.42\,\mathrm{eV} = 1.3 \times 10^{20}\,\mathrm{cm}^{-3}\,. \tag{14}$$

For constructive interference of the two arms of the interferometer, this number needs to be multiplied by a factor of 4. Thus, we can safely conclude that the highest carrier densities approach $10^{20}\,\mathrm{cm}^{-3}$. In the experiment, we use a 0.6 µm thin film of GaAs clad between $\mathrm{Al_{0.3}Ga_{0.7}As}$ barriers, grown by metal-organic vapor phase epitaxy on a GaAs substrate. The sample is glued onto a 1 mm thick sapphire disk, and the GaAs substrate is removed. Finally, a $\lambda/4$-antireflection coating is evaporated. The light emitted by this double heterostructure, held under ambient conditions, is collected by a second reflective microscope objective [21], is spectrally prefiltered by a sequence of four fused-silica prisms, and is sent into a 0.25 m focal length grating spectrometer connected to a liquid-nitrogen-cooled, back-illuminated, UV-enhanced charge-coupled-device (CCD) camera.

Let us first discuss results for single pulses only, i.e., we block one arm of the interferometer. Figure 4 shows spectra at the third harmonic for different pulse intensities I in multiples of I_0, as defined above. For attenuation, we used metallic beam splitters on 100 µm thin fused silica substrates, the dispersion of which has carefully been compensated for by extracavity sequence of four $\mathrm{CaF_2}$ prisms [19]. At low intensity, i.e., for $I = 0.017 \times I_0$, we observe a single maximum around the 300 nm wavelength; this is interpreted as the usual third-harmonic generation which is resonantly enhanced by the GaAs band edge here. With increasing intensity, we find a second maximum emerging at the long wavelength side, which gains more and more weight. At the highest intensity, i.e. $I = 0.779 \times I_0$, tenfold magnification reveals an additional smaller maximum around the 340 nm wavelength. In [11], we have interpreted this overall behavior as a signature of carrier-wave Rabi flopping. Note that the intensities revealing a double-peak structure in the third-harmonic spectrum correspond very well to our simple estimates above, i.e., we estimated a full Rabi flop for an intensity of $I = 0.601 \times I_0$.

In the second set of experiments we study the third-harmonic spectra for excitation with phase-locked pulse pairs with time delay τ, i.e., we open both arms of the interferometer. It is interesting to note that Θ is the same for $\tau = 0$ and for, e.g., τ equal to two optical cycles, because the two optical fields simply add. Yet, the corresponding Rabi frequency is larger for $\tau = 0$. For low intensities (Fig. 5a), i.e., for small Rabi frequencies compared to the light frequency, the third-harmonic spectrum is simply modulated as a function of τ due to interference of the laser pulses within the sample leading to a period

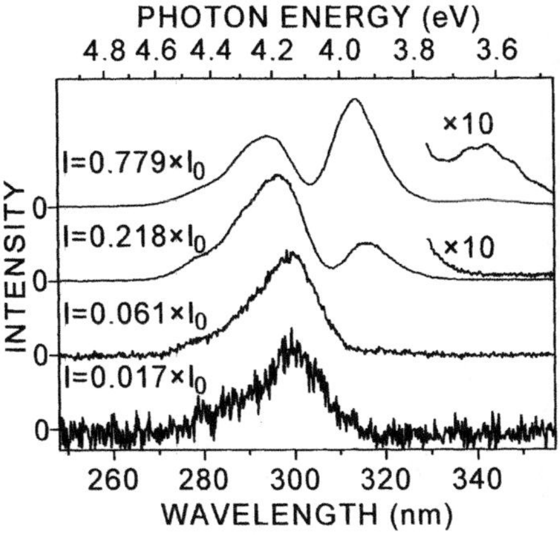

Fig. 4. Experiment: Spectra of light emitted in the forward direction around the third harmonic of the GaAs band-gap frequency. The spectra are shown on a linear scale, vertically displaced and individually normalized (from *top* to *bottom:* maxima correspond to 5664, 439, 34, and 4 counts per second). Excitation with 5 fs pulses. The intensity I of the pulses is indicated. Taken from [11]

of about 2.9 fs. In contrast to this, for higher intensities (Fig. 5b–d), where the Rabi frequency becomes comparable to the light frequency, the shape of the spectra changes dramatically with time delay τ. For $\tau = 0$, e.g., in Fig. 5b, the two pulses simply interfere constructively, and we find the same spectral double maximum structure as in the single pulse experiments (Fig. 4). For larger τ, i.e., after one or two optical cycles, this double maximum disappears and is replaced by one prominent and much larger maximum. For the highest intensity, i.e., for Fig. 5d – which corresponds to an envelope pulse area Θ of more than 4π – the behavior is quite involved with additional fine structure for $|\tau| < 1$ fs. Note that the spectra for $\tau = 0$ nicely reproduce the behavior seen in Fig. 4.

Beside the interference of the laser pulses within the sample, at larger time delays $|\tau|$, one additionally observes interference of the third-harmonic signals corresponding to the two phase-locked pulses on the detector leading to periods around one femtosecond in Fig. 6. It can also be seen from Fig. 6 that the splitting in the spectra gradually approaches zero for large time delays. As reasoned above, this is due to the decreasing electric field, hence decreasing Rabi frequency with increasing $|\tau|$. The smaller splitting at $\tau = 0$ cannot be understood along these lines. It is possibly due to intraband effects, which might lead to a breakdown of the semiconductor band structure in the largest electric fields.

We have also deliberately introduced positive or negative group velocity dispersion by moving one of the extracavity CaF_2 prisms in or out of the beam with respect to the optimum position (Fig. 7). The amount of effectively removed (added) CaF_2 material is given by $-x$ µm ($+x$ µm); 0 µm

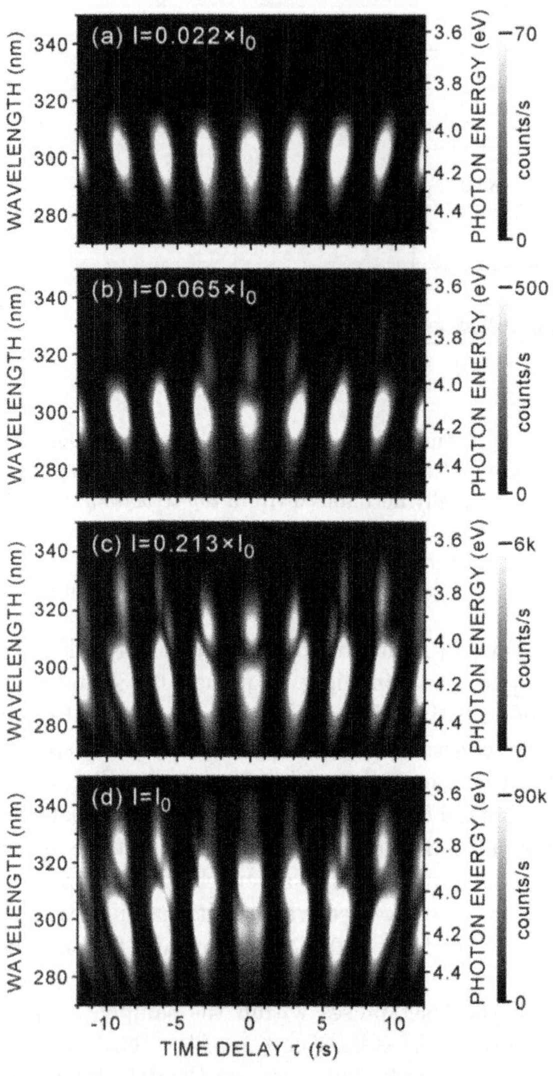

Fig. 5. Experiment: Same as Fig. 4, however, using pairs of phase-locked 5 fs pulses. The signal around the third harmonic of the band gap is depicted versus time delay τ in a gray-scale plot (note the saturated scale on the right-hand side). (**a**)–(**d**) correspond to different intensities I as indicated. I refers to one arm of the interferometer. Taken from [11]

corresponds to the optimum position determined by the field autocorrelation of the laser pulses (Fig. 3). Obviously, this variation leaves the amplitude spectrum of the laser pulses unaffected. We find that one quickly gets out of the regime of carrier-wave Rabi flopping, i.e., both the splitting at $\tau = 0$

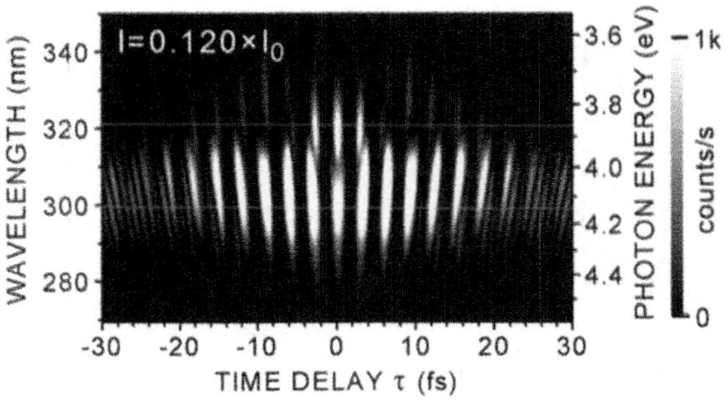

Fig. 6. Experiment: Similar to Fig. 5c, however, for a larger range of time delay τ. Around $\tau = 0$, the interference of the laser pulses within the sample dominates, and we additionally observe interference of the third-harmonic signals corresponding to the two phase-locked pulses on the detector for larger $|\tau|$

and the dependence of the shape on the time delay τ quickly disappear with increasing pulse chirp. This demonstrates that it is not just the large bandwidth of the pulses but that they are short – two optical cycles – which is important for the observation of carrier-wave Rabi flopping.

To test further our assignment of the features seen in the third-harmonic spectra to carrier-wave Rabi flopping, we additionally performed experiments on the corresponding changes in the optical transmission of the sample around the GaAs band gap. One obviously expects a large increase in the sample transmission at the band gap as a result of Rabi flopping. As it is, however, not easy to vary the laser intensity continuously without possibly introducing an additional chirp, we scan the sample position z on the optical axis through the fixed focus of the objective at $z = 0$ (Fig. 8a). We define a differential transmission, $\Delta T/T$, as

$$\frac{\Delta T}{T} = \frac{I_t(z) - I_t(z = -\infty)}{I_t(z = -\infty)}, \tag{15}$$

where $I_t(z)$ is the transmitted light intensity at sample position z. The condition $z = -\infty$ actually corresponds to $z = -20\,\mu\text{m}$ in the experiment, where the profile is so large that we can safely assume that linear optics applies.

Figure 9 shows corresponding results for three different incident light intensities I in units of I_0, as defined above. First, all results are closely symmetrical around $z = 0$, which indicates that changes in absorption dominate. Changes in the refractive index might lead to focusing or defocusing of the beam which would result in asymmetrical dependence on z (similar to the known so-called z-scan technique, e.g., described in [23]). Second, one can see a large increase in transmission for wavelengths shorter than the GaAs band edge (approximately 870 nm) around $z = 0$ (Fig. 9a). $z = 0$ corresponds to the highest intensity in each plot. The maximum around the 670 nm wavelength results from bleaching of the band gap of the $Al_{0.3}Ga_{0.7}As$ barriers of

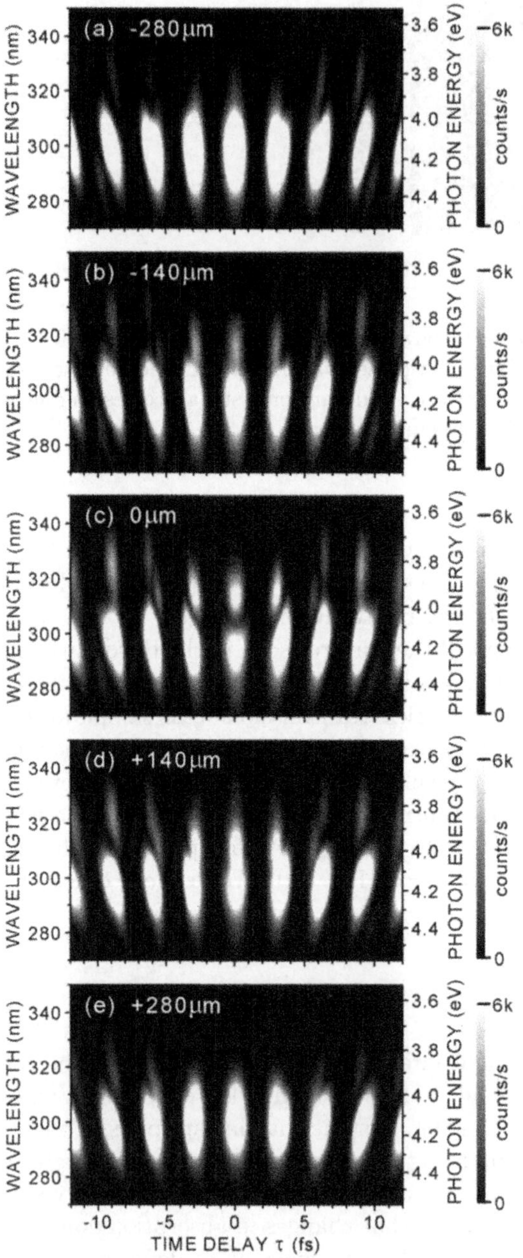

Fig. 7. Experiment: As Fig. 5c, however, the additional/removed amount of CaF_2 material is indicated. This changes the chirp of the pulses but not their amplitude spectrum. (**c**) corresponds to Fig. 5c, i.e., to $I = 0.213 \times I_0$. Taken from [11]

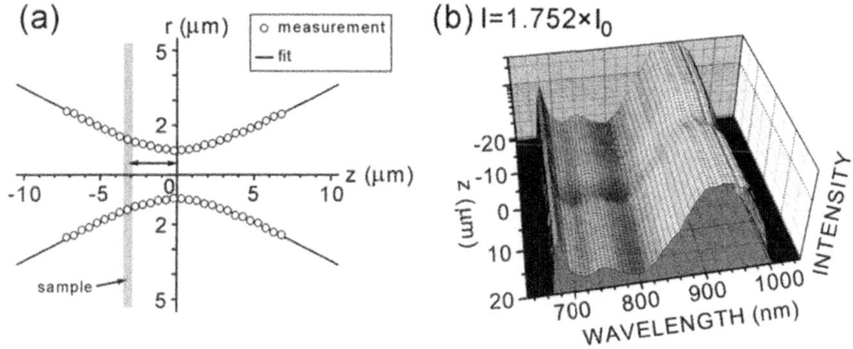

Fig. 8. (**a**) Scheme of the z-scan experiment. The measured beam radii, $r(z)$, are fitted with the formula $r(z) = 0.97\,\mu\text{m}\sqrt{1 + 0.12 \cdot (z/\mu\text{m})^2}$. (**b**) Transmitted light intensity on a logarithmic scale versus sample position z for the highest intensity available $I = 1.752 \times I_0$ (referring to $z = 0$)

the GaAs double heterostructure which accidentally coincides with the pronounced maximum in the laser spectrum (Fig. 3) also around 680 nm. For larger intensity, Fig. 9b, the transmission maximum around $z = 0$ flattens, and we observe pronounced induced absorption for wavelengths longer than the GaAs band edge. On the basis of Rabi flopping, one expects some induced absorption which compensates for the reduced absorption at the band gap (energy conservation). Furthermore, this induced absorption might also be due partly to intraband effects, i.e., to the dynamic Franz–Keldysh effect [13, 14] – consistent with our discussion of Fig. 6 around $\tau = 0$. For the highest intensity (Fig. 9c), this increased absorption becomes the dominating feature throughout most of the spectral range. Note that little if any induced transparency is observed for wavelengths between 780 nm (170 meV above the unrenormalized band gap $E_g = 1.42$ eV) and 700 nm (350 meV above the unrenormalized band gap), whereas the laser spectrum (Fig. 3) still has significant amplitude there. This indicates that these states high up in the band-to-band continuum of GaAs must experience much stronger damping (phase relaxation) and/or energy relaxation than those states near the band gap. We will come back to this point in the theory section.

2.2 Theory of Interband Effects

To describe our experiments, one has to solve the Maxwell equations coupled to the material equations and investigate the light transmitted or emitted in the forward direction. Let us consider wave propagation in the positive z-direction with $\boldsymbol{E}$ and $\boldsymbol{D}$ fields polarized in the x-direction with corresponding components $E(z,t)$ and $D(z,t)$, respectively. The $\boldsymbol{H}$ and $\boldsymbol{B}$ fields are polarized in the y-direction with corresponding components $H(z,t)$ and $B(z,t)$.

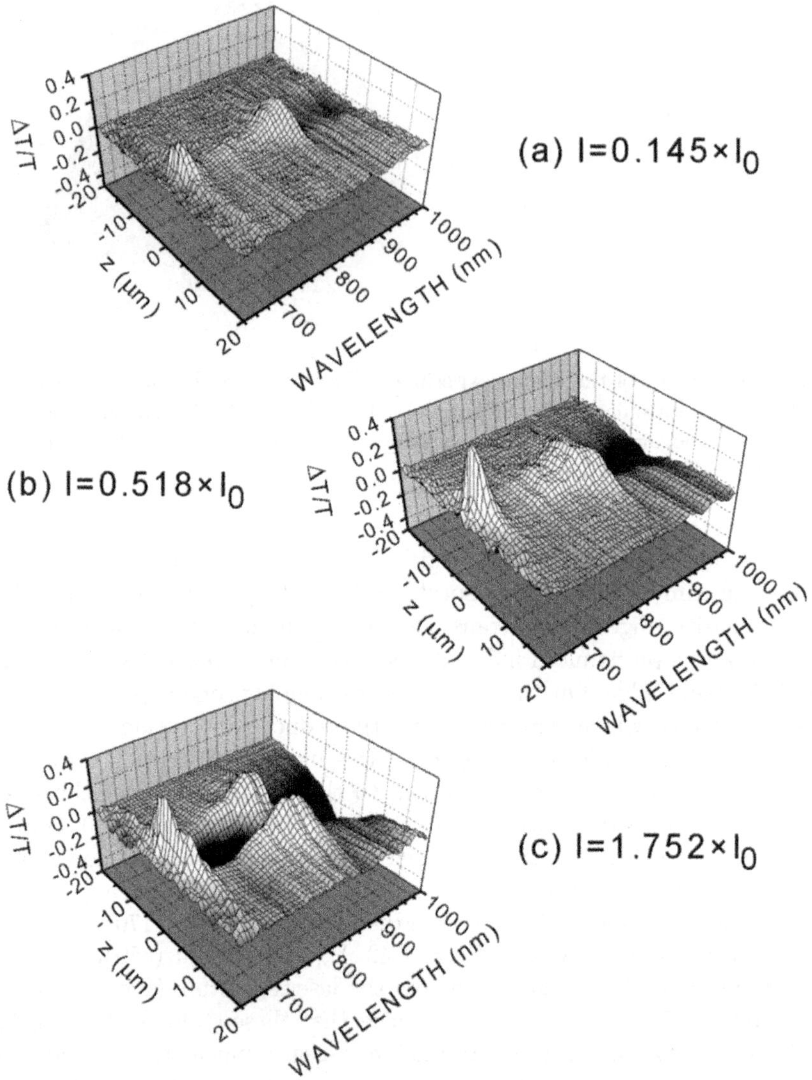

Fig. 9. Differential transmission $\Delta T/T$ as a function of the sample coordinate z for three different incident intensities I (referring to $z = 0$). (**a**) $I = 0.145 \times I_0$, (**b**) $I = 0.518 \times I_0$, (**c**) $I = 1.752 \times I_0$

Under these conditions, Maxwell equations (in SI units) immediately give

$$\frac{\partial E(z,t)}{\partial z} = -\frac{\partial B(z,t)}{\partial t}, \tag{16}$$

$$\frac{\partial H(z,t)}{\partial z} = -\frac{\partial D(z,t)}{\partial t}. \tag{17}$$

In the semiconductor,

$$B(z,t) = \mu_0 H(z,t) \tag{18}$$

and

$$D(z,t) = \varepsilon_0 E(z,t) + P(z,t)\,, \tag{19}$$

with the (real) medium polarization $P(z,t)$. The electric field impinging on the sample from the vacuum on the left, i.e., from $z = -\infty$, is a plane wave and can be written as

$$E(z,t) = \tilde{E}(t - z/c_0)\cos\left[\omega_0(t - z/c_0) + \phi\right] \tag{20}$$

with the vacuum velocity of light $c_0 = 1/\sqrt{\mu_0\varepsilon_0}$, the (real) electric-field envelope $\tilde{E}$, the laser center frequency ω_0, and the carrier–envelope offset (CEO) phase ϕ. Note that ϕ would drop out when using the rotating wave approximation (RWA) and/or the slowly varying envelope approximation (SVEA) [2]. In contrast to this, it is generally important in the carrier-wave regime. Except for the next section, in which we explicitly discuss the dependence on the CEO phase ϕ, we choose $\phi = 0$. The material enters via the polarization P which has to be computed microscopically from the underlying Hamiltonian H. Neglecting the Coulomb interaction of carriers [24], any type of intraband optical processes at this point, phonons and their coupling to the carriers, suppressing spin indexes, and using the dipole approximation for the optical transitions from the valence (v) to the conduction (c) band at wave vector $\boldsymbol{k}$, we have [3]

$$H = \sum_{\boldsymbol{k}} E_{\mathrm{c}}(\boldsymbol{k}) c^{\dagger}_{\mathrm{c}\boldsymbol{k}} c_{\mathrm{c}\boldsymbol{k}} + \sum_{\boldsymbol{k}} E_{\mathrm{v}}(\boldsymbol{k}) c^{\dagger}_{\mathrm{v}\boldsymbol{k}} c_{\mathrm{v}\boldsymbol{k}} - \sum_{\boldsymbol{k}} d_{\mathrm{cv}}(\boldsymbol{k}) E(z,t)\left(c^{\dagger}_{\mathrm{c}\boldsymbol{k}} c_{\mathrm{v}\boldsymbol{k}} + c^{\dagger}_{\mathrm{v}\boldsymbol{k}} c_{\mathrm{c}\boldsymbol{k}}\right). \tag{21}$$

Here, $E_{\mathrm{c,v}}(\boldsymbol{k})$ are the single particle energies of electrons in the conduction and valence bands, respectively (the band structure), and $d_{\mathrm{cv}}(\boldsymbol{k})$ is the (real) dipole matrix element for an optical transition at electron wave vector $\boldsymbol{k}$. Note that in our discussion above, we used $d = d_{\mathrm{cv}}(\boldsymbol{k})$. The creation $c^{\dagger}$ and annihilation c operators create and annihilate crystal electrons in the indicated band (c,v) at the indicated momentum ($\boldsymbol{k}$). The optical polarization is given by

$$P(z,t) = \frac{1}{V}\sum_{\boldsymbol{k}} d_{\mathrm{cv}}(\boldsymbol{k})\left[p_{\mathrm{vc}}(\boldsymbol{k}) + \text{c.c.}\right] + P_{\mathrm{b}}(z,t), \tag{22}$$

where the optical transition amplitudes,

$$p_{\mathrm{vc}}(\boldsymbol{k}) = \langle c^{\dagger}_{\mathrm{v}\boldsymbol{k}} c_{\mathrm{c}\boldsymbol{k}} \rangle, \tag{23}$$

depend on time t as well as parametrically on the propagation coordinate z. As usual, the sum in (22) can be expressed via the combined density of states $D_{\mathrm{cv}}(E)$ as $\sum_{\boldsymbol{k}} \ldots \rightarrow \int D_{\mathrm{cv}}(E) \ldots \mathrm{d}E \rightarrow \sum_n D_{\mathrm{cv}}(E_n) \ldots \Delta E$ (compare with Table 1), which neglects all anisotropies. The background polarization $P_{\mathrm{b}}(z,t) = \varepsilon_0 \chi_{\mathrm{b}}(z) E(z,t) = \varepsilon_0(\varepsilon_{\mathrm{b}}(z) - 1)E(z,t)$ accounts for all high-energy optical transitions not explicitly accounted for in (21) and can be expressed in terms of the background dielectric constant $\varepsilon_{\mathrm{b}}(z)$. The dynamics of $p_{\mathrm{vc}}(\boldsymbol{k})$, as well as those of the occupation numbers in the conduction band,

$$f_{\mathrm{c}}(\boldsymbol{k}) = \langle c_{\mathrm{c}\boldsymbol{k}}^{\dagger} c_{\mathrm{c}\boldsymbol{k}} \rangle , \tag{24}$$

and in the valence band,

$$f_{\mathrm{v}}(\boldsymbol{k}) = \langle c_{\mathrm{v}\boldsymbol{k}}^{\dagger} c_{\mathrm{v}\boldsymbol{k}} \rangle , \tag{25}$$

are easily calculated from the Heisenberg equation of motion for any operator $\mathcal{O}$ according to

$$-\mathrm{i}\hbar \frac{\partial}{\partial t} \mathcal{O} = [H, \mathcal{O}] \, . \tag{26}$$

Employing the usual anticommutation rules, i.e.,

$$\left[c_{\mathrm{c}\boldsymbol{k}}, c_{\mathrm{c}\boldsymbol{k}'}^{\dagger} \right]_{+} = \delta_{\boldsymbol{k}\boldsymbol{k}'} \, , \qquad \left[c_{\mathrm{v}\boldsymbol{k}}, c_{\mathrm{v}\boldsymbol{k}'}^{\dagger} \right]_{+} = \delta_{\boldsymbol{k}\boldsymbol{k}'} \, , \tag{27}$$

and all other anticommutators are zero, leads us to the known Bloch equations for the transition amplitude:

$$\left[\frac{\partial}{\partial t} + \mathrm{i}\Omega(\boldsymbol{k}) \right] p_{\mathrm{vc}}(\boldsymbol{k}) + \left[\frac{\partial}{\partial t} p_{\mathrm{vc}}(\boldsymbol{k}) \right]_{\mathrm{rel}} = \mathrm{i}\hbar^{-1} d_{\mathrm{cv}}(\boldsymbol{k}) E(z,t) \left[f_{\mathrm{v}}(\boldsymbol{k}) - f_{\mathrm{c}}(\boldsymbol{k}) \right] , \tag{28}$$

with the optical transition energy $\hbar\Omega(\boldsymbol{k}) = E_{\mathrm{c}}(\boldsymbol{k}) - E_{\mathrm{v}}(\boldsymbol{k})$, and for the occupation in the conduction band,

$$\frac{\partial}{\partial t} f_{\mathrm{c}}(\boldsymbol{k}) + \left[\frac{\partial}{\partial t} f_{\mathrm{c}}(\boldsymbol{k}) \right]_{\mathrm{rel}} = 2\hbar^{-1} d_{\mathrm{cv}}(\boldsymbol{k}) E(z,t) \, \mathrm{Im} \left[p_{\mathrm{vc}}(\boldsymbol{k}) \right] . \tag{29}$$

Here, we have assumed a real dipole matrix element. $[1 - f_{\mathrm{v}}(\boldsymbol{k})]$ can be interpreted as the occupation of holes and obeys an equation similar to $f_{\mathrm{c}}(\boldsymbol{k})$. The terms with subscript "rel" have been added phenomenologically and describe dephasing and relaxation, respectively. They will be discussed later. Note that the transition amplitude $p_{\mathrm{vc}}(\boldsymbol{k})$ and the occupation factors $f_{\mathrm{c}}(\boldsymbol{k})$ and $f_{\mathrm{v}}(\boldsymbol{k})$ are easily connected to the components of the Bloch vector (u, v, w) mentioned in the introduction via

$$\begin{pmatrix} u \\ v \\ w \end{pmatrix} = \begin{pmatrix} 2\,\mathrm{Re}\left[p_{\mathrm{vc}}(\boldsymbol{k}) \right] \\ 2\,\mathrm{Im}\left[p_{\mathrm{vc}}(\boldsymbol{k}) \right] \\ f_{\mathrm{c}}(\boldsymbol{k}) - f_{\mathrm{v}}(\boldsymbol{k}) \end{pmatrix} . \tag{30}$$

In the modeling presented now, excitation is with $t_{\rm pulse} = 5\,\rm fs$ pulses, the envelope of which, $\tilde{E}(t)$, is sech(t/t_0)-shaped with $t_0 = 1/[2\,\mathrm{arccosh}(\sqrt{2})]t_{\rm pulse}$. Their center frequency is given by $\hbar\omega_0 = E_{\rm g}$. The phase ϕ between the envelope and the carrier wave of the pulses chosen is zero, i.e., the actual optical field is given by $E(t) = \tilde{E}(t)\cos(\omega_0 t)$. The relaxation terms in (28) and (29) are chosen as

$$\left[\frac{\partial}{\partial t}p_{\rm vc}(\boldsymbol{k})\right]_{\rm rel} = -\frac{1}{T_2}p_{\rm vc}(\boldsymbol{k})\,, \text{ with } T_2 = 50\,\text{fs, and } \left[\frac{\partial}{\partial t}f_{\rm c}(\boldsymbol{k})\right]_{\rm rel} = 0\,.$$

For the Maxwell part, a 0.6 µm thin slice of this "material" with additional background dielectric constant $\varepsilon_{\rm b} = 10.9$ is sandwiched between a $\lambda/4$-antireflection coating (with $\lambda = c_0 2.9\,\mathrm{fs}/\sqrt[4]{\varepsilon_{\rm b}}$) on the front side and a semi-infinite substrate with dielectric constant $\varepsilon = \varepsilon_{\rm b}$ on the back. This roughly corresponds to the sample geometry used in the experiment (we will come back to improved models later).

If an antireflection-coated sample is used, the light intensity $\propto nE^2$ in the medium is the same as the light intensity in vacuum. Thus, the electric field and the envelope pulse area inside the sample are smaller by a factor of $\sqrt{n}$. The medium refractive index n is similar to the background refractive index $n_{\rm b}$ (for single two-level systems). Thus, for the convenience of the reader, we quote the incident (vacuum) envelope pulse areas in multiples of $\sqrt{n_{\rm b}}$. For instance, $\Theta = 2\pi\sqrt{n_{\rm b}}$ in vacuum simply corresponds to a 2π envelope pulse area in the semiconductor.

Figure 10 shows results for identical two-level systems with a density of $10^{18}\,\rm cm^{-3}$ and with transition energy $E_{\rm g}$. The single maximum around the third harmonic of the GaAs band gap splits into two maxima which shift symmetrically with respect to the center frequency. In the theory, the energetic separation of the two maxima is given roughly by the envelope pulse area, though it appears to be smaller in the experiment (Fig. 4) by about a factor of 2.

In the following, we discuss the dependence on detuning between the optical transition energy and the laser center photon energy. For clarity, we show solutions of the Bloch equations only (i.e., no propagation effects) at this point. Corresponding results are depicted in Fig. 11. Here, ω denotes the (spectrometer) photon frequency, ω_0 the laser center frequency, and $\Omega = \Omega(\boldsymbol{k}) = \hbar^{-1}[E_{\rm c}(\boldsymbol{k}) - E_{\rm v}(\boldsymbol{k})]$ the transition frequency of one transition within the band. Without band-gap renormalization, it is clear that there are no states below the band-gap energy (dashed horizontal line); nevertheless, we depict these data. Again, the laser center frequency is centered at the band-gap energy, i.e., $\hbar\omega_0 = E_{\rm g}$. The laser spectrum is shown on the right-hand side lower corner as the gray-shaded area. The spectrum for $\hbar\Omega = E_{\rm g}$ is also depicted by the white line. Obviously, it resembles the results shown in Fig. 10 very closely, even though we do not account for propagation effects in Fig. 11. For the small envelope pulse area, $\Theta = 0.5\pi$, we find a single rather narrow maximum around $\omega/\omega_0 = 3$ and $\Omega/\omega_0 = 1$. Its width

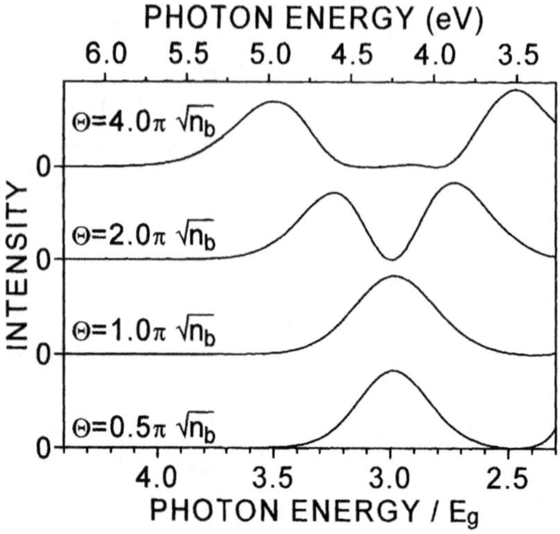

Fig. 10. Theory: Spectra of light emitted in the forward direction around the third harmonic of the transition frequency. The spectra are shown on a linear scale, vertically displaced, and individually normalized. Resonant excitation is with 5 fs sech^2-shaped pulses. The envelope pulse area Θ in front of the sample is indicated. For instance, $\Theta = 2\pi\sqrt{n_b}$, with $n_b = \sqrt{\varepsilon_b}$, corresponds to an envelope pulse area of 2π inside the sample for vanishing reflection losses

correlates with the width of the laser spectrum. This single maximum is nothing but the usual, yet resonantly enhanced, third-harmonic generation. It experiences a constriction for $\Theta = 1.0\pi$, which evolves into a shape that resembles an anticrossing for $\Theta = 2.0\pi$. Here, two separate peaks are observed only in a rather narrow region around $\hbar\Omega = \hbar\omega_0 = E_g$, and for larger $\hbar\Omega$, only a single maximum occurs. We also find that the contribution of large-frequency transitions is by no means small. For example, for $\hbar\Omega = 2\,\text{eV}$ transition energy, the signal is actually larger than for the band gap, i.e., for $\hbar\Omega = 1.42\,\text{eV}$. This trend continues for yet larger pulse areas (see $\Theta = 4.0\pi$ in Fig. 11). There is considerable resonant enhancement (as can be seen from Fig. 11a), but this enhancement becomes less important at large pulse areas because the resonant transitions are completely saturated.

The actual spectra (compare (22)) are integrals over the individual contributions, multiplied by the combined density of states over the relevant range of transition energies. The bands themselves clearly have contributions even at $\hbar\Omega = 5\,\text{eV}$. If one would sum up all these contributions at, e.g., $\Theta = 4.0\pi$ (Fig. 11d), one no longer gets two maxima but rather a single maximum around $\omega/\omega_0 = 3$, which would no longer be in agreement with the experiments. Thus, there must be a reason why the high-energy transitions do not contribute significantly. The reason is that the high-energy transitions have much shorter dephasing times which significantly suppresses their contribution [24]. Another possible reason is that the T_2-approximation is likely to fail

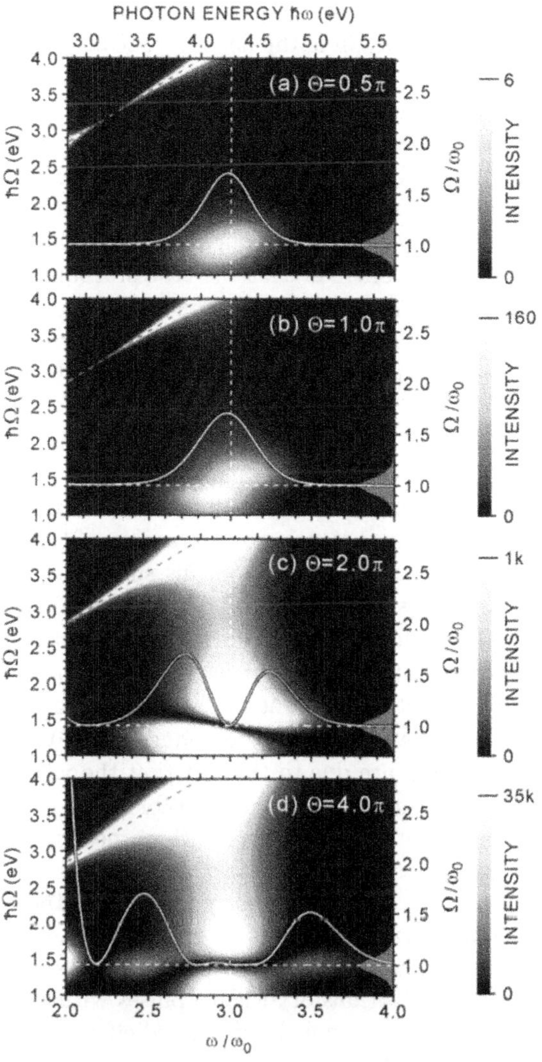

Fig. 11. Theory: Gray-scale plot of intensity as a function of (spectrometer) photon frequency ω and transition frequency Ω. The center frequency ω_0 of the optical pulses (see *gray areas* on the RHS) is centered at the band-gap frequency, i.e., $\hbar\omega_0 = E_g$. The spectrum for a transition right at the band gap, i.e. $\hbar\Omega = E_g$, is highlighted by the *white curve*. The *diagonal dashed line* corresponds to $\Omega = \omega$. Excitation is with sech^2-shaped 5 fs pulses. The envelope pulse area Θ is indicated and increases from (**a**) to (**d**)

under these conditions. For off-resonant excitation (below the resonance), the nonlinear signal arises from the spectral overlap of the laser spectrum and the low-energy tail of an individual Lorentzian line. As the laser spectrum rapidly drops in the experiment (it actually ends above some maximum frequency), this overlap is due solely to the Lorentzian tail of the line. More realistic non-Markovian calculations would lead to a much more rapid decay of the line in frequency-space, as is, e.g., well known from the so-called Urbach tail in semiconductors which is a rapidly decaying exponential rather than a slowly decaying Lorentzian. This behavior can be expressed by frequency-dependent damping, which, in the time domain, is equivalent to a convolution integral, i.e., T_2 has to be replaced by a memory integral. Though no such complete modeling is in reach presently, we can conclude that the states around the semiconductor band edge experience a huge resonant enhancement, whereas states high in the bands make negligible contributions to *nonlinear* optical signals. We will discuss now, however, that these states do have a substantial effect due to their linear optics.

In Fig. 10, we described all higher energy transitions by means of the background dielectric constant [3]. A constant as a function of frequency is equivalent to a $\delta(t)$-response in real time – the higher frequency oscillators (or transitions) are assumed to lead to an instantaneous screening of the external laser field. Remember that this effect is by no means small, i.e., the background dielectric constant in GaAs has a value of 10.9. Is the concept of a background dielectric constant still justified under the aforementioned conditions? A quick estimate shows that the answer is no. The dominant contributions to the background dielectric constant in the model semiconductor GaAs [25,26] stem from the known E_1 and E_2 resonances at 3 eV and 5 eV photon energy, respectively (which are due to the fact that the valence and conduction bands are roughly parallel over a considerable part of the Brillouin zone). The E_1 resonance results in detuning with respect to the band edge (or laser center frequency) of about 1.5 eV, which is equivalent to an oscillation period of about 2.7 fs. This is obviously not instantaneous if compared to the GaAs band-gap period of 2.9 fs, 5 fs optical pulses, and Rabi periods comparable with or even shorter than 2.9 fs. This finite response time of dielectric screening leads to distortions of incident pulses within the sample (if and only if propagation is accounted for) and related effects which altogether modify the experimental results.

To account for these effects, one needs to account for the actual band structure [27] (i.e., one must not employ the effective mass approximation). Difficulties do arise from the fact that the optical dipole matrix element $d_{cv}(\boldsymbol{k})$ depends strongly on the electron wave vector $\boldsymbol{k}$ – which again needs accurate band structure calculations. The dephasing and relaxation processes in the bands are also expected to change strongly with the wave vector. We avoid all these truly challenging problems by taking advantage of the fact that high-energy optical transitions do not contribute significantly anyway to *nonlinear* signals due to a huge resonant enhancement at the band edge (see our discussion above). Thus, we set the occupation of these high-energy transitions to zero and determine the distribution (i.e., the combined density of states) of this ensemble of two-level systems by a fit to the accurately measured linear complex dielectric function [25] (it is quite convenient that this paper actually contains a table of the data points). The corresponding parameters for GaAs and the $Al_{0.3}Ga_{0.7}As$ barriers are summarized in Table 1. The T_1-times drop out of the equations. This fit procedure is advantageous

Table 1. Combined density of states $D_{cv}(E_n)$ (in units of $10^{19}\,cm^{-3}\cdot eV^{-1}$) and dephasing time $T_2(E_n)$ (in units of femtosecond) versus energy E_n (in units of electronvolt) of GaAs and $Al_{0.3}Ga_{0.7}As$, $\Delta E = E_{n+1} - E_n = 0.1\,eV$

	GaAs		$Al_{0.3}Ga_{0.7}As$			GaAs		$Al_{0.3}Ga_{0.7}As$	
E_n	D_{cv}	T_2	D_{cv}	T_2	E_n	D_{cv}	T_2	D_{cv}	T_2
1.5	3	10			3.7	95	10	95	10
1.6	3	10			3.8	95	10	95	10
1.7	4	10			3.9	95	10	95	10
1.8	6	10			4.0	95	10	95	10
1.9	7	10	3	10	4.1	100	10	100	10
2.0	9	10	6	10	4.2	115	10	105	10
2.1	12	10	7	10	4.3	130	10	112	10
2.2	12	10	9	10	4.4	155	10	130	10
2.3	15	10	10	10	4.5	165	10	160	10
2.4	18	10	10	10	4.6	180	9	210	9
2.5	20	10	14	10	4.7	200	9	210	9
2.6	20	10	17	10	4.8	205	8	205	8
2.7	22	10	20	10	4.9	215	8	188	8
2.8	95	10	27	10	5.0	145	8	120	8
2.9	130	10	40	10	5.1	115	8	108	8
3.0	150	10	65	10	5.2	30	8	60	8
3.1	145	10	125	10	5.3	40	8	65	8
3.2	140	10	138	10	5.4	70	8	70	8
3.3	120	10	140	10	5.5	80	8	80	8
3.4	110	10	120	10	5.6	90	8	90	8
3.5	100	10	110	10	5.7	100	8	100	8
3.6	100	10	100	10	5.8	105	8	105	8

because theory was never able to provide a true quantitative match to the measured linear dielectric function anyway [3]; however, the actual quantitative shape of the dielectric function is very important for what follows. The resulting linear dielectric function, fitted by means of 45 two-level systems, is shown in Fig. 12d. The only optical nonlinearity at this point of modeling arises from those two-level transitions at the GaAs ($Al_{0.3}Ga_{0.7}As$) band edge at $E_g = 1.42\,eV$ ($1.83\,eV$) photon energy. As in the previous section, their dipole matrix element is $d = 0.5\,e\,nm$, their density $10^{18}\,cm^{-3}$, and their dephasing time is $T_2 = 50\,fs$. A crucial point in the improved model is that we account for the actual sample geometry, i.e., including all layers of the

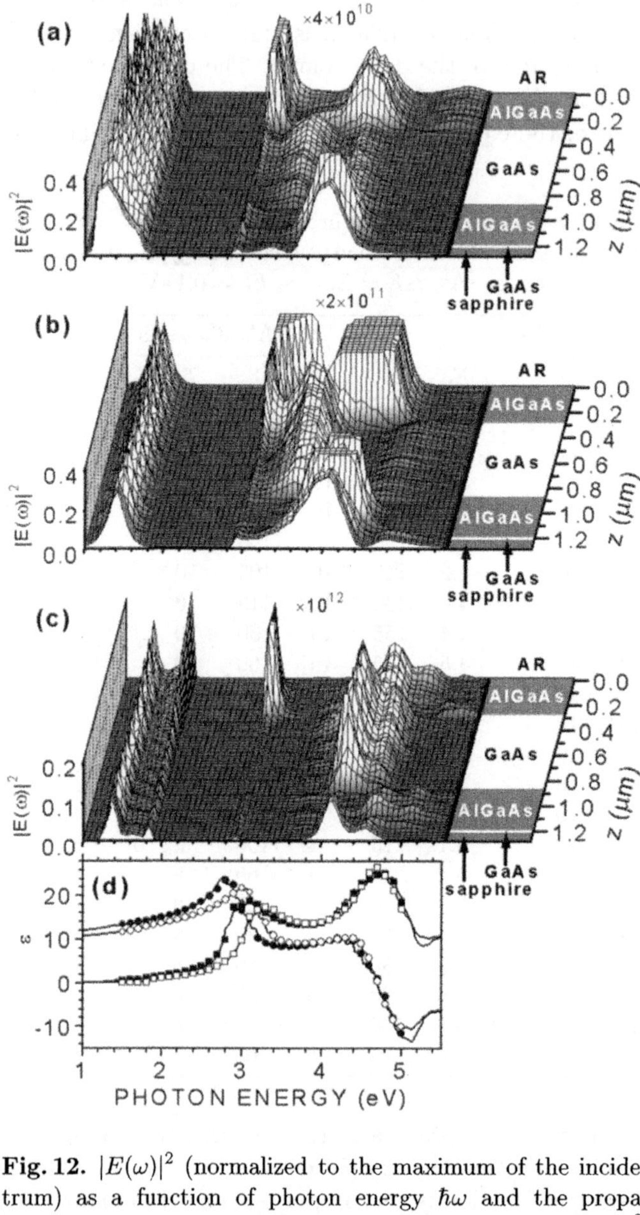

Fig. 12. $|E(\omega)|^2$ (normalized to the maximum of the incident electric field spectrum) as a function of photon energy $\hbar\omega$ and the propagation coordinate z, $\phi = 0$. Note the strong variation as a function of z. (**a**) sinc^2-shaped 5.6 fs pulses, $\tilde{E}_0 = 3.5 \times 10^9$ V/m; the GaAs cap layer thickness is $d_{\text{cap}} = 30$ nm. (**b**) As (**a**), but for sech^2-shaped 5 fs incident optical pulses with $\tilde{E}_0 = 3.5 \times 10^9$ V/m and $d_{\text{cap}} = 10$ nm. (**c**) As (**a**), but for incident pulses which are fitted to the experiment (Fig. 3), $E_0 = 2.5 \times 10^9$ V/m, $d_{\text{cap}} = 10$ nm. (**d**) The real (*circles*) and imaginary (*squares*) parts of the linear dielectric function of GaAs (*full*) and $Al_{0.3}Ga_{0.7}As$ (*open*), respectively, are shown for comparison. The *symbols* are the experimental data taken from [25]; the *full curves* correspond to our modeling

sample. All these aspects together with the fact that we do not employ the slowly varying envelope approximation and do not employ the rotating wave approximation make this a somewhat demanding numerical task. We show in the following, results for optical pulses with $E(t) = \tilde{E}_0 \,\mathrm{sinc}(t/t_0)\cos(\omega_0 t + \phi)$ which roughly resemble the above experiments (see the fit to the measured field autocorrelation function in [19]). The full-width at half-maximum of the intensity envelope is 5.6 fs (i.e., $t_0 = 5.6\,\mathrm{fs}/2.7831$), the center photon energy $\hbar\omega_0 = 1.42\,\mathrm{eV} = E_\mathrm{g}$, and the CEO phase $\phi = 0$.

Calculations for the GaAs double heterostructure used in [11] are shown in Fig. 12. The layer structure is shown on the RHS. While propagating through the sample, the fundamental spectrum becomes significantly distorted, which leads to lengthening of the pulse in time and thus, to a reduction of the field amplitude and the Rabi frequency. The dispersive effects due to the linear dielectric function further enhance this trend. Both effects lead to a reduction of the splitting of the Rabi sidebands around the third harmonic (compare Fig. 12 and Fig. 13). This explains the much (factor of 2) smaller splitting seen in Fig. 4 compared to the simple modeling (see Fig. 10) as well as the tails for large time delays τ in Fig. 6. It also becomes obvious from Fig. 12 that the signal varies very strongly with propagation coordinate z. This is due to the fact that the absorption coefficients (for the third harmonic) of both GaAs and $\mathrm{Al_{0.3}Ga_{0.7}As}$ are around 1/10 nm which can easily be estimated from the corresponding linear dielectric functions shown in Fig. 12d. As a dramatic result, the detected signal does not stem from the 600 nm thick GaAs layer sandwiched between $\mathrm{Al_{0.3}Ga_{0.7}As}$ barriers – which we initially believed [11] – but rather from the thin GaAs cap layer initially employed as an antioxidation layer. The $\mathrm{Al_{0.3}Ga_{0.7}As}$ barriers lead to an (almost) off-resonant nonlinear signal which is expected from the dependence on detuning shown in Fig. 11.

The same overall behavior is also found in sech^2-shaped 5 fs pulses (see Fig. 12b) as well as in pulses corresponding to the experiment (Fig. 12c). For the latter, we have fitted a measured laser spectrum by the sum of three Gaussians and have assumed a constant phase (i.e., no chirp), consistent with the experiment (Fig. 3). Under these conditions, the electric field versus time results from a Fourier transform of the square root of the intensity laser spectrum. For such pulses, the Rabi splitting seen in the emitted light intensity is yet smaller, nicely reproducing the experiment (see Fig. 4). These actual laser pulses are more susceptible to the mentioned propagation effects because the prominent high-energy peak in the laser spectrum (see Fig. 3) is largely reduced as a result of absorption, especially because it accidentally coincides with the band edge of the $\mathrm{Al_{0.3}Ga_{0.7}As}$ barriers, whereas the low-energy peak, which is below the GaAs band edge, remains nearly unaffected. As the short pulses essentially result from beating between the low- and the high-energy peak, selective suppression of the high-energy peak lengthens the pulses substantially, leading to a large reduction in the Rabi energy, and, thus, to smaller observed splitting.

It is rather obvious that one can (in principle) avoid these disturbing propagation effects (which were not clear to us at the time of publication of [11]) by using very thin layers of GaAs without $Al_{0.3}Ga_{0.7}As$ barriers (as they also distort the pulses and also absorb the third harmonic). This also potentially opens the door to significant dependence on the carrier–envelope offset phase, which is sometimes also referred to as the absolute optical phase. In many situations in optics, only the well-known relative optical phase, i.e., the phase between two different beams or pulses, matters. In contrast to this, for few-cycle optical pulses, the CEO phase of a single pulse can significantly influence the outcome of an experiment.

3 Dependence on the Carrier–Envelope Offset Phase

To measure the carrier–envelope offset frequency f_ϕ of pulses directly from a femtosecond laser oscillator, recent work [28, 29] has, for example, used the interference of the fundamental frequency of a laser pulse, which was spectrally broadened by self-phase modulation in a (photonic crystal) fiber of a few millimeter length where the second harmonic was generated with the help of a separate crystal. In [30], the same idea was used, except that the fundamental spectrum did already cover one octave, hence no need for additional broadening. Somewhat similar to this, [31] used the interference of the third harmonic, generated on a silicon wafer surface, with the second harmonic generated in a separate crystal. Related work for amplified pulses [32, 33, 34, 35] addressed the role of the CEO phase ϕ. All the optical nonlinearities used in these and other [36] cases are off-resonant and within the perturbative regime, i.e., an expansion in terms of nonlinear optical susceptibilities is meaningful. Furthermore, in most of these cases, the pulses have to propagate over a considerable distance within the apparatus and, thus, the difference between the phase velocity and the group velocity can change the CEO phase within the measurement setup. A nonperturbative and, hence, distinctly different way to determine the CEO phase would be via X-ray generation in extreme nonlinear optics in atoms [37]. In a recent review [38] on the implications of the CEO phase on metrology [39, 40], the authors state in their outlook

> ... If an experimental technique can be developed that is sensitive to the carrier–envelope phase and works with the direct output of a mode-locked oscillator (i.e., without that amplification that will be necessary for extreme nonlinear optics), it may in turn benefit optical-frequency synthesis because it may create a simpler technique for determining/controlling the comb offset frequency.

We now show that carrier-wave Rabi flopping can exhibit dependence on the CEO phase ϕ, which potentially allows us to determine ϕ. We also show

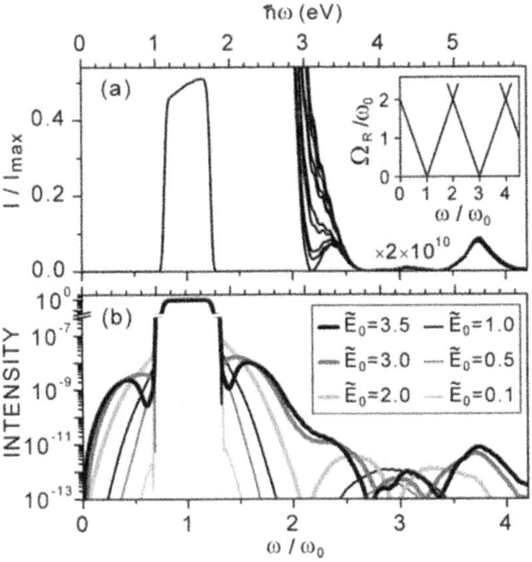

Fig. 13. (**a**) Signal intensity (*linear scale*, normalized to the maximum intensity, I_{max}, of the incident laser spectrum) emitted in the forward direction versus spectrometer frequency ω in units of the laser center frequency ω_0 for different values of the CEO phase ϕ of the 5.6 fs sinc^2 pulses. The GaAs film with $L = 20$ nm thickness on a substrate with $\epsilon_{\mathrm{s}} = \mathrm{const} = (1.76)^2$ has no $\mathrm{Al_{0.3}Ga_{0.7}As}$ barriers on either side, but a front-side antireflection (AR) coating, $\tilde{E}_0 = 3.5 \times 10^9$ V/m. The *inset* illustrates the interference of the different Rabi sidebands as the Rabi frequency $\Omega_{\mathrm{R}} = \hbar^{-1} d\tilde{E}$ increases. (**b**) As (**a**), but signal intensity (normalized) on a logarithmic scale for fixed $\phi = 0$ and for different incident electric-field amplitudes (in units of 10^9 V/m) as indicated. Taken from [27]

that the behavior is distinctly different from the above-mentioned perturbative off-resonant approaches. Given the discussion of the previous section, the idea is quite straightforward. If, for example, the Rabi frequency becomes comparable to the light frequency, the high-energy sidebands of the fundamental and the low-energy sideband of the third harmonic meet at around twice the fundamental laser center frequency (see inset in Fig. 13a). This interference is expected to depend strongly on the CEO phase.

Figure 13a shows spectra of light emitted in the forward direction of a thin layer of GaAs which has no $\mathrm{Al_{0.3}Ga_{0.7}As}$ barriers (in contrast to the previous section!) for various values of the CEO phase ϕ. Samples as such can be produced by molecular-beam epitaxy on GaAs substrates and subsequent selective etching, first, of the GaAs substrate and then of the $\mathrm{Al_{0.3}Ga_{0.7}As}$ etch-stop layer. The meeting of the different Rabi sidebands (see inset) can also be seen in the actual calculations (Fig. 14). Furthermore, it becomes clear from the intensity dependence shown in Fig. 13b that it is *not* simply the interference of the tail of the laser spectrum itself (which is roughly equal

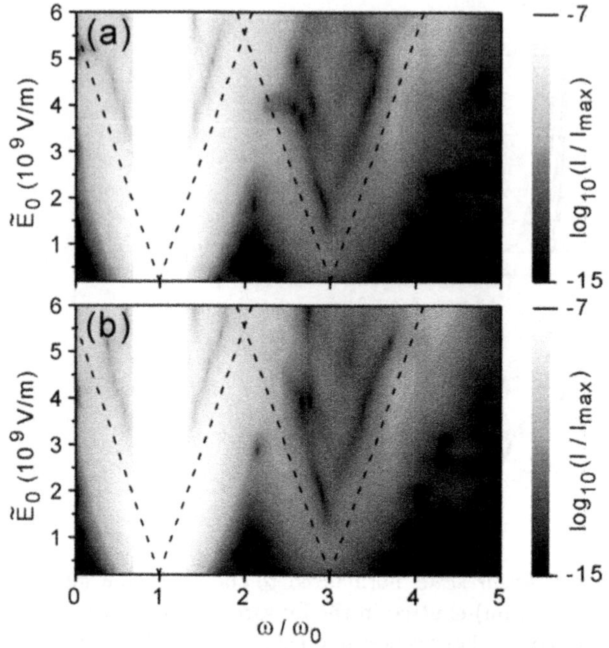

Fig. 14. Gray-scale image of the light intensity emitted in the forward direction versus field amplitude of the incident pulses, corresponding to the *inset* in Fig. 13. Parameters as in Fig. 13. (**a**) $\phi = 0$, (**b**) $\phi = \pi/2$. Taken from [27]

to a box function) with the third-harmonic signal – which would be similar to all the approaches discussed above – but rather the interference of different Rabi sidebands.

Figure 15a depicts the intensity spectra of light emitted in the forward direction versus CEO phase ϕ for a $L = 100$ nm thin GaAs film on a substrate with $\epsilon_s = (1.76)^2$ (e.g., sapphire). Note the dependence on ϕ with large visibility around $\omega/\omega_0 = 2.05$ to 2.25 (this is a 284 meV or 38 nm broad interval) and the period of π (rather than 2π for [28, 29, 30, 31, 36]) resulting from the inversion symmetry of the problem. In other words, the signal does not depend on the sign of the electric field. Figure 15b shows the same for $L = 20$ nm, indicating that one already has some distortions in Fig. 15a due to the finite thickness of the sample as a result of different group and phase velocities. The high-energy transitions do *not* react instantaneously as would be the case for the background dielectric constant. Figure 15c is as Fig. 15b, but introduces a front-side $\lambda/4$-antireflection (AR) coating designed for the fundamental laser frequency ω_0. Note that Fig. 15b and Fig. 15c are shifted horizontally with respect to each other, because the incident optical pulses, and thus also ϕ, are distorted as a result of multiple reflections. Figure 15d is as Fig. 15c, but for a different incident electric-field amplitude $\tilde{E}_0$. This

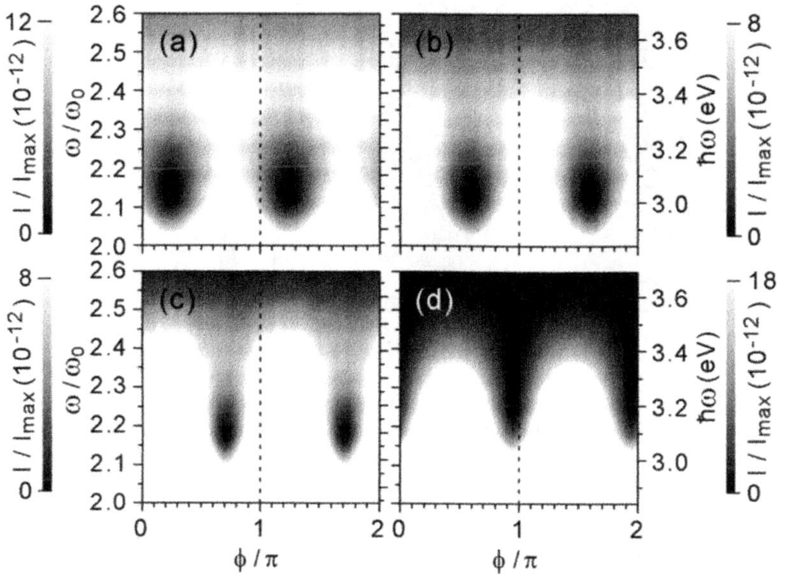

Fig. 15. Gray-scale image of the emitted intensity as a function of ω and ϕ for a thin GaAs film with thickness L without $Al_{0.3}Ga_{0.7}As$ barriers on a substrate with dielectric constant ϵ_s. Excitation with 5.6 fs $sinc^2$ pulses. (**a**) $L = 100$ nm, $\tilde{E}_0 = 3.5 \times 10^9$ V/m, and $\epsilon_s = (1.76)^2$; (**b**) as (**a**), but $L = 20$ nm; (**c**) as (**b**), but with an additional front-side antireflection coating (as in Fig. 13); (**d**) as (**c**), but for an electric-field amplitude of $\tilde{E}_0 = 4.0 \times 10^9$ V/m. Taken from [27]

variation also leads to a horizontal shift, which is both interesting as well as disturbing. It is interesting, on the one hand, because no such intensity dependence occurs in off-resonant perturbative nonlinear optics [28, 29, 30, 31, 36] – pointing out the distinct difference between the two scenarios. It is disturbing, on the other hand, because to use the effect to determine the CEO phase, one needs to calibrate the incident electric-field amplitude, or, more precisely, the Rabi frequency. This is, however, possible via the measured splitting of Rabi sidebands [11]. Finally, it is important to note that the excitation intensities (or field amplitudes) and the absolute emitted signals shown here are comparable to the signals computed for samples corresponding to the experiment (see Fig. 12) in which the nonlinear signals could easily be detected using 5 fs pulses with an 81 MHz repetition rate directly out of a laser oscillator with average powers of the order of merely a few 10 mW.

The square modulus of the electric field at a frequency of $\omega/\omega_0 = 2.14$ as a function of the propagation coordinate z within the GaAs sample and as a function of the CEO phase ϕ is shown in Fig. 16. As discussed above, the difference between the group and phase velocities leads to slipping of the carrier wave underneath the field envelope during propagation and, thus, to

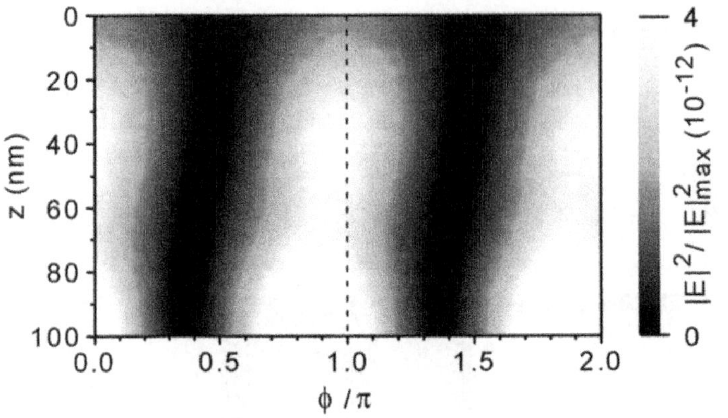

Fig. 16. Gray-scale image of the square modulus of the electric field at $\omega/\omega_0 = 2.14$ as a function of the propagation coordinate z and the CEO phase ϕ (remember that the CEO phase ϕ refers to the *incident* electric field). $\tilde{E}_0 = 4.5 \times 10^9$ V/m, 5.6 fs sinc^2 pulses, and $\epsilon_s = \epsilon_{b,\,\mathrm{GaAs}} = 10.9$. The latter choice suppresses the reflection at the interface between the GaAs film and the substrate

a shift of the maxima/minima to smaller ϕ. For a GaAs film thickness of only $L = 20$ nm, this shift can be neglected.

In an actual experiment using a mode-locked laser, the CEO phase ϕ would vary from pulse to pulse with the CEO frequency f_ϕ, which is generally not identical to the pulse repetition frequency. Thus, the dependence on ϕ could actually be measured by detecting a certain optical frequency, ω, in Fig. 15, e.g., with a photomultiplier tube and sending this signal into a radio-frequency (rf) spectrum analyzer. A period of 2π in Fig. 15 and Fig. 16 would correspond to a peak at the CEO frequency f_ϕ in the rf spectrum. The period of π, which corresponds to Fig. 15 and Fig. 16, is equivalent to a peak at twice the CEO frequency, $2f_\phi$, in the rf spectrum.

4 Conclusions

In conclusion, our resonant nonperturbative few-cycle experiments on carrier-wave Rabi flopping have given access to semiconductor material dynamics on a timescale comparable to only one cycle of light. The characteristic feature of carrier-wave Rabi flopping is the occurrence of sidebands around the third harmonic of the laser center frequency with a splitting comparable to the frequency of light. Though we have explicitly discussed semiconductors, much of the underlying physics is more general and is also expected in other systems, as, e.g., in atoms or molecules. However, we are not currently aware of any corresponding experiments. The interference of the different Rabi sidebands around the fundamental and the third harmonic gives rise to a signal contribution around the second harmonic of the laser center frequency which depends strongly on the carrier–envelope offset phase. This does, however, require suitable, i.e., very thin GaAs samples without $Al_{0.3}Ga_{0.7}As$ barriers to avoid distortions of the incident laser pulses due to the noninstantaneous response of higher energy optical transitions. The anticipated dependence on the carrier–envelope offset phase could in turn be used to determine the carrier–envelope offset phase if one is able to calibrate the Rabi frequency. This would be necessary because – unlike what is known from off-resonant perturbative nonlinear optics – the dependence of the signal on the carrier–envelope offset phase changes with the incident laser intensity, pointing out the distinct difference between resonant and off-resonant nonlinear optics.

Acknowledgements

This work is supported by the DFG Leibniz Award 2000 and by project DFG-We 1497/11-1. We thank Wolfgang Stolz for the high-quality GaAs samples as well as Uwe Morgner and Franz X. Kärtner for intense support in the initial phase of the experiment and for stimulating discussions, especially on the carrier–envelope offset phase. The theoretical part of this work has largely benefited from discussions with the group of Hartmut Haug who early shared their theoretical results concerning state-dependent damping in carrier-wave Rabi flopping with us. During his stay in January/February 2002 in Karlsruhe, we had many stimulating discussions on intraband effects with Ren-Bao Liu.

References

[1] T. Brabec, F. Krausz: Rev. Mod. Phys. **72**, 545 (2000)
[2] P. Meystre, M. Sargent III: *Elements of Quantum Optics*, 2nd ed. (Springer 1991)
[3] W. Schäfer, M. Wegener: *Semiconductor Optics and Transport Phenomena*, Adv. Texts in Phys. (Springer 2002)
[4] R. Binder, S.W. Koch, M. Lindberg, N. Peyghambarian, W. Schäfer: Phys. Rev. Lett. **65**, 899 (1990)
[5] S.T. Cundiff, A. Knorr, J. Feldmann, S.W. Koch, E.O. Göbel, H. Nickel: Phys. Rev. Lett. **73**, 1178 (1994)
[6] H. Giessen, A. Knorr, S. Haas, S.W. Koch, S. Linden, J. Kuhl, M. Hetterich, M. Grün, C. Klingshirn: Phys. Rev. Lett. **81**, 4260 (1998)
[7] A. Schülzgen, R. Binder, M.E. Donovan, M. Lindberg, K. Wundke, H.M. Gibbs, G. Khitrova, N. Peyghambarian: Phys. Rev. Lett. **82**, 2346 (1999)
[8] L. Bányai, Q.T. Vu, B. Mieck, H. Haug: Phys. Rev. Lett. **81**, 882 (1998)
[9] C. Fürst, A. Leitenstorfer, A. Nutsch, G. Tränkle, A. Zrenner: Phys. Status Solidi B **204**, 20 (1997)
[10] S. Hughes: Phys. Rev. Lett. **81**, 3363 (1998)
[11] O.D. Mücke, T. Tritschler, M. Wegener, U. Morgner, F.X. Kärtner: Phys. Rev. Lett. **87**, 057401 (2001)
[12] C. Zener: Proc. R. Soc. Lon. Ser.-A **145**, 523 (1934)

[13] A. P. Jauho, K. Johnsen: Phys. Rev. Lett. **76**, 4576 (1996)
[14] A. H. Chin, J. M. Bakker, J. Kono: Phys. Rev. Lett. **85**, 3293 (2000)
[15] F. Bloch: Z. Phys. **52**, 555 (1929)
[16] G. Wannier: Phys. Rev. **52**, 191 (1937)
[17] V. P. Kalosha, J. Herrmann: Phys. Rev. Lett. **83**, 544 (1999)
[18] R. W. Ziolkowski, J. M. Arnold, D. M. Gogny: Phys. Rev. A **52**, 3082 (1995)
[19] U. Morgner, F. X. Kärtner, S. H. Cho, Y. Chen, H. A. Haus, J. G. Fujimoto, E. P. Ippen, V. Scheuer, G. Angelow, T. Tschudi: Opt. Lett. **24**, 411 (1999)
[20] M. U. Wehner, M. H. Ulm, M. Wegener: Opt. Lett. **22**, 1455 (1997)
[21] The first microscope objective has a focal length of $f = 5.41$ mm and a numerical aperture of NA=0.5 (*Coherent 25-0522*), the second one $f = 13.41$mm and NA=0.5 (*Coherent 25-0555*).
[22] M. Peyghambarian, S. W. Koch, A. Mysyrowicz: *Introduction to Semiconductor Optics* (Prentice Hall, Englewood Cliffs 1993)
[23] M. Sheik-Bahae, A. A. Said, T.-H. Wei, D. J. Hagan, E. W. van Stryland: IEEE J. Quantum Elect. **26**, 760 (1990)
[24] Q. T. Vu, H. Haug, O. D. Mücke, T. Tritschler, M. Wegener, G. Khitrova, H. M. Gibbs: Phys. Rev. Lett. **92**, 217403 (2004)
[25] D. E. Aspnes, S. M. Kelso, R. A. Logan, R. Bhat: J. Appl. Phys. **60**, 754 (1986)
[26] In general, GaAs also shows an off-resonant second-harmonic signal due to its lack of inversion symmetry. This weak signal can be suppressed by proper choice of the light polarization.
[27] O. D. Mücke, T. Tritschler, M. Wegener, U. Morgner, F. X. Kärtner: Phys. Rev. Lett. **89**, 127401 (2002)
[28] D. J. Jones, S. A. Diddams, J. K. Ranka, A. Stentz, R. S. Windeler, J. L. Hall, S. T. Cundiff: Science **288**, 635 (2000)
[29] A. Apolonski, A. Poppe, G. Tempea, C. Spielmann, T. Udem, R. Holzwarth, T. W. Hänsch, F. Krausz: Phys. Rev. Lett. **85**, 740 (2000)
[30] U. Morgner, R. Ell, G. Metzler, T. R. Schibli, F. X. Kärtner, J. G. Fujimoto, H. A. Haus, E. P. Ippen: Phys. Rev. Lett. **86**, 5462 (2001)
[31] M. Mehendale, S. A. Mitchell, J. -P. Likforman, D. M. Villeneuve, P. B. Corkum: Opt. Lett. **25**, 1672 (2000)
[32] P. Dietrich, F. Krausz, P. B. Corkum: Opt. Lett. **25**, 16 (2000)
[33] A. V. Sokolov, D. R. Walker, D. D. Yavuz, G. Y. Yin, S. E. Harris: Phys. Rev. Lett. **87**, 033402 (2001)
[34] G. G. Paulus, F. Grasbon, H. Walther, P. Villoresi, M. Nisoli, S. Stagira, E. Priori, S. De Silvestri: Nature **414**, 182 (2001)
[35] A. Baltuška, T. Fuji, T. Kobayashi: Phys. Rev. Lett. **88**, 133901 (2002)
[36] H. R. Telle, G. Steinmeyer, A. E. Dunlop, J. Stenger, D. H. Sutter, U. Keller: Appl. Phys. B-Lasers O. **69**, 327 (1999)
[37] C. G. Durfee III, A. R. Rundquist, S. Backus, C. Herne, M. M. Murnane, H. C. Kapteyn: Phys. Rev. Lett. **83**, 2187 (1999)
[38] S. T. Cundiff, J. Ye, J. L. Hall: Rev. Sci. Inst. **72**, 3749 (2001)
[39] J. Reichert, M. Niering, R. Holzwarth, M. Weitz, T. Udem, T. W. Hänsch: Phys. Rev. Lett. **84**, 3232 (2000)
[40] T. Udem, S. A. Diddams, K. R. Vogel, C. W. Oates, E. A. Curtis, W. D. Lee, W. M. Itano, R. E. Drullinger, J. C. Bergquist, L. Hollberg: Phys. Rev. Lett. **86**, 4996 (2001)

Dynamics of a Photoexcited Hydrated Electron

Maxim S. Pshenichnikov[1], Andrius Baltuška[2], and Douwe A. Wiersma[1]

[1] Department of Chemistry, Ultrafast Laser and Spectroscopy Laboratory, University of Groningen, The Netherlands
m.s.pshenichnikov@chem.rug.nl
[2] Photonics Institute, Vienna University of Technology, Gusshausstr. 27, 1040 Vienna, Austria

Abstract. Combining photon-echo and frequency-resolved pump-probe techniques with extremely short laser pulses that consist of only few optical cycles, we investigate the dynamics of the equilibrated hydrated electron. The pure dephasing time of the hydrated electron deduced from the photon-echo measurements is approximately 1.7 fs. The shape of the absorption spectrum is excellently described by an extended Lorentzian contour which spectral width is fully determined by the pure dephasing time. The experimental data is successfully described with a model in which the excited state lifetime is approximately 50 fs in water and approximately 70 fs in heavy water. The deuteration effect on the lifetime strongly suggests that OH(OD)-vibrational modes in the first solvation shell act as accepting modes for the energy relaxation.

1 Introduction

Upon injection into a fluid, an electron can be captured in a potential energy well formed by neighboring molecules of the liquid. The first observation of such an electron, known nowadays as the solvated electron, dates back to 1864, when *Weyl* reported a blue color of solutions of metals in ammonia [1]. However, it took more than a half century until *Kraus* [2] suggested that the blue color was caused by bound species: electrons trapped among the ammonia molecules. The solvated electron has been the subject of numerous investigations ever since. In the 1960s, the electron solvated in water – *the hydrated electron* – was discovered through the observation of the extraordinarily broad absorption spectrum spanning from 500 nm to 1000 nm that appeared upon injection of electrons [3, 4].

The vast interest in the hydrated electron from both theoretical and experimental points of view is by no means accidental. The hydrated electron is a transient species in charge-transfer reactions in biology, nonsilicon solar-cell technology [5, 6, 7, 8], and chemistry. Important examples are photosynthesis [9], charge transport through biomembranes [10], and long-distance charge transport in nerves [11]. The hydrated electron is also a key intermediate in radiation chemistry [4] and electrochemistry [5, 6, 7, 8]. In addition, the hydrated electron is a perfect test-ground for various theories of complex

F. X. Kärtner (Ed.): Few-Cycle Laser Pulse Generation and Its Applications, Topics Appl. Phys. **95**, 409–444 (2004)

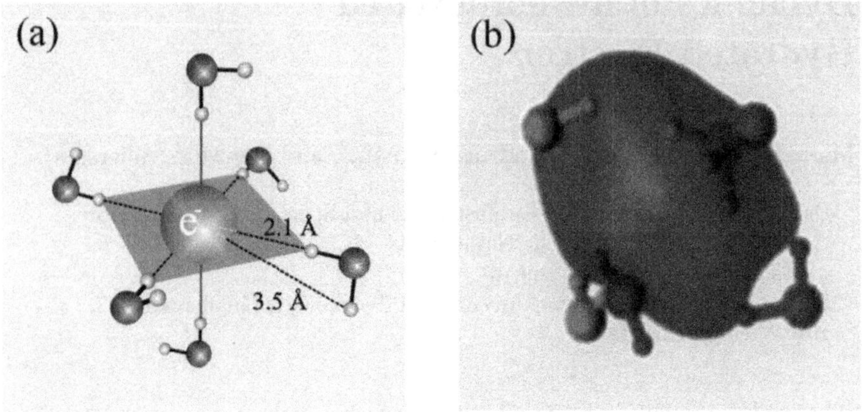

Fig. 1. The structure of the nearest solvation shell of the hydrated electron in glassy (**a**) and liquid water (**b**). The excess electron distribution is shown in the *central area*. Note that the six water molecules are oriented with their OH bonds toward the center of the electron charge distribution. Adapted from [20] and [18], respectively

quantum-mechanical systems. The three-dimensional confinement of a single electron by surrounding water molecules gives rise to discrete quantum states. This closely resembles a quantum dot [12], an entity that has raised immense attention in recent years as a model system for atoms [13, 14, 15] and molecules [16] and holds great promise for use in optical devices [17]. However, the important difference between a quantum dot and a hydrated electron is that in the former case, the potential giving rise to the confinement is static, whereas in the latter case, it changes rapidly in time due to dynamic fluctuations in the liquid surrounding.

Numerous computational studies have been performed to investigate the quantum-mechanical status of the hydrated electron and the microscopic structure of its surroundings. It became clear that the first shell of water around the electron is composed of approximately six molecules with their OH-bonds oriented toward the electron [18, 19] (Fig. 1). A similar result was also found in electron-spin echo measurement on an electron solvated in glassy water [20, 21].

When molecular dynamics simulations matured, they succeeded in reproducing the general features of the absorption band shape, such as its breadth and asymmetry [19, 22, 23, 24, 25], although the transition energies were somewhat overestimated (Fig. 2a). After being shifted toward lower frequencies, the simulated spectra closely resembled the experimental data. According to extensive computational modeling performed by the group of *Rossky* [19,23,24], the absorption spectrum of hydrated electrons is primarily caused by a strongly allowed transition from a roughly spherical localized *s*-like ground state to a triple of *p*-like states that are also bound and localized

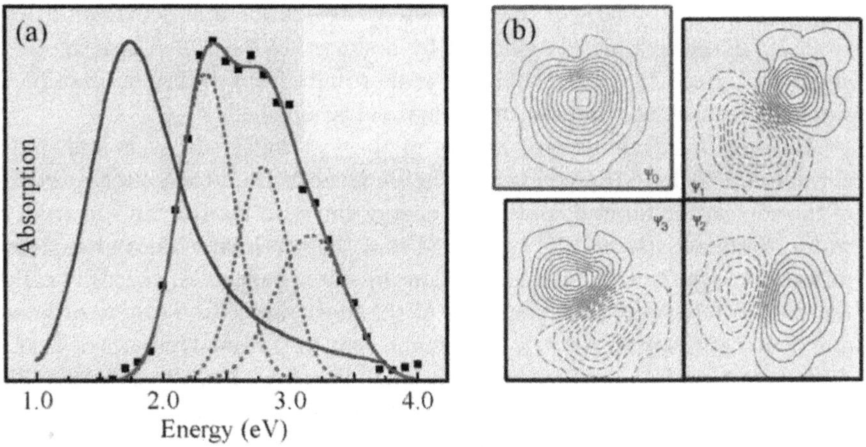

Fig. 2. Overview of the lowest electronic transition in the hydrated electron. (**a**) Absorption spectrum. The smooth *solid curve* shows experimentally measured absorption at room temperature. *Squares* depict the result of quantum molecular simulations (Adapted from [34]). The *dashed curves* correspond to the individual absorption components originating from three nondegenerate s–p transitions. (**b**) Electronic wave function plots for typical ground, s-like, state and lowest three excited, p-like, states. (Reproduced from [24])

(Fig. 2b). The existence of the s-state indicates that, on average, the potential energy well created by the molecules surrounding the electron is close to spherical. However, because of the asymmetries caused by the dynamic nature of liquid water, the potential energy surface does not have a perfect spherical shape, which results in three nondegenerate p-states.

The question of homogeneous vs. inhomogeneous broadening in the optical absorption spectrum of the excess electron in water and other fluids, as well as the explanation of its asymmetry and extraordinary spectral width, has remained a standing problem for over three decades. Numerous attempts have been made to fit the experimental data by various lineshapes [26,27,28,29,30] and superposition of lines [31, 32, 33]. In computer simulations [23], by imposing an ordering of the p-states according to their energy, the resulting absorption contour was decomposed into a superposition of the contributions from separate s–p transitions. Each of the three bands obtained is claimed to be substantially inhomogeneously broadened by different structures of the solvent surroundings. The energy splitting between two adjacent s–p transitions found in these simulations is $\sim 0.4\,\text{eV}$, which constitutes nearly the width of the bands associated with each separate transition.

The dynamic behavior of the hydrated electron, i.e., the energy relaxation after an instantaneous $s \to p$ excitation, has been thoroughly modeled as well [18, 19, 25, 34, 35, 36, 37, 38, 39, 40, 41]. All computer simulations predict that the solvation dynamics are essentially bimodal. The initial decay is responsible for $\sim$ 50% of the total energy relaxation and occurs on a 10 fs to 25 fs timescale which is followed by a slower 130 fs to 250 fs decay. It is generally agreed that the latter timescale results from diffusional motion of water molecules into and out of the first solvation shell.

However, the microscopic nature of the fast initial decay is still under considerable debate. By comparing the frequencies from the power spectrum of the correlation function of the s–p energy gap with the Raman spectrum of water, *Staib* and *Borgis* [25] concluded that the accelerated decay was determined by coupling to hindered rotations of water molecules, generally called *librations.* Based on the dependence of the most rapid timescale in ordinary and deuterated water, *Barnett* et al. found that the initial dynamics of the hydrated electron is caused by *free* rotational diffusion of water molecules [35]. Conversely, *Berg* [36] and the group of *Schwartz* and *Rossky* [37, 39, 40] revealed no isotope effect in the first 25 fs of the solvation dynamics of a hydrated electron. They accordingly inferred that the origin of the initial dynamics is translational. *Park* et al. concluded that the motions of water molecules in the first solvation shell of the hydrated electron are dominantly rotational through repulsion of hydrogen-bonded hydrogen atoms and attraction of dangling hydrogen atoms [18].

The last decade has witnessed numerous studies of the hydrated electron with an $\sim$ 200 fs time resolution [31, 42, 43, 44, 45, 46, 47, 48, 49, 50]. The femtosecond time-resolved studies of hydrated electrons were pioneered by *Migus* et al. [47]. Electrons were generated by multiphoton ionization of neat water and studied by transient absorption of a supercontinuum probe. Later, similar experiments were carried out by several groups evidencing the importance of geminate recombination and pump–probe cross-phase modulation in the recorded transients [31, 43, 45, 46, 48, 49, 50].

In another approach, an already equilibrated hydrated electron is excited from the ground s-state to the p-state using a short pulse, and the resulting solvation dynamics is probed as a function of time with another delayed pulse. Following this route, the group of *Barbara* found decays of $\sim$ 300 fs and $\sim$1.1 ps in a pump–probe experiment [51]. The results were explained using a three-state model, where the fastest decay is caused by the relaxation down to the ground state, giving rise to a not yet equilibrated "hot" ground state. Recently, *Assel* et al. refined this model by including excited-state solvation that took place before relaxation back to the ground state [31].

Another unsolved question surrounding the hydrated electron concerns its excited-state lifetime. Despite the illusive simplicity of the system, the rate of electronic relaxation remains a heavily debated issue. So far MD simulations [41, 52, 53, 54, 55, 56] have not been able to clarify the problem. The predicted lifetime ranges from $\sim$ 120 fs to 1.1 ps, depending on the particular choice of the electron–water pseudopotential and on the model chosen for water molecules. For instance, the initial assumption of "rigid" water molecules and an electron–water pseudopotential [52] led to the prediction of a $\sim$ 1 ps lifetime of the $2p$-state [56]. The implementation of the "flexible" water potential reduced the radiationless relaxation times to ~ 120 fs and ~ 220 fs for an electron solvated in H_2O and D_2O, respectively [53]. Assuming a modified electron–water potential [55], the lifetime of the $2p$-state was inferred at 1.1 ps, with electronic relaxation preceded by a 240 fs solvation process (i.e., rearrangement of the surrounding of the excited state) [41].

Results from femtosecond spectroscopy [31, 42, 44, 51, 57, 58, 59, 60] also remained inconclusive. According to one scenario [31,42,44,57,58], the excited state relaxes with a time constant of ~ 200 fs, thereafter cooling back to the hot ground state at a 1 ps rate. In another scenario [59, 60], $2p$-state solvation takes place on a 300 fs timescale, whereas the excited state decays predominantly with a time constant of 0.7 ps to 1.2 ps.

More recent experiments with a substantially improved time resolution revealed that the early solvation dynamics occurs on a much shorter timescale [59,60,61,62]. In these experiments, it was found that the ~ 300 fs decay is preceded by dynamics on a timescale of less than 50 fs. Interestingly, both experiments with 35 fs pulses [59,60] and with 13 fs pulses [61,62] pointed to the librational nature of this initial solvation dynamics. It became evident, however, that even shorter pulses are required to match the large spectral width of hydrated electron absorption. The rapid progress of state-of-the-art laser technology in the last years has made such pulses available for spectroscopic experiments [63, 64, 65, 66, 67, 68].

In this Chapter, we discuss the latest results on the early dynamics of the hydrated electron obtained with an unprecedented time resolution provided by 5 fs optical pulses. Based on a comparison of photon echo signals from hydrated electrons and from water alone, we derive a 1.7 fs pure dephasing time of hydrated electrons. This value is fully consistent with the lineshape of the absorption spectrum which, it is shown, is overwhelmingly homogeneously broadened. We also present the results of femtosecond frequency-resolved, pump–probe experiments on the hydrated electron, photo-excited from its equilibrated ground state. Our data show that the depopulation of the excited state of the hydrated electron proceeds on a 50 fs timescale, whereas in heavy water, this lifetime increases by a factor $\sqrt{2}$. Finally, we propose a model for a hydrated electron dynamics that successfully accounts for the observed experimental features.

The use of extremely short 5 fs pulses provides obvious advantages to a spectroscopic experiment. In addition to the very high temporal resolution, the broad bandwidth associated with short pulses allows covering an impressive spectral window at once. On the other hand, any experiment with 5 fs pulses is a daunting task. Besides the trivial experimental nuisances such as pulse lengthening during its propagation before and inside the sample because of group velocity dispersion, there are also more fundamental problems to be addressed.

The conventional description of nonlinear signals applicable to multi-cycle pulses becomes questionable for pulses that consist merely of a couple of optical fringes. Clearly, in the latter case, the conventionally employed slowly varying envelope approximation [69, 70, 71, 72], implying that the change of the pulse amplitude in the duration of an optical cycle is negligible compared to the magnitude of the amplitude itself, can no longer be maintained. Furthermore, the phase-matching bandwidth [73, 74] that is limited due to dispersion in the nonlinear medium rapidly gains importance with the broadening of the pulse spectrum. Another point of serious concern is the frequency-dependent variation in the sensitivity of the photodetector employed to register the signal generated in a nonlinear process. In combination, the above listed features of an experiment with broadband pulses result in what is known as a spectral-filter effect [75, 76, 77]. On top of that, artificial lengthening of the observed time dependencies is a direct consequence of the noncollinear geometry employed in spectroscopic experiments. Evidently, if a portion of the signal field is filtered out in frequency and/or the signal is artificially "blurred" in time, this might crucially influence the measured data [78] and subsequently lead to its erroneous interpretation.

To address these issues, we present a comprehensive theoretical analysis in which the frequency- and time-domain formalism of ultrafast nonlinear spectroscopy are thoroughly reexamined. Complete expressions, valid even for single-cycle-pulse applications, are derived for a nonlinear signal in the frequency and time domains. Among others, we show that one does not need to invoke the slowly varying envelope approximation in its aforementioned meaning, i.e., rejecting derivatives of the time-domain electric field. We also assert that the influence of geometric delay smearing does not introduce a significant distortion of the observed traces, provided that the geometry is carefully optimized.

2 Ultrafast Spectroscopy with 5 fs Resolution

In this section, we derive the master equation that describes spectroscopic observables and is valid even for single-cycle optical pulses. Using the frequency-domain framework, we consistently include the effects of phase matching, dispersive pulse broadening, dispersion of third-order nonlinearity, and frequency dependence of the resulting nonlinear signal.

The frequency-domain formalism is then recast in the time-domain that is conventionally used in the description of transient spectroscopy with short light pulses. We subsequently show that, despite the ultrabroad bandwidth associated with 5 fs pulses, the effect of spectral filtering can be disregarded under proper experimental conditions. This allows a straightforward transition from the frequency-domain representation to the time-domain one, with the latter offering a simpler formalism. Most importantly, this simplifies the experimental task by lifting the otherwise unavoidable necessity to frequency-resolve the signals generated by ultrabroadband pulses.

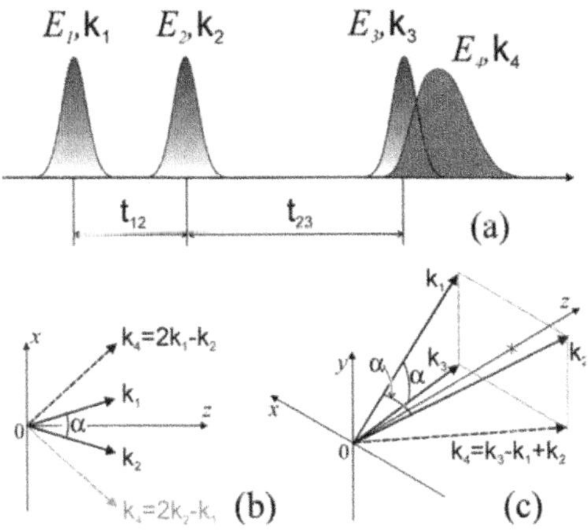

Fig. 3. (**a**) Schematic representation of the pulse sequence in a three-pulse nonlinear spectroscopic experiment. $E_{1,2,3}$ are the input fields, and E_4 is the signal due to the third-order nonlinear process. t_{12} and t_{23} are the delays between pulses E_1–E_2 and E_2–E_3, respectively. (**b**) Self-diffraction (two-pulse photon echo) configuration. Two conjugated signals are emitted in the directions k_4 and k'_4. (**c**) Transient grating in a "box" geometry

2.1 The Formalism

We consider the case of noncollinear geometry where three beams $E_{1,2,3}(z,t)$ intersect at small angles in a nonlinear medium (Fig. 3). The corresponding configurations of two types of noncollinear third-order experiments are depicted in Fig. 3b,c. The self-diffraction (SD, Fig. 1b) and transient grating (TG, Fig. 1c) signals are equivalent to the two- and three-pulse stimulated photon echo signals originating from the systems with phase memory. Focusing conditions of the beams are chosen such that the confocal parameter [73] and the longitudinal beam overlap of the fundamental beams are considerably longer than the interaction length. For simplicity, we assume that neither of the fields is absorbed in the nonlinear medium and that the nonlinear response is purely third order. The input beams induce then third-order nonlinear polarization $P^{\langle 3\rangle}(z,t)$ that serves as a source for the signal field $E_4(z,t)$. By writing both $P^{\langle 3\rangle}(z,t)$ and $E_4(z,t)$ as Fourier superpositions of monochromatic waves, one obtains an equation that governs propagation of the signal wave in the $+z$-direction inside the nonlinear medium [79]:

$$\frac{\partial^2}{\partial z^2}\tilde{E}_4(z,\Omega) + k_{4z}^2(\Omega)\tilde{E}_4(z,\Omega) = -\mu_0\Omega^2\tilde{P}^{\langle 3\rangle}(z,\Omega), \tag{1}$$

where $\tilde{E}_4(z,\Omega)$ and $\tilde{P}^{\langle 3\rangle}(z,\Omega)$ are Fourier transforms of $E_4(z,t)$ and $P^{\langle 3\rangle}(z,t)$ respectively; Ω is the frequency; and $k_{4z}(\Omega)$ is the projection of the wave vector of signal field $k_4^2(\Omega) = \Omega^2\varepsilon_0\mu_0\tilde{\varepsilon}(\Omega)$ onto the z-axis, with $\tilde{\varepsilon}(\Omega)$ the Fourier transform of the complex relative permittivity $\varepsilon(t)$.

To simplify the left part of (1), we write the signal field as a plane wave propagating along the z-axis:

$$\tilde{E}_4(z,\Omega) = \tilde{E}_4(z,\Omega)\mathrm{e}^{\mathrm{i}k_{4z}(\Omega)z}\,, \tag{2}$$

and substitute it in (1):

$$2\mathrm{i}k_{4z}(\Omega)\frac{\partial}{\partial z}\tilde{E}_4(z,\Omega) + \frac{\partial^2}{\partial z^2}\tilde{E}_4(z,\Omega) = -\mu_0\Omega^2\tilde{P}^{\langle 3\rangle}(z,\Omega)\mathrm{e}^{-\mathrm{i}k_{4z}(\Omega)z}\,. \tag{3}$$

We now neglect the second-order derivative over the signal electric field [73, 79]:

$$\left|\frac{\partial}{\partial z}\tilde{E}_4(z,\Omega)\right| \ll \left|2k_{4z}(\Omega)\tilde{E}_4(z,\Omega)\right|\,, \tag{4}$$

which is known as the slowly varying amplitude approximation. Note that the use of the time-domain description of the signal wave propagation results in a second-order differential equation, similar in structure to (3). Unlike (3), though, simplification of the time-domain expression requires rejection of the second-order *temporal* derivative of the envelope, i.e.,

$$\left|\frac{\partial^2}{\partial t^2}E(t)\right| \ll \frac{4\pi}{T_{\mathrm{per}}}\left|\frac{\partial}{\partial z}E(t)\right|\,,$$

where T_{per} is the characteristic period of light oscillation. Such a move implies assumption of the slow envelope variation as a function of time. This condition is not fulfilled for pulses that carry only a few cycles, since the change of the envelope within one optical period is comparable to the magnitude of the envelope itself. *Brabec* and *Krausz* [70], who explored the time-domain approach for the propagation of nearly monocycle pulses, found that rejection of the second-order derivative term is warranted when the phase and the group velocities of light are close to each other. To this point we notice that the application of nonequality (4) to the frequency-domain (3) does not require any assumptions on the change of the temporal envelope altogether. Therefore, nonequality (3) is safe to apply even to monocyclic pulses, provided there is no appreciable linear absorption at lengths comparable to the wavelength:

$$\tilde{E}_4(L,\Omega) = \mathrm{i}\frac{c\mu_0\Omega}{2n_4(\Omega)}\int_0^L \tilde{P}^{\langle 3\rangle}(z,\Omega)\mathrm{e}^{-k_{4z}(\Omega)z}\,\mathrm{d}z\,, \tag{5}$$

where $n_4(\Omega) = \sqrt{\tilde{\varepsilon}(\Omega)}$ is the refractive index for the signal wave and L is the thickness of the nonlinear medium.

To calculate the third-order dielectric polarization induced at frequency Ω by fundamental fields, we should sum over all possible permutations of fundamental frequencies weighted according to the third-order susceptibility [80]:

$$\tilde{P}^{\langle 3\rangle}(z,\Omega) = \iint \mathrm{d}\omega' \, \mathrm{d}\omega'' \tilde{\chi}^{\langle 3\rangle} \left(\omega_{\mathrm{eg}} - \omega', \omega' - \omega'', -\omega_{\mathrm{eg}} + \Omega\right)$$
$$\cdot \tilde{E}_1^*(z,\omega')\tilde{E}_2(z,\omega'')\tilde{E}_3(z,\Omega - \omega'' + \omega')$$
$$\cdot \exp\{\mathrm{i}\left[-k_{1z}(\omega') + k_{2z}(\omega'') + k_{3z}(\Omega - \omega'' + \omega'')\right]z\}$$
$$\cdot \exp\left[-\mathrm{i}\omega'' t_{12} - \mathrm{i}(\Omega - \omega'' + \omega')(t_{12} + t_{23})\right], \qquad (6)$$

where $\tilde{E}_i(z,\omega'')$ is a Fourier transform of $E_i(z,t)$ with their k-vectors represented analogously to (2) and t_{12} and t_{23} are delays between pulses $E_1 - E_2$ and $E_2 - E_3$, respectively. In the SD case (Fig. 1b), t_{23} is set to zero, and t_{12} is scanned, whereas in the TG experiment (Fig. 1c), $t_{12} = 0$ and t_{23} is scanned. Representation of the frequency-dependent nonlinear susceptibility $\tilde{\chi}^{\langle 3\rangle}\left(\omega_{\mathrm{eg}} - \omega', \omega' - \omega'', -\omega_{\mathrm{eg}} + \Omega\right)$ is based on the interaction of the input fields with an electronic transition with the frequency ω_{eg}. The inclusion of the third-order susceptibility due to Raman and two-photon processes is also straightforward. The particular expression for $\tilde{\chi}^{\langle 3\rangle}$ will be discussed below.

To calculate the signal field, one should integrate the signal intensity over the longitudinal coordinate z, according to (5). This can be performed analytically for a low-efficiency nonlinear process ($E_{1,2,3} = \mathrm{const}$), as is usually the case in spectroscopic applications:

$$\tilde{E}_4(\Omega, t_{12}, t_{23}) = \mathrm{i}\frac{c\mu_0 \Omega L}{2n_4(\Omega)} \iint \mathrm{d}\omega' \, \mathrm{d}\omega'' \tilde{\chi}^{\langle 3\rangle} \left(\omega_{\mathrm{eg}} - \omega', \omega' - \omega'', -\omega_{\mathrm{eg}} + \Omega\right)$$
$$\cdot \tilde{E}_1^*(\omega')\tilde{E}_2(\omega'')\tilde{E}_3(\Omega - \omega'' + \omega') \operatorname{sinc}\left[\Delta k_z(\Omega,\omega',\omega'')\frac{L}{2}\right]$$
$$\cdot \exp\left[\mathrm{i}\Delta k_z(\Omega,\omega',\omega'')\frac{L}{2} - \mathrm{i}\omega'' t_{12} - \mathrm{i}(\Omega - \omega'' + \omega')(t_{12} + t_{23})\right]. \qquad (7)$$

The phase mismatch,

$$\Delta k_z(\Omega,\omega',\omega'') = -k_{1z}(\omega') + k_{2z}(\omega'') + k_{3z}(\Omega - \omega'' + \omega'') - k_{4z}(\Omega), \qquad (8)$$

should be calculated for each particular geometry given in Fig. 1b,c.

Equation (7), which will be extensively used throughout this chapter, is valid even for single-cycle optical pulses. The frequency representation allows us to include, in a self-consistent way, dispersive broadening of interacting pulses and frequency dependence of the nonlinear susceptibility. Besides, we avoid the introduction of the carrier frequency [81] the definition of which becomes confusing for few-cycle pulses. We also draw the reader's attention to the Ω-term in front of the integral that follows directly from the Maxwell equations and reflects the fact that higher frequencies are generated more efficiently. It is this term that is responsible for the effect of self-steepening of pulses propagating in optical fibers [82].

The total spectrally resolved signal registered by a quadratic detector is written as

$$\tilde{I}_4(\Omega, t_{12}, t_{23}) = \varepsilon_0 \frac{n_4(\Omega)Q(\Omega)}{c} \left|\tilde{E}_4(\Omega, t_{12}, t_{23})\right|^2, \qquad (9)$$

where $Q(\Omega)$ is the spectral sensitivity of a monochromator–detector combination.

From the point of view of practical application of 5 fs pulses, we now quantify the differences between the complete frequency-resolved signals of TG and SD computed according to (7–9) with $\tilde{\chi}^{(3)} = \text{const}$ and the ideal frequency-resolved TG and SD signals for an instantaneous nonlinear response are [83]

$$I_4^{\text{ideal}}(\Omega, t_{12}, t_{23}) = \left|E_4^{\text{ideal}}(\Omega, t_{12}, t_{23})\right|^2,$$

$$E_4^{\text{ideal}}(\Omega, t_{12}, t_{23}) = \iint \mathrm{d}\omega' \, \mathrm{d}\omega'' \tilde{E}_1^*(z, \omega')\tilde{E}_2(z, \omega'')\tilde{E}_3(z, \Omega - \omega'' + \omega') \cdot \exp\left[-\mathrm{i}\omega'' t_{12} + \mathrm{i}(\omega'' - \omega')(t_{12} + t_{23})\right]. \qquad (10)$$

Comparison of the respective complete and ideal signals provides us with information on the spectral filter effect – a combined influence of the spectral variations in the generation efficiency of the signal field and in its detection. To simulate our experimental conditions, in the calculation of complete SD and TG traces, we included the dispersive properties of a 100 μm layer of water [84, 85] and the impact of the noncollinear beam geometry on the phase mismatch given by (8). The ideal frequency-resolved traces were calculated according to (10). The spectral filters for the SD and TG cases, obtained as the ratios of complete (7–9) versus ideal (10) signals, are presented in Fig. 4. The dashed and dotted curves correspond to TG and SD, respectively, for a flat spectral response of the detector ($Q(\Omega) = \text{const}$). Apparently, both filters are dominated by the Ω^2-dependence that originates from the Ω-term in (5). The curvature of the SD filter compared with that of the TG is somewhat steeper. This reflects the fact that the phase mismatch for SD is greater since SD is intrinsically non-phase-matched geometry [83].

Taking into account the typical real spectral sensitivity of a silicon photodiode, $Q(\Omega)$ (dash-dotted curve in Fig. 4) results in the overall spectral filter for TG depicted by a solid curve. Note that the overall spectral filtering effect is nearly frequency-independent throughout most of the spectrum of a 5 fs pulse (shaded contour in Fig. 4) because the photodetector sensitivity balances the Ω^2-dependence. Therefore, we can disregard the effect of spectral filtering when it is counterweighed by the spectral response of a properly chosen detector. This is an important conclusion for practical nonlinear spectroscopy with 5 fs pulses since it justifies the use of less cumbersome spectrally unresolved detection of TG and SD signals.

Now we demonstrate how to arrive at the conventionally used time-domain description of ultrafast spectroscopy [74]. As we already pointed out,

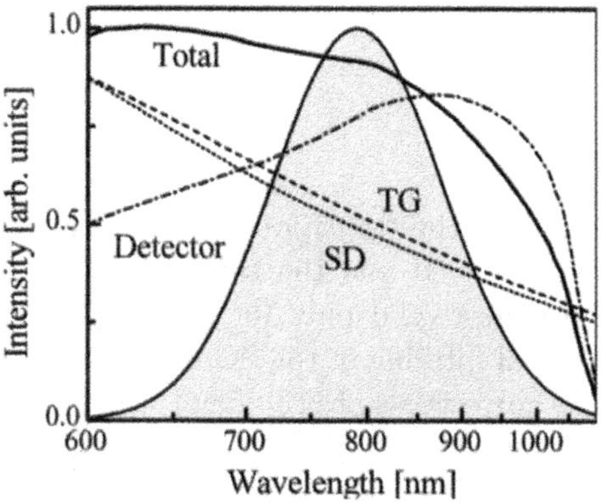

Fig. 4. Spectral filters for two configurations of the photon echo in water. *Shaded contour* represents the spectrum of ideal 5 fs pulses. The spectral filter calculated for self-diffraction is shown by a *dotted line*, and the filter for transient grating is presented by a *dashed line*. The *dash-dotted line* depicts the typical spectral sensitivity of a silicon light detector, $Q(\lambda)$. The spectral filter for transient grating corrected by $Q(\lambda)$ is given by a *solid curve*. The thickness of the water layer is taken as 100 μm, and the intersection angles of the beams are 4°. Note that the overall spectral filter (*solid curve*) is nearly flat up to 900 nm due to the photodetector sensitivity balancing off the more efficient generation of the nonlinear signal at higher frequencies

to match the information obtained in a SD or TG experiment, (9) should be integrated over all frequency components to obtain the total energy of the signal field detected by a quadratic detector. According to Parseval's theorem, the amount of energy carried by the signal is the same whether we compute it in the time domain or in the frequency domain. Therefore, the following formula is a time-domain expression for the same signal:

$$S(t_{12}, t_{23}) = \varepsilon_0 \frac{n(\omega_4)}{2\pi c} \int_{-\infty}^{\infty} \mathrm{d}t \left|E_4(t, t_{12}, t_{23})\right|^2 , \tag{11}$$

where

$$\begin{aligned} E_4(t, t_{12}, t_{23}) = {} & \frac{c\mu_0 L}{2n_4(\omega_4)} \iiint_0^{\infty} \mathrm{d}t_1\, \mathrm{d}t_2\, \mathrm{d}t_3 R(t_1, t_2, t_3) \\ & \cdot E_1^*(t - t_3 - t_2 - t_1) E_2(t - t_{12} - t_3 - t_2) E_3(t - t_{23} - t_{12} - t_3) \\ & \cdot \exp\left[-\mathrm{i}\omega_{\mathrm{eg}}(t_3 - t_2) + \mathrm{i}(\omega_3 + \omega_2 - \omega_1)t_3 + \mathrm{i}(\omega_2 - \omega_1)t_2 - \mathrm{i}\omega_1 t_1\right] \\ & \cdot \exp\left[+\mathrm{i}\omega_2 t_{12} + \mathrm{i}\omega_3(t_{12} + t_{23})\right] \end{aligned} \tag{12}$$

and the so-called nonlinear response function is introduced as a Fourier transform of the nonlinear susceptibility [74]:

$$R(t_1, t_2, t_3) = \iiint \tilde{\chi}^{(3)}(\omega_1, \omega_2, \omega_3) \mathrm{e}^{(-\mathrm{i}\omega_1 t_1 - \mathrm{i}\omega_2 t_2 - \mathrm{i}\omega_3 t_3)}\, \mathrm{d}\omega_1\, \mathrm{d}\omega_2\, \mathrm{d}\omega_3 \,. \tag{13}$$

In (12), we also extracted the oscillations of electrical fields at the optical frequency ω_i:

$$E_i(t) = E_i(t)\mathrm{e}^{-\mathrm{i}\omega_i t}\,. \tag{14}$$

Note that for ultrabroadband optical pulses, the transition between the frequency-domain description formulated by (7–9) and the time-domain representation summarized by (11–13) becomes valid only for a flat spectral filter. In other situations when the spectral filtering of the SG or TG signals does occur (regardless of its cause), the correctness of (12) is not warranted, and one must use the more general (7–9). The distortion of measured time dependencies due to spectral filtering is well known from the example of second-harmonic autocorrelation in $\tilde{\chi}^{\langle 2\rangle}$ crystals with finite thickness [77].

Equation (13) provides the link between the nonlinear response function $R(t_1,t_2,t_3)$ and third-order susceptibility $\tilde{\chi}^{\langle 3\rangle}$. For the former, an extensive formalism of non-Markovian dynamics, based on pathway propagation in Lionville-space [74], has been developed. Here, we restrict ourselves to a simple model of a homogeneously broadened two-level system. In this case, the nonlinear response function is given as

$$R(t_1,t_2,t_3) = \frac{\mu_{\mathrm{eg}}^4 N}{\hbar^4}\exp\left(-\frac{t_1+t_3}{T_2} - \frac{t_2}{T_1}\right), \tag{15}$$

where μ_{eg} is the transition dipole moment; N is concentration; T_1 and T_2 are the population-relaxation and dephasing times, respectively; and

$$T_2^{-1} = (T_2^*)^{-1} + (2T_1)^{-1}\,, \tag{16}$$

where T_2^* is the pure dephasing time. Fourier transformation of (15) yields a well-known result [86]:

$$\begin{aligned}\tilde{\chi}^{\langle 3\rangle}(\omega_{\mathrm{eg}}-\omega',\omega'-\omega'',-\omega_{\mathrm{eg}}+\Omega) &= -\mathrm{i}\frac{\mu_{\mathrm{eg}}^4 N}{\hbar^4}\frac{1}{T_1^{-1}-\mathrm{i}(\omega''-\omega')}\\ &\cdot\left(\frac{1}{T_2^{-1}-\mathrm{i}(\omega_{\mathrm{eg}}-\omega')}+\frac{1}{T_2^{-1}-\mathrm{i}(\omega''-\omega_{\mathrm{eg}})}\right)\\ &\cdot\frac{1}{T_2^{-1}+\mathrm{i}(\omega_{\mathrm{eg}}-\Omega-\omega''+\omega')}.\end{aligned} \tag{17}$$

The second sum term in square brackets in (17) is included to account for the fact that $\tilde{\chi}^{\langle 3\rangle}$ possesses symmetry with respect to ω' and ω'' and the total expression of $\tilde{\chi}^{\langle 3\rangle}$ is the sum of all frequency permutations [73, 79]. Third-order susceptibilities for different four-photon processes like Raman scattering or two-photon absorption can be calculated in a similar fashion. The two-level system can also be dressed in a vibrational manifold to account for coherent excitation of several Frank–Condon transitions [74].

2.2 Ultimate Temporal Resolution of SD and TG Experiments

In this section, we address geometric smearing – the effect deteriorates the temporal resolution of a nonlinear spectroscopic experiment as a direct consequence of employing noncollinear beam geometry. This type of distortion originates from the fact that in a beam inclined at an angle to a plane, different transverse components of a pulse travel different distances before reaching the plane. This means that a fixed delay between two pulses propagating in two intersecting beams changes into a range of delays across the waist of the beams in the intersecting region. The very same idea of yielding a range of delays simultaneously is used in single-shot pulse autocorrelation techniques [87].

The "delay blurring" described above can be of serious concern when dealing with the laser pulses that have duration shorter than 10 fs. This issue has been addressed previously in connection with the temporal resolution of a noncollinear pulse duration measurement via second-harmonic generation [75, 76]. Analogously to [75], here, we evaluate the influence of geometric smearing on the width of self-diffraction and transient grating traces.

For arbitrary pulses and beam profiles, the shape of the resulting traces should be computed numerically by integrating (9) over each transverse component of the beam. For linearly chirped Gaussian pulses with a Gaussian spatial profile, however, these traces can be calculated analytically. Assuming that the nonlinear response of the medium is instantaneous, one can calculate from (10) that the ideal SD or TG trace has a Gaussian intensity profile in time. Its width, τ_0, is broader than the pulse duration by a factor of $\sqrt{3/2}$. The width of the actual signal, τ_{meas}, that has been stretched by geometric smearing, can be expressed by

$$\tau_{\mathrm{meas}}^2 = \tau_0^2 + \beta\delta^2 \,, \tag{18}$$

where β is a scaling constant dependent on the beam geometry employed and δ is the effective delay smearing given by

$$\delta = \frac{\alpha d_{\mathrm{f}}}{2c} \,. \tag{19}$$

Here, d_{f} is the beam diameter in the focal plane, and α is a small intersection angle between the interacting beams (Fig. 1b,c). As reported in [75], the lowest value of δ for Gaussian pulses and beams is 0.4 fs if the central wavelength of the pulse is 800 nm. For beam profiles other than Gaussian, the value of δ becomes larger.

For self-diffraction, the constant β equals 4/3, and for transient grating in the "box" beam arrangement, β takes the value of $\approx 5/3$. The temporal resolution of a self-diffraction experiment is somewhat higher compared to that of transient grating. This is explained by the fact that the smearing in transient grating takes place in the xz- and yz-planes (Fig. 1c) simultaneously. In any case, for intersection angles smaller than 10°, the lengthening of the detected signal does not exceed 10%.

Therefore, the effect of geometric smearing on the generated signals is insignificant even for experiments with pulses as short as 5 fs, provided the intersection angle is kept sufficiently small and the beams are properly focused.

3 Experimental

3.1 Femtosecond Laser System

The femtosecond spectrometer is based on a home-built cavity-dumped Ti:sapphire laser, that has been described in detail elsewhere [64]. A schematic of the setup is presented in Fig. 5. Briefly, the cavity-dumped laser is a standard Ti:sapphire oscillator that incorporates a Bragg cell inside the cavity. This allows us to produce 15 fs pulses at the central wavelength of 790 nm at a desired repetition rate. The energy of the dumped pulses reaches up to 35 nJ and is adjustable by setting the level of rf power applied to the Bragg cell.

To provide adequate time resolution for the study of the ultrafast dynamics of a hydrated electron and broaden the spectral window of our measurements, the output of the cavity-dumped laser is externally compressed to a pulse duration below 5 fs [64]. The 15 fs laser pulses, precompressed by a pair of fused silica prisms, are injected into a single-mode quartz fiber through a microscope objective lens. The white-light continuum resulting from the combined action of self-phase modulation and dispersion in the fiber core is collimated by an off-axis parabolic mirror to avoid chromatic aberrations and the bulk dispersion of a collimating lens. A portion of the white light in the spectral range of 580 nm to 1060 nm (shaded contour in Fig. 6) is then compressed in a state-of-the-art three-stage pulse compressor that includes a pair of 45° quartz prisms, specially designed chirped dielectric mirrors, and thin-film dielectric Gires–Tournois interferometers.

Immediately before performing photon echo spectroscopy on hydrated electrons, the compressor is adjusted to yield the shortest duration of pulses, which are characterized by second-harmonic frequency-resolved gating (SHG FROG) [75]. Because dispersive broadening easily affects 5 fs pulses even as they propagate through air, the FROG characterization is carried out directly at the location of the spectroscopic sample by replacing it with a very thin (~ 10 μm) second-harmonic BBO crystal (EKSMA).

3.2 Transient Grating and Photon Echo Experiments

The experimental arrangement used for photon echo spectroscopy is depicted in Fig. 5. The beam carrying 6 nJ, 5 fs pulses is split into three channels of approximately equal intensity by the 0.5 mm thick beam splitters BS1 and BS2 (pulses E_1, E_2, and E_3 in Fig. 5). Compensating plates (CP) of

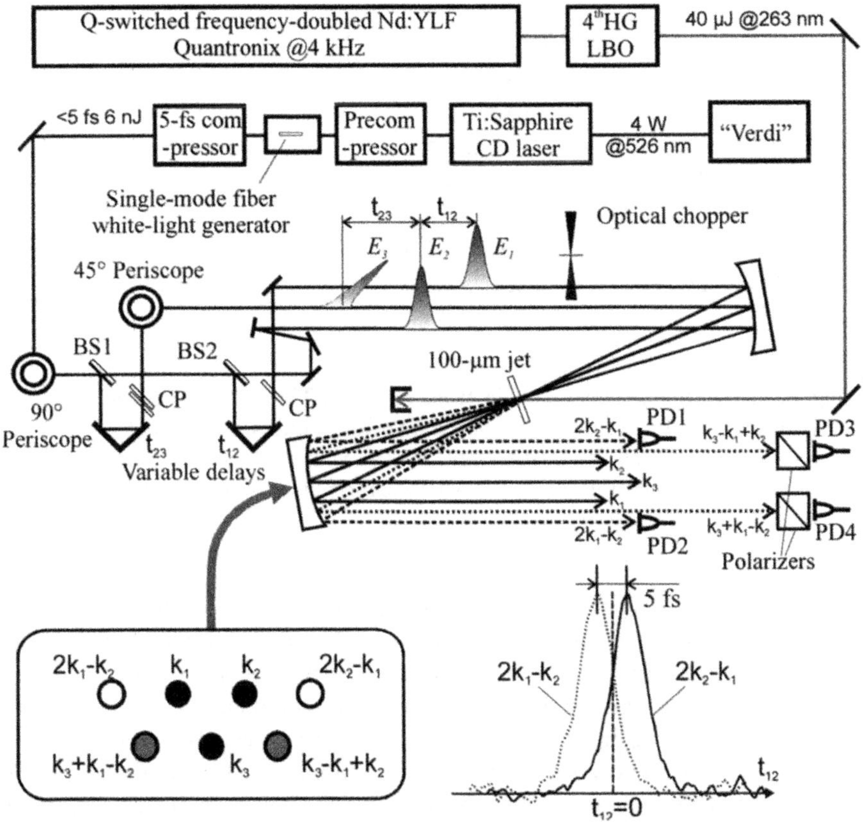

Fig. 5. Schematic of the femtosecond photon echo spectrometer. BS1 is an $R = 30\%$ beam splitter. BS2 is a 50% beam splitter. CPs are compensating plates, and PD1–4 are silicon light detectors. E_1, E_2, and E_3 are femtosecond excitation pulses. t_{12} and t_{23} are the delays between pulses E_1–E_2, and E_2–E_3, respectively. The *bottom-left inset* shows the image of the signal and excitation beam arrangement on the recollimating mirror. The *bottom-right inset* shows the profiles of two-pulse photon echoes measured in two conjugate directions by PD1 and PD2

the same thickness are inserted into the beams to equalize dispersion in all three channels. To minimize distortions in the pulse duration and preserve its spectral content, silver mirrors overcoated by a thin protective layer are used throughout. To match the reflectivity of the beam splitters, the initial horizontal polarization of the laser beam is turned 90° by a mirror periscope. The optical polarization of one of pulses, E_3, is further turned by 45° with respect to the polarization of the pulses, E_1 and E_2, by another periscope to facilitate photo echo measurements in the parallel and orthogonal polarization directions. Two independent optical delay lines, t_{12} and t_{23}, are employed to fabricate desired sequences of the three pulses.

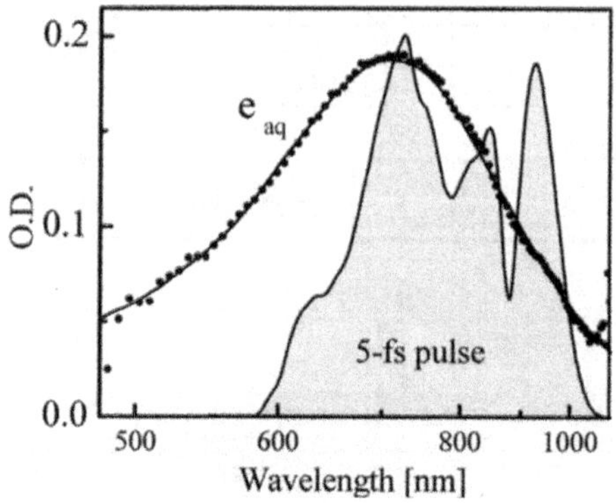

Fig. 6. Absorption spectrum of electrons in water. The *solid line* is the absorption of hydrated electrons produced by electron beams in bulk water (adopted from [32]). The *solid dots* are measured with a femtosecond white-light continuum upon photoionization of potassium ferrocyanide with 263 nm pulses. The *shaded contour* shows the spectrum of the 5 fs pulses

The beams are focused into the sample – a 100 μm water jet, in which the hydrated electrons are generated – and recollimated behind it by spherical mirrors with the radii of curvature $R = -250$ mm. A small incidence angle on the spherical mirror is chosen to prevent astigmatism of the beams within the intersecting region. The intersecting angles between the incident beams are kept at $\sim 4°$. The waist of the focused beam is ~ 30 μm in diameter. The phase-matching geometry of the laser beams is explained in the bottom-left inset in Fig. 5, presenting the enlarged image of the beam configuration on the recollimating mirror. The advantages of employing such a beam arrangement have been discussed previously [88].

The two- and three-pulse stimulated echo signals are detected simultaneously behind the sample by silicon photodiodes PD1–4 equipped with built-in amplifiers. The photodiode signals are processed by lock-in amplifiers, digitized, and stored in the computer memory.

The ability to determine the exact overlap of pulses E_1–E_3 in time with a high degree of precision presents a considerable experimental challenge and is vital for the photon echo measurements with 5 fs pulses that occupy merely 1.5 μm in space. Fortunately, the overall symmetry of the configuration employed makes it possible to find $t_{12} = 0$ and $t_{23} = 0$ accurately without employing additional means.

The two-pulse photon echo traces are measured as a function of the delay between the pulses E_1 and E_2. A typical result of a t_{12} scan is depicted in the bottom-right inset in Fig. 5. Here, the solid and the dashed curves represent the contours obtained in the two conjugate directions, monitored by photodiodes PD1 and PD2, respectively. Since these two-pulse photon echo traces are intrinsically symmetrical around $t_{12} = 0$, such a scan allows finding the precise overlap of pulses E_1 and E_2. Similarly, the precise location of $t_{23} = 0$, i.e., the overlap between pulses E_2 and E_3, can be verified by a t_{12} scan of the three-pulse echo signals, which for coinciding pulses E_2 and E_3, correspond to the signals depicted in the inset in Fig. 5.

The transient grating scans, a variety of three-pulse photon echo spectroscopy, are performed by scanning the delay t_{23} between the time-coincident pair of excitation pulses E_1–E_2 and the third (probe) pulse E_3. The symmetry of the incident beam arrangement implies that in this case two identical signals are emitted in the $k_3 + k_2 - k_1$ and $k_3 - k_2 + k_1$ directions. This also serves as a sensitive indicator to monitor continuously the perfect time overlap of the excitation pulses. By rotating polarizing cubes in front of PD3–4, we record the components of the transient grating signal parallel and perpendicular to the polarization of the pulses E_1 and E_2.

3.3 Generation of Hydrated Electrons

Hydrated electrons are generated by photoionizing a small amount of potassium ferrocyanide [60, 89, 90, 91] added to water, with the quadrupled output of a Nd:YLF laser (263 nm). Potassium ferrocyanide was obtained from Merck. Water and heavy water of HPLC grade were purchased from Aldrich and used without further purification. The polished sapphire nozzle (Kiburtz) ensured good quality of the jet and, therefore, no addition to the solution of the chemical substances stabilizing the jet surfaces [60] was needed. All measurements were carried out at room temperature.

The repetition rate of the YLF laser and, consequently, the repetition rate of the entire spectrometer was set at 4 kHz. The UV pulses are focused directly into the intersecting region of pulses E_1, E_2, and E_3 in the water jet. To avoid noticeable variation of the electron concentration within the interaction area of the femtosecond pulses, the spot size of the UV beam on the water jet is approximately two times larger than the waist of the three other beams. The triggering of the Nd:YAG laser is synchronized with the cavity dumping of the Ti:sapphire laser, so that the UV pulse precedes the femtosecond pulse by $\sim$ 200 ns. The absorption spectrum of hydrated electrons has been measured in the spectral region of 480 nm to 1100 nm with uncompressed white-light pulses and obtained as a difference in optical density of the water jet in the presence of and without UV radiation. The typical absorption spectrum (Fig. 6, open circles) has a peak value of optical density O.D. ≈ 0.2 at around 720 nm. The data of our absorption measurement coincide very well with the absorption spectrum, known from the literature [32] of hydrated electrons (Fig. 6, solid line) that have been directly injected into a volume of water.

3.4 Pump–Probe Experiments

Pump–probe experiments are carried out as follows. Two replicas of a 5 fs input pulse, separated in time by a delay τ, intersect in a 100 μm water jet containing the sample. Behind the sample, the probe beam is dispersed through a 1/8 m monochromator (CVI). The pump–probe signal ΔT (i.e., variations in the probe transmission with and without pump) is measured by a digital lock-in amplifier that is referenced to a mechanical chopper (~ 1 kHz) inserted in the pump beam. The differential absorption signal is computed as the $-\Delta T/T$ ratio, where T is the probe beam intensity. Orthogonal polarizations of pump and probe beams are used to avoid unwanted heterodyning of the signal by the light scattered from the pump pulse in the sample. All measurements are carried out at room temperature at a 5 kHz repetition rate. The contribution of pure solvent to the pump–probe signal ($< 10\%$ of the peak signal) is measured independently by blocking the UV preparation pulses and subtracted from the corresponding transient spectra of hydrated electrons.

4 Results and Discussion

4.1 Pure Dephasing Time of Hydrated Electrons

The method of generating hydrated electrons employed provides us with a unique opportunity to compare the photon echo signal from hydrated electrons with that measured in pure water, that is, in the absence of photoionizing radiation. Since electronic hyperpolarizability [92, 93, 94] heavily dominates the overall water response [95, 96], treating the nonlinearity as nearly instantaneous on the timescale of our pulses is well justified. Therefore, the signals obtained from pure water correspond to the ultimate instrument response of the spectrometer. This instrument function, among its other merits, automatically accounts for pulse duration, mode size, and spatial as well as spectral filtering in the detection. Therefore, the differences in the shape of photon echo traces recorded with and without UV radiation provide us with direct information on electronic dephasing of hydrated electrons.

The two-pulse photon echo signals from water and hydrated electrons are shown as solid circles in Fig. 7a,b, respectively. A minute difference in the widths of these two traces suggests that the electronic dephasing of photoexcited hydrated electrons is extremely fast. To fit to the experimental data (solid curves in Fig. 7a,b), we used the formalism developed in Sect. 2. The precise pulse parameters were obtained from independent FROG characterization, as described in Sect. 3.1. For water (Fig. 7a), we assumed an instantaneous response function (15) $R(t_1, t_2, t_3) \propto \delta(t_1)\delta(t_2)\delta(t_3)$ or, in other words, the frequency-independent, third-order susceptibility. For the hydrated electron (Fig. 7a), the complete (15) was used with the dephasing time T_2 as the fitting parameter. The population lifetime T_1 was considered much longer than any relevant experimental timescale, including pulse duration (see Sect. 4.3). Therefore, the difference between the full electronic dephasing time, T_2, and the pure dephasing time, T_2^*, which are connected by (16), is negligible.

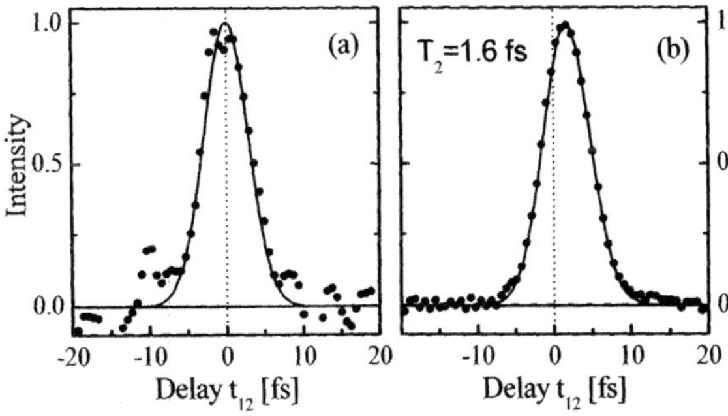

Fig. 7. Results of two-pulse photon echo experiments on water alone (**a**) and hydrated electrons (**b**). *Circles* represent experimental data points, and *solid curves* show fits obtained according to the procedure described in the text

The finite population lifetime of the electrons in the excited state causes the delay of the echo trace in Fig. 7b (its shift further away from $t_{12} = 0$ compared to the data in Fig. 7a). The best fit to the experimental data yields the dephasing time of $T_2 = 1.6\,\mathrm{fs}$. Note that this value is reasonably close to theoretical estimates derived from a model based on a Gaussian wave packet approximation for the bath [38,97,98]. The addition of any appreciable amount of inhomogeneity immediately results in pulling the echo maximum away from zero and the appearance of noticeable asymmetry in the trace.

Evidently, the 1.6 fs dephasing time should manifest itself in the absorption lineshape that has to be substantially homogeneously broadened [74]. Here, we stress that the use of a standard Lorentzian lineshape is not warranted for the spectra with widths comparable to the central frequencies. Instead, a more general relation should be used [72, 73, 99]:

$$\begin{aligned}\sigma_{\mathrm{A}}(\omega) &\propto \omega \operatorname{Im}\left[\chi^{(1)}(\omega)\right] \\ &\propto \omega \operatorname{Im}\left(\frac{1}{(\omega_{\mathrm{eg}} - \omega) - \mathrm{i}T_2^{-1}} + \frac{1}{(\omega_{\mathrm{eg}} + \omega) + \mathrm{i}T_2^{-1}}\right) \\ &\propto \frac{4\omega^2 T_2^{-2}}{(\omega_{\mathrm{eg}}^2 - \omega^2)^2 + 4\omega^2 T_2^{-2}} \,. \end{aligned} \tag{20}$$

In (21), σ_{A} is the absorption cross section, and ω_{eg} is the transition frequency. In a narrow absorption band, i.e., $T_2^{-1} \ll \omega_{\mathrm{eg}}$, one can use the approximation $\omega \cong \omega_{\mathrm{eg}}$ which immediately gives the conventional Lorentzian contour:

$$\sigma_A(\omega) \propto \frac{T_2^{-2}}{(\omega_{\mathrm{eg}} \bar{} \omega)^2 + T_2^{-2}} \,. \tag{21}$$

Rejection of the second term of the sum in (20), resulting in (21), amounts to application of the so-called rotating-wave approximation (RWA) [73, 74]. The difference between the two lineshapes, given by (20) and (21), mostly

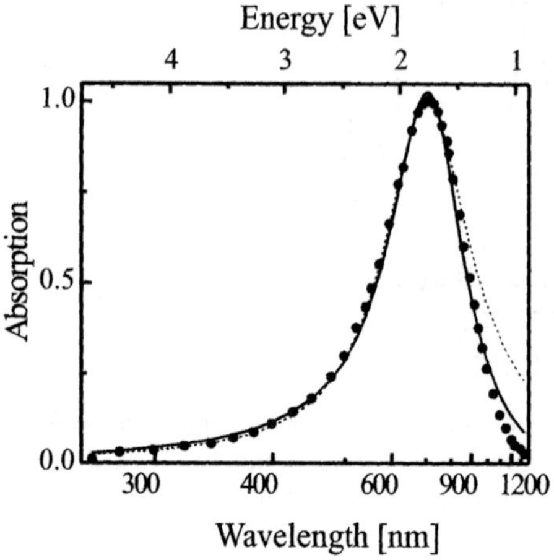

Fig. 8. Fit of absorption spectrum of electrons in water. The *circles* show the measured absorption of hydrated electrons produced by electron beams in bulk water (adopted from [32]). The *solid curve* shows the fit by homogeneously broadened line-shape calculated according to (20). The *dotted line* depicts a Lorentzian line-shape given by (21)

affects the asymptotic behavior in the low-frequency spectral wings. Besides, unlike the pure Lorentz function (21), that produces a centrosymmetric contour, the lineshape described by (20) is essentially asymmetrical. The latter contour has a more abrupt red wing and a prolonged blue wing. The need to account for the experimentally observed asymmetry precluded the use of a single Lorentzian lineshape in past attempts to model the absorption spectrum of hydrated electrons. Consequently, a collection of spectral lines [30, 31, 33] or a combined lineshape [26, 27, 28, 29, 32] was required for a reasonable fit.

Employing the more general relation for a homogeneously broadened lineshape given by (20), we obtained a fit (solid curve in Fig. 8) to the absorption spectrum of electrons injected into bulk water at 298 K [32]. Evidently, the whole absorption spectrum can be excellently reproduced by a homogeneously broadened lineshape. The dephasing time $T_2 = 1.7$ fs, deduced from the fit of the spectrum, agrees perfectly with that obtained from the photon echo experiment. This leaves no room for doubt about the homogeneous nature of spectral broadening of the hydrated electron absorption band.

Having established that the absorption spectrum of a hydrated electron is homogeneously broadened, we address the results of MD simulations from which the conclusion of inhomogeneous broadening of each s–p band was drawn [19, 23, 24]. It is clear that the stipulated nondegeneracy of p-states is a direct and natural consequence of a not entirely spherically symmetrical solvent cavity. However, the decomposition of the absorption spectrum into three bands seems to be somewhat artificial. As the energy of a given

p-state fluctuates in time from the rearrangement of the surrounding solvent molecules, this state might well become the highest or the lowest among the three p-states. According to the proposed ordering of the excited states by energy, one should then reassign the transition to this p-state as belonging to a different absorption subband. As a consequence of this artificial reassignment, the width of the Gaussian-like subbands revealed by these simulations becomes nearly equal to the separation of the band centers (vide supra). If the energy-dependent order were dropped, the absorption band associated with each s–p transition eventually would have the same width as the whole absorption spectrum of the solvated electron. This means that, depending on the current precise solvent surroundings, the optical transition to a designated p-state can take place anywhere across the whole absorption spectrum. The dynamic fluctuations of the solvent cavity causes constant "migration" of the respective s–p absorption bands within the common envelope.

We state that the seeming confusion with the assignment of the three absorption subbands arose from the averaging over an ensemble of solvent configurations [19, 23, 24] rather than averaging of the same configuration evolving in time. The same averaging of multiple static cavity "snapshots" led to the conclusion of the inhomogeneous broadening of each absorption line [19, 23, 24].

To the best of our knowledge, there has been no experimental evidence supporting the hypothesis of inhomogeneous line broadening in the absorption of an electron in water. Both picosecond [100] and femtosecond [31] transient pump–probe spectra revealed no presence of hole-burning behavior. Our recent attempts to imprint a spectral hole in the hydrated-electron absorption line employing a 15 fs excitation pulse centered around 800 nm, and a broadband 5 fs readout pulse likewise resulted in uniform and immediate bleaching of the whole absorption contour, suggesting that the latter is overwhelmingly homogeneously broadened. The lack of asymmetry in the two-pulse photon echo signal (Fig. 8b) is other strong evidence in favor of overwhelming homogeneous broadening. Besides, in the TG experiments, we observed no quantum beats which are usually associated with the coherent excitation of several transitions (vide infra).

Therefore, based on the current and previously accumulated experimental data, we conclude that the absorption spectrum of a hydrated electron is predominantly homogeneously broadened. An important question now should be raised about the physical origin of the very rapid dephasing associated with a very broad absorption band. Hardly any nuclear motion on such a short timescale should be expected. However, electronic dephasing is not necessarily the consequence of rapid fluctuation of local structures, nor does it mean that a large amount of energy has to be dissipated by the bath within a few femtoseconds. It is well known that the difference in the frequencies of (harmonic) ground- and excited-state potentials influences electronic dephasing through so-called quadratic electron–phonon coupling [101, 102, 103, 104, 105]. This mechanism, it also has been shown, leads to prevalently homogeneous broadening of absorption lines [106]. Since multi-phonon interactions are involved in nonlinear electron–phonon coupling, the latter explains large breadths of absorption spectra in a relatively modest width of spectral density of solute–solvent fluctuations [74].

To conclude this section, we suggest that the absorption band of equilibrated hydrated electrons is primarily *homogeneously* broadened with the corresponding dephasing time $T_2 \cong 1.7\,\mathrm{fs}$. The asymmetry of the spectrum is explained by the frequently neglected dependence of the optical absorption cross section on frequency. The mechanism responsible for the extraordinary width of the absorption spectrum is most probably quadratic electron–phonon coupling due to the appreciable difference in the steepness of the ground- and excited-state potentials.

4.2 Transient Grating Spectroscopy

The early part of TG transients of a hydrated electron is shown in Fig. 9 for parallel polarizations of an excitation pulse pair and the probe pulse. The signal has a sharp peak around zero which is followed by a prominent recurrence at $\sim 40\,\mathrm{fs}$. Subsequently, the signal decays on an $\sim 150\,\mathrm{fs}$ timescale.

Although the 5 fs pulses have enough spectral bandwidth to excite more than one of the *p*-states at once, there is no indication of quantum beats that are associated with the presence of several vibronic transitions [74]. From the calculated splitting between the different *s*–*p* subbands [19, 23, 24], one expects quantum beats with an $\sim 10\,\mathrm{fs}$ period. As we discussed in Sect. 4.1, the total spectrum consists of the sum of three bands with approximately the same width as the total spectrum and nearly identical central frequencies. As a result, no quantum beats can be observed under these circumstances.

Next, we focus on the origin of the narrow peak around zero delay in the TG signal. If the polarizations of the excitation pulse pair and the probe pulse are orthogonal (Fig. 9, solid dots), the peak around zero vanishes. The second intriguing feature of the transient is that its amplitude is precisely equal to that obtained with collinear polarizations. Obviously, the nonzero part of the difference between the TG signals with parallel and orthogonal polarizations is confined to the region of the pulse overlap. Note that this is highly uncommon; usually the signal obtained with parallel polarizations is stronger than that obtained with orthogonal polarizations because of polarization anisotropy [107], that is, only after some time, the amplitudes of the two signals become equal as a consequence of rotational diffusion processes which randomize the preferential orientation of dipole moments induced by the first pulse(s).

Clearly, the situation of the hydrated electron is entirely different. As we have shown in Sect. 4.1, the absorption spectrum of a hydrated electron is primarily homogeneously broadened. The underlying mechanism is that

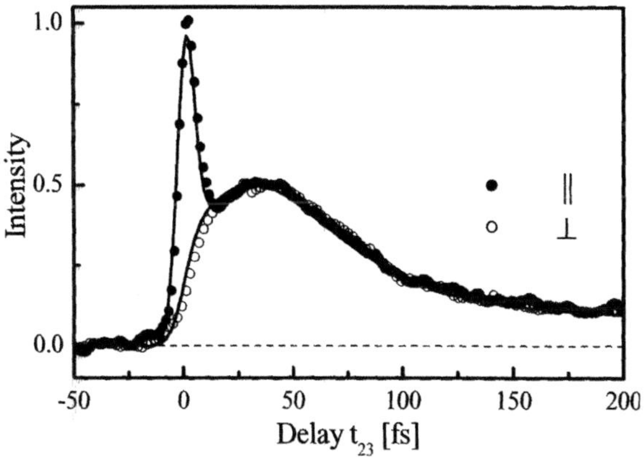

Fig. 9. Transient grating signals obtained from the hydrated electron in H_2O. *Solid dots* and *open circles* represent experimental data measured with the parallel and perpendicular polarization of excitation pulses, respectively. *Solid curves* depict the fits calculated, as described in Sect. 4.3. Note that the nonzero difference between two signals is confined to the region where the pulses overlap in time

the charge distribution in the ground state rapidly samples many configurations with different orientations. The timescale at which the sampling takes place is shorter than the duration of 5 fs excitation pulses. A hydrated electron experiences a variety of possible orientations of the neighboring water molecules *during* the excitation process. Therefore, the orientations of the excited *p*-state are completely randomized by the time the excitation is completed. As a result, no anisotropy can be observed after the excitation pulse pair ends.

We conclude that, in the case of an aqueous electron, the transient anisotropy disappears in less than 5 fs. The important consequence is that when the polarizations of the pump pulses and the probe pulse are perpendicular, the coherent coupling between them is extremely ineffective. Therefore, if the contributions with irregular time ordering (as, for instance, E_1–E_3–E_2) are omitted, the total number of permutations is decreased by a factor of 2 [88]. Hence, the TG signal near $t_{23} = 0$ for perpendicular polarizations is expected to be approximately one-fourth of the signal with parallel polarizations. This value is very close to that observed in the experiment (Fig. 9). The recurrence (or delayed response) of the TG signal at $\sim$ 40 fs will be explained in the next section as a direct consequence of the extremely short population lifetime.

4.3 Transient Pump–Probe Spectroscopy

Representative transient absorption spectra of hydrated electrons in water at different pump–probe delays are presented in Fig. 10 (solid dots). A dominant feature of the transient spectra in Fig. 10 at initial pump–probe delays is an absorption bleach contour (negative signal) that stretches through most of

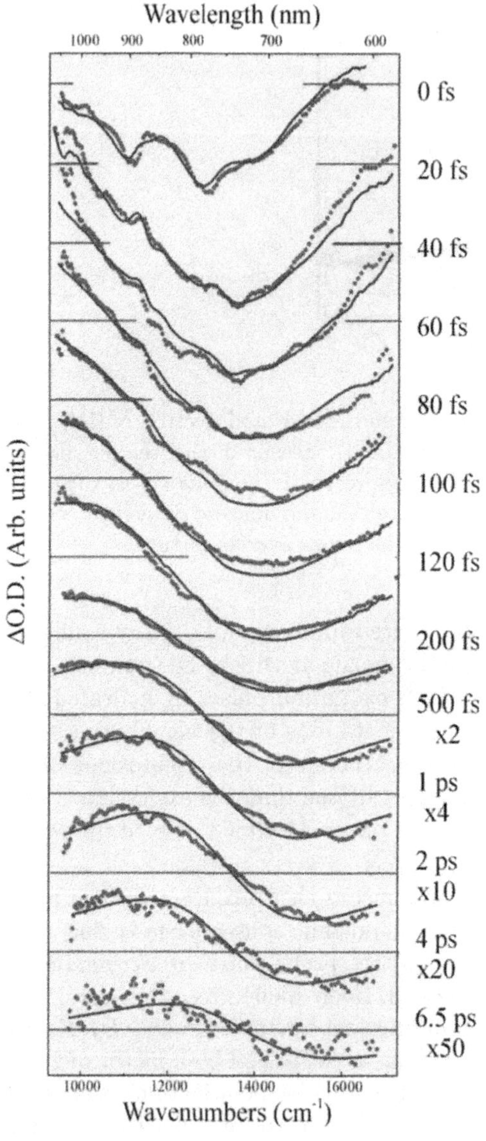

Fig. 10. Differential absorption signals. Experimental data and results of numerical simulations are depicted by *solid dots* and *curves*, respectively. The delay between pump and probe pulses is indicated on the *right*. *Horizontal lines* show the level

the covered spectral window. Around 720 nm, the bleach decreases by a factor of ~ 2 during the first 100 fs. The high-frequency wing (below 700 nm) of the bleaching contour becomes noticeably flatter in this time window. On the same timescale, a sharp rise in induced absorption (positive signal) at low frequencies is observed. All these processes are consistent with a rapid (within 100 fs) depopulation of the excited state, resulting in a loss of excited-state emission/absorption and simultaneous development of hot ground-state absorption. No signature of p-state solvation or intraband dynamics was found on this timescale, even with 5 fs resolution in the experiments. Had such processes existed on this timescale, they would manifest themselves through a redshift of the emission. However, no such redshift of the emission was detected.

After $\sim$ 100 fs, no drastic changes in the transient spectra are observed; the induced absorption contour moves gradually toward the center of the steady-state absorption of the hydrated electron with chararcteristic times of $\sim$ 300 fs and 1.1 ps, corroborating previous reports [31, 57, 58].

Figure 11a outlines a simple, yet realistic model to explain the experimental results of energy relaxation. The first pulse promotes the electron from the initial $1s$ to the $2p$-state. The population relaxation of the $2p$-state occurs with the time constant T_1 after which the electron finds itself in the "hot" ground state (denoted as $1s'$) due to rearrangement of the first solvation shell. Subsequently, the contribution from the redshifted induced absorption builds up, whereas the contributions from the $2p$-state fluorescence and absorption decrease correspondingly (Fig. 11b). The latter flattens the blue flank of the pump–probe signal (Fig. 11b). Finally, the energy flows from the first solvation shell further into the water bath leading to thermalization of the hydrated electron.

The pump–probe signal is therefore a sum of four terms [74], which originate from bleaching of the $1s$-state, induced absorption from the $1s'$-state, and fluorescence and induced absorption from the $2p$-state, respectively (Fig. 2b):

$$\begin{aligned} -\Delta T/T(\Omega) \propto & -\sigma_{sp}\left(\Omega, \Gamma_{sp}, \omega_{sp}\right) + a_{s'}(\tau)\sigma_{s'p'}\left[\Omega, \Gamma_{s'p'}(\tau), \omega_{s'p'}(\tau)\right] \\ & - a_p(\tau)\sigma_{sp}\left(\Omega, \Gamma_{sp}, \omega_{sp}\right) + a_p(\tau) a_{pc}\sigma_{pc}\left(\Omega, \Gamma_{pc}, \omega_{pc}\right), \end{aligned} \tag{22}$$

where $a_p(\tau)$ and $a_{s'}(\tau)$ are the respective populations of the $2p$- and $1s'$-states, a_{pc} is the absorption cross section of the $2p$–c transition, and the σs denote the transition spectral shapes. The condition of population conservation dictates that $a_p(\tau) + a_{s'}(\tau) = 1$. The crucial difference between the current model and the previous ones [31, 58] is that solvation and/or intra-p-band dynamics play no role, in accord with our experimental observations on this timescale.

In general, expressions for the σs are quite complex and include a twofold integral over the applied electric fields and the third-order susceptibility, as given by (7). To highlight the main features of the pump–probe signal, we derived a simplified expression that is valid when the pump–probe pulse duration is much shorter than the population relaxation time:

$$\sigma(\Omega, \Gamma, \omega) = \frac{\Gamma_{sp}}{\Gamma} \frac{4\Omega^2\Gamma^2}{(\omega^2 - \Omega^2)^2 + 4\Omega^2\Gamma^2}. \tag{23}$$

Note that upon deriving (23), one *cannot* invoke the rotating-wave approximation because the central frequency ω is of the same order of magni-

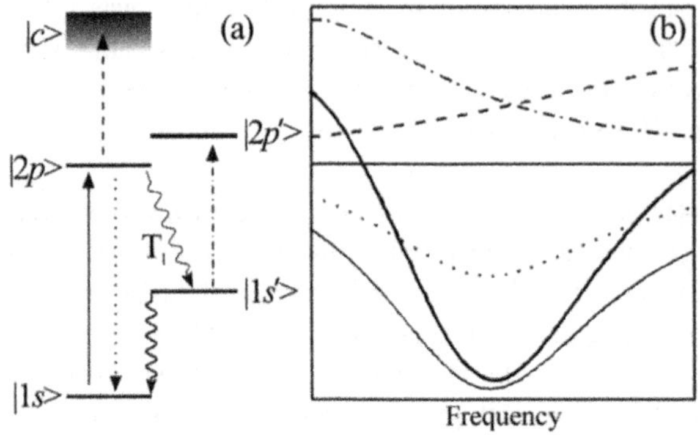

Fig. 11. Model of hydrated electron dynamics (**a**) and processes (**b**) that contribute to the pump–probe signal (*thick solid curve*): bleaching of the 1*s*-state (*thin solid curve*), emission from the 2*p*-state (*dotted line*), and absorption from the 2*p*-state to continuum states (*dashed curve*), and absorption from the "hot" $1s'$-state (*dash-dotted line*). T_1 denotes the lifetime of the 2*p*-state

tude as the characteristic half-width Γ [108]. As follows from (22), the total pump–probe signal is proportional to a linear combination of lineshapes of the form given by (23) (Fig. 11b). Remarkably, it does not depend on the particular shape or chirp of the excitation pulses, as long these pulses remain much shorter than the population relaxation time T_1 of the 2*p*-state. At delay times considerably longer than T_1, $a_p(\tau) \to 0$. As a result, (22) can be reduced to the first two terms that account for the interplay between the 1*s*-state bleach and $1s'$-state induced absorption. Apparently, as the hydrated electron becomes fully thermalized, the central frequency $\omega_{s'p'}$ and half-width $\Gamma_{s'p'}$ of the s'–p' transition should approach their steady-state values ω_{sp} and Γ_{sp}, respectively.

The fit to the data proceeds as follows. The width Γ_{sp} is set at 3300 cm^{-1}, which corresponds to dephasing time $T_2 = 1.7$ fs, as measured in photon echo experiments (see Sect. 4.1). The width, the central frequency, and the time-dependent amplitude of the $1s'$–$2p'$ contour are treated as local fitting parameters (i.e., specific for each delay time), whereas the rest of the variables are global (i.e., common for all delays). For the pump–probe spectra measured at short time delays (≤ 100 fs), we used the full expressions for the lineshape functions, similar to (7), to be able to account for the actual amplitude and phase properties of the laser pulse. The fits of individual pump–probe spectra obtained using the model are depicted as solid curves in Fig. 10. Clearly, the overall fit quality is excellent, which supports the proposed model. We can even reproduce reasonably well the wiggles clearly seen at short delays that originate from irregularities in the pulse spectrum and phase.

Figure 12 shows the populations of 2*p* and $1s'$ states as a function of the pump–probe delay time derived from the global fit. The monoexponential fit of the 2*p*-state population yields an $\sim$ 50 fs lifetime. Correspondingly,

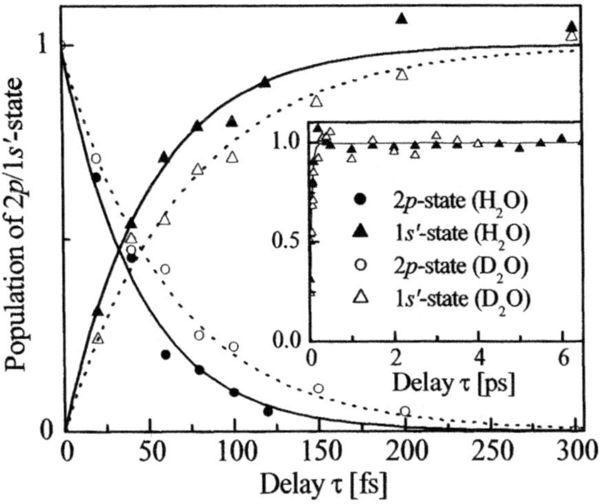

Fig. 12. Population of the $2p$-state (*circles*) and $1s'$-states (*triangles*) as deduced from the fit to experimental data (Fig. 1). Exponential fits with time of 50 fs and 70 fs are shown for water (*solid curve*) and heavy water (*dashed curve*), respectively

the population of the $1s'$-state increases with a 50 fs time constant. After the initial buildup, $a_{s'}(\tau)$ stays close to unity (Fig. 12, inset), which is fully consistent with overall population conservation. This feature originates solely from the fit, is not assumed a priori, and therefore, presents a powerful check of the correctness of the model. Note that the extraction of a 50 fs lifetime was not possible from the data obtained in the previous pump–probe studies [31,42,44,57,58,60], due to the lack of adequate temporal resolution (typically $\sim$ 150 fs).

To gain more insight into the physical processes underlying hydrated electron dynamics, we examine the impact of deuteration on the spectro-temporal behavior of the pump–probe signal (Fig. 12, open symbols). The lifetime of the $2p$-state and, accordingly, the buildup time of the $1s'$-state of the electron in heavy water is $\sim$ 70 fs which is longer by a factor of $\approx \sqrt{2}$ than that found in H_2O. The subsequent dynamics, however, are nearly identical in both solvents [61, 108].

The difference in relaxation times can be explained by the fact that the moments of inertia of H_2O and D_2O differ precisely by a factor of 2. This immediately leads us to the conclusion that the maximum in TG signals is caused by *overdamped* librational motion of the water molecules. The specific dependence on deuteration rules out translations because the total mass of a water molecule changes only by a factor of 18/20.

The extremely short excited-state lifetime allows the following interpretation of the results of a transient-grating (TG) experiment on the hydrated electron (Fig. 9) [61, 108]. The signal peak delayed by $\sim$ 40 fs in H_2O and $\sim$ 60 fs in D_2O has been assigned before to librational wave-packet dynamics [59, 61, 62] and/or non-Condon (time-dependent variations of a transition

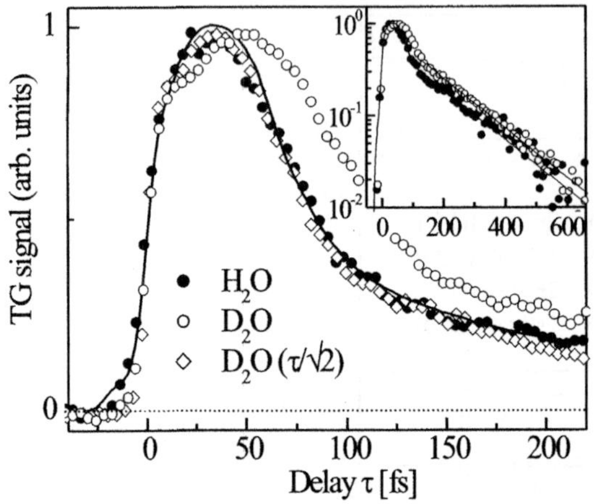

Fig. 13. Comparison of transient grating signals obtained from electrons hydrated in H_2O (*solid dots*) and D_2O (*open circles*). All data are given for perpendicular polarizations of excitation pulses. The delay axis for D_2O has been compressed by a factor of $\sqrt{2}$ to highlight the isotopic effect (*diamonds*). The *inset* shows the same data on a logarithmic scale

dipole moment) effects [108]. In the framework of the proposed model, the retarded TG response has a perfectly natural explanation as interplay among different contributions. At early times, the loss in $2p$-state fluorescence due to population relaxation is counteracted by a simultaneous drop in $2p$-state absorption. On the other hand, induced $1s'$-state absorption does not substantially contribute to the total TG signal as it is formed mostly outside the spectral region of the pulses and detector. The combined effect of all these factors leads to an *increase* in the TG signal with time. Only after ~ 50 fs, when the contribution from $1s'$-state absorption begins to dominate, is there a sharp decrease in the TG signal. However, after ~ 100 fs, the $2p$-state is almost depopulated, and only two contributions ($1s$-state bleaching and hot $1s'$-absorption) donate to the TG signal. This results in leveling of the signal that further decays with a time constant of ~ 150 fs within the dynamic range of detection.

The results of the TG experiment on an electron solvated in heavy water are shown in Fig. 13. The TG signals are similar to those for normal water, but the maximum is delayed to longer times (~ 60 fs). However, if the timescale of the D_2O TG data is compressed by $\sqrt{2}$, the TG signal becomes virtually indistinguishable from that for H_2O (Fig. 13, diamonds). As we discussed, this is a direct consequence of the $2p$-state lifetime that changes by a factor of $\sqrt{2}$ in H_2O and D_2O. The fact that both TG signals behave similarly on the long timescale indicates that librational motions of water molecules are overtaken by those that exhibit no isotopic effect (e.g., translations).

5 Hydrated-Electron Dynamics: The Microscopic Picture

According to our experiments, the relaxation dynamics of a hydrated electron consists of three stages: 1. a rapid, 50 fs depopulation of the excited state, which corresponds to transfer of the energy of $\sim 8500\,\mathrm{cm}^{-1}$ to the surrounding water molecules; 2. a slow, 1 ps equilibration of the ground state; and 3. an intermediate phase between the first two processes. The early delay region of up to about 100 fs is dominated by librational motions of water molecules, which is manifested by a $\sqrt{2}$ difference in the timescales for H_2O and D_2O. The energy released to the solvent by ground-state thermalisation amounts to $\sim 1/3$ of the initial s–p transition energy. The features of pump–probe spectra and TG signals in the range of delays from zero to 60 fs point out the presence of a short-lived excited state.

The microscopic picture underlying the early time dynamics of the hydrated electron that has emerged from our experiments is schematically presented in Fig. 14. In this cartoon, only four of the approximately six water molecules in the first solvation shell around the electron are depicted for the sake of simplicity. The relevant mean size of the electron cloud in the ground $|g\rangle$ state was matched to the value reported by the group of *Rossky* (1.9 Å) [24], and the distances between the electron and the water molecules were taken from [20].

Before excitation, the electron finds itself in the s-like ground state in equilibrium with surrounding water molecules. Upon excitation, the electron makes a transition to the excited p-state wave function that is elongated in a particular direction. Since the excitation pulses are very short, the water molecules have no time to react, and, therefore, they still preserve their compact geometry.

From the fit to transient absorption spectra, we extract that $\sim 8500\,\mathrm{cm}^{-1}$ of the energy deposited by photoexcitation is dissipated with a time constant of 50 fs due to population relaxation from the initially excited $2p$-state. Such a fast relaxation process can be explained only by assuming that a conical intersection exists between the excited and ground states coupled to several vibrational modes [109]. For the latter, the most likely candidates are the OH-stretching vibrational modes of the six nearest water molecules in the Kevan geometry [20]. Upon electron excitation to the p-state, the combined system of an "electron coupled to six OH modes" tends to lower its energy by elongating the octahedral structure (Jahn–Teller-like instability). Our estimates show that the energy gain due to cavity elongation can be as high as 1 eV, in good agreement with the measured value. After cavity deformation, a non-Born–Oppenheimer transition [109] takes place from the $2p$- to the $1s$-state within several vibrational periods (~ 10 fs). The observed $\sqrt{2}$-dependence of the lifetime points to a dominant role for the OH-stretching mode in the radiationless relaxation process of the excited state. Note that MD

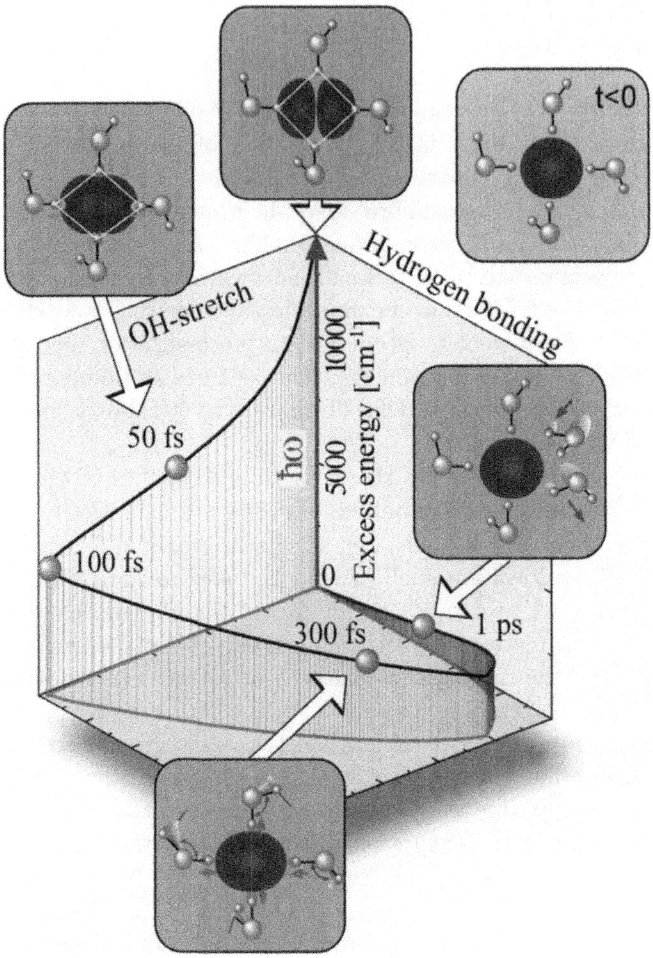

Fig. 14. Artist's impression of hydrated electron dynamics. The schematic trajectory of energy relaxation is depicted as a function of two separate solvent coordinates. At $t = 0$, the hydrated electron is photoexcited from the origin of the reference frame (photon energy is shown by a *vertical arrow*). Positions at different times following the excitation are indicated by *circles*. *Insets* show configurations of the charge distribution of the electron (*contour plot*) and surrounding water molecules

simulations based on the assumption of an *averaged* water–electron potential cannot account for the suggested relaxation path.

As revealed by the fit results of the pump–probe data (Fig. 12), by approximately 100 fs after excitation, most of the electrons already returned to the (hot) ground state. This corresponds to the "collapse" of the size of the hydrated electron, which now occupies a more compact hot s-state, leaving a void in the solvent. The more compact configuration of the electron changes the character of the librational motion from forcibly driven by the electron expansion to a more random one. Accordingly, the "distance" covered by the electron along the librational coordinate stops increasing, and the "propagation" along the translational coordinate begins. During this phase, the relaxation process does not exhibit any appreciable isotopic effect. There are two timescales in this process: $\sim$ 300 fs and $\sim$ 1 ps (Fig. 13). The former time is reasonably close to the population relaxation time of the OH-stretching mode of $\sim$ 260 fs [100]. During this phase, the water molecules in the first solvation shell deploy the accumulated excess energy into a collective-type translational motion. It is likely that the existing hydrogen bond network [110, 111, 112] helps to speed up the energy transfer. After approximately 300 fs, the energy is released in the form of local heating (i.e., a rise in the local temperature) and spreads further away from the electron. Finally, after about 5 ps, practically a full equilibration of the hydrated electron is achieved.

6 Conclusions

Photon echo, transient grating and pump–probe spectroscopy on the hydrated electron performed with the best time resolution available to date has provided a powerful insight into the microscopic processes that underlie solvation dynamics.

To solve the nontrivial fundamental issues related to nonlinear spectroscopy, the optical pulses that consist of 2.5 optical cycles, we developed a general formalism describing the signal field generated in both the time- and frequency-domain. The frequency-domain representation is more powerful since it allows a consistent account of a variety of effects, such as phase mismatch, self-steepening, and dispersive pulse broadening. Additionally, the use of the frequency-domain formalism removed the necessity to invoke a number of approximations, for example, the slowly varying envelope approximation. The formulation derived also avoids the use of parameters that are ill-defined for broadband optical pulses, for instance, the carrier frequency of the pulse. Equations (7–10) constitute the backbone of the general description of the third-order nonlinear experiments. Importantly, these equations remain valid and can be directly applied, even for single-cycle pulses.

We have developed a general procedure for calculating the spectral-filter effect. Such a routine should be employed to optimize the experimental configuration for any third-order spectroscopic experiment that uses laser pulses shorter than 10 fs. We have demonstrated that the ability to defeat the damaging role of the spectral-filter effect legitimizes a transition to the time-domain formulation typically employed for multicycle pulses.

Both two-pulse photon echo and TG experiments, involving two and three femtosecond pulses, respectively, have been performed on equilibrated hydrated electrons. By comparing two-pulse echo signals from hydrated electrons and from water alone, we have derived the pure dephasing time of ~ 1.7 fs. We have shown that the absorption band of equilibrated solvated electrons in water is predominantly homogeneously broadened, and we succeeded in modeling the whole absorption spectrum of hydrated electrons by a single homogeneous lineshape. The symmetrical Lorentzian lineshape typically employed has been abandoned in favor of a more general expression for a homogeneously broadened line. Importantly, in the lineshape used in our fit, the long-puzzling issue of the asymmetry of the absorption spectrum found a natural explanation. The TG experiments on the hydrated electron with perpendicular polarziations of excitation pulses have revealed that the induced anisotropy decays within the duration of the pulses (~ 5 fs).

From the fit of the pump–probe spectra, we inferred a ~ 50 fs lifetime of the *p*-state of the hydrated electron. The lifetime increases by a factor of $\sqrt{2}$ in heavy water. This fast relaxation process, it is suggested, is associated with a conical intersection between the ground and excited state caused by strong coupling between the electron and OH-vibrational modes of water molecules in the first solvation shell. The next stage of electron thermalization is determined by the vibrational relaxation of the OH-stretching modes of water molecules ($\sim$300 fs). Finally, the hydrogen bond network of water molecules is fully equilibrated on an ~ 1 ps timescale.

Electrons in polar liquids continue to be a vast experimental and theoretical field in which many intriguing questions remain to be answered. There are still many conflicting and not well-understood issues concerning electron equilibration [48, 113], energy relaxation of photoexcited electrons in fluids [31,60], the nature and number of bound localized and unbound states of electrons trapped in a solvent cavity [23, 44], etc. New, emerging spectroscopic techniques could help clarify many mysteries of this attractive and challenging system. We foresee great possibilities for the application of femtosecond infrared spectroscopy [114]. By temporally and spectrally resolving the transient dynamics of the OH-bond that absorbs infrared, one would obtain an invaluable direct insight into the motions of solvent molecules and how they respond to the photoexcitation of a solvated electron.

References

[1] W. Weyl: Pogg. Ann. **123**, 350 (1864)
[2] C. A. Kraus: J. Am. Chem Soc. **30**, 1323 (1908)
[3] J. W. Boag, E. J. Hart: Nature **197**, 45 (1963)
[4] E. J. Hart, J. W. Boag: J. Am. Chem Soc. **84**, 4090 (1962)
[5] U. Bach, D. Lupo, P. Comte, J. E. Moser, F. Weissrtel, J. Salbeck, H. Spreitzer, M. Gretzel: Nature **395**, 583 (1998)
[6] Borja, P. K. Dutta: Nature **362**, 43 (1993)
[7] O. Khaselev, J. A. Turner: Science **280**, 425 (1998)
[8] Sykora, J. R. Kincaids: Nature **387**, 162 (1997)
[9] M. H. B. Stowell, T. M. McPhillips, D. C. Rees, S. M. Soltis, E. Abresch, G. Feher: Science **276**, 812 (1997)

[10] G. Steinberg-Yfrach, P. A. Liddell, S. C. Hung, A. L. Moore, D. Gust, T. A. Moore: Nature **385**, 239 (1997)
[11] H. Schindelin, C. Kisker, J. L. Schlessman, J. B. Howard, D. C. Rees: Nature **387**, 370 (1997)
[12] A. P. Alivisatos: Science **271**, 933 (1996)
[13] L. P. Kouwenhoven, T. H. Oosterkamp, M. W. S. Danoesastro, M. Eto, D. G. Austing, T. Honda, S. Tarucha: Science **278**, 1788 (1997)
[14] P. L. McEuen: Science **278**, 1729 (1997)
[15] D. R. Stewart, D. Sprinzak, C. M. Marcus, C. I. Duruz, J. S. H. Jr: Science **278**, 1784 (1997)
[16] T. H. Oosterkamp, T. Fujisawa, W. G. v. d. Wiel, K. Ishibiashi, R. V. Hijman, S. Tarucha, L. P. Kouwenhoven: Nature **395**, 873 (1998)
[17] J. Hasen, L. N. Pfeiffer, A. Pinczuk, S. He, K. W. West, B. S. Dennis: Nature **390**, 54 (1997)
[18] I. Park, K. Cho, S. Lee, K. Kim, J. D. Joannopoulos: Comp. Mater. Sci. **21**, 291 (2001)
[19] B. J. Schwartz, P. J. Rossky: J. Comp. Phys. **101**, 6902 (1994)
[20] L. Kevan: Accounts Chem. Res. **14**, 138 (1981)
[21] O. Mishima, H. E. Stanley: Nature **396**, 329 (1998)
[22] C. Romero, C. D. Jonah: J. Comp. Phys. **90**, 1877 (1988)
[23] P. J. Rossky, J. Schnitker: J. Phys. Chem. **92**, 4277 (1988)
[24] J. Schnitker, K. Motakabbir, P. J. Rossky, R. Friesner: Phys. Rev. Lett. **60**, 456 (1988)
[25] A. Staib, D. Borgis: J. Comp. Phys. **103**, 2642 (1995)
[26] A. Banerjee, J. Simons: J. Comp. Phys. **68**, 415 (1978)
[27] W. M. Bartczak, M. Hilczer, J. Kroh: J. Phys. Chem. **91**, 3834 (1987)
[28] I. Carmichael: J. Phys. Chem. **84**, 1076 (1980)
[29] T. Kajiwara, K. Funabashi, C. Naleway: Phys. Rev. A **6**, 808 (1972)
[30] L. Kevan, B. C. Webster: *Electron Solvent and Anion Solvent Interactions* (Elsevier, Amsterdam 1976)
[31] M. Assel, R. Laenen, A. Laubereau: J. Phys. Chem. A **102**, 2256 (1998)
[32] F. Y. Jou, G. R. Freeman: J. Phys. Chem. **83**, 2383 (1979)
[33] R. Lugo, P. Delahay: J. Comp. Phys. **57**, 2122 (1972)
[34] B. J. Schwartz, P. J. Rossky: J. Comp. Phys. **101**, 6917 (1994)
[35] R. B. Barnett, U. Landman, A. Nitzan: J. Comp. Phys. **90**, 4413 (1989)
[36] M. A. Berg: J. Comp. Phys. **110**, 8577 (1999)
[37] P. Graf, A. Nitzan, G. H. F. Diercksen: J. Phys. Chem. **100**, 18916 (1996)
[38] B. J. Schwartz, E. R. Bittner, O. V. Prezhdo, P. J. Rossky: J. Comp. Phys. **104**, 5942 (1996)
[39] B. J. Schwartz, P. J. Rossky: J. Phys. Chem. **99**, 2935 (1995)
[40] B. J. Schwartz, P. J. Rossky: J. Mol. Liq. **65/66**, 23 (1995)
[41] B. J. Schwartz, P. J. Rossky: J. Comp. Phys. **105**, 6997 (1996)
[42] J. C. Alfano, P. K. Walhout, Y. Kimura, P. F. Barbara: J. Comp. Phys. **98**, 5996 (1993)
[43] Hertwig, H. Hippler, A. N. Unterreiner, P. Vöhringer: Ber. Bunsen Phys. Chem. **102**, 805 (1998)
[44] Y. Kimura, J. C. Alfano, P. K. Walhout, P. F. Barbara: J. Phys. Chem. **98**, 3450 (1994)
[45] F. H. Long, H. Lu, K. B. Eisenthal: Phys. Rev. Lett. **64**, 1469 (1990)

[46] M. C. Messmer, J. D. Simon: J. Phys. Chem. **94**, 1220 (1990)
[47] A. Migus, Y. Gauduel, J. L. Martin, A. Antonetti: Phys. Rev. Lett. **58**, 1559 (1987)
[48] C. Pupin, T. Goulet, D. Houde, J. P. Jay-Gerin: J. Phys. Chem. A **101**, 4351 (1997)
[49] A. Reuther, A. Laubereau, D. N. Nikogosyan: J. Phys. Chem. **100**, 16794 (1996)
[50] X. Shi, F. H. Long, H. Lou, K. B. Eisenthal: J. Phys. Chem. **100**, 11903 (1996)
[51] P. J. Reid, C. Silva, P. K. Walhout, P. F. Barbara: Chem. Phys. Lett. **228**, 658 (1994)
[52] R. N. Barnett, U. Landman, A. Nitzan: J. Comp. Phys. **93**, 8187 (1990)
[53] E. Neria, A. Nitzan, R. N. Barnett, U. Landmann: Phys. Rev. Lett. **67**, 1011 (1991)
[54] O. V. Prezhdo, P. J. Rossky: Phys. Rev. Lett. **81**, 5294 (1998)
[55] K. Toukan, A. Rahman: Phys. Rev. B **31**, 2643 (1985)
[56] F. J. Webster, J. Schnitker, M. S. Friedrichs, R. A. Friesner: Phys. Rev. Lett. **66**, 3172 (1991)
[57] A. Hertwig, H. Hippler, A. N. Unterreiner: Phys. Chem. Chem. Phys. **1**, 5633 (1999)
[58] M. Assel, R. Laenen, A. Laubereau: Chem. Phys. Lett. **317**, 13 (2000)
[59] C. Silva, P. K. Walhout, K. Yokoyama, P. F. Barbara: Phys. Rev. Lett. **80**, 1086 (1998)
[60] K. Yokoyama, C. Silva, D. H. Son, P. K. Walhout, P. F. Barbara: J. Phys. Chem. **102**, 6957 (1998)
[61] M. F. Emde, A. Baltuška, A. Kummrow, M. S. Pshenichnikov, D. A. Wiersma: Phys. Rev. Lett. **80**, 4645 (1998)
[62] A. Kummrow, M. F. Emde, A. Baltuška, M. S. Pshenichnikov, D. A. Wiersma: J. Phys. Chem. **102**, 4172 (1998)
[63] A. Baltuška, Z. Wei, M. S. Pshenichnikov, D. A. Wiersma: Opt. Lett. **22**, 102 (1997)
[64] A. Baltuška, Z. Wei, R. Szipcs, M. S. Pshenichnikov, D. A. Wiersma: Appl. Phys. B-Lasers O. **65**, 175 (1997)
[65] T. Kobayashi, A. Shirakawa, T. Fuji: IEEE J. Sel. Top. Quant. **7**, 525 (2001)
[66] T. Kobayashi, A. Shirakawa, H. Matsuzawa, H. Nakanishi: Chem. Phys. Lett. **321**, 385 (2000)
[67] M. Nisoli, S. D. Silvestri, R. Szipucs, K. Ferencz, C. Spielmann, S. Sartania, F. Krausz: Opt. Lett. **22**, 522 (1997)
[68] M. Nisoli, S. Stagira, S. D. Silvestri, O. Svelto, S. Sartania, Z. Cheng, M. Lenzner, C. Spielmann, F. Krausz: Appl. Phys. B-Lasers O. **65**, 189 (1997)
[69] L. Allen, J. H. Eberly: *Optical Resonance and Two-Level Atoms* (Dover, New York 1987)
[70] T. Brabec, F. Krausz: Phys. Rev. Lett. **78**, 3282 (1997)
[71] P.N. Butcher, D. Cotter: *The Elements of Nonlinear Optics* (Cambridge Univ. Press, Cambridge 1990)
[72] K. Shimoda: *Introduction to Laser Physics*, 2nd ed. (Springer, Berlin, Heidelberg 1991)
[73] R. W. Boyd: *Nonlinear Optics* (Academic, San Diego 1992)
[74] S. Mukamel: *Principles of Nonlinear Optical Spectroscopy* (Oxford University Press, New York 1995)

[75] A. Baltuška, M. S. Pshenichnikov, D. A. Wiersma: IEEE J. Sel. Top. Quant. **35**, 459 (1999)
[76] G. Taft, A. Rundquist, M. M. Murnane, I. P. Christov, H. C. Kapteyn, K. W. DeLong, D. F. Fittinghoff, M. A. Krumbuguel, J. N. Sweetser, R. Trebino: IEEE J. Sel. Top. Quant. **2**, 575 (1996)
[77] A. M. Weiner: IEEE J. Quantum Elect. **QE-19**, 1276 (1983)
[78] M. Pshenichnikov, A. Baltuška, R. Szipocs, D. A. Wiersma: Sub-5 fs pulses: generation, characterisation, and experiments, in T. Elsaesser, J. G. Fujimoto, D. A. Wiersma, et al. (Eds.): Ultrafast Phenomena XI (Springer, Berlin, Heidelberg 1998)
[79] Y. R. Shen: *The Principles of Nonlinear Optics* (Wiley, New York 1984)
[80] M. Schubert, W. Wilhelmi: *Nonlinear Optics and Quantum Electronics* (Wiley, New York 1986)
[81] S. A. Akhmanov, V. A. Vysloukh, A. S. Chirkin: *Optics of Femtosecond Laser Pulses* (American Institute of Physics, New York 1992)
[82] G. P. Agrawal: *Nonlinear Fiber Optics*, 2nd ed. (Academic, San Diego 1995)
[83] R. Trebino, K. W. DeLong, D. N. Fittinghoff, J. Sweetser, B. R. M. A. Krumbgel, a. D. J. Kane: Rev. Sci. Inst. **68**, 3277 (1997)
[84] The International Association for the Properties of Water and Steam, Erlangen: *Release on the Refractive Index of Ordinary Water Substance as a Function of Wavelength, Temperature and Pressure* (1997) URL `http//www.iapws.org/relguide/rindex.pdf`
[85] A. H. Harvey, J. S. Gallagher, J. M. L. Sengers: J. Phys. Chem. **27**, 761 (1998)
[86] N. Bloembergen: *Nonlinear Optics* (Benjamin, New York 1965)
[87] G. R. Fleming: Chemical applications of ultrafast spectroscopy (1986)
[88] W. P. de Boeij, M. S. Pshenichnikov, D. A. Wiersma: J. Phys. Chem. **100**, 11806 (1996)
[89] M. S. Matheson, W. A. Mulac, J. Rabani: J. Phys. Chem. **67**, 2613 (1963)
[90] M. Shirom, G. Stein: J. Comp. Phys. **55**, 3372 (1971)
[91] M. Shirom, G. Stein: J. Comp. Phys. **55**, 3379 (1971)
[92] D. McMorrow: Opt. Commun. **86**, 236 (1991)
[93] D. McMorrow, W. T. Lotshaw: Chem. Phys. Lett. **174**, 85 (1990)
[94] D. McMorrow, W. T. Lotshaw, G. A. Kenney-Wallace: IEEE J. Quantum Elect. **24**, 443 (1988)
[95] E. W. Castner, Y. J. Chang, Y. C. Chu, G. E. Walrafen: J. Comp. Phys. **102**, 653 (1994)
[96] S. Palese, L. Schilling, R. J. D. Miller, P. R. Staver, W. T. Lotshaw: J. Phys. Chem. **98**, 6308 (1994)
[97] E. Neria, A. Nitzan: J. Comp. Phys. **99**, 515 (1993)
[98] O. V. Prezhdo, P. J. Rossky: Phys. Rev. Lett. **81**, 5294 (1998)
[99] A. E. Siegman: *Lasers* (Univ. Science Books, Mill Valley 1986)
[100] D. Huppert, P. Avouris, P. M. Rentzepis: J. Phys. Chem. **82**, 2282 (1978)
[101] M. Aihara: Phys. Rev. B **25**, 53 (1982)
[102] B. A. Grishanin, V. M. Petnikova, V. V. Shuvalov: J. Appl. Spect. **47**, 1278 (1987)
[103] B. A. Grishanin, V. M. Petnikova, V. V. Shuvalov: J. Appl. Spect. **47**, 1309 (1987)
[104] M. Toutounji, G. J. Small, S. Mukamel: J. Comp. Phys. **109**, 7949 (1998)
[105] M. Toutounji, G. J. Small, S. Mukamel: J. Comp. Phys. **110**, 1017 (1999)

[106] I. S. Osad'ko, M. V. Stashek, M. A. Michailov: Laser Phys. **6**, 175 (1996)
[107] A. C. Albrecht: in G. Porter (Ed.): *Progress in Reaction Kinetics*, vol. 5 (Pergamon Press, Oxford 1970) p. 301
[108] A. Baltuška, M. F. Emde, M. S. Pshenichnikov, D. A. Wiersma: J. Phys. Chem. **103**, 10065 (1999)
[109] A. Kuhl, W. Domcke: J. Comp. Phys. **116**, 263 (2002)
[110] S. W. Benson, E. D. Siebert: J. Am. Chem Soc. **114**, 4269 (1992)
[111] J. L. Green, A. R. Lacey, M. G. Sceats: J. Phys. Chem. **90**, 3958 (1986)
[112] M. F. Kropman, H. K. Nienhuys, S. Woutersen, H. J. Bakker: J. Phys. Chem. **105**, 4622 (2001)
[113] L. Turi, P. Holpar, E. Keszei: J. Phys. Chem. **101**, 5469 (1997)
[114] C. Chudoba, E. T. J. Nibbering, T. Elsaesser: Phys. Rev. Lett. **81**, 3010 (1998)

Index

Topics in Applied Physics

78 **High-Power Diode Lasers**
By R. Diehl (Ed.) 2000, 260 Figs. XIV, 416 pages

79 **Frequency Measurement and Control**
Advanced Techniques and Future Trends
By A. N. Luiten (Ed.) 2001, 169 Figs. XIV, 394 pages

80 **Carbon Nanotubes**
Synthesis, Structure, Properties, and Applications
By M. S. Dresselhaus, G. Dresselhaus, Ph. Avouris (Eds.) 2001, 235 Figs. XVI, 448 pages

81 **Near-Field Optics and Surface Plasmon Polaritons**
By S. Kawata (Ed.) 2001, 136 Figs. X, 210 pages

82 **Optical Properties of Nanostructured Random Media**
By Vladimir M. Shalaev (Ed.) 2002, 185 Figs. XIV, 450 pages

83 **Spin Dynamics in Confined Magnetic Structures I**
By B. Hillebrands and K. Ounadjela (Eds.) 2002, 166 Figs. XVI, 336 pages

84 **Imaing of Complex Media with Acoustic and Seismic Waves**
By M. Fink, W. A. Kuperman, J.-P. Montagner, A. Tourin (Eds.) 2002, 162 Figs. XII, 336 pages

85 **Solid–Liquid Interfaces**
Macroscopic Phenomena – Microscopic Understanding
By K. Wandelt and S. Thurgate (Eds.) 2003, 228 Figs. XVIII, 444 pages

86 **Infrared Holography for Optical Communications**
Techniques, Materials, and Devices
By P. Boffi, D. Piccinin, M. C. Ubaldi (Eds.) 2003, 90 Figs. XII, 182 pages

87 **Spin Dynamics in Confined Magnetic Structures II**
By B. Hillebrands and K. Ounadjela (Eds.) 2003, 179 Figs. XVI, 321 pages

88 **Optical Nanotechnologies**
The Manipulation of Surface and Local Plasmons
By J. Tominaga and D. P. Tsai (Eds.) 2003, 168 Figs. XII, 212 pages

89 **Solid-State Mid-Infrared Laser Sources**
By I. T. Sorokina and K. L. Vodopyanov (Eds.) 2003, 263 Figs. XVI, 557 pages

90 **Single Quantum Dots**
Fundamentals, Applications, and New Concepts
By P. Michler (Ed.) 2003, 181 Figs. XII, 352 pages

91 **Vortex Electronis and SQUIDs**
By T. Kobayashi, H. Hayakawa, M. Tonouchi (Eds.) 2003, 259 Figs. XII, 302 pages

92 **Ultrafast Dynamical Processes in Semiconductors**
By K.-T. Tsen (Ed.) 2004, 190 Figs. XI, 400 pages

93 **Ferroelectric Random Access Memories**
Fundamentals and Applications
By H. Ishiwara, M. Okuyama, Y. Arimoto (Eds.) 2004, Approx. 200 Figs. XIV, 288 pages

94 **Silicon Photonics**
By L. Pavesi, D.J. Lockwood (Eds.) 2004, 262 Figs. XVI, 397 pages

95 **Few-Cycle Laser Pulse Generation and Its Applications**
By Franz X. Kärtner (Ed.) 2004, 209 Figs. XIV, 448 pages

Printing: Krips bv, Meppel
Binding: Litges & Dopf, Heppenheim